Jellyfish Blooms: Causes, Consequences, and Recent Advances

Developments in Hydrobiology 206

*Series editor*

K. Martens

# Jellyfish Blooms: Causes, Consequences, and Recent Advances

*Proceedings of the Second International Jellyfish Blooms Symposium, held at the Gold Coast, Queensland, Australia, 24–27 June, 2007*

*Editors*

Kylie A. Pitt[1] & Jennifer E. Purcell[2]

[1]*Australian Rivers Institute and Griffith School of Environment, Griffith University, Australia*
[2]*Western Washington University, Shannon Point Marine Center, Washington, USA*

***Reprinted from Hydrobiologia, Volume 616 (2009)***

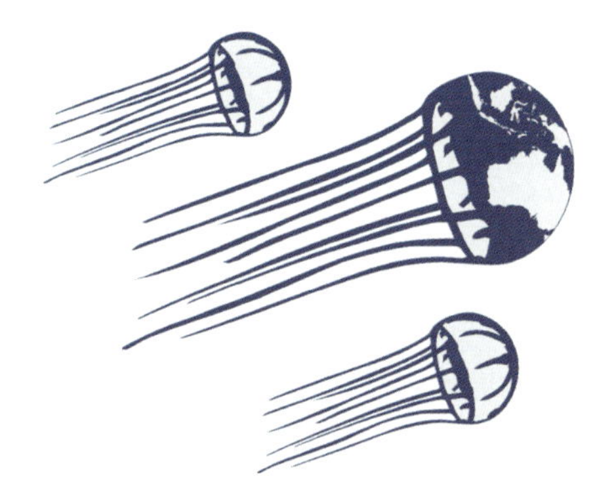

**Library of Congress Cataloging-in-Publication Data**

A C.I.P. Catalogue record for this book is available from the Library of Congress.

ISBN: 978-1-4020-9748-5

---

Published by Springer,
P.O. Box 17, 3300 AA Dordrecht, The Netherlands

Cite this publication as Hydrobiologia vol. 616 (2009).

---

**Cover illustration:** An aggregation of the giant jellyfish, *Nemopilema nomurai*, in a set net in Kyoto Prefecture, Japan. Photo by Yomiuri Shinbun.

---

***Printed on acid-free paper***

# TABLE OF CONTENTS

## EVOLUTION AND MOLECULAR APPROACHES



## BENTHIC STAGES OF THE LIFE HISTORY



## GENERAL BIOLOGY AND ECOLOGY

Hydrobiologia (2009) 616:1–5
DOI 10.1007/s10750-008-9599-2



# Preface

Published online: 30 September 2008


Jellyfish are among the most conspicuous animals in the ocean. They are fascinating ecologically as they are major consumers of zooplankton, are widely distributed, and most have complex life histories and demographics. They also have societal and economic significance as they clog cooling water intakes of coastal industries and ships, interfere with fisheries and deter holidaymakers from enjoying a swim at the beach. Although it is not a ubiquitous phenomenon, there is little doubt that some species of jellyfish are flourishing in parts of the world. The causes of jellyfish blooms are equivocal, but have been related to overfishing, climate change, eutrophication, and the increasing amount of artificial marine structures that provide suitable habitat for the benthic stage of the life history.

The Second International Jellyfish Blooms Symposium was held on the Gold Coast, Queensland, Australia, in June 2007 and followed the highly successful inaugural symposium that was held in Alabama, USA in 2000. In the intervening years, some remarkable changes have occurred. Most notable has been the population explosion of the 'giant jellyfish' (>1-m diameter), *Nemopilema nomurai*, in East Asian waters. The ctenophore, *Mnemiopsis leidyi*, infamous for its invasion of the Black Sea, has also recently appeared in northern European waters. Techniques for studying jellyfish have also advanced. In particular, molecular techniques are showing an increasing diversity of cryptic species of jellyfish and are being developed as tools to screen the hulls of ships and other infrastructure for invasive polyps. Acoustic techniques are being refined and are increasingly being used to quantify abundances, distributions, and movements of medusae. Natural tracers, such as stable isotopes and fatty acids more frequently are being used to complement traditional gut content analyses to examine trophic relationships. The polyp stage of the life history also recently has gained greater recognition because of its potential contribution to the development of blooms.

Thus, the 61 delegates from 17 nations who gathered in Australia had much to discuss! Professor Daniel Pauly, best known for his theory of 'fishing down marine food webs' opened the symposium by discussing how commercial fishing may contribute to jellyfish blooms and by encouraging jellyfish researchers to consider using applied fisheries techniques and to communicate better with ecosystem modelers. Professor Bill Hamner and Dr. Mike Dawson presented the plenary talk for the session on Diversity and discussed the phylogeography of jellyfish blooms. Dr. Jenny Purcell led the session on Trophic Ecology and encouraged researchers of jellyfish to start thinking on large, rather than local,

Guest editors: K. A. Pitt & J. E. Purcell
Jellyfish Blooms: Causes, Consequences, and Recent Advances

Participants of the Second International Jellyfish Blooms Symposium held at the Gold Coast, Queensland, Australia, 24–27 June 2007

scales. The final session on the Ecology of the Benthic Polyps was led by Professor Gerhard Jarms who emphasized that an understanding of the dynamics of jellyfish blooms required better knowledge of the demographics of the cryptic benthic polyps.

The papers presented in this volume are a subset of the 60+ talks and posters presented at the meeting. They provide a good representation of the diversity of issues discussed and hopefully will stimulate researchers worldwide to continue research into how and why jellyfish blooms occur.

The guest editors and conference organisers wish to thank the sponsors of the symposium, specifically the Ian Potter Foundation, CSIRO Land and Water, Griffith University, the Australian Rivers Institute (Griffith University), and James Cook University. The guest editors are also indebted to the numerous referees for their detailed and insightful reviews of manuscripts and, of course, to the contributing authors who stoically endured the numerous queries and requests of the guest editors.

Kylie Pitt (conference organiser and guest editor)
Jenny Purcell (guest editor)
Jamie Seymour (conference organiser)

## Complete list of conference presentations

Oral presentations

Jellyfish in marine ecosystems. D. Pauly. University of British Columbia, Canada.

The phylogeography of jellyfish blooms. W. M. Hamner[1], M. N. Dawson[2]. [1]University of California, Los Angeles, USA; [2]University of California, Merced, USA.

Phylogeography of an invasive jellyfish indicates multiple independent invasions and intense founder effects. K. M. Bayha[1], T. Bolton[2], W. M. Graham[1,3]. [1]Dauphin Island Sea Lab, USA; [2]The Flinders University of South Australia, Australia; [3]University of South Alabama, USA.

Molecular analysis of the genus *Physalia* (Cnidaria: Siphonophora) in New Zealand. D. R. Pontin, R. H. Cruickshank. Lincoln University, New Zealand.

Mitochondrial genomes of two jellyfishes, *Aurelia aurita* and *Chrysaora quinquecirrha.* D.-S. Hwang, J.-S. Lee. Hanyang University, South Korea.

Patterns in pelagic cnidarian blooms in the North Atlantic. M. J. Gibbons[1], A. J. Richardson[2]. [1]University of the Western Cape, Republic of South Africa; [2]CSIRO Marine and Atmospheric Research and University of Queensland, Australia.

Population dynamics and dispersal of *Pelagia noctiluca* in the Adriatic Sea. A. Malej[1], V. Malačič[1], A. Malej Jr[2], M. Malej[2], A. Ramšak[1], K. Stopar[1]. [1]National Institute of Biology; [2]University of Primorska, Slovenia.

Jellyfish biodiversity in Taiwan, western North Pacific Ocean. W. T. Lo[1], P. H. Ho[2]. [1]National Sun Yat-Sen University; [2]National Taiwan Ocean University, Taiwan, Republic of China.

Jellyfish appearances in Korean waters in 2006. W. D. Yoon, H. J. Gal, D. Lim, S. J. Chang. National Fisheries Research and Development Institute, Korea.

Ecological niche conservatism in box-jellyfishes. B. Bentlage. University of Kansas, USA.

An overview of the Rhopaliophora (Cnidaria, Medusozoa) research in Indian waters. G. Kanagaraj[1], P. Samapthkumar[1], A. C. Morandini[2]. [1]Annamalai University, India; [2]Universidade Federal do Rio de Janeiro, Brazil.

Reconstructing the invasion history of two independent introductions of the ctenophore *Mnemiopsis leidyi*. K. M. Bayha[1], G. R. Harbison[2], J. H. McDonald[3], P. M. Gaffney[4]. [1]Dauphin Island Sea Lab, USA; [2]Woods Hole Oceanographic Institute, USA; [3]University of Delaware, USA.

Phylogeny of Rhopaliophora and the evolution of the paraphyletic Rhizostomae. I. Straehler-Pohl. University of Hamburg, Germany.

Gelatinous zooplankton trophic ecology in the 21st Century: utilizing broad-scale and predictive approaches. J. E. Purcell. Western Washington University, USA.

Estimation of trophic level of giant jellyfish, *Nemopilema nomurai*, using stable isotope ratio. M. Toyokawa. National Institute of Fisheries Research, Japan.

Vertical migration of medusae in the open waters of south Adriatic Sea during 96 h sampling period (July 2003). D. Lučić[1], A. Benović[1], M. Morović[2], I. Onofri[3]. [1]University of Dubrovnik, Croatia; [2]Institute of Oceanography and Fisheries, Croatia; [3]University of Split, Croatia.

Links between dissolved organic matter excretion by gelatinous zooplankton and bacterial metabolism. R. H. Condon[1], D. K. Steinberg[1], T. C. Bouvier[2], P. A. del Giorgio[3]. [1]Virginia Institute of Marine Science, USA; [2]Université de Montpellier, France; [3]Université du Québec à Montréal, Canada.

The advection of scyphomedusae in the central Baltic Sea—Modelling, field data and possible consequences for trophic interactions. K. Barz[1], H.-J. Hirche[1], H.-H. Hinrichson[2]. [1]Alfred Wegener Institute for Polar and Marine Research, Germany; [2]Leibniz Institute of Marine Science, Germany.

Ontogeny of association behaviour with jellyfish in jack mackerel *Trachurus japonicus* revealed by rearing experiment using *Aurelia aurita* and underwater observation in relation to *Nemopilema nomurai*. R. Masuda. Kyoto University, Japan.

Bioenergetics in gelatinous zooplankton—the effect of prey species on energetic parameters. L. F. Møller, P. Thor, P. Tiselius. Göteborg University, Sweden.

Development, biological regulation, and fate of *Mnemiopsis leidyi* blooms in the York River estuary, USA. R. H. Condon, D. K. Steinberg. Virginia Institute of Marine Science, USA.

Jellyfish blooms that coincide with nutrient pulses can generate red tides. K. A. Pitt[1], M. J. Kingsford[2], D. Rissik[3], K. Koop[4]. [1]Griffith University, Australia; [2]James Cook University, Australia; [3]New South Wales Department of Natural Resources, Australia; [4]New South Wales Department of Environment and Climate Change, Australia.

The polyps of the Scyphozoa and their impact on jellyfish blooms. G. Jarms. University of Hamburg, Germany.

The life-cycle of the deep sea jellyfish *Atorella octagonos* (Cnidaria: Scyphozoa: Coronatae). A. C. Morandini[1], S. N. Stampar[2], H. M. Pacca[2],

F. L. da Silveira[2], G. Jarms[3]. [1]Universidade Federal do Rio de Janeiro, Brazil; [2]Universidade de São Paulo, Brazil; [3]University of Hamburg, Germany.

Morphological plasticity in polyps of *Aurelia* sp. in response to developmental temperature and food availability, and its subsequent effects on ephyrae: an experimental approach. L. M. Chiaverano[1], W. M. Graham[1,2]. [1]University of South Alabama, USA; [2]Dauphin Island Sea Lab, USA.

Thermal decline and density of population as factors for inducing strobilation and budding in moon jellyfish, *Aurelia aurita*. N. L. Lim, E. Y. Koh, W. D. Yoon, D. Lim. National Fisheries Research and Development Institute, Korea.

Seasonal changes in the density of *Aurelia aurita* polyps in the innermost part of Tokyo Bay. H. Ishii, K. Katsukoshi. Tokyo University of Marine Science and Technology, Japan.

Effects of changing environmental conditions on polyp populations and production of ephyrae. S. Holst. University of Hamburg, Germany.

Seasonal lunar spawning periodicity in offshore cubozoans. T. J. Carrette, J. Seymour. James Cook University, Australia.

Visual ecology in two species of box jellyfish. A. Underwood, J. Seymour. James Cook University, Australia.

Movement patterns in the Australian box jellyfish *Chironex fleckeri*: passive drifters or active locomotors? M. R. Gordon, R. Jones, J. E. Seymour. James Cook University, Australia.

Behavioural responses of zooplankton to the presence of predatory jellyfish. E. F. Carr, K. A. Pitt. Griffith University, Australia.

A comparison of the influence of symbiotic and non-symbiotic jellyfish on the nutrient cycling of coastal waterways. E. J. West, K. A. Pitt, D. T. Welsh. Griffith University, Australia.

Life cycle of the jellyfish *Lychnorhiza lucerna* Haeckel, 1880 (Scyphozoa, Rhizostomeae). A. Schiariti[1,2], M. Kawahara[3], S. Uye[3], H. W. Mianzan[1,2]. [1]Instituto Nacional de Investigación y Desarrollo Pesquero, Argentina; [2]Concejo Nacional de Investigaciones Cientificas y Ténicas, Argentina; [3]Hiroshima University, Japan.

Real-time field measurements of *Aurelia aurita* using a Self-Contained Underwater Velocimetry Apparatus (SCUVA). K. Katija, J. O. Dabiri. California Institute of Technology, USA.

The role of water column structure on jellyfish aggregation and reproduction in shallow coastal waters. W. M. Graham[1,2], J. E. Higgins III[1], K. Park[1,2]. [1]Dauphin Island Sea Lab, USA; [2]University of South Alabama, USA.

New approach to the understanding of a marine ecosystem using acoustic survey of jellyfish aggregation. G. A. Colombo[1], A. Benović[2], M. Acha[1], D. Lučić[2], T. Makovec[3], V. Onofri[2], A. Malej[3], H. Mianzan[1]. [1]Instituto Nacional de Investigación y Desarrollo Pesquero, Argentina; [2]University of Dubrovnik, Croatia; [3]National Institute of Biology, Slovenia.

Mechanism and significance of *Aurelia* aggregations in Uwa Bay, Japan. N. Fujii, Y. Nanjo, J. Ooyama, H. Takeoka. Ehime University, Japan.

Enhancement of jellyfish (*Rhopilema esculentum* Kishinouye, 1891) in Liaodong Bay, China. J. Dong, L. Jiang, K. Tan, C. Ye, P. Li, B. Wang. Liaoning Ocean and Fisheries Science Research Institute, China.

Recent bloom of the giant jellyfish *Nemopilema nomurai* in the East Asian Marginal Seas: A brief summary. S. Uye. Hiroshima University, Japan.

Mass appearance of the giant jellyfish, *Nemopilema nomurai*, along Japanese coasts and countermeasures. H. Iizumi[1], O. Kato[1], T. Watanabe[2]. [1]Japan Sea National Fisheries Research Institute, Fisheries Research Agency, Japan; [2]National Research Institute of Fisheries Engineering, Japan.

Seasonal change of oocyte size of giant jellyfish, *Nemopilema nomurai*. M. Toyokawa[1], A. Shimizu[1], K. Sugimoto[2], K. Nishiuchi[3], T. Yasuda[4], N. Iguchi[5]. [1]National Research Institute of Fisheries Science, Fisheries Research Agency, Japan; [2]Fukui Prefectural Fisheries Experimental Station, Japan; [3]Seikai National Fisheries Research Institute, Fisheries Research Agency, Japan; [4]Yasuda Fisheries Office, Japan; [5]Japan Sea National Fisheries Research Institute, Fisheries Research Agency, Japan.

Toxic components in the venom of the Lion's mane jellyfish *Cyanea* sp. H. Helmholz, C. Ruhnau, S. Lassen, R. Pepelnik, A. Prange. GKSS Research Centre, Germany.

Poster presentations

Kinematical properties of the jellyfish *Aurelia* sp. T. Bajcar[1], V. Malačič[2], A. Malej[2], B. Širok[1]. [1]University of Ljubljana, Slovenia; [2]National Institute of Biology, Slovenia.

Prospects of molecular systematics for Cubozoa. B. Bentlage[1], A. G. Collins[2], P. Cartwright[1]. [1]University of Kansas, USA; [2]NOAA's National Systematics Lab and National Museum for Natural History, Washington, DC, USA.

Nematocysts and morphometrics of lion's mane jellyfishes, genus *Cyanea*, supporting the validity of *C. nozakii* and *C. annaskala* in Australian waters. R. H. Condon[1], M. D. Norman[2], K. M. Bayha[3]. [1]Virginia Institute of Marine Science, USA; [2]Museum of Victoria, Australia; [3]Dauphin Island Sea Lab, USA.

Life cycle of a sea anemone parasite on *Aurelia* sp. from Veliko Jezero (Mljet, Croatia). I. D'Ambra, W. M. Graham. Dauphin Island Sea Lab and University of South Alabama, USA.

Preliminary observations on the diet and feeding of *Chrysaora hysoscella* from Walvis Bay Lagoon, Namibia. B. Flynn, M. J. Gibbons. University of the Western Cape, Republic of South Africa.

Seasonal variations in abundance, biomass and trophic role of *Aurelia aurita* s.l. medusae in a brackish water lake, Japan. C. H. Han, M. Kawahara, S. Uye. Hiroshima University, Japan.

Complete mitochondrial genomes of two jellyfish, *Aurelia* sp. 1 (like *A. aurita*) and *Dactylometra quinquecirrha*. D.-S. Hwang[1], J.-S. Ki[1], Y. S. Kang[2], K. Shin[3], J.-S. Lee[1]. [1]Hanyang University, South Korea; [2]National Fisheries Research and Development Institute, South Korea; [3]Korea Ocean Research and Development Institute, South Korea.

Biological characteristics of benthic polyp and planktonic juvenile stages of the giant jellyfish *Nemopilema nomurai*. M. Kawahara, S. Uye. Hiroshima University, Japan.

Importance of jellyfish molecular analysis for global expansion and a parallel high throughput DNA-chip detection. J.-S. Ki, D.-S. Hwang, J.-S. Lee. Hanyang University, South Korea.

Population studies of *Aurelia aurita* from Southampton Water and Horsea Lake, UK. C. H. Lucas. National Oceanography Centre Southampton, University of Southampton, U.K.

Characterization of microsatellite markers for three species of invasive hydrozoans in the San Francisco Estuary, CA, USA. M. H. Meek, M. R. Baerwald, B. P. May. University of California, Davis, USA.

The influence of ecological and physical factors on the settlement and viability of the moon jelly (Scyphozoa; *Aurelia* sp.) in the Northern Gulf of Mexico. M.-E. C. Miller, W. M. Graham. Dauphin Island Sea Lab and University of South Alabama, USA.

Population structure of the scyphomedusae *Chrysaora lactea* and *Lychnorhiza lucerna* (Cnidaria: Scyphozoa: Discomedusae) in an estuarine area on southeastern Brazil. A. C. Morandini[1], S. N. Stampar[2], F. L. da Silveira[2]. [1]Universidade Federal do Rio de Janeiro, Brazil; [2]Universidade de São Paulo, Brazil.

*Scolionema* or *Gonionemus* (Cnidaria: Hydrozoa: Limnomedusae): morphology and life cycle helping to solve taxonomical problems. A. C. Morandini[1], S. N. Stampar[2], V. B. Tronolone[2]. [1]Universidade Federal do Rio de Janeiro, Brazil; [2]Universidade de São Paulo, Brazil.

Stable isotopes emphasise the importance of large and emergent zooplankton to the diet of jellyfish. K. A. Pitt[1], A.-L. Clement[1,2], R. M. Connolly[1], D. Thibault-Botha[2]. [1]Griffith University, Australia; [2]Aix-Marseille Université, France.

Recruitment to benthic phase of planula and growth of scyphozoan polyps (*Aurelia aurita*). K. Shin, W. J. Lee, D. H. Son. Korea Ocean Research and Development Institute, Korea.

Effects of warming and low dissolved oxygen upon survival and asexual reproduction of scyphozoan polyps (*Aurelia aurita* s.l.). M. Takao, S. Uye. Hiroshima University, Japan.

Interactions between fishes and invasive jellyfish in the upper San Francisco Estuary, USA. A. P. Wintzer, P. B. Moyle. University of California, Davis, USA.

A new enzyme-assay for PLA2 activity in jellyfish venom based on phosphorus detection using HPLC-CC-ICP-MS. A. Zimmermann[1,2], H. Helmholz[1], D. Pröfrock[1], A. Prange[1]. [1]GKSS Research Centre, Germany; [2]University of Hamburg, Germany.

Hydrobiologia (2009) 616:7–10
DOI 10.1007/s10750-008-9619-2

JELLYFISH BLOOMS

# Obituary: Francesc Pagès (1962–2007)

**Josep-Maria Gili · José Luis Acuña · Humberto E. González**

Published online: 24 October 2008

Francesc Pagès was born in Barcelona on 6th July 1962. He started studies on gelatinous plankton as an undergraduate student of biology at the University of Barcelona, in 1989, and maintained his enthusiasm for these organisms throughout his professional career. Francesc's first project was a careful study of the Mediterranean gelatinous fauna under the supervision of Josep-Maria Gili, and this would later become the topic of his PhD thesis. It was surprising that a person of his age could have such a clear professional interest. Those were fantastic years, when he learned the taxonomy and ecology of jellyfish and siphonophores. That period was especially hard for Francesc because he had to balance his passion and inclination for gelatinous species with the demands of his undergraduate studies.

Guest editors: K. A. Pitt & J. E. Purcell
Jellyfish Blooms: Causes, Consequences, and Recent Advances

J.-M. Gili (✉)
Instituto de Ciencias del Mar (CSIC), Passeig Maritim de la Barceloneta, 37–49, Barcelona 08003, Spain
e-mail: gili@cmima.csic.es

J. L. Acuña
University of Oviedo, Oviedo, Asturias, Spain

H. E. González
University Austral of Chile, Valdivia and COPAS Center of Oceanography, Concepción, Chile

After graduation, he started a long pilgrimage around the world studying gelatinous zooplankton. His first cruise was in Guinea Bissau on board the fishing vessel Lulu. Back in Spain he started a study of the southeast Atlantic fauna, which became the main subject of his PhD thesis. During those 5 years, he participated in several cruises aboard oceanographic and fishing vessels off Namibia and South Africa. These expeditions allowed Francesc to produce the most complete account of pelagic cnidarians in that area. His studies of cnidarian fauna later were continued in the Antarctic and Mediterranean, and in Japan, although his Japanese studies remain unfinished due to his unexpected passing. In the south Atlantic, he developed another of his main subjects of interest, the relationship between zooplankton distribution and the physical environment.

He next turned to the Antarctic Ocean, as a postdoctoral fellow in the Alfred Wegener Institute in

Bremerhaven, Germany. During those years, he participated in cruises to Antarctica, without losing his interest in gelatinous plankton in other regions of the world. After this period, he accepted an invitation from the Japan Society for the Promotion of Science to study the Japanese siphonophoran fauna. During a 1-year visit to the Seto Marine Biological Laboratory at Shirahama-cho, he initiated what would become his favourite item of research during the last years of his life, exploration of the gelatinous fauna of the deep ocean by means of oceanographic submersibles. In 2000, he returned to Barcelona to start his own research group focused on the study of gelatinous zooplankton. Two years later he was appointed to a staff position in the Institute of Marine Sciences at Barcelona (CSIC), while initiating research within the European Union Project Eurogel. His purpose was to study the causes of jellyfish swarms in European waters. Meanwhile, he participated actively in deep dives aboard the submersible Johnson-Sea-Link II in the Gulf of Maine, California, Sargasso Sea, Sagami Bay in Japan, and the Norwegian fjords.

The infectious enthusiasm of Francesc for the wonderful marine gelatinous world had profound impact on the international community. He visited Chile on several occasions where he participated in projects and seminars. His cooperation with Chilean scientists broadened our knowledge on the biology and ecology of gelatinous zooplankton in the Humboldt Current System, which he liked to compare with the Benguela Current. He disseminated his motivation to several under- and post-graduate students who started their theses on groups such as chaetognaths and siphonophores.

Francesc Pagès passed away on 5 May 2007 on his return from Plymouth, where he was organising a workshop for the Hydrozoan Society. He was one of the most active members of the Society and played an important role as a bridge between the specialists and colleagues around the world. During the preparation of the meeting, he displayed his usual vitality and enthusiasm, and his death came as a shock to all of us. He was moving to a new research position at the Mediterranean Institute for Advanced Studies in the Balearic Islands, which represented an exciting scientific opportunity and a landmark in his personal life.

Francesc was an avid reader. He was well known in second-hand book stores, where he would find an original jellyfish illustration or a precious old book that he had been pursuing for quite a long time. Francesc was a great scholar, with vast knowledge on zooplankton biology and ecology, and was always willing to engage in conversation with colleagues, friends, and students. He never dismissed other people's opinions, particularly when these came from very young or inexperienced colleagues, and he always was prepared to ask questions, to listen and to learn in a gentle, humble tone that we will always remember.

During his short but intense career, Francesc's achievements were impressive, and he obtained a worldwide reputation in his field, the biology of gelatinous zooplankton. Fully devoted to his family and his work, he sailed all the oceans of the Earth sharing his experiences and knowledge with scientists, sailors, naturalists, and friends. Wherever he went he made friends, and he was an excellent ambassador for his Institute and for his country and culture. His passion for the oceans, and particularly for the organisms that inhabit them, was unparalleled. He was kind, affectionate, and generous. His capacity to assimilate information and to generate knowledge always surprised us, and we regarded him as a master. We have lost a young scientist whom we will never forget—a tireless, hard-working, and honest colleague, and far more than a friend.

## Selected publications of Francesc Pagès

Rossi, S., M. Youngbluth, F. Pagès, X. Garrofé & C. Jacoby, 2008. Lipid biomarkers and trophic links between seston, crustacean zooplankton and the siphonophore *Nanomia cara* in Georges Basin and Oceanographer Canyon (NW Atlantic). Scientia Marina 72: 403–416.

Pagès, F., J. Martín, A. Palanques, P. Puig & J.-M. Gili, 2007. High occurrence of the elasipodid holothurian Penilpidia ludwigi (von Marenzeller, 1893) in bathyal sediment traps moored in a western Mediterranean submarine canyon. Deep-Sea Research I 54: 2170–2180.

Pagès, F., J. Corbera & D. J. Lindsay, 2007. Piggybacking pycnogonids and parasitic narcomedusae on *Pandea rubra* (Anthomedusae, Pandeidae). Plankton and Benthos Research 2: 83–90.

Hosia, A. & F. Pagès, 2007. Unexpected new species of deep-water Hydroidomedusae from Korsfjorden, Norway. Marine Biology 151: 177–184.

Gili, J.-M., S. Rossi, F. Pagès, C. Orejas, P. González & W. Arntz, 2006. A new trophic link between the pelagic and benthic systems on the Antarctic Shelf. Marine Ecology Progress Series 322: 43–49.

Fuentes, V. & F. Pagès, 2006. Description of *Jubanyella plemmyris* gen. nov. et sp. nov. (Cnidaria: Hydrozoa: Narcomedusae) from a specimen stranded off Jubany Antarctic station, with a new diagnosis for the family Aeginidae. Journal of Plankton Research 28: 959–963.

Bouillon, J., C. Gravili, F. Pagès, J. Gili & F. Boero, 2006. An introduction to Hydrozoa. Mémoires du Muséum d'Histoire Naturelle, 194 Paris, 591 pp.

Pagès, F., P. R. Flood & M. Youngbluth, 2006. Gelatinous zooplankton net-collected in the Gulf of Maine and adjacent submarine canyons: New species, new family (Jeanbouillonidae), taxonomic remarks and some parasites. Scientia Marina 70: 363–379.

Bouillon, J., M. D. Medel, F. Pagès, J.-M. Gili, F. Boero & C. Gravili, 2004. Fauna of the Mediterranean Hydrozoa. Scientia Marina 68(Supp. 2): 5–438.

Masó, M., E. Garcés, F. Pagès & J. Camp, 2003. Drifting plastic debris as a potential vector for dispersing harmful algae bloom (HABs) species. Scientia Marina 67: 107–111.

Pagès, F. & P. R. Pugh, 2002. Fuseudoxid: The elusive sexual stage of the calycophoran siphonophore *Crystallophyes amygdalina* (Clausophyidae, Crystallophyinae). Acta Zoologica 83: 328–336.

Bouillon, J., F. Pagès & J.-M. Gili, 2001. New species of benthopelagic hydromedusae from the Weddell Sea. Polar Biology 24: 839–845.

Graham, W. M., F. Pagès & W. M. Hamner, 2001. A physical context for the aggregations of gelatinous zooplankton. A review. Hydrobiologia 451: 199–212.

Pagès, F., H. E. González, M. Ramón, M. Sobarzo & J.-M. Gili, 2001. Gelatinous zooplankton assemblages associated with water masses in the Humboldt Current System, and potential predatory impact by *Bassia bassensis* (Siphonophora: Calycophorae). Marine Ecology Progress Series 210: 13–24.

Pagès, F., 2000. Biological associations between barnacles and jellyfish with special emphasis on the ectoparasitism of *Alepas pacifica* (Lepadomorpha) on *Diplulmaris malayensis* (Scyphozoa). Journal of Natural History 34: 2045–2056.

Gili, J-M., F. Pagès, J. Bouillon, A. Palanques, P. Puig, S. Heussner, A. Calafat, M. Canals & A. Monaco, 2000. A multidisciplinary approach to the understanding of hydromedusan populations inhabiting Mediterranean submarine canyons. Deep-Sea Research I 47: 1513–1533.

Pagès, F. & C. Orejas, 1999. Medusae, siphonophores and ctenophores of the Magellan region. Scientia Marina 63(Suppl. 1): 51–57.

Gili, J-M., J. Bouillon, F. Pagès, A. Palanques & P. Puig, 1999. Submarine canyons as habitat of singular plankton populations: Three new deep-sea hydroidomedusae in the western Mediterranean. Zoological Journal of the Linnean Society 125: 313–329

Pagès, F., P. R. Pugh & V. Siegel, 1999. Discovery of an Antarctic epiplanktonic medusa in the Mediterranean. Journal of Plankton Research 12: 2431–2435.

Gili, J.-M., J. Bouillon, F. Pagès, A. Palanques, P. Puig & S. Heussner, 1997. Origin and biogeography of the deep-water Mediterranean Hydromedusae including the description of two new species collected in submarine canyons of Northwestern Mediterranean. Scientia Marina 62: 113–134.

Pagès, F. & J. Bouillon, 1997. Redescription of *Paragotoea bathybia* Kramp 1942 (Hydroidomedusae:Corymorphidae) with a new diagnosis for the genus *Paragotoea*. Scientia Marina 61: 487–493.

Pugh, P. R. & F. Pagès, 1997. A re-description of *Lensia asymmetrica* Stepanjants 1970 (Siphonophorae, Diphyidae). Scientia Marina 61: 153–161.

Pagès, F., 1997. The gelatinous zooplankton in the pelagic system of the Southern Ocean: A review. Annales de l Institut Océangrapique, Paris 73: 139–158.

Pagès, F. & S. B. Schnack-Schiel, 1996. Distribution patterns of the mesozooplankton, principally siphonophores and medusae, in the vicinity of the Antarctic Slope Front (eastern Weddell Sea). Journal of Marine Systems 9: 231–248.

Pagès, F., M. G. White & P. G. Rodhouse, 1996. Abundance of gelatinous carnivores in the nekton community of the Antarctic Polar Front in summer 1994. Marine Ecology Progress Series 141: 139–147.

Pagès, F., H. E. González & S. R. González, 1996. Diet of the gelatinous zooplankton in Hardangerfjord (Norway) and potential predatory impact by *Aglantha digitale* Trachymedusae). Marine Ecology Progress Series 139: 69–77.

Pugh, P. R. & F. Pagès, 1995. Is *Lensia reticulata* a diphyine species (Siphonophorae, Calycophora, Diphyidae)? A redescription. Scientia Marina 59: 181–192.

Pagès, F., P. R. Pugh & J.-M. Gili, 1994. Macro- and megaplanktonic cnidarians collected in the eastern part of the Weddell Gyre during summer 1979. Journal of the Marine Biological Association of the United Kingdom 74: 873–894.

Pagès, F. & F. Kurbjeweit, 1994. Vertical distribution and abundance of mesoplanktonic medusae and siphonophores in the Weddell Sea, Antarctica. Polar Biology 14: 243–251.

Pagès, F. & J.-M. Gili, 1992. Influence of the thermocline on the vertical migration of medusae during a 48 h sampling period. South African Journal of Zoology 27: 50–59.

Pagès, F., J.-M. Gili & J. Bouillon, 1992. Planktonic cnidarians of the Benguela Current: Station data. Scientia Marina 56(Suppl. 1): 144.

Pagès, F., 1992. Mesoscale coupling between planktonic cnidarian distribution and water masses during a temporal transition between active upwelling and abatement in the northern Benguela ecosystem. South African Journal of Marine Science 12: 42–52.

Gili, J.-M., F. Pagès & X. Fusté, 1991. Mesoscale coupling between spatial distribution of planktonic cnidarians and hydrographic features along the Galician coast (Northwestern Iberian Peninsula). Scientia Marina 55: 419–426.

Pagès, F., J. Bouillon & J.-M. Gili, 1991. Four new species of hydromedusae (Cnidaria, Hydrozoa) from Namibian waters. Zoologica Scripta 20: 98–107.

Pagès, F. & J.-M. Gili, 1991. Effects of large-scale advective processes on gelatinous zooplankton populations in the northern Benguela ecosystem. Marine Ecology Progress Series 75: 205–215.

Sabatés, A., J.-M. Gili & F. Pagès, 1989. Relationship between zooplankton distribution, geographic characteristics and hydrographic patterns off the Catalan coast (western Mediterranean). Marine Biology 103: 153–159.

Pagès, F., J.-M. Gili & J. Bouillon, 1989. The siphonophores (Cnidaria, Hydrozoa) of Hansa Bay, Papua New Guinea. Indo-Malayan Zoology 6: 133–144.

Gili, J.-M., A. Sabatés, F. Pagès & J. D. Ros, 1988. Small-scale distribution of a cnidarian population in the Western Mediterranean. Journal of Plankton Research 10: 385–401.

## A complete bibliography of Francesc Pagès can be found at:

Youngbluth, M., 2007. Marine Biology Research 3: 272–274.

Hydrobiologia (2009) 616:11–21
DOI 10.1007/s10750-008-9582-y

JELLYFISH BLOOMS

# The growth of jellyfishes

**M. L. D. Palomares · D. Pauly**

Published online: 15 October 2008


**Abstract** To date, a disparate array of concepts and methods have been used to study the growth of jellyfish, with the result that few generalities have emerged which could help, e.g., in predicting growth patterns in unstudied species. It is shown that this situation can be overcome by length-frequency analysis (LFA), applied to jellyfish bell diameter (i.e., "length") frequency data. A selection of LFA methods (ELEFAN, Wetherall plots and length-converted catch curves, all implemented in the FiSAT software) is applied here to 34 sets of bell diameter frequency data of jellyfish. This led to the estimates of parameters of the von Bertalanffy growth function (VBGF), which, especially in its seasonal form, was found to fit the available size-frequency data reasonably well. We also obtained numerous estimates of mortality, useful for modeling the life history of jellyfish. Finally, by scaling their asymptotic weight ($W_\infty$, a parameter of the VBGF) to the weight they would have if they had the same water content as fish, we show that most jellyfish grow at the same rate as small fishes (guppies and anchovies). As in fish, the VBGF parameters $K$ and $W_\infty$, when plotted in a double logarithmic ("auximetric") plot, tend to cluster into ellipsoid shapes, which increase in area when shifting from species to genera, families, etc. If validated by subsequent studies, auximetric plots for jellyfish would provide a powerful tool for testing comparative hypotheses on jellyfish life history.

**Keywords** Von Bertalanffy growth function · Length-frequency analysis · ELEFAN · FiSAT · Natural mortality · Water content

Guest editors: K. A. Pitt & J. E. Purcell
Jellyfish Blooms: Causes, Consequences, and Recent Advances

M. L. D. Palomares (✉) · D. Pauly
The Sea Around Us Project, Fisheries Centre, University of British Columbia, 2202 Main Mall, Vancouver, BC, Canada V6T 1Z4
e-mail: m.palomares@fisheries.ubc.ca

D. Pauly
e-mail: d.pauly@fisheries.ubc.ca

## Introduction

Much more attention is recently being devoted to jellyfish than previously, possibly because of some spectacular outbreaks, such as for example in the Benguela ecosystem (Lyman et al., 2006). Also, targeted fisheries catches of jellyfish are increasing (Kingsford et al., 2000). As a result, it can be expected that studies of the life history and ecology of jellyfish are likely to be intensified. Most of these studies require estimates of individual growth (i.e., the relationship between size and age), which then allows mortality rates to be inferred.

Jellyfishes differ from most other aquatic metazoans, notably fishes, in having very high water content. This feature (which they share with salps), along with their externalized anatomy, render direct comparisons of growth rates with other groups, e.g., fishes, difficult. This has led to situations where the authors of papers on jellyfish growth and related topics have not attempted to apply the methods and concepts used successfully in other branches of marine biology, i.e., in studies of the growth of fish. Rather, we find in papers on life history of jellyfishes, a confusing mix of methods, with a preference for percent and/or instantaneous growth rates ("this jellyfish grew 2% per day, while that grew 5%"), which cannot be used for comparisons within species and even less for comparisons between species.

This is a situation similar to that prevailing in the 1970s with tropical fish, and with squids and shrimps, about which little was known and a number of contradictory inferences were made (see Longhurst & Pauly, 1987, Chap. 9). The growth of all three of these groups was tackled, as suggested in Kingsford et al. (2000), using length-frequency analyses (LFA; see Longhurst & Pauly, 1987; and contributions in Pauly & Morgan, 1987). In the process, an understanding of their growth emerged, and their life history is now understood, and their mortality rates readily inferred, as required, e.g., for their incorporation in ecosystem models (Pauly et al., 2008).

We propose to do this for jellyfish in three steps. First, we show that frequency distributions of bell diameters, habitually collected by jellyfish scientists, provide an appropriate basis for standard LFA. Second, we show that the von Bertalanffy Growth Function (VBGF; von Bertalanffy, 1957), commonly used in fisheries research, can describe the growth of jellyfishes, especially when seasonal growth oscillations are explicitly taken into account. Finally, we show that the asymptotic weight ($W_\infty$) of jellyfishes, when scaled to the same water content as in fishes, leads to growth curves that resemble those of small fishes, e.g., guppies and anchovies. This becomes particularly visible on auximetric plots (plots of log $K$ vs. log $W_\infty$), which allow for growth comparisons between widely disparate groups.

## Materials and methods

### Growth parameter estimation

Bell diameter frequency distributions of 34 jellyfish populations were assembled from the published

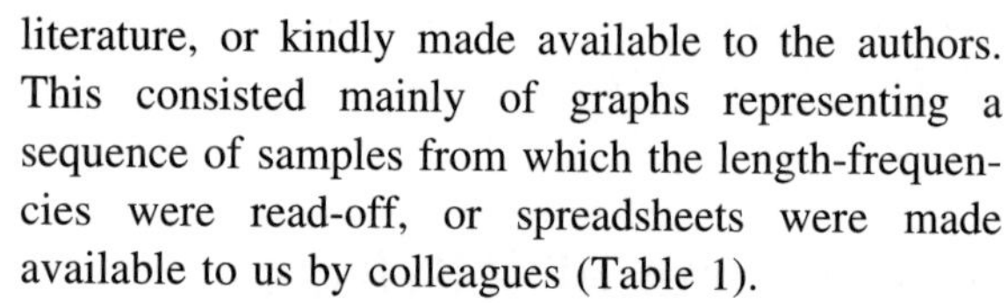

literature, or kindly made available to the authors. This consisted mainly of graphs representing a sequence of samples from which the length-frequencies were read-off, or spreadsheets were made available to us by colleagues (Table 1).

These were arranged sequentially in time, and fitted with the VBGF using a non-parametric, robust approach known as ELEFAN (Pauly, 1987, 1998a), implemented in the FiSAT software package (Gayanilo et al., 1995, available at www.fao.org/fi/statist/fisoft/fisat).

The bell-diameter frequency measurements were analyzed using the ELEFAN routine of FiSAT, i.e., a routine for fitting the von Bertalanffy growth function (VBGF) to length-frequency data (Gayanilo et al., 1995). This routine, although non-parametric, was preferred over more sophisticated approaches (e.g., MULTIFAN; Fournier et al., 1998), because it makes less demand on the underlying data, and especially because it can accommodate any unknown number of cohorts originating of the same year (Pauly, 1987), a feature common in jellyfish (Kingsford et al., 2000).

The standard version of the VBGF has the form:

$$L_t = L_\infty \cdot \left(1 - e^{-K\cdot(t-t_0)}\right) \quad (1)$$

where $L_t$ is the length at age $t$, $L_\infty$ is the asymptotic length, i.e., the mean length the animal would reach if they were to grow indefinitely (and similar to the size of the largest specimens), $K$ is a coefficient of dimension $t^{-1}$, and $t_0$ is a parameter setting the origin of the curve on the $X$-axis.

As it is obvious that the growth of jellyfish, like that of any poikilothermic organism, tends to fluctuate seasonally (Longhurst & Pauly, 1987; Pauly, 1998b), a form of the VBGF (Somers, 1988) was also used that accounts for seasonality, i.e.,

$$L_t = L_\infty \cdot \left(1 - e^{-[K\cdot(t-t_0)+S_{ts}-S_{t0}]}\right) \quad (2)$$

where $S_{ts} = \mathrm{CK}/2\pi \cdot (\sin 2\pi \cdot (t - t_s))$, $S_{t0} = (\mathrm{CK}/2\pi \sin 2\pi \cdot (t_0 - t_s))$, and $L_t$, $L_\infty$, $K$, $t$, and $t_0$ are as defined in Eq. 1.

In Eq. 2, $C$ is the intensity of the (sinusoid) growth oscillations of the growth curve and $t_s$ is the onset of the first oscillation relative to $t = 0$. A "winter point" (WP), can also be defined (from WP $= t_s + 0.5$ year) as the period of the year when growth is slowest. This is usually during the coldest month of the year, i.e., in February in the northern, and July in the southern

**Table 1** Summary of jellyfish (Scyphozoa and Hydrozoa) and comb jelly (Tentaculata: Ctenophora) stocks with size-frequency distributions obtained from various published and unpublished sources

| Species (class) | Locality (number of stocks) | b | a | *N* (*r*) | Water content (%) | Source |
|---|---|---|---|---|---|---|
| *Aequorea aequorea* (Forskål, 1775) (Hydrozoa) | Namibian shelf (1) | 2.017 | 1.10* | 2475 (0.770) | (96) | Brierley et al. (2001) and Buecher et al. (2001) |
| *Aurelia aurita* (Scyphozoa) | Kertinge Nor, Fyn Island, Denmark (1); Tokyo Bay, Japan (3); Black Sea (2); Kiel Bight, Germany (1); Urazoko Bay, Japan (1); Kagoshima Bay, Japan (1); Wadden Sea, Netherlands (1), Vagsbøpollen, Norway (1), Topeng Bay, Taiwan (1); Tomales Bay, USA (1); Big Jelly Lake, Koror, Palau (1) | 2.79; 2.32 | 0.0556; 0.290 | 52 (0.990); n.a. | 97.6 | Chen (2002), Dawson & Martin (2001), Hamner & Jenssen (1974), Ishii & Båmstedt (1998), Ishii & Tanaka (2001), Miyake et al. (1997), Mutlu (2001), Möller (1980), Olesen et al. (1994), Omori et al. (1995), van der Veer & Oorthuysen (1985), Weisse & Gomoiu (2000), and Yasuda (1971) |
| *Catostylus mosaicus* (Scyphozoa) | Australia (4) | | | | (96) | Pitt & Kingsford (2000) and Pitt & Kingsford (2003) |
| *Chiropsalmus* sp. (Scyphozoa) | Australia (1) | | | | (96) | Gordon et al. (2004) |
| *Chrysaora hysoscella* (Linnaeus, 1766) (Scyphozoa) | Namibian shelf (3) | 2.706; 2.896 | 0.100*; 0.060* | 40 (0.970); 635 (0.970) | (96) | Brierley et al. (2001); Buecher et al. (2001) |
| *Chrysaora melanaster* Brandt, 1838 (Scyphozoa) | USA (1) | | | | (96) | Brodeur et al. (2002) |
| *Cotylorhiza tuberculata* (Macri, 1778) (Scyphozoa) | Vlyho Bay, Ionian Island, Greece (1) | 3.100 | 0.080 | 11 (0.935) | 95.8 | Kikinger (1992) |
| *Cyanea* sp. (Scyphozoa) | Niantic River Estuary, USA (1) | 3.340 | 0.0306* | n.a.; (0.982) | (96) | Brewer (1989) |
| *Halecium petrosum* Stechow, 1919 (Hydrozoa) | Tossa de Mar, Spain (1) | | | | (96) | Coma et al. (1992) |
| *Halecium pusillum* Sars, 1856 (Hydrozoa) | Tossa de Mar, Spain (1) | | | | (96) | Coma et al. (1992) |
| *Mastigias* cf. *papua etpisoni* Dawson 2005 (Scyphozoa) | Ongeim'l Tketau, Mecherchar Island, Koror, Palau (1) | | | | (96) | Unpublished data from Lori J. Bell (Coral Reef Research Foundation), Michael N Dawson (Coral Reef Research Foundation and University of California, Merced), Laura E. Martin (Coral Reef Research Foundation and University of California, Merced) and Sharon Patris (Coral Reef Research Foundation). |

**Table 1** continued

| Species (class) | Locality (number of stocks) | b | a | $N$ ($r$) | Water content (%) | Source |
|---|---|---|---|---|---|---|
| *Mnemiopsis leidyi* A. Agassiz, 1865 (Tentaculata) | Narragansett Bay, USA (1) | 2.636 | 0.0303* | n.a.; (0.96)0 | 96.6 | Deason (1982) |
| *Periphylla periphylla* (Péron & Lesueur, 1810) (Scyphozoa) | Lurefjorden, Norway (1) | 2.870 | 0.148 | n.a.; (0.950) | 96.0 | Fosså (1992) |
| *Phyllorhiza punctata* (Scyphozoa) | Paraná State, Brazil; Laguna Joyuda, Puerto Rico (4) | 2.689; 2.625 | 0.152; 84.7* | 22 (0.995); 31; (0.966) | (96) | García & Durbin (1993), García (1990), and Haddad & Noqueira (2006) |

Here, a and b refer to the y-intercept and the slope of log transformed length weight relationship of the form $W = aL^b$, respectively, while $N$ refers to the number of individuals considered in the regression analyses, and $r$ is the square of the regression's coefficient of determination. Values of the length–weight relationship coefficient a with * were adjusted to weight in grams and bell diameter in cm. Water content values in brackets refer to the mean water content of 96% reported by Larson (1986)

hemisphere, corresponding to WP = 0.1, and WP = 0.7, respectively (Pauly, 1987, 1998a).

Although the $C$-parameter, which defines the amplitude of growth oscillations up to $C = 1$, when a complete halt of growth is induced (at WP), can be estimated from detailed, well-sampled length frequency data (as can WP, see Pauly, 1987, 1998a), Eq. 2 was fitted to the available length-frequency data with fixed values of $C = 0.25$, 0.5, and 0.75 for populations sampled from sub-tropical, temperate and boreal localities, respectively. This is justified by the observation that $C$ in fishes and aquatic invertebrates correlates strongly with the summer–winter difference in temperature in various environments ($\Delta T$), and usually takes a value of 1 when $\Delta T = 10$°C, i.e., in temperate, and boreal waters (Pauly, 1987, 1998a). Tropical samples were treated without seasonality using Eq. 1. This fixed the temperature-induced oscillations at predictable levels, and reduced the number of free parameters to one (i.e., $K$, with $L_\infty$ fixed see below) or two ($K$ and $L_\infty$).

In some cases, when the available length-frequency samples were too sparse to use ELEFAN, another LFA method was used, the Wetherall Plot (Wetherall, 1986; Gayanilo et al., 1995). This method requires a single sample representative of a population, which is generally approximated by adding up a number of samples collected at different times.

The version of the Wetherall plot used here consists of successive mean lengths ($L_{mean}$) computed from successive cut-off lengths ($L_i$), minus the $L_i$ (i.e., $L_{mean} - L_i$) being plotted against $L_i$. This results in a series of points trending downward, which can be fitted with a linear regression of the form $Y = a + b \cdot X$, with $L_\infty = a/-b$) and $Z/K = (1 + b)/(-b)$, where $Z$ is the instantaneous rate of total mortality (Gayanilo et al., 1995; Pauly, 1998a). The cases where we used Wetherall plots for LFA can be identified in Table 2 as they provide only estimates of asymptotic size and $Z/K$, which is equivalent to $M/K$ in unexploited populations (see below).

In a few more cases, length at (presumed) age or growth increment data were available in the literature. These were fitted with a non-seasonal version of the VBGF (Eq. 1) through direct non-linear fitting. Such cases can be identified in Table 2 as they provide only estimates of asymptotic size and $K$.

ELEFAN, the Wetherall plot, and the direct non-linear fitting of the VBGF all assume that shrinkage

**Table 2** Summary of von Bertalanffy growth parameters estimated for 34 populations of jellyfishes (Scyphozoa and Hydrozoa) and comb jellies (Ctenophora) from analyses of length-frequency distributions, growth increment, and age-at-length data

| Species | Locality (temp in °C) | Sample size ($L_{min} - L_{max}$, cm) | $L_\infty$ (cm) | $W_\infty$ (WW, g) | $W_\infty$ (norm) | $K$ (year$^{-1}$) | $C$ | WP | $M$ ($M/K$) | Data source(s) |
|---|---|---|---|---|---|---|---|---|---|---|
| *Aequorea aequorea* | Namibian shelf, Namibia | 3396 (4–10) | 11.1 | 141 | 22.6 | 0.87 | 0.50 | 0.7 | 2.09 | Buecher et al. (2001); Brierley et al. (2001) |
| *Aurelia aurita* | NW Black Sea (17.0) | 1909 (1.5–10.5) | 11.6 | 69 | 6.6 | 0.62 | 0.25 | 0.1 | 2.03 | Weisse & Gomoiu (2000) |
| *Aurelia aurita* | Kertinge Nor, Denmark | 300 (0–6) | 11.5 | 67 | 6.4 | 1.80 | – | – | – | Olesen et al. (1994) |
| *Aurelia aurita* | Kiel Bight, Germany | n.a. (1–19.7) | 25.2 | 482 | 46.3 | 3.11 | – | – | – | Möller (1980) |
| *Aurelia aurita* | Urazoko Bay, Japan (17.5) | 3631 (1–32) | 34.5 | 1079 | 103.6 | 0.45 | 0.50 | 0.1 | 1.35 | Yasuda (1971) |
| *Aurelia aurita* | Tokyo Bay, Japan (19.4) | 2915 (5–29) | 31.4 | 848 | 81.4 | 1.50 | 0.50 | 0.1 | 5.38 | Ishii & Tanaka (2001) |
| *Aurelia aurita* | Tokyo Bay, Japan (17.5) | 3101 (1–31) | 35.5 | 1161 | 111.4 | 0.86 | 0.50 | 0.1 | 2.95 | Omori et al. (1995) |
| *Aurelia aurita* | Tokyo Bay, Japan | 272 (0–30) | 31 | 821 | 78.8 | 0.54 | 0.50 | 0.1 | 1.03 | Kinoshita et al. (2006) |
| *Aurelia aurita* | Kagoshima Bay, Japan | n.a. (0–23) | 27.0 | 575 | 55.2 | 3.83 | – | – | – | Miyake et al. (1997) |
| *Aurelia aurita* | Dutch Wadden Sea | 1094 (2–30) | 37.4 | 1326 | 127.3 | 2.90 | 0.50 | 0.1 | 10.3 | van der Veer & Oorthuysen (1985) |
| *Aurelia aurita* | Vagsbøpollen, Norway (11.9) | 1228 (0.5–11.5) | 12.0 | 75 | 7.2 | 3.70 | 0.50 | 0.1 | 6.61 | Ishii & Bamstedt (1998) |
| *Aurelia aurita* | Topeng Bay, Taiwan | 2051 (1–30) | 33.4 | 993 | 95.3 | 0.74 | 0.25 | 0.1 | 1.74 | Chen (2002) |
| *Aurelia aurita* | Black Sea, Turkey (17) | 973 (1.5–31.5) | 35 | 1119 | 107.4 | 0.68 | 0.25 | 0.1 | 2.33 | Mutlu (2001) |
| *Aurelia aurita* | Tomales Bay, USA | 31 (0–17) | 19.4 | 250 | 24.0 | 2.93 | – | – | – | Hamner & Jenssen (1974) |
| *Aurelia* sp. | Big Jellyfish Lake, Koror, Palau (30.0) | 2474 (2–38) | 33.2 | 978 | 93.9 | 0.98 | 0.25 | 0.1 | 1.74 | Dawson & Martin (2001) |
| *Catostylus mosaicus* | Botany Bay, Australia | n.a. (15–25) | 31.2 | 1867 | 298.7 | 2.28 | – | – | – | Pitt & Kingsford (2000) |
| *Catostylus mosaicus* | Botany Bay, Australia | n.a. (15–25) | 28.2 | 1376 | 220.1 | 2.14 | – | – | – | Pitt & Kingsford (2000) |
| *Catostylus mosaicus* | Botany Bay, Australia | 3409 (1.5–28) | 37.0 | 3083 | 493.2 | 0.60 | 0.50 | 0.7 | 4.25 | Pitt & Kingsford (2003) |
| *Catostylus mosaicus* | Lake Illawarra, Australia | 5710 (1–31) | 35.5 | 2727 | 436.3 | 1.50 | 0.50 | 0.7 | 4.46 | Pitt & Kingsford (2003) |
| *Chiropsalmus* sp. | Four Mile Beach, Australia (28.6) | 1627 (0.5–12.5) | 14.8 | 210 | 33.6 | 1.50 | – | – | 5.99 | Gordon et al. (2004) |
| *Chrysaora hysoscella* | Namibian shelf | 1954 (7.5–62.5) | 68.8 | 10993 | 1758.8 | 1.10 | 0.25 | 0.1 | (3.12) | Buecher et al. (2001) |
| *Chrysaora hysoscella* | Namibian shelf | 283 (10.5–55.5) | 60.8 | 7777 | 1244.3 | – | – | – | (2.26) | Brierley et al. (2001) |
| *Chrysaora hysoscella* | Namibian shelf | 2240 (7.5–62.5) | 68.2 | 10725 | 1716.0 | 4.30 | 0.25 | 0.7 | 4.83 | Buecher et al. (2001); Brierley et al. (2001) |
| *Chrysaora melanaster* | SE Bering Sea, USA | 304 (1.5–55) | 56.3 | 6252 | 1000.4 | – | – | – | (3.04) | Brodeur et al. (2002) |
| *Cotylorhiza tuberculata* | Vlyho Bay, Ionian Sea, Greece (21.5) | 3500 (2–35) | 39.0 | 6845 | 1150.0 | 0.73 | – | – | – | Kikinger (1992) |
| *Cyanea* sp. | Niantic River Estuary, USA | 704 (1.5–13.5) | 14.2 | 214 | 34.2 | 2.30 | 0.50 | 0.1 | 5.97 | Brewer (1989) |
| *Halecium petrosum* | Mediterranean Sea, Spain | 2918 (1–19) | 19.2 | 427 | 68.3 | 0.85 | 0.25 | 0.1 | 1.17 | Coma et al. (1992) |
| *Halecium pusillum* | Mediterranean Sea, Spain | 695 (1–27) | 28.1 | 920 | 147.2 | 1.20 | 0.25 | 0.1 | 1.73 | Coma et al. (1992) |

**Table 2** continued

| Species | Locality (temp in °C) | Sample size ($L_{min} - L_{max}$, cm) | $L_\infty$ (cm) | $W_\infty$ (WW, g) | $W_\infty$ (norm) | $K$ (year$^{-1}$) | $C$ | WP | $M$ ($M/K$) | Data source(s) |
|---|---|---|---|---|---|---|---|---|---|---|
| *Mastigias* cf. *papua etpisoni* | Mecherchar Island, Palau | 160343 (0–21) | 23.7 | 2115 | 338.4 | – | – | – | (11.7) | Unpublished data (see Table 1) |
| *Mnemiopsis leidyi* | Narragansett Bay, RI, USA | 501 (0–4) | 4.3 | 1 | 0.2 | – | – | – | (1.14) | Deason (1982) |
| *Periphylla periphylla* | Lurefjorden, Norway (7.0) | 438 (5.5–13.5) | 13.5 | 262 | 41.9 | – | – | – | (1.19) | Fosså (1992) |
| *Phyllorhiza punctata* | Laguna Joyuda, Puerto Rico (28.0) | 191 (1–50) | 57.5 | 7960 | 1273.6 | 2.30 | 0.25 | 0.1 | 4.69 | García (1990, Fig. 4) |
| *Phyllorhiza punctata* | Laguna Joyuda, Puerto Rico (28.0) | 3667 (2.5–38.5) | 40.2 | 3048 | 487.7 | 3.00 | – | – | – | García (1990, Table 1) |
| *Phyllorhiza punctata* | Laguna Joyuda, Puerto Rico (28.0) | n.a. (3–24) | 38.3 | 2681 | 428.9 | 4.69 | – | – | – | García & Durbin (1993) |

Here, $L_\infty$ and $W_\infty$ are the asymptotic bell diameter (in cm) and weight (in grams), respectively, while $K$ is the growth coefficient of the von Bertalanffy growth equation. The unitless parameter $C$ refers to the amplitude and WP is the winter point, i.e., the coldest point of the year when growth is assumed to stop for a seasonal von Bertalanffy growth curve. Finally, $M$ is the natural mortality expressed in year$^{-1}$. Note that in cases where length-frequency distributions cover only a few samples, and/or are not adapted to the VBGF, the Wetherall method was used, and thus only providing an estimate of $L_\infty$ and $M/K$

does not occur. We believe that this is a second-order effect (but see below).

## Mortality estimation

In fishery biology, mortality rates are usually expressed as instantaneous rates, defined by $Z$ in:

$$N_{i+1} = N_i \cdot e^{-Z \cdot (t_{i+1} - t_i)} \quad (3)$$

where $N_{i+1}$ is the population size at time $t_{i+1}$, $N_i$ is the starting population size, and $Z$ is the total mortality rate, which is the sum of natural mortality, $M$, and fishing mortality, $F$. As the jellyfish populations reported here are not exposed to fishing mortality, we have $Z = M$ throughout.

Length-frequency data can be used to infer mortality rates if certain assumptions are met pertaining to the quality of the data, notably, that the samples should be representative of the population in the juvenile and adult phases (Wetherall, 1986; Wetherall et al., 1987). With the Wetherall plot, mortality itself cannot be estimated, but the ratio $M/K$ can.

Another method for estimation of mortality from length-frequency samples are length-converted catch curves (Gayanilo et al., 1995; Pauly 1987, 1998a), which require growth parameters, but which, contrary to the Wetherall plot, explicitly accounts for seasonal growth (Gayanilo et al., 1995; Pauly, 1998a). We abstain here from describing this method in detail. Suffice to say that it results in a plot of the logs of numbers (corrected for the effect of non-linear growth) reflecting population abundances by age, and whose descending slope is an estimate of $Z$ (Pauly, 1998a).

## Scaling jellyfish weight for water content

The length–weight relationships in Table 1 (i.e., $W = aL^b$), as obtained from pairs of live wet weights ($W$) in g and bell diameters in cm ($L$), were used to convert our estimates of $L_\infty$ into estimates of $W_\infty$. In cases where length–weight relationships were not available for a given species, length–weight relationships of other species in the same genus or higher taxa were used. In cases where several length–weight relationships were available, the average of $W_\infty$ was calculated for each length–weight relationship that was used. The parameters of length–weight

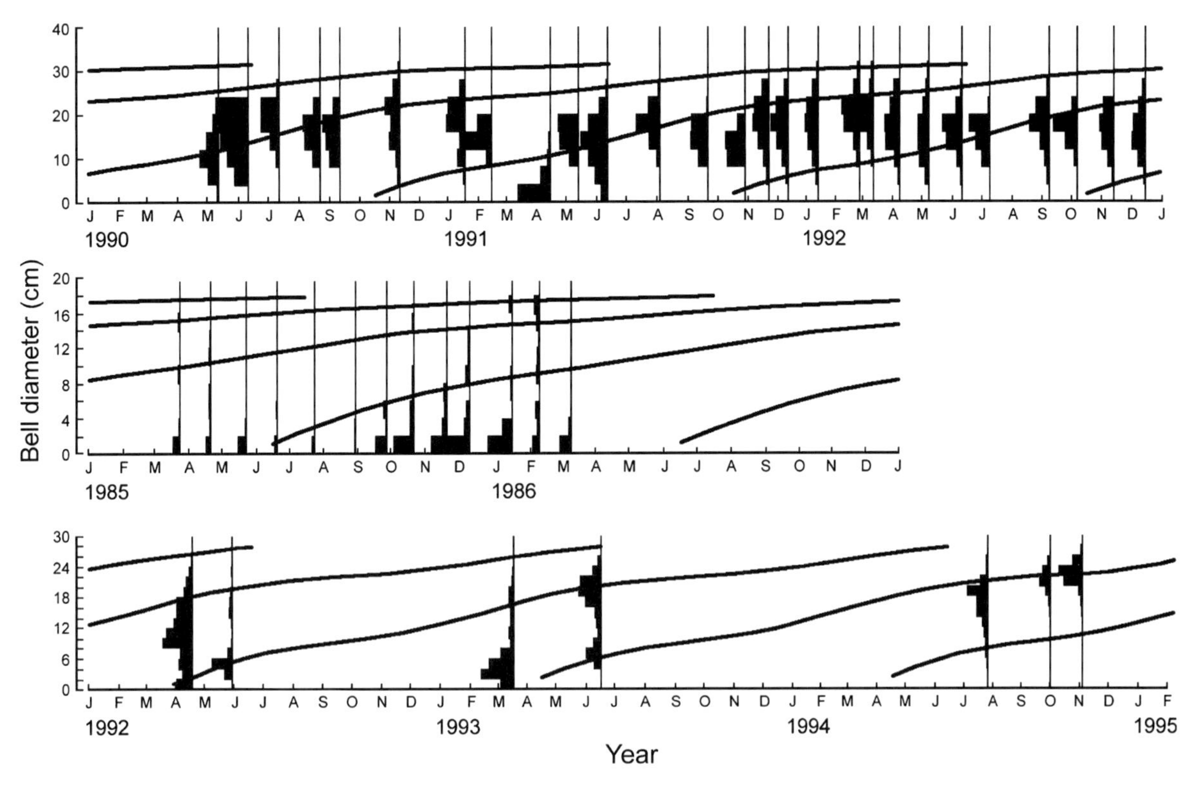


**Fig. 1** Jellyfish growth curve fitting with ELEFAN. Top panel: *Aurelia aurita* from Tokyo Bay, Japan in 1990–1992 (Omori et al., 1995), with $L_\infty = 35.5$ and $K = 0.86$ year$^{-1}$ for fixed values of $C = 0.5$ and WP = 0.1. Central panel: *Halecium petrosum* from the Mediterranean Sea, Spain, sampled from 1985 to 1986 (Coma et al., 1992) with $L_\infty = 19.2$ and $K = 0.85$ year$^{-1}$ for fixed values of $C = 0.25$ and WP = 0.1. Bottom panel: First 3 of the 6-year samples of *Catostylus mosaicus* from Botany Bay, Australia between March 1990 and February 1998 (Pitt and Kingsford, 2003) with $L_\infty = 37.0$ and $K = 0.60$ year$^{-1}$ for fixed values of $C = 0.5$ and WP = 0.7. These three growth curves were selected from thousands of alternatives using a search algorithm in FiSAT (see Fig. 5 and text)

relationships are usually estimated with great precision, and the uncertainty in the $W_\infty$ values discussed here is mainly due to the water content of various species of jellyfish not being known with precision.

The reported water content of jellyfish species (see Table 1) was obtained from available reports in the scientific literature. In cases where no water content estimate was available, the average water content of jellyfish (96%) estimated by Larson (1986) was used. Fish water content were obtained from data in FishBase (www.fishbase.org), based mainly on Bykov (1983), which yielded for 530 fish species, a mean water content, mainly for muscle tissue, of 75 ± 5.13%.

The re-scaled $W_\infty$ values of jellyfish were then calculated, i.e., the asymptotic weight they would have if they had the same water content as fish. This was achieved by multiplying the original value of $W_\infty$ by the ratio of the % dry weights, i.e., in most cases 25/4 = 6.25 (see Table 1).

## Comparing the growth of jellyfish with fish and invertebrates

The parameter $K$ of the VBGF tends to vary inversely with the parameter $W_\infty$. Thus, if an organism is small, it will have a high value of $K$, and vice versa. This makes it difficult to compare "growth" in different organisms using a single number. On the other hand, a bivariate "auximetric plot" can be used to compare the parameters $K$ and $W_\infty$ in different populations of the same species or between species. Such plots thus allow various inferences on the likely range of growth parameters of organisms, which are related to organisms whose positions on such auximetric plot is known and thus for which the value of

$K$ can be inferred from the $W_\infty$ value (which itself can be inferred from the maximum reported weight in a population).

## Results

Table 2 presents the key results of growth and mortality studies we performed on the 34 populations of jellyfishes in Table 1. The available size-frequency data were sufficient for the use of ELEFAN in 30 of these (as illustrated in Fig. 1), though Wetherall plots were used in all cases to provide estimates of asymptotic length (see Fig. 2 for an example). Mortality was then obtained from a catch curve (see Fig. 3 for an example).

Table 3 presents the values $L_m/L_\infty$ that we were able to assemble, $L_m$ being the mean length at first maturity in a given population. The mean value of $L_m/L_\infty$, is 0.43 (Table 3), which is lower than the mean value reported from fishes, i.e., around 0.6–0.7 (Froese & Binohlan, 2000). Figure 4 presents an auximetric plot with the growth parameters in Table 2.

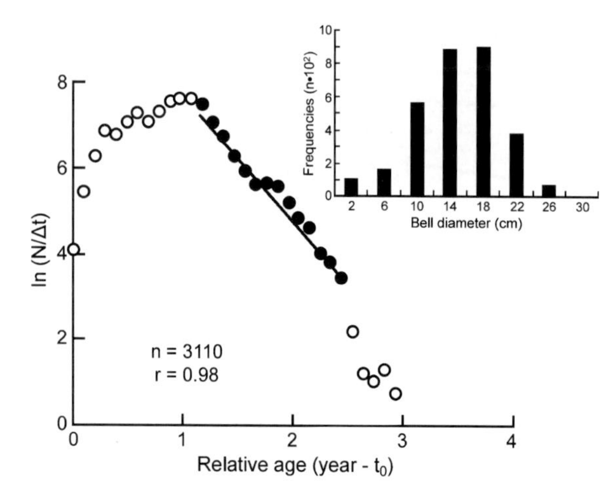


**Fig. 3** Example of a catch curve analysis (see methods section) applied to a cumulative sample of bell diameters of *Aurelia aurita* in Tokyo Bay, Japan (inset; from Omori et al., 1995; Fig. 4) obtained between May 1990 and December 1992. Using the von Bertalanffy growth parameters $L_\infty = 35.5$ and $K = 0.86$ year$^{-1}$ for fixed values of $C = 0.5$ and WP $= 0.1$ (see Fig. 1, top panel) yields an estimate of $Z = 2.95$ year$^{-1}$, which may correspond to an estimate of P/B (see text and Table 2)

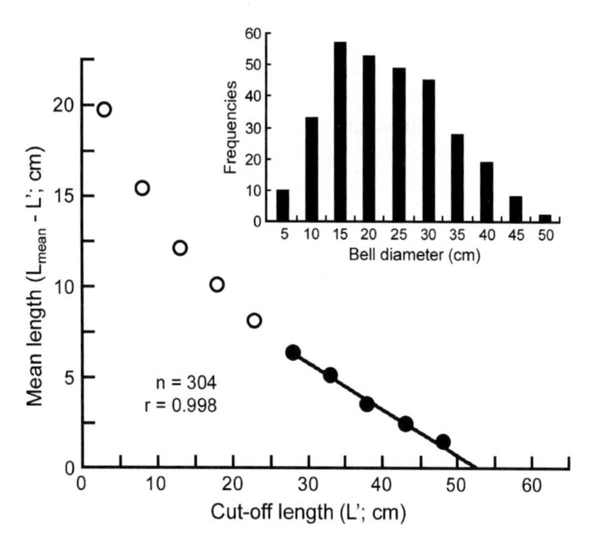


**Fig. 2** Example of a Wetherall Plot (see methods section) applied to a cumulative sample of bell diameters of *Chrysaora melanaster* from the Bering Sea, USA (inset; from Brodeur et al., 2002, Fig. 8) obtained in September of 1996, 1997, and 1999. Only the lengths assumed to be fully retained by the gear (straight section of graph; bell diameters >25 cm) are used for the regression, which yields $L_\infty = 56.3$ cm and $M/K = 2.86$. Note that the points are weighted by cumulative sample size, which give a crucial importance to the selection of the 1st point (see text)

## Discussion

This paper adapted various concepts from length-based fish population dynamics to jellyfish, and the literature cited above discusses the various assumptions and pitfalls.

Here, we will mention only an important feature of the Wetherall plot, i.e., that on statistical grounds, it should be estimated from a regression where each point is weighted by the number of animals in the sample (Wetherall et al., 1987). This results in the first point included in the regression (representing the smallest length fully retained by the sampling gear) having a strong impact on the estimation of results ($L_\infty$ and $Z/K$). Moreover, the points are not strictly independent, thus violating an assumption of regression analysis. Thus, the Wetherall method should be considered mainly as a useful heuristic.

Our conclusion regarding growth is that many papers on jellyfish biology present size frequency data, which can be fitted with the VBGF (see Table 2 for a subset of that literature). We suspect, moreover,

**Table 3** Summary of available length at first maturity ($L_m$) data for five species of scyphozoan jellyfishes with computed $L_m/L_\infty$ ratio ($L_\infty$ values, i.e., asymptotic bell diameter, from Table 2)

| Species | Locality | $L_m$ | $L_m/L_\infty$ | Remarks (source) |
|---|---|---|---|---|
| *Aurelia aurita* | Tokyo Bay, Japan | 8.75 | 0.246 | Seasonal (Omori et al., 1995) |
| *Aurelia aurita* | Vagsbøpollen, Norway | 9.40 | 0.783 | Only summer months (Ishii & Båmstedt, 1998) |
| *Catostylus mosaicus* | Botany Bay, Australia | 13.0 | 0.351 | Seasonal (Pitt & Kingsford 2000) |
| *Chiropsalmus* sp. | Four Mile Beach, Australia | 5.50 | 0.371 | M from 5 samples winter to spring (Gordon et al., 2004) |
| *Cyanea* sp. | Niantic River Estuary, USA | 8.15 | 0.574 | Seasonal (Brewer, 1989) |
| *Phyllorhiza punctata* | Laguna Joyuda, Puerto Rico | 15.0 | 0.261 | Seasonality due to rainfall (García, 1990) |
| Means | – | – | 0.431 | This study |

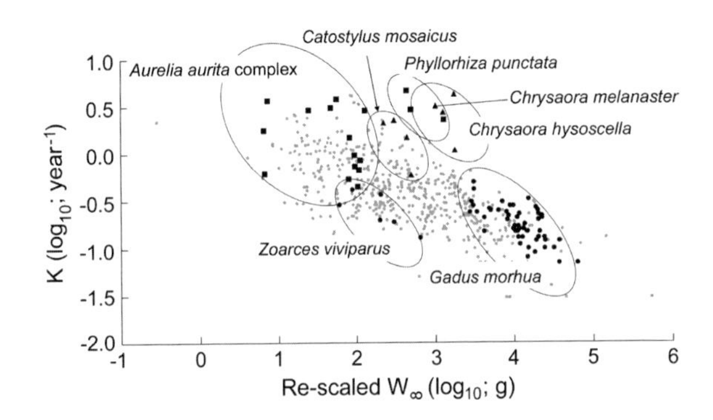


**Fig. 4** Auximetric plot of ($\log_{10}$) $K$ against re-scaled values of ($\log_{10}$) $W_\infty$ for the major groups of jellyfishes, on a background of grey dots representing fishes (in addition to two highlighted species, *Zoarces viviparus* and *Gadus morhua*). As might be seen, the *Aurelia aurita* complex and *Catostylus mosaicus* resemble small fishes in their growth pattern, but *Chrysaora* spp. and *Phryllorhiza punctata* (and other species in Table 2) may grow faster (higher $K$ for a given $W_\infty$) than fishes. This result might need to be validated using more growth parameter sets. Also, it is very sensitive to the assumed ratio between the water contents of jellyfish and fish (see text)

that the VBGF is not only a convenient mathematical function for describing the growth of jellyfish, but also that it does so because their respiratory physiology makes this growth function, derived from physiological considerations, the model of choice (von Bertalanffy, 1957; Longhurst & Pauly, 1987; Pauly, 1998b).

The crucial step in estimating the parameters of the VBGF does not consist of the estimation of asymptotic size, for which the maximum size in a field sample usually provides a good approximation, nor with the parameter from seasonal growth, which can be approximated from first principles. Rather, the crucial parameter is $K$. How well this parameter is estimated can be assessed by plots such as that shown in Fig. 5, which are a standard feature of the ELEFAN procedure.

Our estimate of the mean value of $M/K$ in jellyfish is 3.03, about two times higher than the values reported for fishes, which usually range between 1 and 2 (Beverton & Holt, 1956; Pauly, 1998a). This high value of M/K may be due to, at least in some cases, shrinkages of the bells of jellyfish (Hamner & Jenssen, 1974), which could have biased the (fixed) interrelationships of number, size, and age, which are assumed in LFA. Note that when $K$ is underestimated by ELEFAN or other LFA, $M$ is also underestimated (and conversely for overestimation), for which reason our estimates of $M/K$ should be robust.

The auximetric plot in Fig. 4, finally, suggests that some jellyfish, once account is taken of their high water contents, have growth patterns similar to small and very small fishes, such as guppies and anchovies (*Aurelia aurita* (Linnaeus, 1758) complex, *Catostylus mosaicus* (Quoy & Gaimard, 1824)). Others (*Phyllorhiza punctata* (von Lendenfeld, 1884), *Chrysaora* spp.) may grow faster than fishes (i.e., have higher values of $K$ for their value of $W_\infty$). However, the accuracy of the position of an organism on an auximetric plot depends on the accuracy of the growth parameters, and in the case of jellyfish, on correct conversion to standard water content. Because of this, we consider the results of this study to be preliminary.

We are, however, encouraged by the observation that, as in fish (here exemplified by *Gadus morhua* Linnaeus, 1758 and *Zoarces viviparous* (Linnaeus, 1758), the different populations in a given species appear to form ellipsoid clusters on an auximetric plot (see www.fishbase.org for more). Genera and higher taxa can be expected, as well, to form such

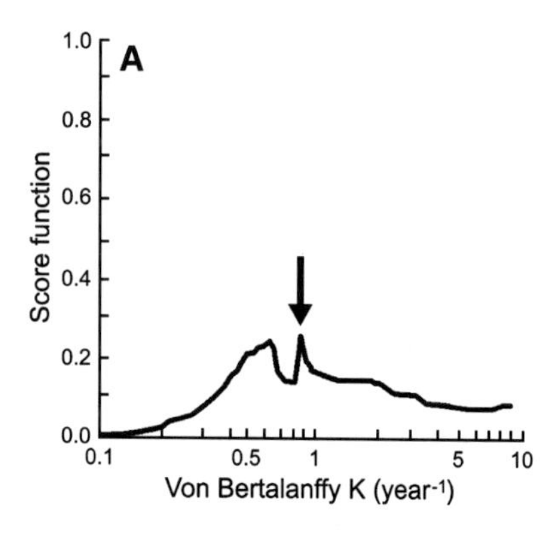

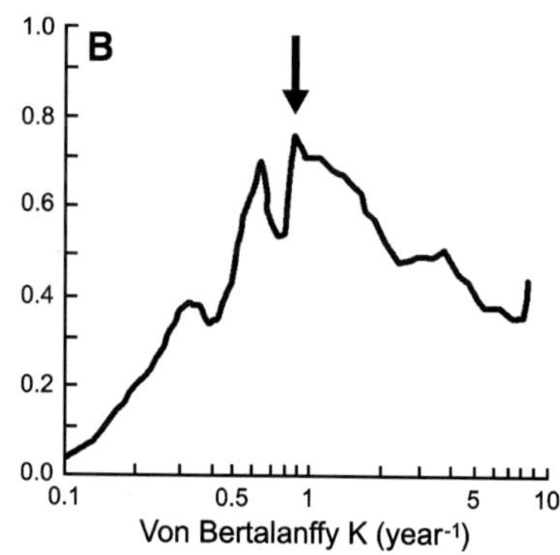

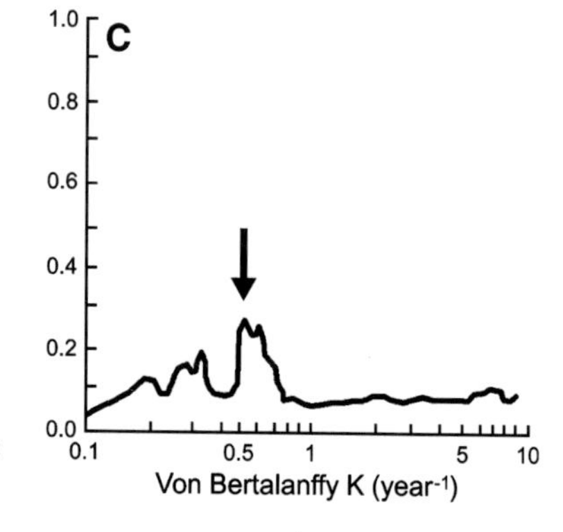


**Fig. 5** Examples of response surface of the goodness-of-fit estimator of ELEFAN, as used to estimate $K$ (and to assess the uncertainty associated with the point estimate) when the other parameters of the seasonally oscillating VBGF ($L_\infty$, $C$ and WP) are known or assumed (see arrows). Here, the panel A pertains to the top of Fig. 1, i.e., to *Aurelia aurita* from Tokyo Bay, and the best fitted $K$ value is not very distinct from adjacent values; hence, the best estimate of $K$ (= 0.86 year$^{-1}$) is highly uncertain. Panel B, corresponding to the middle panel of Fig. 1, i.e., to *Halecium petrosum* from the Mediterranean, suggests that $K$ is well estimated, as the response surface exhibits a sharp peak close to 0.85 year$^{-1}$. Panel C, representing the bottom panel of Fig. 1, i.e., *Catostylus mosaicus* in Botany Bay, shows $K$ (= 0.60 year$^{-1}$) to be more reliably estimated

clusters, albeit larger ones. This suggests that the large cluster we drew for the *Aurelia aurita* complex would, indeed, include more than one species.

**Acknowledgements** We would like to thank Ms. Christine Dar (SeaLifeBase Project, Philippines) for help with assembling the data in the correct format for VBGF analysis and for extracting and encoding in SeaLifeBase, over a short period of time, the life history and ecological information on jellyfishes. We would also like to thank Dr. Laura E. Martin (Coral Reef Research Foundation and University of California, Merced) for sending us unpublished jellyfish size-frequency data, which extended the coverage of the analysis presented here.

## References

Beverton, R. J. H. & S. J. Holt, 1956. A review of methods for estimating rates in exploited fish populations, with special reference to sources of bias in catch sampling. Rapports et Procès-verbaux des Réunions du Conseil International de l'Exploration de la Mer 140: 67–83.

Brewer, R. H., 1989. The annual pattern of feeding, growth, and sexual reproduction in *Cyanea* (Cnidaria: Scyphozoa) in the Niantic River Estuary, Connecticut. Biological Bulletin 176: 272–281.

Brierley, A. S., B. E. Axelsen, E. Buecher, C. A. J. Sparks, H. Boyer & M. J. Gibbons, 2001. Acoustic observations of jellyfish in the Namibian Benguela. Marine Ecology and Progress Series 210: 55–66.

Brodeur, R. D., H. Sugisaki & G. L. J. Hunt, 2002. Increases in jellyfish biomass in the Bering Sea: Implications for the ecosystem. Marine Ecology and Progress Series 233: 89–103.

Buecher, E., C. Sparks, A. Brierley, H. Boyer & M. Gibbons, 2001. Biometry and size distribution of *Chrysaora hysoscella* (Cnidaria, Scyphozoa) and *Aequorea aequorea* (Cnidaria, Hydrozoa) off Namibia with some notes on their parasite *Hyperia medusarum*. Journal of Plankton Research 23: 1073–1080.

Bykov, V. P., 1983. Marine Fishes: Chemical Composition and Processing Properties. Amerind Publishing Co. Pvt. Ltd., New Delhi.

Chen, E. L., 2002. Population Dynamics and Feeding of the Moon Jellyfish (*Aurelia aurita*) in Tapeng Bay, Southwestern Taiwan. National Sun Yat-Sen University, Taiwan.

Coma, R., I. Llobet, M. Zabala, J. Gili & R. G. Hughes, 1992. The population dynamics of *Halecium petrosum* and *Halecium pusillum* (Hydrozoa, Cnidaria), epiphytes of *Halimeda tuna* in the northwestern Mediterranean. Scientia Marina 56: 161–169.

Dawson, M. N. & L. E. Martin, 2001. Geographic variation and ecological adaptation in *Aurelia* (Scyphozoa, Semaeostomeae): some implications from molecular phylogenetics. Hydrobiologia 451: 259–273.

Deason, E. E., 1982. *Mnemiopsis leidyi* (Ctenophora) in Narragansett Bay, 1975–79: abundance, size composition and estimation of grazing. Estuarine Coastal and Shelf Science 15: 121–134.

Fosså, J. H., 1992. Mass occurrence of *Periphylla periphylla* (Schyphozoa, Coronatae) in a Norwegian fjord. Sarsia 77: 237–251.

Fournier, D. A., J. Hampton & J. R. Sibert, 1998. MULTIFAN-CL: a length-based, age-structured, model for fisheries stock assessment, with application to south Pacific albacore (*Thunnus alalunga*). Canadian Journal of Fisheries and Aquatic Science 55: 2105–2116.

Froese, R. & C. Binohlan, 2000. Empirical relationships to estimate asymptotic length, length at first maturity and length at maximum yield per recruit in fishes, with a simple method to evaluate length frequency data. Journal of Fish Biology 56: 758–773.

García, J. R., 1990. Population dynamics and production of *Phyllorhiza punctata* (Cnidaria: Scyphozoa) in Laguna

Joyuda, Puerto Rico. Marine Ecology Progress Series 64: 243–251.
Garcia, J. R. & E. Durbin, 1993. Zooplanktivorous predation by large scyphomedusae *Phyllorhiza punctata* (Cnidaria: Scyphozoa) in Laguna Joyuda. Journal of Experimental Marine Biology and Ecology 173: 71–93.
Gayanilo, J. F., P. Sparre & D. Pauly, 1995. FAO/ICLARM Stock Assessment Tools (FiSAT) User's Guide. Report No. 8. FAO, Rome.
Gordon, M., C. Hatcher & J. Seymour, 2004. Growth and age determination of the tropical Australia cubozoan *Chiropsalmus* sp. Hydrobiologia 530/531: 339–345.
Haddad, M. A. & M. J. Noqueira, 2006. Reappearance and seasonality of *Phyllorhiza punctata* von Lendenfeld (Cnidaria, Scyphozoa, Rhizostomeae) medusae in southern Brazil. Revista Brasileira de Zoologia 23: 824–831.
Hamner, W. M. & R. M. Jenssen, 1974. Growth, degrowth and irreversible cell differentiation in *Aurelia aurita*. American Zoologist 14: 833–849.
Ishii, H. & F. Tanaka, 2001. Food and feeding of *Aurelia aurita* in Tokyo Bay with an analysis of stomach contents and a measurement of digestion times. Hydrobiologia 451: 311–320.
Ishii, H. & U. Båmstedt, 1998. Food regulation of growth and maturation in a natural population of *Aurelia aurita* (L.). Journal of Plankton Research 20: 805–816.
Kikinger, R., 1992. *Cotylorhiza tuberculata* (Cnidaria: Scyphozoa) – life history of a stationary population. Marine Ecology (Berlin) 13: 333–362.
Kingsford, M. J., K. A. Pitt & B. M. Gillanders, 2000. Management of jellyfish fisheries, with special reference to the order Rhizostomeae. Oceanography and Marine Biology: An Annual Review 38: 85–156.
Kinoshita, J., J. Hiromi & Y. Yamada, 2006. Abundante and biomass of scyphomedusae, *Aurelia aurita* and *Chrysaora melanaster*, and Ctenophora, *Bolinopsis mikado*, with estimates of their feeding impact on zooplankton in Tokyo Bay, Japan. Journal of Oceanography 62: 607–615.
Larson, R. J., 1986. Water content, organic content, and carbon and nitrogen composition of medusae from the northeast pacific. Journal of Experimental Marine Biology and Ecology 99: 107–120.
Longhurst, A. & D. Pauly, 1987. Ecology of Tropical Oceans. Academic Press, San Diego.
Lyman, C. P., M. J. Gibbons, B. E. Axelsen, C. A. J. Sparks, J. Coetzee, B. G. Heywood & A. S. Brierley, 2006. Jellyfish overtake fish in a heavily fished ecosystem. Current Biology 16: R492–R493.
Miyake, H., K. Iwao & Y. Kakinuma, 1997. Life history and environment of *Aurelia aurita*. South Pacific Studies 17: 273–285.
Möller, H., 1980. Population dynamics of *Aurelia aurita* medusae in Kiel Bight, Germany (FRG). Marine Biology 60: 123–128.
Mutlu, E., 2001. Distribution and abundance of moon jellyfish (*Aurelia aurita*) and its zooplankton food in the Black Sea. Marine Biology 138: 329–339.
Olesen, N. J., K. Frandsen & H. U. Riisgard, 1994. Population dynamics, growth and energetics of jellyfish *Aurelia aurita* in a shallow fjord. Marine Ecology Progress Series 105: 9–18.
Omori, M., H. Ishii & A. Fujinaga, 1995. Life history strategy of *Aurelia aurita* (Cnidaria, Scyphomedusae) and its impact on the zooplankton community of Tokyo Bay. ICES Journal of Marine Science 52: 597–603.
Pauly, D., 1987. A review of the ELEFAN system for analysis of length-frequency data in fish and aquatic invertebrate. In Pauly, D. & G. R. Morgan (eds), Length-Based Models in Fisheries Research. ICLARM Conference Proceedings 13. ICLARM, Manila, Philippines: 7–34.
Pauly, D., 1998a. Beyond our original horizons: the tropicalization of Beverton and Holt. Reviews in Fish Biology and Fisheries 8: 307–334.
Pauly, D., 1998b. Why squids, though not fish, may be better understood by pretending they are. South African Journal of Marine Science 20: 47–58.
Pauly, D., S. Libralato, L. Morissette & M. L. D. Palomares, 2008. Jellyfish in ecosystems, online databases and ecosystem models. Proceedings of the Second International Jellyfish Blooms Symposium, Australia, June 2007. Hydrobiologia. doi:10.1007/s10750-008-9583-x.
Pauly, D. & G. R. Morgan (eds), 1987. Length-based Methods in Fisheries Research. ICLARM Conference Proceedings 13. ICLARM, Manila, Philippines.
Pitt, K. A. & M. J. Kingsford, 2000. Reproductive biology of the edible jellyfish *Catostylus mosaicus* (Rhizostomeae). Marine Biology 137: 791–799.
Pitt, K. A. & M. J. Kingsford, 2003. Temporal and spatial variation in recruitment and growth of medusae of the jellyfish, *Catostylus mosaicus* (Scyphozoa: Rhizostomeae). Marine and Freshwater Research 54: 117–125.
Somers, I. F., 1988. On a seasonally oscillating growth function. Fishbyte 6: 8–11.
van der Veer, H. W. & W. Oorthuysen, 1985. Abundance, growth and food demand of the scyphomedusa *Aurelia aurita* in the western Wadden Sea. Netherlands Journal of Sea Research 19: 38–44.
von Bertalanffy, L., 1957. Quantitative laws in metabolism and growth. Quarterly Review of Biology 32: 217–231.
Weisse, T. & M. Gomoiu, 2000. Biomass and size structure of the scyphomedusa *Aurelia aurita* in the northwestern Black Sea during spring and summer. Journal of Plankton Research 22: 223–239.
Wetherall, A., 1986. A new method for estimating growth and mortality parameters from length frequency data. Fishbyte (ICLARM/The WorldFish Center) 4(1): 12–14.
Wetherall, A., J. J. Polovina & S. Ralston, 1987. Estimating growth and mortality in steady-state fish stocks from length-frequency data. In Pauly, D. & G. R. Morgan (eds), Length-based Models in Fisheries Research. ICLARM Conference Proceedings 13. ICLARM, Manila, Philippines: 53–74.
Yasuda, T., 1971. Ecological studies on the jelly-fish, *Aurelia aurita* in Urazoko Bay, Fukui Prefecture – 4. Monthly change in the bell-length composition and breeding season. Bulletin of the Japanese Society of Scientific Fisheries 37: 364–370.

Hydrobiologia (2009) 616:23–50
DOI 10.1007/s10750-008-9585-8

JELLYFISH BLOOMS

# Extension of methods for jellyfish and ctenophore trophic ecology to large-scale research

**Jennifer E. Purcell**

Published online: 27 September 2008


**Abstract** Science has rapidly expanded its frontiers with new technologies in the 20th Century. Oceanography now is studied routinely by satellite. Predictive models are on global scales. At the same time, blooms of jellyfish and ctenophores have become problematic, especially after 1980. Although we have learned a great deal about gelatinous zooplankton ecology in the 20th Century on local scales, we generally have not scaled-up to estimate the extent, the causes, or effects of large blooms. In this age of global science, research on gelatinous zooplankton needs to utilize large-scale approaches and predictive equations. Some current techniques enable jellyfish populations (aerial, towed cameras), feeding (metabolic rates, stable isotopes), and dynamics (predictive modeling) to be studied over large spatial and temporal scales. I use examples of scyphomedusae (*Aurelia* spp., *Cyanea capillata*, *Chrysaora quinquecirrha*) and *Mnemiopsis leidyi* ctenophores, for which considerable data exist, to explore expanding from local to global scales of jellyfish trophic ecology. Regression analyses showed that feeding rates of *Aurelia* spp. (FR in copepods eaten medusa$^{-1}$ d$^{-1}$) generally could be estimated ±50% from in situ data on medusa wet weight (WW) and copepod density; temperature was not a significant factor. FR of *C. capillata* and *C. quinquecirrha* were similar to those of *Aurelia* spp.; the combined scyphomedusa regression underestimated measured FR of *C. quinquecirrha* and *Aurelia* spp. by 50% and 180%, respectively, and overestimated measured FR of *C. capillata* by 25%. Clearance rates (CR in liters cleared of copepods ctenophore$^{-1}$ d$^{-1}$) of *M. leidyi* were reduced in small containers ($\leq$20 l), and a ratio of container-volume to ctenophore-volume of at least 2,500:1 is recommended for feeding experiments. Clearance rates were significantly related to ctenophore WW, but not to prey density or temperature, and estimated rates within 10–159%. Respiration rates of medusae and ctenophores were similar across habitats with greatly ambient different temperatures (10–30°C), and can be predicted from regressions using only mass. These regressions may permit estimation of feeding effects of gelatinous predators without exhaustive collection of feeding data in situ. I recommend that data on feeding and metabolism of jellyfish and ctenophores be entered in a database to allow generalized predictive relationships to be developed to promote inclusion of these important predators in ecosystem studies and models.

**Keywords** Metabolism · Scaling · Feeding · Global · Model · Methods · Review

Guest editors: K. A. Pitt & J. E. Purcell
Jellyfish Blooms: Causes, Consequences, and Recent Advances

J. E. Purcell (✉)
Shannon Point Marine Center, Western Washington University, 1900 Shannon Point Rd., Anacortes, WA 98221, USA
e-mail: purcelj3@wwu.edu

## Introduction

During the 20th Century, we have learned a great deal about the ecology of the predaceous gelatinous zooplankton, including jellyfish (scyphomedusae, cubomedusae, and hydromedusae), siphonophores, and ctenophores. They are ubiquitous in the world's oceans and estuaries, living from the surface to the greatest depths. They affect the food web from microplankton (e.g., Colin et al., 2005) to whales (Purcell et al., in press). Since they can consume large quantities of ichthyoplankton and zooplankton, their potential importance as both predators and competitors of fish is of particular interest to humans.

Blooms of jellyfish and ctenophores have been problematic in coastal water, especially since the 1980s (reviewed in Purcell et al., 2001b, 2007). When great abundances occur, jellyfish can interfere with fishing, kill fish in aquaculture enclosures, clog power and desalination plant water intakes, and cause health concerns for swimmers (Purcell et al., 2007). *Mnemiopsis leidyi* A. Agassiz ctenophores have caused great damage to fisheries by competing with fish for zooplankton, and eating fish eggs and larvae in the Black Sea, where they were accidentally introduced in the early 1980s. The ctenophores spread to the Azov, Caspian, Marmara, and Mediterranean seas, and recently (2006), were discovered in the North and Baltic seas (e.g., Boersma et al., 2007). Generally, jellyfish and ctenophore blooms are detrimental to human enterprise.

Jellyfish and ctenophore blooms occur over broad regions, such as in the Black, Azov, and Caspian seas (e.g., Shiganova et al., 2003), the Mediterranean Sea (Bernard et al., 1988; Goy et al., 1989), Gulf of Mexico (Graham et al., 2003a, b), the Seto Inland Sea of Japan (Uye et al., 2003), and the East Asian Marginal Seas (Uye, 2008). In spite of their importance and our increased knowledge, few attempts have been made to estimate population trends or the effects of jellyfish blooms on plankton food webs. In order to study the effects of jellyfish blooms, researchers need to utilize large-scale methods for estimating jellyfish and ctenophore size, abundances, and their predation effects.

In large-scale research, some error is inevitable, which is against our training for accuracy and precision as scientists. Nevertheless, atmospheric and oceanographic scientists now routinely study the Earth from satellite data. For example, estimation of sea surface chlorophyll *a* (Chl *a*) is derived from algorithms based on properties of light reflected from the sea surface, as measured by satellite (SeaWiFS). Empirical measurements of Chl *a* from ocean water, as typically measured by fluorescence, show significant deviation from the satellite estimates, but this widely used method provides global estimates of production (e.g., Marrari et al., 2006).

Because of large sizes, fragility, and non-dispersed distributions, many gelatinous zooplankton species present problems both for field sampling and laboratory experiments, which have limited research efforts on them as compared with the more robust crustaceans (e.g., Raskoff, 2003). The relatively abundant data for copepods have enabled development of algorithms for predicting their feeding, growth, fecundity, and mortality rates in relation to Chl *a*, temperature, and size (e.g., Hansen et al., 1997; Hirst & Bunker, 2003; Bunker & Hirst, 2004; Hirst & Kiørboe, 2002; Hirst et al., 2003). Unfortunately, such data are much more limited for gelatinous species than for copepods, and few predictive algorithms have been developed (see Palomares & Pauly, 2008).

In this article, I review recent use of large-scale methods of data collection for jellyfish and ctenophore size and abundances, and explore developing algorithms to estimate feeding effects so that local-scale knowledge can be expanded to large-scale research. These recommendations are intended to promote research on gelatinous zooplankton by utilizing standard methods and existing knowledge.

## Large-scale techniques to determine jellyfish and ctenophore population sizes

### Net-sampling

Data on the abundances of jellyfish and ctenophores are basic to research on their ecological importance. Gelatinous zooplankton presents many challenges for sampling (Raskoff, 2003). The traditional method of quantitative sampling of zooplankton and nekton by nets with flow-meters and preservation in formalin is inappropriate for many jellyfish and ctenophores that are large, sparsely or unevenly distributed, or delicate.

Net sampling is often adequate for small, abundant hydromedusae and calycophoran siphonophores (e.g., Pagès et al., 1996a, b; Hosia & Båmstedt, 2007), and some robust ctenophores, specifically, *Pleurobrachia* spp., *Mertensia ovum* (Fabricius), *Beroe* spp., and with care, *Mnemiopsis leidyi* (e.g., Purcell, 1988; Siferd & Conover, 1992; Shiganova et al., 2003). Large species require a large sampling volume and larger nets; semi-quantitative sampling of scyphomedusae and large hydromedusae is possible with fish and shrimp trawls and seines (e.g., Brodeur et al., 1999, 2002, 2008a, b; Graham, 2001; Purcell, 2003). A single method is typically adequate for only one type within a mixture of taxa (e.g., scyphomedusae, hydromedusae, and ctenophores).

National and state fisheries services usually have annual stock surveys that sample with large trawls, cover large regions, and have been conducted for decades. Such stock surveys have provided invaluable data on jellyfish populations, when their numbers or biomass have been documented from the by-catch. Important contributions include data on *Chrysaora melanaster* Brandt in the eastern Bering Sea (Brodeur et al., 1999, 2002, 2008a), *Chrysaora quinquecirrha* (Desor) and *Aurelia aurita* (Linné) in the Gulf of Mexico (Graham, 2001), *Chrysaora hysoscella* Eschscholtz, *A. aurita*, and *Cyanea capillata* (Linné) in the North Sea (Lynam et al., 2004, 2005), and *Nemopilema nomuri* (Kishinouye) around Japan (Uye, 2008). It would be virtually impossible for individual researchers to sample over the extensive spatial and temporal scales of government-sponsored fisheries programs. For example, annual surveys in the Bering Sea comprised 356 stations over 27 years (Brodeur et al., 2008a). This sampling is semi-quantitative because fisheries do not target jellyfish and fish catch is standardized only as Catch Per Unit Effort (CPUE). Unfortunately, not all fisheries sampling quantifies jellyfish by-catch. Thus, fish trawls are reasonable for sampling large, robust gelatinous species. This sampling is inadequate for small and delicate species, which pass through the large meshes or are destroyed.

### Other animals as samplers

An ingenious method of determining the large-scale distribution of gelatinous species has been by use of their predators as samplers (Link & Ford, 2006). Many fish eat gelatinous species (e.g., Arai, 2005), and gut analyses routinely are performed on commercial fish species during the annual surveys. Link & Ford (2006) used a large-scale dataset that showed a long-term (1980–2000) increase of ctenophores in the fresh stomach contents of spiny dogfish, *Squalus acanthius* Linnaeus, off the U. S. North Atlantic coast. This method does not yet yield quantitative data on jellyfish or ctenophore abundance as well as feeding rates, but could be improved with knowledge of coincident abundances and digestion times of gelatinous species in the predators' stomachs (Arai et al., 2003).

### Satellite and electronic tracking, and acoustic sampling

Use of predators as samplers might allow location of gelatinous organisms by satellite. Leatherback turtles feed almost exclusively on gelatinous zooplankton and can be routinely tracked by satellite tags (Benson et al., 2007); areas where their tracks are concentrated may indicate jellyfish aggregations. Collaborations between sea turtle and jellyfish researchers would produce important data for both (e.g., Houghton et al., 2006; Witt et al., 2007).

Various attempts have been made to tag jellyfish with fish tags, with limited success. Difficulty arises because the tags sink the jellyfish and migrate out of the gelatinous tissue. Recent successful studies show movements of *Chironex fleckeri* Southcott cubomedusae (Seymour et al., 2004; Gordon & Seymour, 2008). Time-at-depth recorders (TDRs) were glued to the large, rigid swimming bells of these medusae. A TDR was attached by a cable tie to *Chrysaora hysoscella* medusae, which enabled their vertical movements to be tracked (Hays et al., 2008). As tags become increasingly smaller and less expensive, such methods should become more widely applicable.

Acoustics routinely are used to estimate fish abundance, and can estimate jellyfish population abundances as well (e.g., Båmstedt et al., 2003; Brierley et al., 2004; Lynam et al., 2006; Kaartvedt et al., 2007; Colombo et al., 2003, 2008). The most extensive work has been in the Namibian Benguela Current, where distributions and biomass of *Chrysaora hysoscella* and *Aequorea forskalea* Peron & Lesueur jellyfish, Cape horse mackerel, and clupeids were estimated with multifrequency acoustics

(Lynam et al., 2006). Validation of acoustic sampling for robust *P. periphylla* Peron & Lesueur showed that small medusae were missed by this method, but that large ones were detected individually (Båmstedt et al., 2003). Difficulties with acoustical methods would be encountered for flaccid species that do not reflect the acoustic signals well and associated fish confound the acoustic signals. Depth-discrete net sampling should be used to determine the species, sizes, and relative abundances of the fish and jellyfish components of the scattering layers.

Continuous plankton recorder (CPR) surveys

Extensive CPR sampling has been conducted for decades in the North Sea and North Atlantic, and part of the sample analysis includes counting nematocysts. Attrill et al. (2007) documented a positive relationship between jellyfish (nematocysts) in the North Sea and climatic factors (the North Atlantic Oscillation Index, NAOI) during 1958–2000. Gelatinous records from the CPR were used to identify favorable leatherback turtle habitat (Witt et al., 2007). Analysis of cnidarian abundance in CPR data showed that seasonal and decadal patterns, and cnidarian relationships with climate and food indicators differed in the shelf and oceanic regions of the North Atlantic Ocean from 1946 to 2005 (Gibbons & Richardson, 2008). Limitations of the CPR data are that they are from surface waters only and the identities of the nematocyst-bearers are unknown; however, CPR data are collected on vessels of opportunity over vast ocean regions, and the data provide a mostly untapped source of data on cnidarians. Currently, molecular analysis of CPR samples (Kirby & Lindley, 2005) is aiding in identification of soft tissues (Cnidaria and Chordata; P. Licandro, SAHFOS, personal communication).

Video surveys

A towed video-recording system has been used to quantify scyphomedusae and ctenophores relative to environmental conditions (depth, temperature, salinity, dissolved oxygen, and chlorophyll) in the Gulf of Mexico (Purcell et al., 2001a; Graham et al., 2003b). Densities and distributions relative to the physical conditions can be measured vertically as well as over long horizontal distances. Densities estimated with the video system agreed well (<40% difference) with those from a Tucker trawl for medusae >15 cm in diameter (Graham et al., 2003b).

Remotely Operated Vehicles (ROVs) have been used for semi-quantitative sampling of gelatinous species (e.g., Raskoff, 2001; Båmstedt et al., 2003; Raskoff et al., 2005). Because ROVs allow study at great depths, most work has been on deep-living species. Both horizontal and vertical transecting has been conducted. Methods such as nearest-neighbor distances (Mackie & Mills, 1983) and apparent size vs. visible volume (Båmstedt et al., 2003) have been used to estimate densities. Paired lasers allow size calculation. If only the duration of viewing is known, the relative abundances of organisms can be determined (e.g., Raskoff et al., 2005, in press). ROV abundance estimates of *P. periphylla* medusae were roughly twice those from a WP3 net (0.8 $m^2$ mouth area; Båmstedt et al., 2003).

Ocean-surface surveys

When in situ sampling is not possible, visual observations from the surface can yield data on large-scale distributions of large jellyfish (Sparks et al., 2001; Doyle et al., 2007). Doyle et al. (2007) counted large scyphomedusae along regular ferry routes across the Irish and Celtic seas, which provided relative distributions and abundances among species and years. The distances from the ship that jellyfish were visible in different sea-states were determined to ensure comparability of counts. Surface data could be compared with concurrent trawl data in order to convert surface counts to estimated jellyfish abundance in the water column.

Shore-based surveys

Some of the longest records of jellyfish occurrence have been from shore-based surveys. Daily counts from a pier in Chesapeake Bay for 30 years showed that *Chrysaora quinquecirrha* scyphomedusae were most abundant in years of low freshwater input and warm spring temperatures (Cargo & King, 1990). Sting reports from beaches can provide long-term and large-scale records, such as for *Pelagia noctiluca* (Forskal) in the Mediterranean Sea (Bernard et al., 1988). Jellyfish strandings along the Irish and Celtic seacoasts provided data on the relative distributions, abundances, seasonality, and inter-annual variation among scyphomedusan species (Doyle et al., 2007; Houghton et al., 2007). Above-water video has been used to track in situ aggregations of *Aurelia aurita* jellyfish (Fuji et al., 2007), and this technique could be applied for beach stranding surveys.

### Aerial surveys

Near-surface jellyfish aggregations and large jellyfish can be quantified from aerial surveys (e.g., Purcell et al., 2000; Houghton et al., 2006). The numbers of *Aurelia labiata* Chamisso and Eysenhardt aggregations showed great inter-annual variation in Prince William Sound, Alaska, where between 28 and 770 occurred in 1995–1998 (Purcell et al., 2000). Flight-path and targets were recorded by use of a hand-held GPS connected to a laptop computer with a flight log program. The sizes and numbers of the aggregations allowed estimation of surface areas. Details of the aerial methodology are in Brown et al. (1999). Individual *Cyanea capillata* medusae also were visible from the plane. Densities of jellyfish could be estimated if aerial data were combined with in situ sampling of jellyfish densities in the aggregations (e.g., Uye et al., 2003). Aerial surveys can cover large areas at low cost in comparison with sea-going surveys. Fish schools, marine vertebrates, and birds can also be quantified by aerial surveys (Brown et al., 1999; Houghton et al., 2006).

### Modeling of jellyfish population dynamics and ecosystem effects

Key objectives are to understand the causes of variation in jellyfish and ctenophore population sizes, to predict future population sizes, and to estimate their trophic importance. Inter-annual variation in jellyfish occurrence and spatial distribution in relationship to climatic variables have shown associations of large populations with high temperature and salinity for *Pelagia noctiluca* in the Mediterranean Sea (Goy et al., 1989; Molinero et al., 2005) and *Chrysaora quinquecirrha* medusae in Chesapeake Bay (Cargo & King, 1990; Brown et al., 2002; Decker et al., 2007). Generalized additive models (GAM) allow non-linear analysis of variables, which showed the largest populations of *Chrysaora melanaster* medusae occurred in years with moderate temperatures and ice cover in the Bering Sea; biotic variables, such as zooplankton, and fish biomass, were also incorporated into the models (Brodeur et al., 2008a). While such models identify possible causes of past jellyfish blooms, they also enable prediction of abundances in future conditions (Goy et al., 1989; Cargo & King, 1990; Decker et al., 2007).

To my knowledge, similar models have not been developed yet for any ctenophore species; however, Kremer (1976) used an energetics model for *Mnemiopsis leidyi* to predict seasonal population dynamics from the measurements of clearance, metabolic, reproduction, and assimilation rates. The model used temperature and zooplankton abundance as forcing functions to estimate population changes. The model later was coupled with a deterministic simulation model for Naragansett Bay, in which zooplankton biomass was not forced (Kremer & Kremer, 1982).

Several ecosystem models have incorporated jellyfish or ctenophores (e.g., Baird & Ulanowicz, 1989; Oguz et al., 2001; Oguz, 2005a, b; Ruzicka et al., 2007; reviewed in Pauly et al., 2008). Generally, such efforts suffer from insufficient information on jellyfish biomass and biology (Pauly et al., 2008). Prediction of the responses of jellyfish and ctenophore populations to the multiple changes occurring in the global ocean makes obtaining the necessary data for such modeling studies of great importance.

## Use of feeding data to estimate jellyfish and ctenophore predation on large scales

The diets and predation rates of many gelatinous species have been detailed since the 1970s. Although previously thought to be generalists, most species show various degrees of selectivity (reviewed in Purcell, 1997). Knowledge of such dietary differences is necessary for understanding the roles of jellyfish and ctenophores in the food web. In addition to mesozooplankton and ichthyoplankton, they eat microplankton (e.g., Stoecker et al., 1987a, b; Sullivan & Gifford, 2004; Colin et al., 2005), gelatinous species (reviewed in Purcell, 1997), and emergent zooplankton (Pitt et al., 2008a).

Stable isotope and fatty acid analyses are being used to follow the transfer of the organic matter through the food webs and to understand trophic relationships of gelatinous species (reviewed in Pitt et al., 2008b). Stable isotopes showed that *Catostylus mosaicus* (Quoy and Gaimard) medusae heavily utilized emergent zooplankton, and would be important contributors to benthic–pelagic coupling (Pitt et al., 2008a). Stable isotopes show that *Aurelia* spp. have a lower trophic level than other scyphomedusae (Kohama et al., 2006; Brodeur et al., 2008b), indicating use of microplankton by this genus, which blooms in eutrophic waters around the world. Similarly, *Mnemiopsis leidyi* ctenophores bloom in eutrophic waters, and the diets of young ctenophores < 1 cm length contain high percentages of microplankton (Sullivan & Gifford, 2004; Rapoza et al., 2006), although stable isotope analysis of large *M. leidyi* did not indicate extensive consumption of microplankton (Montoya et al., 1990). The interactions of pelagic cnidarians and ctenophores with microplankton communities generally have been seldom studied (except Pitt et al., 2007, 2008b, c), and this may be especially important for the problem species *Aurelia* spp. and *M. leidyi* in eutrophic waters. Neither stable isotope nor fatty acid analyses provide quantitative feeding rates; however, feeding with $^{14}$C-labeled prey can provide information about the amount of C assimilated (Pitt et al., 2008c).

Feeding rates (FR) of gelatinous species have been estimated by several methods, including prey removal in laboratory containers, in situ gut contents with digestion times, and metabolic rates to indicate minimum (reviewed in Purcell, 1997). Containers generally reduce FR of even small species, and metabolic rates yield minimum consumption estimates; hence, the gut-content method usually gives the highest feeding estimates. While the gut-content method may be most representative of FR on mesozooplankton in situ, it is very labor intensive and time-consuming, and inaccurate for species eating microplankton.

Few studies compare results obtained by the different methods to estimate feeding. FR of *Pleurobrachia* sp. ctenophores estimated by the gut content method always were higher than estimates by the clearance method (in 1,300-l mesocosms; Sullivan & Reeve, 1982). Comparisons for several siphonophore species showed that gut-content estimates were higher than metabolic estimates for large species, but similar for small species, and that clearance rates generally gave the lowest FR estimates (Mackie et al., 1987). In situ ingestion (gut-contents) was similar to laboratory ingestion at low prey densities (5 $l^{-1}$); specific ingestion in situ (2% $d^{-1}$) was similar to specific metabolism (3% $d^{-1}$) for the small siphonophore *Sphaeronectes gracilis* (Claus) (Purcell & Kremer, 1983). Specific rations of the scyphomedusan *Linuche unguiculata* (Swartz), as estimated by gut contents and feeding experiments, were within a factor of two (Kremer, 2005).

Several generalizations result from previous studies (Purcell, 1997). One is that the main prey of most jellyfish and ctenophore species is copepods. A second is that feeding rates increase in proportion to predator size and prey density. A third is that digestion times are inversely correlated with temperature. Therefore, I hypothesize that the feeding rates can be predicted by multiple regressions of predator size, prey densities, and temperature.

## Scyphomedusae

I tested the above hypothesis for scyphomedusae by use of raw data from previous studies on the numbers of copepods in field-collected medusae, medusa size (wet weight (WW) in g), prey density (PD in copepods $m^{-3}$), and temperature (T in °C), in combination with digestion times and medusa size conversions in those publications, to calculate feeding rates (FR in copepods eaten medusa$^{-1}$ $d^{-1}$). I restricted the analyses to studies in which individual medusae were collected by dip net or by SCUBA divers. I added 1 to all gut content data so that zero feeding would not be lost from the analyses. All data, except temperature, were $\log_{10}$ transformed, after which the data met normal-distribution and constant-variance assumptions of the analyses. Outliers greater than 2 standard deviations were identified by studentized residuals and removed from each dataset. Pearson's product moment correlations tested for correlations among all variables. Collinearity was evaluated by means of VIF and Durbin-Watson (D-W) statistics; VIF values $\approx$ 1 and D-W values $\approx$ 2 were acceptable. First, I tested *Aurelia* spp. from habitats differing in medusa size (0.005–1,139 g wet weight), prey densities (388–74,222 copepods $m^{-3}$), and temperature (9–31°C). Data for *Aurelia* spp. were from July 1998 and 1999 in Prince William Sound, Alaska (PWS), March–April 1991 in Southampton water, United Kingdom (UK), August 1991 and May–6 August in the Inland Sea, Japan (ISJ), and May 1997 in Ngermeaungel Lake,

Koror, Palau (NLK); previous analyses are in Purcell (2003), C. H. Lucas (unpublished), Uye & Shimauchi (2005), and Dawson & Martin (2001), respectively. Then, I compared *Aurelia* spp. with two other species, *Chrysaora quinquecirrha* from June –August 1987, June–September 1988 and 1989, August 1990, and July 1991 in Chesapeake Bay, USA (CB; Purcell, 1992) and *Cyanea capillata* from July 1998 and 1999 in PWS (Purcell, 2003).

*Aurelia* spp.

Pearson correlations showed that all variables (FR, T, WW, and PD) were correlated in nearly every combination (Table 1). This is reasonable biologically because after ephyrae are produced in early spring, temperature, prey density, and jellyfish size increase. Temperature was not variable in two cases (PWS and NLK). Where temperature varied (UK, ISJ, and *Aurelia* spp. combined), it was positively correlated with WW and PD; WW was positively correlated with PD. FR generally had much stronger correlations with the other variables (T, WW, and PD) than correlations among those variables.

Because T, WW, and PD logically could affect medusa feeding rates, I conducted multiple regression analyses (Table 2). VIF and D-W statistics showed that multicollinearity existed when all predictor variables (T, WW, and PD) were tested against the dependent variable (FR). Temperature either did not

**Table 1** Pearson product moment correlations of medusa wet weight (WW), prey density (PD), and temperature (T) with the feeding rate (FR; numbers of copepods eaten medusa$^{-1}$ d$^{-1}$)

| Pair of variables | Pearson's correlation | |
|---|---|---|
| | *R* | *P* |
| $FR_{PWS}$ vs. $WW_{PWS}$ | 0.750 | $1.20 \times 10^{-11}$ |
| **$FR_{PWS}$ vs. $PD_{PWS}$** | **0.121** | **0.364** |
| **$WW_{PWS}$ vs. $PD_{PWS}$** | **0.247** | **−0.154** |
| $FR_{UK}$ vs. $WW_{UK}$ | 0.456 | $8.20 \times 10^{-5}$ |
| $FR_{UK}$ vs. $PD_{UK}$ | 0.348 | 0.003 |
| $FR_{UK}$ vs. $T_{UK}$ | 0.329 | 0.006 |
| $WW_{UK}$ vs. $PD_{UK}$ | 0.236 | 0.051 |
| $WW_{UK}$ vs. $T_{UK}$ | 0.225 | 0.062 |
| $PD_{UK}$ vs. $T_{UK}$ | 0.832 | $8.84 \times 10^{-19}$ |
| $FR_{ISJ}$ vs. $WW_{ISJ}$ | 0.604 | $5.02 \times 10^{-8}$ |
| $FR_{ISJ}$ vs. $PD_{ISJ}$ | 0.521 | $5.21 \times 10^{-6}$ |
| $FR_{ISJ}$ vs. $T_{ISJ}$ | 0.367 | 0.002 |
| $WW_{ISJ}$ vs. $PD_{ISJ}$ | 0.399 | <0.001 |
| $WW_{ISJ}$ vs. $T_{ISJ}$ | 0.472 | $4.90 \times 10^{-5}$ |
| $PD_{ISJ}$ vs. $T_{ISJ}$ | 0.551 | $1.15 \times 10^{-6}$ |
| $FR_{NLK}$ vs. $WW_{NLK}$ | 0.755 | $7.75 \times 10^{-28}$ |
| **$FR_{NLK}$ vs. $PD_{NLK}$** | **−0.126** | **0.131** |
| **$WW_{NLK}$ vs. $PD_{NLK}$** | **0.015** | **0.862** |
| $FR_{AUR}$ vs. $WW_{AUR}$ | 0.550 | $2.67 \times 10^{-28}$ |
| **$FR_{AUR}$ vs. $PD_{AUR}$** | **−0.042** | **0.435** |
| $FR_{AUR}$ vs. $T_{AUR}$ | 0.184 | <0.001 |
| **$WW_{AUR}$ vs. $PD_{AUR}$** | **0.089** | **0.101** |
| $WW_{AUR}$ vs. $T_{AUR}$ | 0.354 | $1.70 \times 10^{-11}$ |
| $PD_{AUR}$ vs. $T_{AUR}$ | 0.469 | $4.92 \times 10^{-20}$ |

**Table 1** continued

| Pair of variables | Pearson's correlation | |
|---|---|---|
| | *R* | *P* |
| **$FR_{CYA}$ vs. $WW_{CYA}$** | **−0.0231** | **0.775** |
| $FR_{CYA}$ vs. $PD_{CYA}$ | 0.463 | $1.14 \times 10^{-9}$ |
| $WW_{CYA}$ vs. $PD_{CYA}$ | −0.178 | 0.026 |
| $FR_{CHR}$ vs. $WW_{CHR}$ | 0.547 | $1.59 \times 10^{-31}$ |
| $FR_{CHR}$ vs. $PD_{CHR}$ | 0.207 | $4.13 \times 10^{-5}$ |
| **$FR_{CHR}$ vs. $T_{CHR}$** | **−0.026** | **0.614** |
| $WW_{CHR}$ vs. $PD_{CHR}$ | 0.109 | 0.032 |
| $WW_{CHR}$ vs. $T_{CHR}$ | −0.145 | 0.004 |
| **$PD_{CHR}$ vs. $T_{CHR}$** | **−0.058** | **0.258** |
| $FR_{SCY}$ vs. $WW_{SCY}$ | 0.543 | $7.77 \times 10^{-69}$ |
| **$FR_{SCY}$ vs. $PD_{SCY}$** | **−0.028** | **0.406** |
| $FR_{SCY}$ vs. $T_{SCY}$ | 0.130 | <0.001 |
| $WW_{SCY}$ vs. $PD_{SCY}$ | −0.098 | 0.004 |
| $WW_{SCY}$ vs. $T_{SCY}$ | 0.081 | 0.016 |
| $PD_{SCY}$ vs. $T_{SCY}$ | 0.376 | $5.91 \times 10^{-31}$ |
| $DT_{AUR}$ vs. $T_{AUR}$ | −0.588 | 0.002 |
| $DT_{AUR}$ vs. $WW_{AUR}$ | −0.408 | <0.05 |
| $T_{AUR}$ vs. $WW_{AUR}$ | 0.548 | 0.004 |

*R* = correlation coefficient; *P*, probability; $P > 0.05$ are not significant (marked in bold). Data on *Aurelia* spp. are from Prince William Sound, Alaska (PWS; Purcell, 2003), Southampton, United Kingdom (UK; Lucas, unpublished), the Inland Sea, Japan (ISJ; Uye & Shimauchi, 2005), and Palau (NLK; Dawson & Martin, 2001). AUR, combined *Aurelia* spp.; CHR, *Chrysaora quinquecirrha* (from Purcell, 1992); CYA, *Cyanea capillata* (from Purcell, 2003). SCY, combined scyphozoan species. Digestion times (DT) also were tested against WW and T

**Table 2** *Aurelia* spp., *Cyanea capillata*, *Chrysaora quinquecirrha*: Results of multiple linear regression analyses evaluating the relationships of medusa wet weight (WW), prey density (PD), and temperature (T) with the feeding rate (numbers of copepods eaten medusa$^{-1}$ d$^{-1}$; FR)

| Predator (number examined) location | Wet weight (WW in g) | | Prey density (PD in number m$^{-3}$) | | Temperature (T in °C) | | FR | Multiple $R^2$ and $F$ | $P$ and SE overall | Predictive equation |
|---|---|---|---|---|---|---|---|---|---|---|
| | Range | $t$ and $P$ | Range | $t$ and $P$ | Range | $t$ and $P$ | Range | | | |
| *Aurelia labiata* (58) PWS | 3.4–2,466 | $t = 6.926$<br>$P < 0.001$ | 388–10,211 | $t = 5.867$<br>$P < 0.001$ | 14 | ND | 20–2,183 | $R^2 = 0.587$<br>$F_{2, 55} = 39.01$ | <0.001<br>SE 0.424 | $Log_{10}FR = 0.620{*}Log_{10}WW + 0.791{*}Log_{10}PD - 1.172$ |
| *Aurelia aurita* (70) UK | 0.005–5 | $t = 5.537$<br>$P < 0.001$ | 482.6–7,705 | $t = 1.837$<br>$P = 0.106$ | 9.1–10.3 | $t = 0.936$<br>$P = 0.353$ | 4–150 | $R^2 = 0.406$<br>$F_{3, 66} = 22.18$ | <0.001<br>SE 0.302 | $Log_{10}FR = 0.265{*}Log_{10}WW + 0.181{*}Log_{10}PD + 0.840$ |
| *Aurelia aurita* (68) ISJ | 48–1,440 | $t = 8.451$<br>$P < 0.001$ | 830–13,990 | $t = 3.288$<br>$P = 0.014$ | 16.2–24.8 | $t = 0.626$<br>$P = 0.950$ | 312–96,576 | $R^2 = 0.705$<br>$F_{3, 64} = 77.703$ | <0.001<br>SE 0.283 | $Log_{10}FR = 1.189{*}Log_{10}WW + 0.346{*}Log_{10}PD - 0.314$ |
| *Aurelia aurita* (144) NLK | 3–1,139 | $t = 22.674$<br>$P < 0.001$ | 2,556–74,222 | $t = 2.825$<br>$P = 0.005$ | 31 | ND | 34–28,631 | $R^2 = 0.787$<br>$F_{2, 141} = 260.97$ | <0.001<br>SE 0.285 | $Log_{10}FR = 0.802{*}Log_{10}WW - 0.153{*}Log_{10}PD + 2.11$ |
| *Aurelia* spp. (340) combined | 0.005–1,139 | $t = 35.613$<br>$P < 0.001$ | 388–74,222 | $t = 3.481$<br>$P = 0.01$ | 9.1–31 | $t = 0.218$<br>$P = 0.827$ | 4–96,576 | $R^2 = 0.809$<br>$F_{3, 336} = 714.29$ | <0.001<br>SE 0.504 | $Log_{10}FR = 0.643{*}Log_{10}WW + 0.178{*}Log_{10}PD + 1.208$ |
| *Chrysaora quinquecirrha* (386) CB | 0.007–146 | $t = 12.052$<br>$P < 0.001$ | 400–232,218 | $t = 8.757$<br>$P < 0.001$ | 22.9–29.1 | $t = 5.076$<br>$P < 0.001$ | 1–17,011 | $R^2 = 0.455$<br>$F_{4, 381} = 106.55$ | <0.001<br>SE 0.359 | $Log_{10}FR = 0.367{*}Log_{10}WW + 0.258{*}Log_{10}PD + 1.447$ |
| *Cyanea capillata* (156) PWS | 1.4–1,642 | $t = 5.702$<br>$P < 0.001$ | 203–10,211 | $t = 6.255$<br>$P < 0.001$ | 14 | ND | 12–5,148 | $R^2 = 0.284$<br>$F_{2, 152} = 30.41$ | <0.001<br>SE 0.475 | $Log_{10}FR = 0.389{*}Log_{10}WW + 0.670{*}Log_{10}PD - 0.512$ |
| Scyphomedusae (882) combined | 0.005–1,642 | $t = 30.296$<br>$P < 0.001$ | 203–232,218 | $t = 9.533$<br>$P < 0.001$ | 9.1–31 | $t = 11.800$<br>$P < 0.001$ | 1–96,576 | $R^2 = 0.672$<br>$F_{3, 878} = 599.34$ | <0.001<br>SE 0.486 | $Log_{10}FR = 0.501{*}Log_{10}WW + 0.512{*}Log_{10}PD + 0.145$ |

Large medusae with < 10 prey were omitted from these analyses (numbers examined are in parentheses). $t$ = $t$ statistic; $P$ = probability, where $P > 0.05$ are not significant; $R^2$ = coefficient of determination; $F$, F statistic; SE, standard error; PWS, Prince William Sound, Alaska (Purcell, 2003); UK, Southampton, United Kingdom (Lucas, unpublished); ISJ, Inland Sea of Japan (Uye & Shimauchi, 2005); NLK, Palau (Dawson & Martin, 2001); CB, Chesapeake Bay (Purcell, 1992). ND, no data

vary or was not significant (Table 2), and multicollinearity was resolved by the removal of temperature from the predictive regressions.

Feeding rates on copepods by *Aurelia* spp. showed great variability, but were strongly correlated with medusa size and prey density, but not temperature (Table 2, Fig. 1). WW had stronger effects on FR in all cases than PD. In the Inland Sea of Japan (ISJ), the copepods were mainly (85.6%) cyclopoids (*Oithona* spp.), and medusae there, especially large ones, showed greater feeding than other locations where calanoid copepods predominated (80–100%) (Fig. 1). Southampton (UK) had the smallest medusae, the coolest temperatures, and lowest feeding. Although the Palau marine lake (NLK) had higher prey densities than Japan (ISJ), medusa feeding was lower in NLK, presumably because of smaller medusa size and warmer temperatures.

The reliability of predicting FR is of key importance. The predicted residual error sum of squares (PRESS) statistic indicates how well a regression predicts new data, with small values (>0) being best (SPSS, 1997). The PRESS statistics indicated that all local regressions would be reasonable predictors of feeding; however, the combined equation (PRESS = 87.1) was a relatively worse predictor (Table 3). In order to test the predictions by the regressions vs. the measured feeding rate data, I entered WW and PD from each dataset into their own (local) and combined regressions (Table 3). Measured *Aurelia* spp. data in local regressions underestimated feeding by 14–45%. The combined regression overestimated measured FR for PWS medusae by 92%. FR of UK and NLK medusae were predicted well by the combined regression; however, FR of the ISJ medusae estimated by the combined regression was only one-fifth of the measured rate. This poor result for ISJ was probably due to the different prey available (small *Oithona* spp. in ISJ but mostly calanoids in PWS, UK, and NLK). Therefore, except for ISJ, where prey differed dramatically from other habitats, the combined *Aurelia* spp. regression estimated feeding rates as well as did the regressions derived from local data.

I also tested measured FR data for *Aurelia* spp. that had not been used to develop the regressions ('novel') against values calculated from the regressions equations. Results for *A. aurita* in Taiwan (Lo & Chen, 2008) differed depending on the net mesh

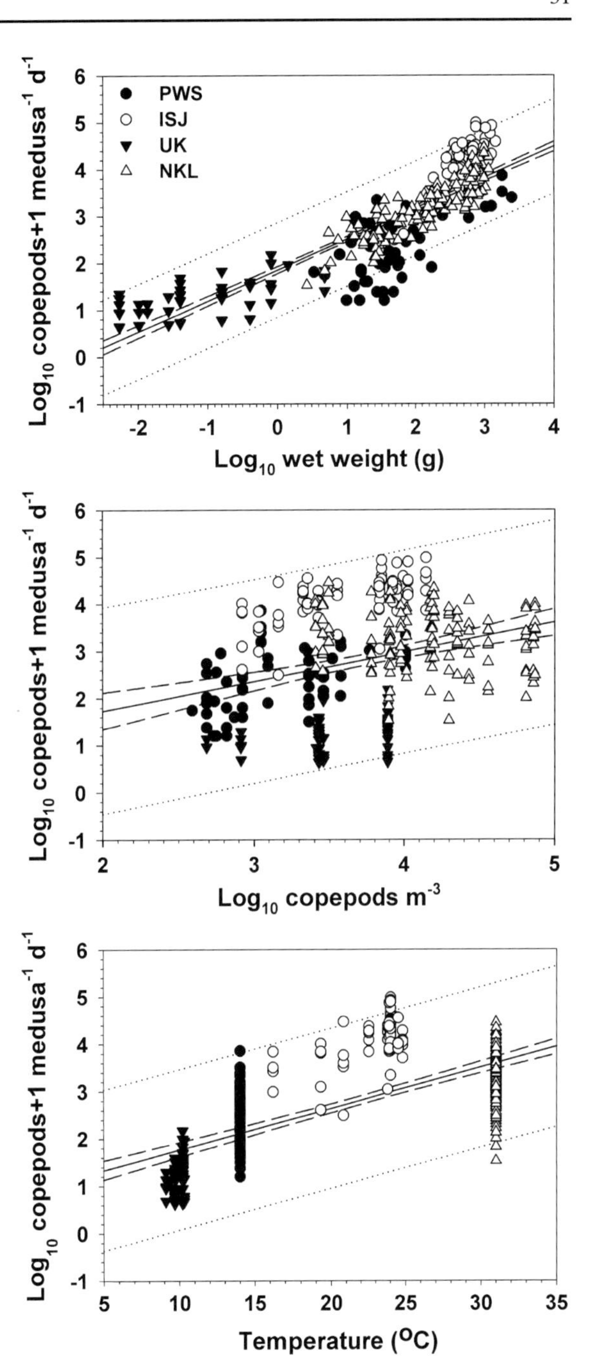


**Fig. 1** Feeding rates ($\log_{10}$ number of copepods + 1 eaten $d^{-1}$) of individual scyphomedusae (*Aurelia* spp.) from field gut contents vs. medusa wet weight (top), prey density (middle), and temperature (bottom). Medusae were collected from Prince William Sound, Alaska (PWS; Purcell, 2003), the Inland Sea, Japan (ISJ; Uye & Shimauchi, 2005), Southampton waters, United Kingdom (UK; C.-H. Lucas, unpublished), and Ngermeaungel Lake, Koror, Palau (NLK; Dawson & Martin, 2001). Lines are: solid, linear regressions; dashed, 95% confidence intervals; dotted, prediction errors

**Table 3** Parameters (means ± standard error) used in regression analyses of feeding rates (FR in copepods $medusa^{-1}$ $d^{-1}$) and clearance rate (CR in liters cleared $ctenophore^{-1}$ $d^{-1}$). Mesh refers to plankton netting

| Species and location | T (°C) | WW (g) | PD (number $m^{-3}$) | Mesh (μm) | PRESS | Feeding rate (FR) or clearance rate (CR) ratio (measured/local or combined) | | | | | Original reference |
|---|---|---|---|---|---|---|---|---|---|---|---|
| | | | | | | Measured | Local or species | | Combined | | |
| | | | | | | FR or CR | FR or CR | Ratio | FR or CR | Ratio | |
| *Aurelia labiata* PWS | 14 | 197.4 ± 63.9 | 2,488.3 ± 378.6 | 243 | 10.9 | 640.4 ± 144.7 | 440.7 ± 66.7 | 1.45 | 1,229.4 ± 209.3 | 0.52 | Purcell (2003) |
| *A. labiata* Oregon (6) | ND | 251.3 ± 102.6 | 164.1 | 202 | – | 729.7 ± 297.9 | 219.6 ± 20.3 | 3.32 | 2,699.4 ± 257.3 | 0.27 | Suchman et al. (2008) |
| *Aurelia aurita* UK | 9.9 ± 0.05 | 0.36 ± 0.12 | 4,703.6 ± 351.4 | 200 or 212 | 6.6 | 21.4 ± 3.0 | 17.9 ± 1.5 | 1.20 | 23.5 ± 5.0 | 0.91 | Lucas (unpublished) |
| *A. aurita* ISJ | 22.9 ± 0.3 | 553.7 ± 36.8 | 5,691.3 ± 484.9 | 100 | 5.6 | 20,928.4 ± 2,434.6 | 18,403.8 ± 1,462.2 | 1.14 | 4,134.6 ± 187.5 | 5.06 | Uye & Shimauchi (2005) |
| *A. aurita* NLK | 31 | 338.6 ± 27.3 | 22,842.2 ± 1,910.3 | 80 | 11.9 | 3,734.0 ± 403.6 | 2,967.0 ± 216.9 | 1.26 | 3,330.8 ± 174.9 | 1.12 | Dawson & Martin (2001) |
| *A. aurita* Taiwan | 22.3 | 96.0 | 5,854 | 100 | – | 1,193 | – | – | 1,433.1 | 0.83 | Lo & Chen (2008) |
| | 22.3 | 96.0 | 653 | 330 | | 1,193 | | | 963.1 | 1.24 | |
| *Chrysaora quinquecirrha* | 27.0 ± 0.1 | 12.5 ± 0.9 | 30,021.7 ± 1,790.0 | 64 | 50.4 | 1,162.8 ± 90.0 | 793.1 ± 27.8 | 1.46 | 773.3 ± 37.3 | 1.50 | Purcell (1992) |
| *Chrysaora fuscescens* (30) | 8.4 ± 0.2 | 489.6 ± 42.0 | 6,308.1 ± 1,074.9 | 202 | – | 401.7 ± 60.6 | 2,313.4 ± 142.9 | 0.17 | 2,391.2 ± 263.7 | 0.17 | Suchman et al. (2008) |
| *Cyanea capillata* PWS | 14 | 144.1 ± 19.8 | 1,637.8 ± 141.7 | 243 | 35.8 | 377.2 ± 48.8 | 209.2 ± 11.7 | 1.80 | 501.8 ± 28.0 | 0.75 | Purcell (2003) |
| *Mnemiopsis leidyi* 20–55 l | 21 | 3.1 ± 0.4 | 42.7 ± 3.3 | – | 6.9 | 9.0 ± 1.0 | 9.2 ± 1.1 | 1.04 | 6.6 ± 0.6 | 1.16 | Kremer & Reeve (1989) |
| *M. leidyi* 90 l | 22.5 | 16.8 ± 1.2 | 37.8 ± 1.3 | – | 6.0 | 17.9 ± 1.9 | 14.6 ± 0.9 | 1.22 | 20.5 ± 1.4 | 0.90 | Decker et al. (2004) |
| *M. leidyi* 1,000 l | 25 | 10.7 ± 0.8 | 17.4 ± 2.1 | – | 0.4 | 45.0 ± 5.1 | 40.2 ± 2.6 | 1.10 | 17.0 ± 1.0 | 2.59 | Purcell (unpublished) |

Ratios of measured feeding rates or clearance rates to those calculated from the local regressions (for *Aurelia aurita*) or species regressions (for species other than *A. aurita*). Abbreviations are as in Table 1. PRESS Statistic indicates how well regression predicts new data, with small values being best. ND, no data

size used for zooplankton data. Use of mean values of in situ variables, including 100-μm copepod data, in the combined *Aurelia* spp. regression underestimated the measured FR by 20%, and use of 330-μm copepod data overestimated measured FR by 24% (Table 3). The mean FR of *Aurelia labiata* medusae off the Oregon coast was 3.3 times that predicted by the PWS regression, and only 27% of that predicted by the combined *Aurelia* spp. regression. Those differences may be due to the low FR of PWS *A. labiata* relative to the *Aurelia* spp. data generally, and to the relatively high rates of the combined regression, as influenced by ISJ data. These comparisons show that novel FR data only sometimes compare very well with the combined *Aurelia* spp. calculated FR, and that net mesh-size affects the FR estimates.

*Other scyphomedusae*

Feeding rates of *Chrysaora quinquecirrha* and *Cyanea capillata* medusae were also variable, but very similar to FR of *Aurelia* spp. (Fig. 2). Data for *C. quinquecirrha* medusae overlapped the *Aurelia* spp. data by WW, although FR of *C. capillata* medusae were somewhat lower than *Aurelia* spp. at the same sizes (Fig. 2), possibly because prey are digested more rapidly by *C. capillata*. Data for ISJ *A. aurita* noticeably differed from the others due to high FR on small *Oithona* spp. vs. PD; prey available and eaten in PWS and CB were mostly calanoids. As for *Aurelia* spp., the variables (T, WW, PD, and FR) were all correlated, except that FR of *C. quinquecirrha* was not correlated with T (Table 1). Therefore, T was removed from the predictive equation (Table 2), which reduced the $R^2$ (0.455–0.419) and prediction (PRESS values 50.4–53.3) to some extent. WW was more important in determining FR than PD for *C. quinquecirrha* but similar for FR of *C. capillata*.

When *Chrysaora quinquecirrha* and *Cyanea capillata* FR data were combined with the *Aurelia* spp. data, the variables were all correlated, except that FR of scyphomedusae combined was not correlated with PD (Table 1). The overall fit of the scyphomedusa FR regression was reduced ($R^2 = 0.672$), although the overall regression and all variables were highly significant ($P < 0.001$; Table 2). Removal of T from the predictive regression eliminated multicollinearity and a failed constant variance assumption.

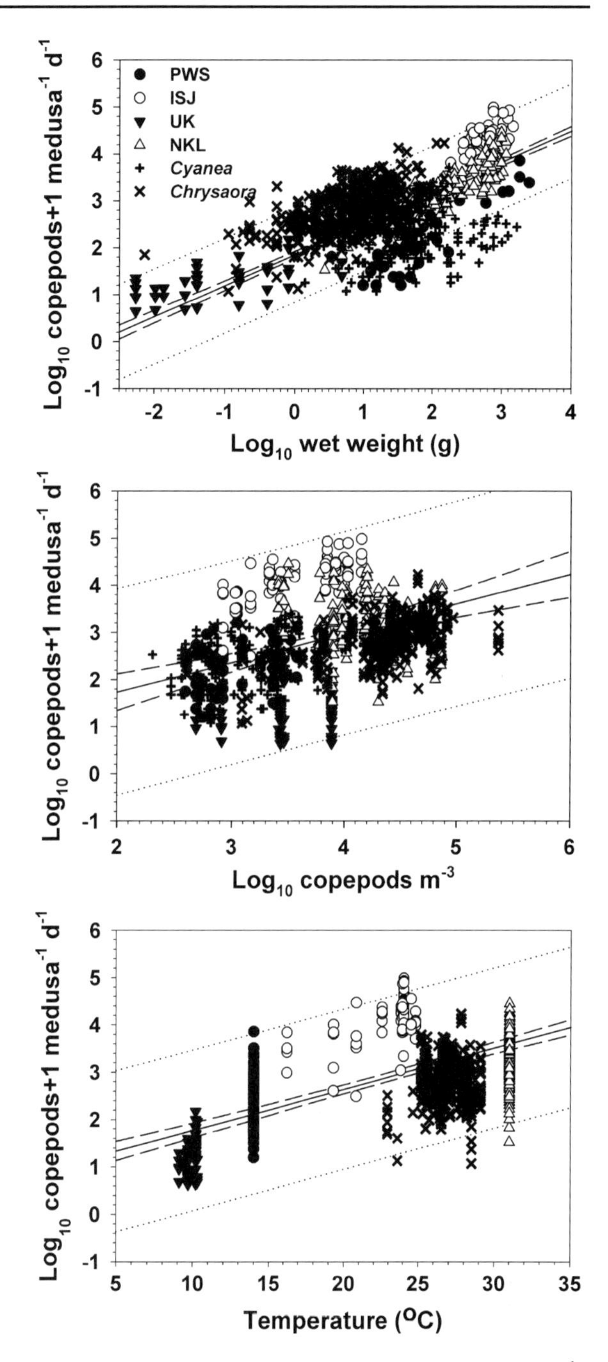


**Fig. 2** Feeding rates ($\log_{10}$ number of copepods + 1 eaten $d^{-1}$) by individual scyphomedusae vs. medusa wet weight (top), prey density (middle), and temperature (bottom) as in Fig. 1 for *Aurelia* spp., with the addition of *Chrysaora quinquecirrha* from Chesapeake Bay (Purcell, 1992) and *Cyanea capillata* from Prince William Sound, Alaska (Purcell, 2003). Location abbreviations for *Aurelia* spp. and lines are as in Fig. 1

In order to test FR predicted by the species regressions and the combined scyphomedusa regression against the measured FR data, I entered WW and

**Table 4** Conditions of experiments to measure digestion rates of scyphomedusae feeding on copepods

| Medusa species | Diameter (mm) | WW (g) | Copepod species | Size (mm) | T (°C) | Digestion time (h) | Reference |
|---|---|---|---|---|---|---|---|
| *Aurelia aurita* | 39 | 3.7 | *Calanus finmarchicus* (Gunner) | 2 | 10 | 2.14–2.51 | Båmstedt & Martinussen (2001) |
| *A. aurita* | 11–14 | 0.1–0.2 | *C. finmarchicus* | 2 | 10 | 5.4–7.7 | Martinussen & Båmstedt (2001) |
| *A. aurita* | 13–15 | 0.15–0.23 | *Pseudocalanus* sp. | 0.76–1 | 9.5–10 | 3.21–6.26 | Martinussen & Båmstedt (1999, 2001) |
| *A. aurita* | 5 | 0.01 | *Pseudocalanus* sp. | 0.76 | 4.5 | 3.63 | Martinussen & Båmstedt (2001) |
| *A. aurita* | 100 | 54.2 | *Acartia omorii* Bradford | 0.5 | 19 | 1 | Uye & Shimauchi (2005) |
| *A. aurita* | 100 | 54.2 | *Oithona davisae* Ferrari and Orsi | 0.5 | 19 | 1 | Uye & Shimauchi (2005) |
| *A. aurita* | 60 | 12.6 | *Acrocalanus* sp. | 0.5 | 30 | 0.71 | Dawson & Martin (2001) |
| *A. aurita* | 50 | 7.5 | Mixed | 0.5 | 22.5 | 1.6 | Lo & Chen (2008) |
| *A. aurita* | 160 | 207.8 | Mixed | 0.5 | 22.5 | 2.05 | Lo & Chen (2008) |
| *A. aurita* | 30 | 1.7 | Calanoids | 0.75 | 7 | 3.5 | Sullivan et al. (1994) |
| *A. aurita* | 80 | 28.6 | Calanoids | 0.5 | 4 | 3.85 | Matsakis & Conover (1991) |
| *A. aurita* | 166 | 230.9 | *Oithona* sp. | 0.5 | 22 | 0.95 | Ishii & Tanaka (2001) |
| *A. labiata* | 110.6 | 72.3 | Calanoids | 1 | 14 | 3 | Purcell (2003) |
| *Cyanea capillata* | 65.5–71.5 | 38.6–49.3 | *C. finmarchicus* | 1.5 | 9.5 | 1.5–2.0 | Martinussen & Båmstedt (1999) |
| *C. capillata* | 102.6 | 135.6 | Calanoids | 1 | 14 | 2 | Purcell (2003) |
| *Chrysaora quinquecirrha* | 25–126 | 1.1–170.4 | *Acartia tonsa* Dana | 1 | 20–27 | 2.5–4.7 | Purcell (1992); $Y = 10.86 - 0.31T$ |

Data were used in regression analyses (Fig. 3). WW, wet weight; T, temperature

PD from each dataset into their own (local) and the combined regressions (Table 3). The combined *Aurelia* spp. data in the combined scyphomedusa regression gave mean FR of 2,061.1 ± 133.3 copepods medusa$^{-1}$ d$^{-1}$, as compared with the measured FR of 5,899.6 ± 661.8 copepods medusa$^{-1}$ d$^{-1}$; thus, the scyphomedusa regression underestimated feeding by *Aurelia* spp. by 180%. Measured *Chrysaora quinquecirrha* and *Cyanea capillata* data in local regressions underestimated feeding by 46 and 80%, respectively. The combined regression overestimated measured FR for *C. quinquecirrha* medusae by 50% and underestimated measured FR for *C. capillata* medusae by 25%. I also tested novel measured FR data for *Chrysaora fuscescens* Brandt against FR calculated from the *C. quinquecirrha* regression and the combined scyphomedusa regression (Table 3). Both regressions predicted FR of *C. fuscescens* poorly, overestimating measured FR by nearly sixfold.

*Digestion times*

In order to calculate feeding rates from gut contents, digestion times (DT) need to be measured. Rates for *Aurelia* spp. have been measured repeatedly (reviewed in Martinussen & Båmstedt, 2001; Hansson et al., 2005). I developed a multiple regression equation for *Aurelia* spp. using DT data measured at ambient temperatures (Table 4). I did not use DT measured at experimentally altered temperatures (Martinussen & Båmstedt, 2001), which may affect the rates. I did not consider the possible effects of prey number and size, which affect DT of small *A. aurita* (Martinussen & Båmstedt, 1999).

DT of *Aurelia* spp. were negatively and significantly correlated with both T and WW (Table 1, Fig. 3). WW were positively correlated with T, which must be due to an experimental artifact; the smallest medusae and no large medusae were tested at cold temperatures (Tables 1, 4). Multiple linear regression

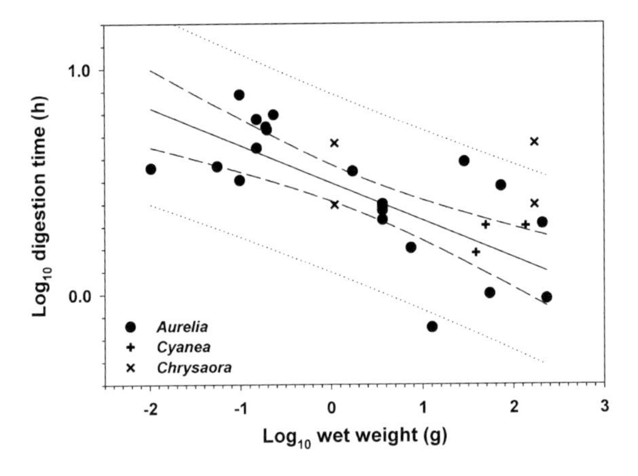


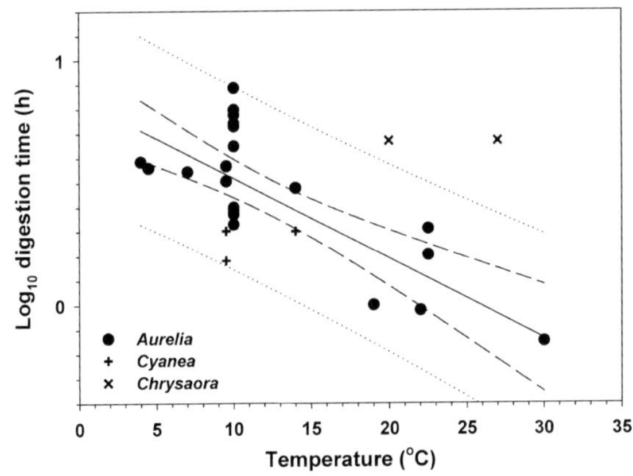


**Fig. 3** Digestion times ($\log_{10}$ h) for scyphomedusae eating copepods vs. medusa wet weight (top) and temperature (bottom) for *Aurelia* spp. Points for *Chrysaora quinquecirrha* and *Cyanea capillata* are shown for comparison but are not included in the regression. Data and sources are in Table 4. Lines are as in Fig. 1

of $\log_{10}$ –transformed data passed assumptions of the analysis and showed no multicollinearity. DT were significantly and negatively related to temperature and jellyfish size ($F_{2,\ 21} = 23.272$; $P < 0.001$; $R^2 = 0.689$) (Fig. 3). There was a significant effect of temperature (T; $t = -3.305$; $P = 0.003$), with digestion by jellyfish of similar size (12.6–28.6 g WW) ranging between 0.71 h at 30°C and 3.85 h at 4°C. This is equivalent to a $Q_{10}$ of 2.08. Rapid digestion at warm temperatures may have been exacerbated by small prey sizes at those locations (Dawson & Martin, 2001; Ishii & Tanaka, 2001; Uye & Shimauchi, 2005; Lo & Chen, 2008). Jellyfish size also was significant (WW; $t = -2.680$; $P = 0.014$); digestion of copepods by jellyfish < 0.3 g WW (<15 mm diameter) required very long times (5–8 h) relative to larger jellyfish. Digestion times (DT in h) for *Aurelia* spp. jellyfish could be predicted according to the following equation: $\log_{10} DT = 0.745 - (0.0943 * \log_{10} WW) - (0.0211 * T)$. DT for *Cyanea capillata* and *Chrysaora quinquecirrha* were similar to those for *Aurelia* spp. medusae of similar sizes (Fig. 3), although DT for *C. capillata* were shorter, and DT for *C. quinquecirrha* were longer than those for *Aurelia* spp. medusae at similar temperatures (Fig. 3).

## Ctenophores

Most studies of feeding rates of *Mnemiopsis leidyi* (called *M. mccradyi* in some publications) ctenophores have been in experimental containers from 4 to 1,000 l volume (reviewed in Purcell et al., 2001b; see also Kremer & Reeve, 1989; Decker et al., 2004; Purcell & Decker, 2005). Clearance rates (CR in liters cleared of prey ctenophore$^{-1}$ d$^{-1}$) usually are presented relative to ctenophore size (Table 5). I reanalyzed raw data for ctenophores (mostly lobates > 1 cm) feeding on copepods from earlier publications (Kremer & Reeve, 1989; Purcell et al., 2001b; Decker et al., 2004). I did not include data for cydippid larvae feeding on nauplii and microplankton (Stoecker et al., 1987a; Sullivan & Gifford, 2004; Finenko et al., 2006). I only used data measured at ambient temperatures because adjustment to new temperatures might affect feeding. CR measured in small containers seem low relative to those measured in large containers, and probably are not representative of in situ rates (Purcell et al., 2001b). This has not been tested directly, and probably depends on ctenophore size. Therefore, I tested *M. leidyi* CR measured in 3.5–1,000-l containers vs. ctenophore size, prey density, and temperature with Pearson product moment correlations and regressions (Tables 3, 5, 6, Figs. 4, 5).

CR on copepods were strongly correlated with ctenophore size (WW) in the three studies separately and combined (Table 5, Fig. 4). Prey densities (PD) were not significantly correlated with CR in any study. CR increased with container volume (CV). Experimental conditions were co-correlated in the combined analysis. CV was correlated with WW, because containers (3.5–55 l) were chosen according to ctenophore size and differed between datasets. PD was correlated with CV and T, because very high (100 and 200 copepods l$^{-1}$) prey densities were used

**Table 5** Pearson product moment correlations of *Mnemiopsis leidyi* ctenophore wet weight (WW in g), prey density (PD in copepods $l^{-1}$), temperature (T in °C), and container volume (CV in liters) with the clearance rate (CR in liters cleared ctenophore$^{-1}$ d$^{-1}$)

| Pair of variables | Pearson's correlation | |
|---|---|---|
| | *R* | *P* |
| $CR_{90}$ vs. $WW_{90}$ | 0.579 | $3.42 \times 10^{-10}$ |
| **$CR_{90}$ vs. $PD_{90}$** | **−0.067** | **0.511** |
| $CR_{90}$ vs. $T_{90}$ | 0.194 | 0.054 |
| **$WW_{90}$ vs. $PD_{90}$** | **0.018** | **0.857** |
| $WW_{90}$ vs. $T_{90}$ | 0.440 | $5.23 \times 10^{-6}$ |
| **$T_{90}$ vs. $PD_{90}$** | **0.058** | **0.568** |
| $CR_{1000}$ vs. $WW_{1000}$ | 0.410 | 0.01 |
| **$CR_{1000}$ vs. $PD_{1000}$** | **−0.312** | **0.068** |
| **$CR_{1000}$ vs. $CV_{1000}$** | **−0.0386** | **0.826** |
| $WW_{1000}$ vs. $CV_{1000}$ | 0.438 | 0.008 |
| **$WW_{1000}$ vs. $PD_{1000}$** | **−0.179** | **0.304** |
| $PD_{1000}$ vs. $CV_{1000}$ | −0.351 | 0.038 |
| $CR_{55}$ vs. $WW_{55}$ | 0.865 | $8.45 \times 10^{-64}$ |
| **$CR_{55}$ vs. $PD_{55}$** | **0.122** | **0.080** |
| $CR_{55}$ vs. $CV_{55}$ | 0.424 | $1.61 \times 10^{-10}$ |
| $WW_{55}$ vs. $CV_{55}$ | 0.217 | 0.002 |
| $WW_{55}$ vs. $PD_{55}$ | 0.369 | $3.76 \times 10^{-8}$ |
| $PD_{55}$ vs. $CV_{55}$ | −0.051 | 0.461 |
| $CR_{MN}$ vs. $WW_{MN}$ | 0.565 | $2.91 \times 10^{-30}$ |
| $CR_{MN}$ vs. $PD_{MN}$ | −0.166 | 0.002 |
| $CR_{MN}$ vs. $CV_{MN}$ | 0.459 | $2.70 \times 10^{-19}$ |
| $CR_{MN}$ vs. $T_{MN}$ | 0.601 | $4.84 \times 10^{-35}$ |
| $WW_{MN}$ vs. $CV_{MN}$ | 0.139 | 0.010 |
| **$WW_{MN}$ vs. $PD_{MN}$** | **−0.053** | **0.331** |
| $T_{MN}$ vs. $WW_{MN}$ | 0.378 | $4.54 \times 10^{-13}$ |
| $PD_{MN}$ vs. $CV_{MN}$ | −0.245 | $3.33 \times 10^{-6}$ |
| $T_{MN}$ vs. $PD_{MN}$ | −0.363 | $3.94 \times 10^{-12}$ |
| $T_{MN}$ vs. $CV_{MN}$ | −0.0231 | 0.775 |

*R*, correlation coefficient; *P*, probability; $P > 0.05$ are not significant (marked in bold). Data are from 90-l containers (90; Decker et al., 2004), 1,000-l mesocosms (1,000; Purcell, unpublished), ≤55-l containers (55; Kremer & Reeve, 1989). MN, combined *Mnemiopsis*

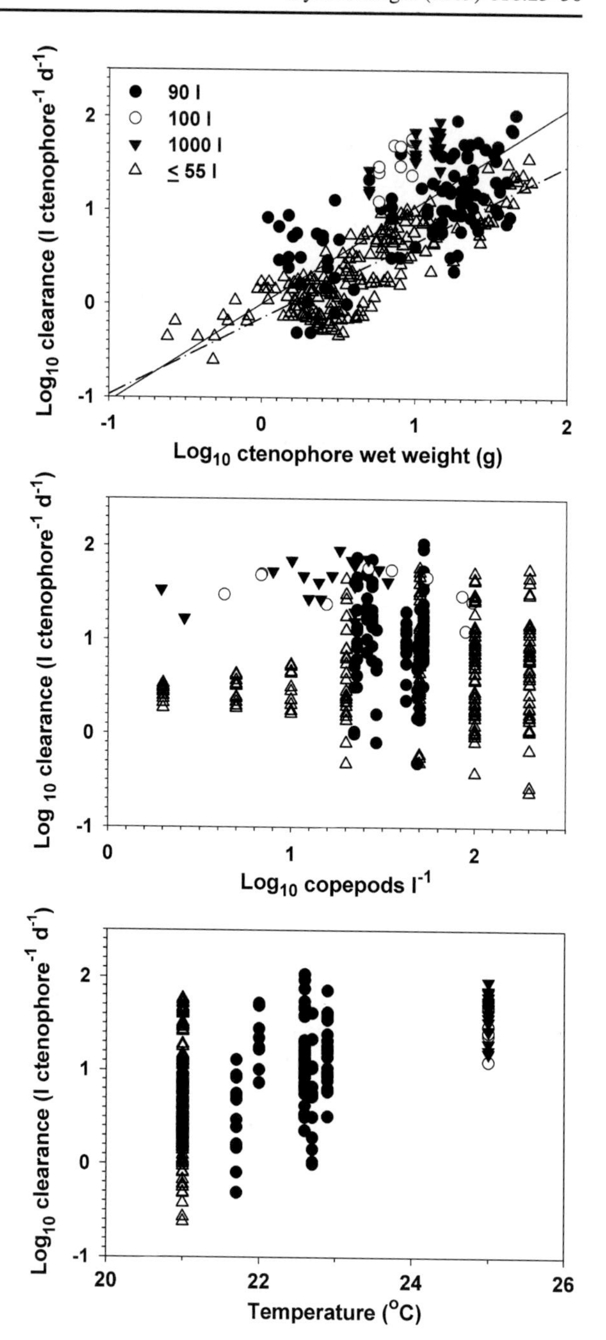


**Fig. 4** Clearance rates ($\log_{10}$ liters cleared ctenophore$^{-1}$ d$^{-1}$) for *Mnemiopsis leidyi* feeding on copepods vs. wet weight (top), prey density (middle), and temperature (bottom). Laboratory experiments were conducted in containers of 90-l (Decker et al., 2004), 100- and 1,000-l (Purcell, unpublished), and 3.5–55-l (Kremer & Reeve, 1989). Regression lines shown vs. wet weight are dot-dash for ≤55-l and solid for others combined

only in 3.5–55-l containers, while PD were lowest and T highest in 1,000-l containers. For subsequent regression analyses, I removed PD (not significant) and T, which was over a small range (21–25°C) and was correlated with other experimental conditions.

Because of the strong correlations between CR and CV, additional analyses were made for *Mnemiopsis leidyi*. Toonen & Chia (1993) recommended a ratio of

**Table 6** Clearance rates (CR in liters cleared ctenophore$^{-1}$ d$^{-1}$) of *Mnemiopsis leidyi* in relation to container size

| Ctenophore size (mm) | Mean or median clearance rate (number) | | | | | | F or H statistic | *P* |
|---|---|---|---|---|---|---|---|---|
| | Container volume (l) | | | | | | | |
| | 3.5 | 20 | 40 | 55 | 90 | 1,000 | | |
| 10–14 | – | 1.85 (37) *a* | 4.21 (12) *b* | – | 6.70 (3) *b* | – | $H_2 = 27.729$ | <0.001 |
| 15–20 | 1.60 (10) *a* | – | – | 3.70 (25) *b* | 3.15 (25) *b* | – | $H_2 = 9.861$ | 0.007 |
| 20–30 | 6.50 (12) | 3.64 (5) *a* | 5.58 (3) | 9.36 (13) *b* | 4.86 (9) | 16.74 (3) *b* | $H_5 = 16.221$ | 0.006 |
| 30–40 | 8.20 (19) *a* | 9.15 (6) *a* | – | 24.15 (9) *b* | 8.32 (6) *a* | – | $F_{3,36} = 22.440$ | <0.001 |
| 40–50 | 8.43 (9) *a* | – | – | 30.80 (6) *b* | 8.16 (8) *a* | 41.77 (9) *b* | $H_3 = 23.000$ | <0.001 |

Statistical tests were one way ANOVA (F statistic) or Kruskal-Wallis one way ANOVA on Ranks (H statistic) by ctenophore size. Numbers of replicates are in parentheses. Different letters (*a, b*) indicate significantly different groups determined by multiple comparison procedures (Dunn's Method). *P*, probability; $P > 0.05$ are not significant

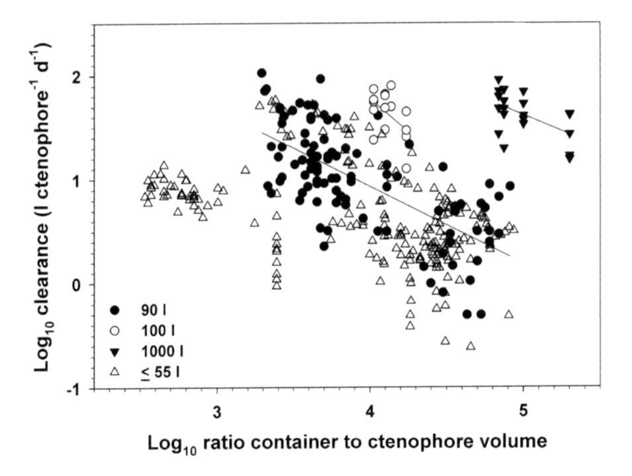


**Fig. 5** Effect of container size (3.5–1,000 l) on clearance rates ($\log_{10}$ l cleared ctenophore$^{-1}$ d$^{-1}$) for *Mnemiopsis leidyi* feeding on copepods in laboratory experiments (as in Fig. 4). Clearance rates are plotted against the $\log_{10}$ ratios of container volume to ctenophore volume. The regression line for containers ≤55 l is omitted for clarity

container volume to jellyfish volume of ≥15,000:1 for the small, ambush predator, *Proboscidactyla flavicirrata* (Brandt), otherwise, feeding by the hydromedusan was affected. Comparison of the ratios of CV to ctenophore volume (~WW) showed that when ratios were <2,500:1, CR were reduced (Fig. 5). CR of ctenophores of similar size in containers of different sizes (3.5–1,000 l) were greater in the larger containers ($P \leq 0.006$), and greatest in 1,000-l (Table 6). Therefore, for subsequent regression analyses of CR, I removed data from 3.5- and 4-l containers, which all had low volume ratios. The remaining data had volume ratios of 2,500 to 200,000. I also removed data from experiments using 200 prey l$^{-1}$ (from Kremer & Reeve, 1989), which was a much higher PD than that used in the other studies.

The individual and combined regressions of WW on CR were strong (Table 7). Thus, clearance rates of *Mnemiopsis leidyi* feeding on copepods in situ can be estimated from data on ctenophore size. The reliability of predicting CR is of great importance; therefore, to test predictions of the regressions vs. the measured CR, I entered WW from each dataset into their own (local) and the combined regression (Table 3). Measured *M. leidyi* CR matched CR from the 20–55-l regression (+2%) and underestimated CR in 90- and 1,000-l regressions by 10–22%. The combined *M. leidyi* CR regression underestimated measured CR in 20–55-l by 16% and CR in 1,000-l by 159%; CR in 90-l containers were overestimated by 10%. The PRESS statistics indicated that all local regressions would be good predictors of feeding, while the combined equation (PRESS = 28.8) was a relatively worse predictor.

## Use of metabolic rates to estimate jellyfish and ctenophore predation on large scales

Respiration and excretion are basic physiological processes that are related to body mass, temperature, and activity for all animals. They have been used to estimate the minimum food requirements and ingestion for some gelatinous species (e.g., Ishii & Tanaka, 2006). Although metabolic rates yield low feeding estimates because they usually are measured on unfed animals and also lack estimates for growth (but see

**Table 7** Clearance rate (CR; liters cleared ctenophore$^{-1}$ d$^{-1}$) equations for *Mnemiopsis leidyi* ctenophores calculated from regression analysis (last equation, combined)

| Container volume (l) (number examined) | Wet weight (WW in g) | | CR | $R^2$ and $F$ | $P$ and SE | Predictive equation |
|---|---|---|---|---|---|---|
| | Range | $t$ and $P$ | Range | | | |
| 20–55 (126) | 0.25–24.2 | $t = 19.876$ | 0.4–58.1 | $R^2 = 0.761$ | $P < 0.001$ | $Log_{10}CR = 0.945*Log_{10}WW + 0.416$ |
| | | $P < 0.001$ | | $F_{1, 124} = 395.054$ | SE 0.232 | |
| 1,000 (16) | 5.0–14.5 | $t = 2.918$ | 15.6–72.3 | $R^2 = 0.476$ | $P = 0.003$ | $Log_{10}CR = 0.843*Log_{10}WW + 0.715$ |
| | | $P = 0.003$ | | $F_{1, 14} = 12.722$ | SE 0.158 | |
| 90 (78) | 1.1–44.0 | $t = 4.839$ | 1.0–74.6 | $R^2 = 0.589$ | $P < 0.001$ | $Log_{10}CR = 0.699*Log_{10}WW + 0.361$ |
| | | $P < 0.001$ | | $F_{1, 88} = 108.741$ | SE 0.275 | |
| Combined (220) | 0.25–44.0 | $t = 23.496$ | 0.4–74.6 | $R^2 = 0.717$ | $P < 0.001$ | $Log_{10}CR = 0.766*Log_{10}WW + 0.423$ |
| | | $P < 0.001$ | | $F_{1, 247} = 552.072$ | SE 0.281 | |

$t$, $t$ statistic; $P$, probability where $P > 0.05$ are not significant; $R^2$, coefficient of determination; $F$, F statistic; SE, standard error

Møller & Riisgård, 2007), they have the advantage of being measured in laboratory containers with fewer artifacts than feeding rates. I hypothesize that respiration rates can be predicted by multiple regressions of predator size and temperature (e.g., Uye & Shimauchi, 2005; Ishii & Tanaka, 2006), and thus, feeding can be estimated from metabolic rates across species.

I tested this hypothesis for scyphomedusae by comparing published regressions of respiration rates (RR) measured at ambient temperatures for *Aurelia aurita, A. labiata, Chrysaora quinquecirrha* (converted from excretion by use of the atomic ratio (11.6:1) of oxygen respired to nitrogen excreted), and *Cyanea capillata* (Table 8). Medusa mass was standardized to carbon (C) by published conversions (Table 9). I entered RRs at the minimum and maximum sizes at experimental temperatures for each regression (1 point for each temperature and size; Fig. 6) in a multiple regression to predict respiration rate (ml $O_2$ medusa$^{-1}$ d$^{-1}$) from medusa mass (g C) and temperature (°C). The regression for *Aurelia* spp. was strong ($R^2 = 0.954$; $P < 0.001$), with mass being significant, but not temperature (Table 10). The regression for scyphomedusan species was equally strong ($R^2 = 0.951$; $P < 0.001$), with respiration rates of *C. quinquecirrha* and *C. capillata* coinciding with those of *Aurelia* spp. (Fig. 6); again, mass was significant, but temperature was not (Table 10). The regressions were recalculated without temperature for the predictive equations (Table 10). PRESS statistics indicated strong predictability of the regressions (*Aurelia* spp. 1.206; scyphomedusae 1.432). Respiration rates of *Aurelia* spp. medusae of equal mass were similar across ambient temperatures from 10 to 30°C, in marked contrast to published increases determined in the laboratory (e.g., $Q_{10} = 2.9$; Fig. 7); $Q_{10}$ of the combined *Aurelia* spp. regression was only 1.67. Respiration of *C. quinquecirrha* increased somewhat with temperature. It was unclear if temperature in the *C. capillata* experiment was adjusted to 15°C (Larson, 1987).

I also developed a multiple regression equation for respiration rate vs. mass and temperature for *Mnemiopsis leidyi* ctenophores (2 sizes) from published respiration equations at ambient temperatures (Table 8). I did not include the regression from Pavlova & Minkina (1993), which gave very low rates compared with the others. I used data only from freshly collected ctenophores from Kremer (1982). The combined regression for *M. leidyi* was strong ($R^2 = 0.874$; $P < 0.001$), with mass but not temperature being significant (Table 10, Fig. 6). The regression was recalculated without temperature for the predictive equation (Table 10). The PRESS statistic (2.279) indicated that the regression would be a good predictor of respiration. Respiration rates of *M. leidyi* ctenophores of equal mass (g C) showed a greater sensitivity to temperature than scyphomedusae (Fig. 7); however, temperature was not significant in the multiple regression (Table 10). Experimental temperatures in Miller (1970) differed by 0 to 8°C from ambient; his data suggest that respiration rates may be reduced when ctenophores

**Table 8** Respiration rates (RR) of scyphomedusae and ctenophores measured at near-ambient temperatures and used in multiple regression analyses

| Species | Weight (g) Range | $R^2$ | T (°C) | Metabolic rate equation | Reference |
|---|---|---|---|---|---|
| *Aurelia labiata* | ~0.01–7 DW | 0.98 | 10 | (10°C) $\mathrm{Log_{10}RR}$ (μl $h^{-1}$) = $0.2*\mathrm{log_{10}DW(mg)}^{0.92}$ | Larson (1987) |
| | ~0.01–3 DW | 0.97 | 15 | (15°C) $\mathrm{Log_{10}RR}$ (μl $h^{-1}$) = $0.39*\mathrm{log_{10}DW(mg)}^{0.91}$ | |
| *Aurelia labiata* | 1.8–26.6 WW | | 10 | RR (μmol g $WW^{-1}$ $h^{-1}$) = 0.297 | Rutherford & Thuesen (2005) |
| *Aurelia aurita* | ~10–1,000 WW | 0.982 | 30 | RR (μl $h^{-1}$) = $1.084*WW^{1.059}$ | Dawson & Martin (2001) |
| *Aurelia aurita* | 11–1,330 WW | 0.83 | 20, 28 | RR (ml $d^{-1}$) = $0.0765*2.8^{(T-20)/10}*WW^{(-1.038)}$ | Uye & Shimauchi (2005) |
| | | 0.97 | | | |
| *Aurelia aurita* | 2.33–17.4 DW | 0.72 | 15, 20, 24 | RR (ml $h^{-1}$) = $10^{(0.11T-3.1)}*DW^{1.093}$ | Ishii & Tanaka (2006) |
| | | 0.94 | | | |
| *Aurelia aurita* | ~0.03–900 DW | 0.99 | 15 | RR (μl $d^{-1}$) = $4.8*DW(mg)^{1.01}$ | Møller & Riisgård (2007) |
| *Chrysaora quinquecirrha* | 0.01–2.8 DW | 0.80 | 18–28 | RR (ml $h^{-1}$) = $(11.6)*0.974*\mathrm{Log_{10}DW} + 0.021*T + 0.134$ | Nemazie et al. (1993) |
| *Cyanea capillata* | 0.2–10 DW | 0.98 | 10 | (10°C) $\mathrm{Log_{10}RR}$ (μl $h^{-1}$) = $0.47*\mathrm{Log_{10}DW(mg)}^{1.0}$ | Larson (1987) |
| | 0.1–1 DW | 0.99 | 15 | (15°C) $\mathrm{Log_{10}RR}$ (μl $h^{-1}$) = $0.72*\mathrm{Log_{10}DW(mg)}^{1.04}$ | |
| *Mnemiopsis leidyi* | 0.1–7.0 WW | 0.94–0.99 | 4 (7) | (4°C) RR (ml $d^{-1}$) = $0.0352*WW^{0.76}$ | Miller (1970) |
| | | | 7 (7) | (7°C) RR (ml $d^{-1}$) = $0.0634*WW^{0.88}$ | |
| | | | 12 (10) | (12°C) RR (ml $d^{-1}$) = $0.0811*WW^{0.79}$ | |
| | | | 16 (8) | (16°C) RR (ml $d^{-1}$) = $0.1157*WW^{0.84}$ | |
| | | | 21.5 (23) | (21.5°C) RR (ml $d^{-1}$) = $0.1024*WW^{0.87}$ | |
| | | | 24 (22) | (24°C) RR (ml $d^{-1}$) = $0.1127*WW^{0.84}$ | |
| | | | 29 (27) | (29°C) RR (ml $d^{-1}$) = $0.1350*WW^{0.83}$ | |
| *M. leidyi* | 0.04–1.0 DW | ND | 10–24 | RR (μg atoms g $DW^{-1}$ $h^{-1}$) = $0.79*e^{0.13T}$ | Kremer (1977) |
| *M. leidyi* | 1.5–28 WW | ND | 22 | RR (μg atoms g $DW^{-1}$ $h^{-1}$) = 8.7*DW | Kremer (1982) |
| *M. leidyi* | 0.007–0.391 DW | 0.78 | 18–27 | RR (ml $h^{-1}$) = $(13.3) \times 1.952DW^{0.742}$ | Nemazie et al. (1993) |
| *M. leidyi* | | | 20–21 | RR (ml $h^{-1}$) = $0.045*DW^{0.584}$ | Finenko et al. (1995) |
| *M. leidyi* starved | 0.05–3 WW | | 22 | RR (μl $h^{-1}$) = $4.57*WW^{0.83}$ | Anninsky et al. (1998) |
| *M. leidyi* fresh | 0.05–3 WW | | 22 | RR (μl $h^{-1}$) = $6.73*WW^{0.83}$ | Anninsky et al. (1998) |
| *M. leidyi* | 0.0015–2.46 WW | | 24 | RR (ml $h^{-1}$) = $0.0042*WW^{0.776}$ | Finenko et al. (2006) |

T, experimental (ambient) temperature in °C; DW, dry weight; WW, wet weight. Excretion rates were converted to respiration by the O:N atomic ratios of 11.6 for *Chrysaora quinquecirrha* and 11.3 for *Mnemiopsis leidyi* (in Nemazie et al., 1993 and Kremer, 1977). Rates were standardized to ml $O_2$ $ind^{-1}$ $d^{-1}$ by the conversions 1 ml $O_2$ = 1.42 mg $O_2$ = 44.88 μmol $O_2$ = 89.76 μg atoms $O_2$. $R^2$, coefficient of determination

**Table 9** Biometric conversions and $Q_{10}$s for scyphomedusae and ctenophores, with ambient salinities

| Species | $Q_{10}$ | DW%WW | C%DW | Salinity | Reference |
|---|---|---|---|---|---|
| *Aurelia aurita* | 2.8 | 3.6 | 3.7 | >30 | Uye & Shimauchi (2005) |
| *Aurelia aurita* | ND | ND | 3.9 | ND | Ishii & Tanaka (2006) |
| *Aurelia aurita* | 3.1 | ND | 5 | 20.0 | Møller & Riisgård (2007) |
| *Aurelia labiata* | 2.9 | 3.8 | 4 | 28–30 | Larson (1986) |
| *Cyanea capillata* | 3.4 | 4.2 | 13 | 28–30 | Larson (1986) |
| *Chrysaora quinquecirrha* | 1.6 | 0.95 | 11.1 | 6–12 | Nemazie et al. (1993) |
| *Mnemiopsis leidyi* | 3.7 | 3.4 | 1.7 | 31 | Kremer (1977) |
| *Mnemiopsis leidyi* | 3.4 | ND | 4.2 | 18 | Finenko et al. (1995) |
| *Mnemiopsis leidyi* | 3.4 | 0.95 | 5.1 | 6–12 | Nemazie et al. (1993) |

DW, dry weight; WW, wet weight; C, carbon. ND, no data

were cooled, and conversely, increased when warmed (Fig. 7, Table 8). $Q_{10}$ calculated from the *M. leidyi* combined equation = 1.33, which is less than the published $Q_{10}$s ($\geq$ 3.4; Table 9).

The respiration rates for medusae and ctenophores can be used to estimate predation rates. The minimum daily carbon ingestion (MDCI) can be calculated by multiplying the daily respiration rate by the respiratory quotient (RQ = 0.8). The MDCI can be converted to numbers of prey ingested from prey carbon mass when the prey types are known (e.g., ICES, 2000). Estimates of predation effects on prey populations can be made from these data and in situ prey densities. Thus, estimates of predation by gelatinous species in situ can be made from laboratory respiration or excretion measurements, in combination with field data on predator mass and density, prey type, densities, and temperature.

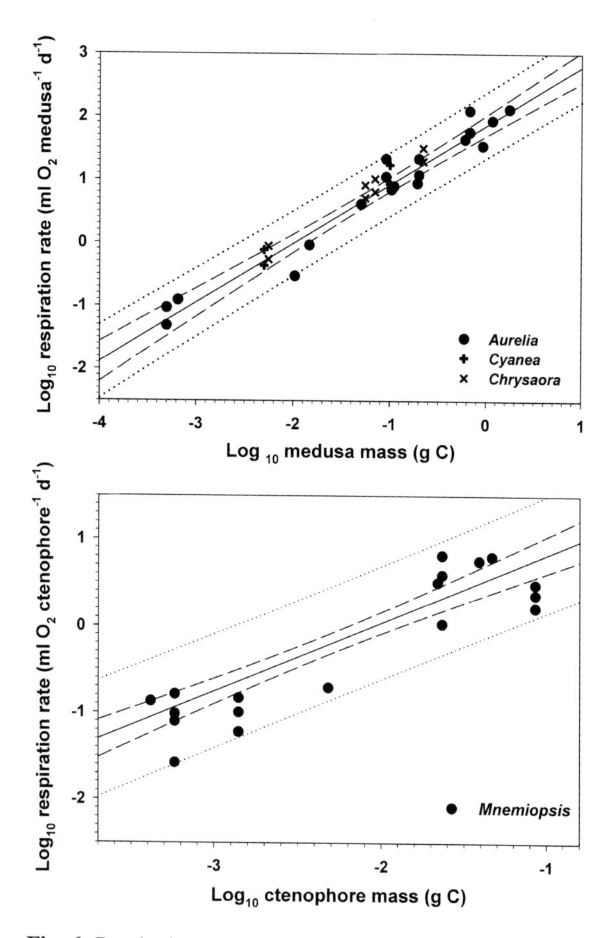


**Fig. 6** Respiration rates measured at ambient temperatures from published regressions (Table 8) against mass. Top for scyphomedusae *Aurelia* spp., *Cyanea capillata,* and *Chrysaora quinquecirrha.* Bottom for *Mnemiopsis leidyi* ctenophores. Lines are as in Fig. 1

## Discussion

### General comments and suggestions

Any estimation of the importance of jellyfish relies first on the determination of their abundance and biomass. Generally, sampling effort is limited by logistics, and the method chosen is assumed to be adequate. Very few studies evaluate the efficacy of any method of estimating jellyfish abundance. *Aurelia aurita* densities in aggregations determined by echo sounder were much lower than those in net (0.8–1.6 m mouth) tows (Toyokawa et al., 1997). Towed-camera estimates of *A. aurita* abundance compared very well against Tucker trawl estimates (Graham et al., 2003a, b). Densities of robust medusae (*Periphylla periphylla*) sampled by several nets and trawls, an ROV, and acoustics were compared by

**Table 10** Regression analyses evaluating the relationships of jellyfish (*Aurelia* spp., *Cyanea capillata*, and *Chrysaora quinquecirrha*) or ctenophore (*Mnemiopsis leidyi*) mass (in g carbon, C) and ambient temperature (T) with respiration rates (RR)

| Predator | $n$ | Weight (g carbon) | | Temperature (°C) | | $R^2$ and $F$ statistics | $P$ and SE overall | Predictive equation |
|---|---|---|---|---|---|---|---|---|
| | | Range | $t$ and $P$ | Tested | $t$ and $P$ | | | |
| *Aurelia* spp. combined | 19 | 0.0005–1 | $t = 12.270$ $P < 0.001$ | 10–30 | $t = -0.608$ $P = 0.552$ | 0.954 $F_{1,17} = 345.74$ | $P < 0.001$ SE 0.245 | $\text{Log}_{10}\text{RR (ml O}_2\text{ d}^{-1}) = 0.936 * \text{Log}_{10}\text{C} + 1.862$ |
| Scyphomedusae combined | 31 | 0.0005–1.7 | $t = 22.268$ $P < 0.001$ | 10–30 | $t = 0.338$ $P = 0.738$ | 0.951 $F_{1,29} = 563.80$ | $P < 0.001$ SE 0.212 | $\text{Log}_{10}\text{RR (ml O}_2\text{ d}^{-1}) = 0.935 * \text{Log}_{10}\text{C} + 1.907$ |
| *Mnemiopsis leidyi* combined | 18 | 0.0014–0.02 | $t = 9.935$ $P < 0.001$ | 4–29 | $t = 0.998$ $P = 0.334$ | 0.874 $F_{1,16} = 102.879$ | $P < 0.001$ SE 0.308 | $\text{Log}_{10}\text{RR (ml O}_2\text{ d}^{-1}) = 0.871 * \text{Log}_{10}\text{C} + 1.686$ |

$t$, $t$ statistic; $P$, probability where $P > 0.05$ are not significant; $R^2$, coefficient of determination; $F$, F statistic; SE, standard error. Data from regressions in Table 8

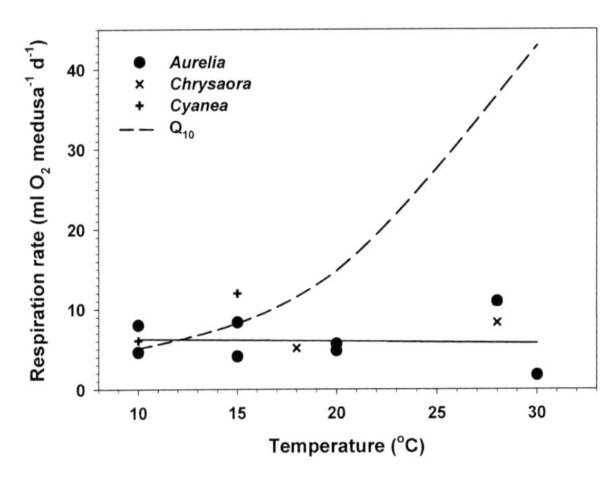


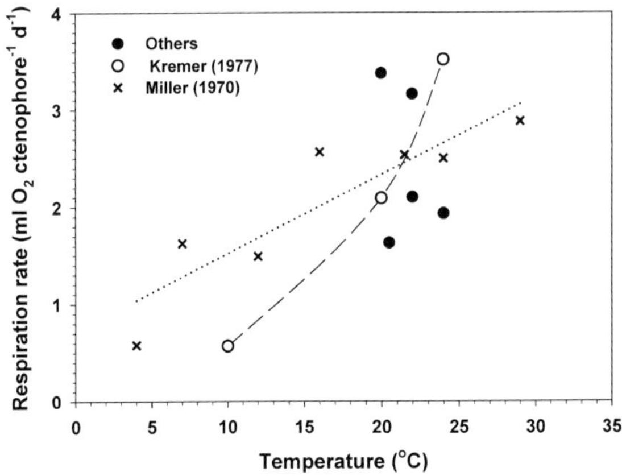


**Fig. 7** Respiration rates against ambient/experiment temperatures from regressions in Table 8. Top for scyphomedusae (*Aurelia* spp., *Cyanea capillata*, and *Chrysaora quinquecirrha*) of equal sizes (0.6–0.7 g C). The dashed line shows predicted respiration for *Aurelia* spp. assuming a $Q_{10}$ of 2.9 (Larson, 1987). Solid line is for all data points. Bottom for *Mnemiopsis leidyi* ctenophores of equal sizes (~0.02 g C). Dashed line is for Kremer (1977) and dotted line is linear regression for Miller (1970)

Båmstedt et al. (2003). Evaluation of all of the reviewed methods is important for meaningful estimates of jellyfish ecosystem effects. If a semi-quantitative method is employed (e.g., surface surveys), efforts should be made to determine what portion of the population is being sampled, and ideally, develop an index to convert the method to abundance/biomass estimates.

Sizes of gelatinous species and conversions among mass units (WW, DW, C) have been determined repeatedly (see Larson, 1986; ICES, 2000). The relationships are very consistent (Table 9), and probably do not need to be measured in every location. One difficulty in mass conversions is that DW increases with salinity, and hence, conversions involving DW and dried tissues (e.g., C) differ depending on ambient salinity, as emphasized by Nemazie et al. (1993) and Hirst & Lucas (1998). Thus, use of DW should be avoided, and necessary conversions from DW should be from specimens from similar ambient salinities.

For gelatinous zooplankton to be included in ecosystem models, data on diet and trophic level, population and individual biomass, as well as growth (production) need to be collected (see Pauly et al., 2008). Dietary data already have been published for many common species. Stable isotopes can yield new insights into trophic interactions. Population biomass and growth data generally have been incompletely collected on depth, spatial, and temporal scales. More in situ data are needed on the polyp, ephyra, and planula stages, specifically, when and where do the various stages occur, and the dates of strobilation in relation to environmental variables. Depth-specific

data are important. Greater emphasis on such data would greatly improve our understanding of gelatinous species in the ocean's ecosystems.

In this review, I focused on scyphomedusae for several reasons. First, large-scale sampling techniques work best or only for large species. Second, reports of problem blooms of scyphomedusae have increased in recent decades. Third, abundant data are available for temperate semaeostome scyphomedusae, especially *Aurelia* spp. Semaeostomes are the predominant scyphomedusae in cool coastal waters, but rhizostome scyphomedusae can predominate in tropical waters. Comparatively few studies exist on rhizostome ecology (but see Larson, 1991; Graham et al., 2003a; Pitt & Kingsford, 2003; Pitt et al., 2005, 2007; Uye, 2008; West et al., 2008). Rhizostomes are also of particular interest because of problem blooms (e.g., Graham et al., 2003a; Uye, 2008), and the use of some species as human food (Omori & Nakano, 2001). Millions of the preferred species, *Rhizostoma esculentum* Kishinouye, are reared and released in Chinese waters annually for later harvest (Dong et al., 2008) with little understanding of the ecological effect (Liu & Bi, 2006). Because of their large sizes and complex feeding structures, rhizostome medusae are excellent candidates for estimation of consumption from respiration rates (e.g., Uye, 2008). Rhizostomes are stronger swimmers than semaeostomes (D'Ambra et al., 2001); therefore, their feeding rates and metabolic demands probably are greater and will require analyses separate from the semaeostomes.

Although scyphomedusae form conspicuous blooms and may predominate as predators in summer, the other gelatinous taxa should not be neglected. There are now approximately 840 recognized species of hydromedusae (Bouillon & Boero, 2000), as compared with only 190 species of scyphomedusae (Arai, 1997), 20 species of cubomedusae (Mianzan & Cornelius, 1999), 200 species of siphonophores (Pugh, 1999), and 150 species of ctenophores (Mianzan, 1999). Only a small fraction of these many species have been studied. The small hydromedusae and fragile ctenophores, in particular, often go unnoticed; however, they are ubiquitous, can occur in high densities and biomass in coastal waters, are important predators (e.g., Pagès et al., 1996a; Purcell & Arai, 2001; Costello & Colin, 2002; Hansson et al., 2005; Hosia & Båmstedt, 2007), and need further study. Because of the great morphological differences between scyphomedusae and hydromedusae, it is unlikely that hydromedusan feeding could be predicted by use of the semaeostome scyphomedusa regressions herein. Similarly, because of the great differences among hydromedusan species, algorithms would need to be developed that group species of similar morphology, feeding behavior, and diet, such as for anthomedusae and for leptomedusae. Among ctenophores, only coastal ctenophores, *Pleurobrachia* spp. and *Mnemiopsis leidyi*, have been studied relatively well because of their abundance and the ability to sample them with plankton nets. Because of their different feeding methods, a different feeding algorithm probably would be necessary for cydippid ctenophores (e.g., *Pleurobrachia* spp.) than for lobate ctenophores (e.g., *M. leidyi*).

The above analyses of feeding and metabolic rates of *Aurelia* spp. medusae from disparate habitats and *Mnemiopsis leidyi* ctenophores show that predictive algorithms can be developed. The ecology of *Mnemiopsis leidyi* is the same in its native (American Atlantic coasts) and introduced (Black Sea region) waters (e.g., reviewed in Kremer, 1994; Purcell et al., 2001b; Shiganova et al., 2003). The regressions herein could be used to predict its predators' effects in different habitats. This approach recently was used to estimate the predation impact of *Chrysaora melanaster* in the Bering Sea from the metabolic rates of *Cyanea capillata* (Brodeur et al., 2002), and of *M. leidyi* in Danish waters from previously determined clearance rates (Riisgård et al., 2007). The ecologies of tropical and sub-tropical jellyfish, including coronates (Kremer, 2005), rhizostomes (Uye, 2008; West et al., 2008), cubomedusae (Gordon & Seymour, 2008), hydromedusae, siphonophores, and ctenophores (Kremer et al., 1986), generally have been studied less than temperate semaeostome scyphomedusae and *M. leidyi*; therefore, additional data on feeding and metabolic rates of those groups probably are needed before generalized algorithms are developed.

## Specific comments on use of feeding data to estimate jellyfish and ctenophore predation

There is inherently greater variability among species in feeding rates (Figs. 1, 2, 4) than in metabolic rates (Figs. 6, 7) because of the differences in predator morphology, nematocysts, and behavior, as well as prey morphology and behavior (reviewed in Purcell, 1997). In addition to the inherent variation among species, our inability to precisely sample the prey population in which the predators fed contributes to variation in the data. The gut contents (prey medusa$^{-1}$) used here from various studies could have influenced variability because of different methodologies used. Medusae usually were dipped from the surface, but some were collected at depth by divers (Dawson & Martin, 2001). Preservation time differed from 0 to 45 min, which would affect the numbers of recognizable prey. Some studies preserved whole medusae (Purcell, 2003; Uye & Shimauchi, 2005; Lo & Chen, 2008), but others rinsed out gastric pouch contents (Dawson & Martin, 2001; Lucas, unpublished), which could affect the numbers of prey retrieved. Plankton nets with different mesh-size (200 or 212 μm in Lucas, unpublished; 243 μm in Purcell, 2003; 10 and 100 μm in Uye & Shimauchi, 2005; 80 μm in Dawson & Martin, 2001) or a pump and 64 μm mesh (Purcell, 1992) were used to sample available prey, which strongly affected estimates of prey density, and subsequent utility of the regressions. Use of zooplankton densities from 100-μm and 330-μm net samples gave different results in the *Aurelia* spp. regression (Table 3).

Despite obvious morphological dissimilarities among the semaeostome species tested here, different prey populations, and different methodologies, FR on copepod prey were reasonably well predicted (generally within a factor of 2) over wide ranges of predator size, prey density, and temperature (Figs. 1 and 2, Tables 2 and 3). The similarity in FR among four scyphozoan species suggests that FR of other species may be estimated from the predictive equation; however, the poor match for *Chrysaora fuscescens* measured and calculated FR indicates that caution is necessary. Only species of similar size and habitat may be appropriate. The greatest divergence among the data was for *Aurelia aurita* in Japan, where small *Oithona* spp. copepods were the predominant prey, while calanoids predominated in the other locations. Therefore, assessment of the prey available may be especially important when choosing which feeding regression to apply.

The present analyses only considered copepods, which are the most abundant prey in most situations; however, gelatinous species eat a variety of zooplankton taxa. Substitution of combined zooplankton taxa consumed by a predator instead of copepods m$^{-3}$ in the regressions should also approximate total consumption. Alternatively, increasing consumption on copepods as calculated by the regressions by the percentages of other prey would approximate total consumption.

Although temperature did not significantly affect FR of *Aurelia aurita* medusae in different habitats, it significantly affected *Chrysaora quinquecirrha* FR in different seasons in Chesapeake Bay. Warm temperature could increase medusa swimming and digestion rates, as well as prey activity; therefore, although prey capture could increase, the prey in the gut contents may remain similar across temperatures because of more rapid digestion.

The length of time required for digestion of copepod prey decreased with *Aurelia* spp. medusa size and temperature. The effect of size mainly was due to the long times for ephyrae and very small medusae to digest large copepods. This regression could be used in combination with gut contents to calculate the feeding rates of *Aurelia* spp. medusae throughout their range of habitats.

### Specific comments on use of metabolic rates to estimate jellyfish and ctenophore predation

Data compiled here show that jellyfish respiration scales with mass with an exponent of ~1 (Fig. 6, Tables 8, 10). This is in agreement with previous conclusions for jellyfish and pelagic animals in general (e.g., Glazier, 2006); however, the empirical relationships for ctenophores were closer to 0.75 than 1 (Fig. 6, Tables 8, 10).

These analyses showed that respiration rates of scyphomedusae and *Mnemiopsis leidyi* ctenophores measured at or near ambient temperatures did not change with temperature in accordance with experimentally measured $Q_{10}$s (Fig. 7, Table 9). Although respiration rates increase when temperature is raised to measure $Q_{10}$ and increase with ambient seasonal warming within a habitat, respiration rates in locations differing in ambient temperature did not reflect the laboratory-determined $Q_{10}$s. $Q_{10}$ of the combined *Aurelia* spp. data was only 1.67. Dawson & Martin (2001) noted that the respiration rates of tropical *A. aurita* were similar to those of temperate *A. aurita*, even though the ambient temperatures were very different, and that temperature adaptation was

common among other animals. Metabolic rates of *Chrysaora quinquecirrha* medusae also increased to some extent with temperature ($Q_{10} = 1.67$) in Chesapeake Bay (Nemazie et al., 1993). Metabolic rates of *M. leidyi* ctenophores increased seasonally with temperature (combined data $Q_{10} = 1.33$), which is less than published $Q_{10}$s (Table 9). Similarly, metabolic rates of ctenophores from Biscayne Bay, Florida increased little with temperatures from 10 to 28°C (Baker, 1973, shown in Kremer, 1977). I conclude that the respiration rates of *Aurelia* spp., *Chrysaora* spp., and *Cyanea* spp. scyphomedusae, and *M. leidyi* ctenophores can be predicted from most habitats with the above regressions using mass, and that adjustment for temperature by $Q_{10}$ determined from experimentally changed temperatures may misestimate metabolic rates. It is important to measure metabolic rates of the organisms at their ambient temperatures.

The prior feeding condition of the specimens also affects their metabolic rates. Variation in prior acclimation duration and feeding in the experiments used here contributed to their different results. Specimens were acclimated for one to several hours before measuring metabolism in most studies. Some studies used newly collected specimens to reflect rates in situ (e.g., Nemazie et al., 1993), while others explicitly tested the effects of food (e.g., Kremer, 1982; Møller & Riisgård, 2007). High levels of food in the laboratory increased the metabolism of *Aurelia aurita* by 3.5 times (Møller & Riisgård, 2007), but that may not be representative of metabolic rates in situ. Metabolic rates of newly collected, lightly fed, and heavily fed specimens of a small siphonophore species showed that the newly collected and lightly fed rates were identical, while the heavily fed rates were higher (Purcell & Kremer, 1983). Therefore, I conclude that the metabolic rates that most resemble in situ rates are those measured on newly collected specimens at ambient temperatures.

A weakness of using metabolic rates to estimate ingestion is that metabolic rates usually do not account for requirements for growth or reproduction, and thus are underestimates (see Møller & Riisgård, 2007). Growth rates of scyphomedusae in situ were about 7% WW $d^{-1}$ (Schneider, 1989; Omori et al., 1995; Lucas, 1996; Uye & Shimauchi, 2005). Therefore, increasing the metabolic rates by 7% for WW, by 0.2% for DW, and by 0.015% for C over basal rates for trophic estimates may be appropriate. Maximal in situ growth of *Chrysaora quinquecirrha* was 60% diameter $d^{-1}$ (= 25% C $d^{-1}$; Olesen et al., 1996). Adjustment for growth would depend on food availability, and would be time- and location-specific.

## Conclusions

The above algorithms would allow estimation of feeding effects, generally within a factor of two, without extensive collection of in situ data on jellyfish or ctenophore feeding. The combined regressions predicted feeding and metabolic rates nearly as well as the local regressions. That seems like a reasonable level of uncertainty, given that all other biological measurements probably have the same or greater errors. Population data for *Aurelia* spp., *Chrysaora quinquecirrha*, *Cyanea capillata*, and *Mnemiopsis leidyi* densities, mean individual mass, zooplankton densities, and water temperature could be used to estimate feeding and respiration rates and consequent effects on the zooplankton population. In general, estimation of consumption by the metabolic regressions probably has less error than estimation from the feeding regressions. These algorithms should be tested for other species to determine how broadly they can be applied. New algorithms should be developed for other key gelatinous taxa, and analyses conducted on combined data from other species. Although these methods are approximate, and Arai (1997) cautions against such extrapolation, it is important that gelatinous species be included in ecosystem studies and models that now are conducted on regional to global scales (Pauly et al., this volume). These methods offer alternatives to when limited person-power, resources, and time do not permit exhaustive in situ collection of jellyfish feeding data. I briefly summarize recommendations for trophic research methods for gelatinous predators:

- Determine densities and size distributions (mass) of the gelatinous species.
- Sample small ctenophores and hydromedusae as well as large scyphomedusae. Three types of sampling may be necessary—nets as small 0.5-m-diameter can be sufficient for hydromedusae, short tows of soft-mesh plankton net with a non-draining cod-end improve sampling for

ctenophores, and plankton nets larger than 1-m-diameter are required for scyphomedusae.

- Test the accuracy of the various large-scale sampling methods against quantitative methods, and develop conversions to make methods as quantitative as possible.
- Use a fine-mesh net ($\sim 100$ μm) for zooplankton sampling.
- Report temperature and salinity.
- Use the gut-content method to estimate the feeding rates on mesozooplankton.
- For clearance rate experiments use high container-to-specimen volume ratios, at least 2,500:1.
- Use natural prey, not *Artemia* sp. nauplii, in feeding experiments.
- Use ambient temperature for all feeding, digestion, and metabolic experiments.
- Conduct metabolic experiments on newly collected specimens for rates that reflect natural food conditions.
- Do not convert metabolic rates by use of $Q_{10}$ values measured at experimentally manipulated temperatures.
- Before sampling, examine the data criteria of a central database and submit data to a central database after publication.
- Develop algorithms among taxa that can be used to predict gelatinous predator effects on large scales.

**Acknowledgments** I thank Drs K. A. Pitt and J. Seymour for the opportunity to speak at the Second International Jellyfish Blooms Symposium. I especially thank Drs. M. N Dawson, L. E. Martin, W.-T. Lo, C. H. Lucas, S.-I. Uye, M. B. Decker, P. Kremer, M. R. Reeve, and C. L. Suchman for use of raw data, and A. G. Hirst, P. Kremer, M. N Dawson, and an anonymous reviewer for suggestions. Studies in Chesapeake Bay were funded by NOAA grants NA86AA-D-SG006 and NA890AA-D-SG063 to the University of Maryland Sea Grant Program, and NSF grant no. DEB-9412113 to the TIES group and by NSF grant no. OCE-9633607 to the author. The research in Alaska was funded by the *Exxon Valdez* Oil Spill Trustee Council as part of the Alaska Predator Ecosystem eXperiment (APEX Projects 96163A, 97163A, 98163A, 98163S, and 99163S). This review was partially prepared while the author was supported by grants from the National Science Council and the Ministry of Education of the Republic of China to Dr. W.-T. Lo [NSC91-2621-Z 110-001, 94-C030220 (Kuroshio project)]. I dedicate this article to Hoyt Purcell, my champion.



## References

Anninsky, B. E., Z. A. Romanova, G. I. Abolmasova, A. C. Gucu & A. E. Kideys, 1998. The ecological and physiological state of the ctenophore *Mnemiopsis leidyi* (Agassiz) in the Black Sea in autumn 1996. In Ivanov, L. I. & T. Oguz (eds), Ecosystem Modeling as a Management Tool for the Black Sea. Kluwer Academic Publishers, Dordecht: 249–262.

Arai, M. N., 1997. Coelenterates in pelagic food webs. In den Hartog, J. C. (ed.), Proceeding of the 6th International Conference on Coelenterate Biology. Publication of the National Natuurhistorisch Museum, Leiden: 1–9.

Arai, M. N., 2005. Predation on pelagic coelenterates: a review. Journal of the Marine Biological Association of the United Kingdom 85: 523–536.

Arai, M. N., D. W. Welch, A. L. Dunsmuir, M. C. Jacobs & A. R. Ladouceur, 2003. Digestion of pelagic Ctenophora and Cnidaria by fish. Canadian Journal of Fisheries and Aquatic Sciences 60: 825–829.

Attrill, M. J., J. Wright & M. Edwards, 2007. Climate-related increases in jellyfish frequency suggest a more gelatinous future for the North Sea. Limnology and Oceanography 52: 480–485.

Baird, D. & R. E. Ulanowicz, 1989. The seasonal dynamics of the Chesapeake Bay ecosystem. Ecological Monographs 59: 329–364.

Baker, L. D., 1973. The ecology of the ctenophore *Mnemiopsis mccradyi* Mayer, in Biscayne Bay, Florida, 131 pp. M.S. Thesis, University of Miami.

Båmstedt, U., S. Kaartvedt & M. Youngbluth, 2003. An evaluation of acoustic and video methods to estimate the abundance and vertical distribution of jellyfish. Journal of Plankton Research 25: 1307–1318.

Benson, S. R., P. H. Dutton, C. Hitipeuw, B. Samber, J. Bakarbessy & D. Parker, 2007. Post-nesting migrations of leatherback turtles (*Dermochelys coriacea*) from Jamursba-Medi, Bird's Head Peninsula, Indonesia. Chelonian Conservation and Biology 6: 150–154.

Bernard, P., F. Couasnon, J.-P. Soubiran & J.-F. Goujon, 1988. Surveillance estivale de la méduse *Pelagia noctiluca* (Cnidaria, Scyphozoa) sur les côtes Méditerranéennes Françaises. Annales de l'Institut océanographique, Paris 64: 115–125.

Boersma, M., A. M. Malzahn, W. Greve & J. Javidpour, 2007. The first occurrence of the ctenophore *Mnemiopsis leidyi* in the North Sea. Helgoland Marine Research 61: 153–155.

Bouillon, J. & F. Boero, 2000. The Hydrozoa: a new classification in the light of old knowledge. Thalassia Salentina 24: 3–296.

Brierley, A. S., B. E. Axelsen, D. C. Boyer, C. P. Lynam, C. A. Didcock, H. J. Boyer, C. A. J. Sparks, J. E. Purcell & M. J. Gibbons, 2004. Single-target echo detections of jellyfish. ICES Journal of Marine Science 61: 383–393.

Brodeur, R. D., M. B. Decker, L. Ciannelli, J. E. Purcell, N. A. Bond, P. J. Stabeno, E. Acuna & G. L. Hunt Jr., 2008a. The rise and fall of jellyfish in the Bering Sea in relation to climate regime shifts. Progress in Oceanography 77: 103–111.

Brodeur, R. D., C. E. Mills, J. E. Overland & J. D. Shumacher, 1999. Evidence for a substantial increase in gelatinous zooplankton in the Bering Sea, with possible links to climate change. Fisheries Oceanography 8: 296–306.

Brodeur, R. D., C. L. Suchman, D. C. Reese, T. W. Miller & E. A. Daly, 2008b. Spatial overlap and trophic interactions between pelagic fish and large jellyfish in the northern California current. Marine Biology 154: 649–659.

Brodeur, R. D., S. Sugisaki & G. L. Hunt Jr., 2002. Increases in jellyfish biomass in the Bering Sea: implications for the ecosystem. Marine Ecology Progress Series 233: 89–103.

Brown, C. W., R. R. Hood, Z. Li, M. B. Decker, T. Gross, J. E. Purcell & H. Wang, 2002. Forecasting system predicts presence of sea nettles in Chesapeake Bay. EOS, Transactions of the American Geophysical Union 83: 321, 325–326.

Brown, E. D., S. M. Moreland, B. L. Norcross & G. A. Borstad, 1999. Estimating forage fish and seabird distribution and abundance using aerial surveys: survey design and uncertainty. In Cooney, R. T. (ed.), Sound Ecosystem Assessment (SEA)-An Integrated Science Plan for the Restoration of Injured Species in Prince William Sound. *Exxon Valdez* Oil Spill Restoration Project Final Report (Restoration Project 99320T), Anchorage: 131–172.

Bunker, A. J. & A. G. Hirst, 2004. Fecundity of marine planktonic copepods: global rates and patterns in relation to chlorophyll a, temperature and body weight. Marine Ecology Progress Series 279: 161–181.

Cargo, D. G. & D. R. King, 1990. Forecasting the abundance of the sea nettle, *Chrysaora quinquecirrha*, in the Chesapeake Bay. Estuaries 13: 486–491.

Colin, S. P., J. H. Costello, W. M. Graham & J. Higgins III, 2005. Omnivory by the small cosmopolitan hydromedusa *Aglaura hemistoma*. Limnology and Oceanography 50: 1264–1268.

Colombo, G. A., A. Benović, A. Majej, D. Lučić, T. Makovec, V. Onofri, M. Achal, A. Madirolas & H. Mianzan, 2008. Acoustic survey of a jellyfish-dominated ecosystem (Mljet, Croatia). Hydrobiologia. doi:10.1007/s10750-008-9587-6.

Colombo, G. A., H. Mianzan & A. Madirolas, 2003. Acoustic characterization of gelatinous-plankton aggregations: four case studies from the Argentine continental shelf. ICES Journal of Marine Science 60: 650–657.

Costello, J. H. & S. P. Colin, 2002. Prey resource utilization by co-occurring hydromedusae from Friday Harbor, Washington, USA. Limnology and Oceanography 47: 934–942.

D'Ambra, I., J. H. Costello & F. Bentivegna, 2001. Flow and prey capture by the scyphomedusa *Phyllorhiza punctata* von Lendenfeld, 1884. Hydrobiologia 451 (Developments in Hydrobiology 155): 223–227.

Dawson, M. N. & L. E. Martin, 2001. Geographic variation and ecological adaptation in *Aurelia* (Scyphozoa, Semaeostomeae): some implications from molecular phylogenetics. Hydrobiologia 451 (Developments in Hydrobiology 155): 259–273.

Decker, M. B., D. L. Breitburg & J. E. Purcell, 2004. Effects of low dissolved oxygen on zooplankton predation by the ctenophore, *Mnemiopsis leidyi*. Marine Ecology Progress Series 280: 163–172.

Decker, M. B., C. W. Brown, R. R. Hood, J. E. Purcell, T. F. Gross, J. Matanoski & R. Owens, 2007. Development of habitat models for predicting the distribution of the scyphomedusa, *Chrysaora quinquecirrha*, in Chesapeake Bay. Marine Ecology Progress Series 329: 99–113.

Dong, J., L.-x. Jiang, K.-f. Tan, H.-y. Liu, J. E. Purcell, P.-j. Li & C.-c. Ye., 2008. Stock enhancement of the edible jellyfish (*Rhopilema esculentum* Kishinouye) in Liaodong Bay, China: a review. Hydrobiologia. doi:10.1007/s10750-008-9592-9.

Doyle, T. K., J. D. R. Houghton, S. M. Buckley, G. C. Hays & J. Davenport, 2007. The broad-scale distribution of five jellyfish species across a temperate coastal environment. Hydrobiologia 579: 29–39.

Finenko, G. A., G. I. Abolmasova & Z. A. Romanova, 1995. Intensity of the nutrition, respiration and growth of *Mnemiopsis mccradyi* in relation to grazing conditions. Biologia Morya 21: 315–320 (in Russian).

Finenko, G. A., A. E. Kideys, B. E. Anninsky, T. A. Shiganova, A. Roohi, M. R. Tabari, H. Rostami & S. Bagheri, 2006. Invasive ctenophore *Mnemiopsis leidyi* in the Caspian Sea: feeding, respiration, reproduction and predatory impact on the zooplankton community. Marine Ecology Progress Series 314: 171–185.

Fuji, N., A. Fukushima, Y. Naojo & H. Takeoka, 2007. Aggregation of Aurelia aurita in Uwa Sea, Japan. In Tanabe, S., H. Takeoka, T. Isobe & Y. Nishibe (eds), Chemical Pollution and Environmental Changes. Frontiers Science Series, Vol. 48. Universal Academy Press, Inc., Tokyo: 379–381.

Gibbons, M. J. & A. J. Richardson, 2008. Patterns in pelagic cnidarian abundance in the North Atlantic. Hydrobiologia (in press).

Glazier, D. S., 2006. The ¾-power law is not universal: evolution of isometric, ontogenetic metabolic scaling in pelagic animals. BioScience 56: 325–332.

Gordon, M. & J. Seymour, 2008. Quantifying movement patterns in the tropical Australian Chirodropid *Chironex fleckeri* using ultrasonic telemetry. Hydrobiologia (in press).

Goy, J., P. Morand & M. Etienne, 1989. Long-term fluctuations of *Pelagia noctiluca* (Cnidaria, Scyphomedusa) in the western Mediterranean Sea. Prediction by climatic variables. Deep-Sea Research 36: 269–279.

Graham, W. M., 2001. Numerical increases and distributional shifts of *Chrysaora quinquecirrha* (Desor) and *Aurelia aurita* (Linné) (Cnidaria: Scyphozoa) in the northern Gulf of Mexico. Hydrobiologia 451 (Developments in Hydrobiology 155): 97–111.

Graham, W. M., D. L. Martin, D. L. Felder, V. L. Asper & H. M. Perry, 2003a. Ecological and economic implications of the tropical jellyfish invader, *Phyllorhiza punctata* von Lendenfeld, in the northern Gulf of Mexico. Biological Invasions 5: 53–69.

Graham, W. M., D. L. Martin & J. C. Martin, 2003b. In situ quantification and analysis of large jellyfish using a novel video profiler. Marine Ecology Progress Series 254: 129–140.

Hansen, P. J., P. K. Bjørnsen & B. W. Hansen, 1997. Zooplankton grazing and growth: scaling within the

2–2,000-μm body size range. Limnology and Oceanography 42: 687–704.
Hansson, L. J., O. Moeslund, T. Kiørboe & H. U. Riisgård, 2005. Clearance rates of jellyfish and their potential predation impact on zooplankton and fish larvae in a neritic ecosystem (Limfjorden, Denmark). Marine Ecology Progress Series 204: 117–131.
Hays, G. C., T. K. Doyle, J. D. R. Houghton, M. K. S. Lilley, J. D. Metcalfe & D. Righton, 2008. Diving behaviour of jellyfish equipped with electronic tags. Journal of Plankton Research 30: 325–331.
Hirst, A. G. & A. J. Bunker, 2003. Growth of marine plankton copepods: global rates and patterns in relation to chlorophyll a, temperature, and body weight. Limnology and Oceanography 48: 1988–2010.
Hirst, A. G. & T. Kiørboe, 2002. Mortality of marine planktonic copepods: global rates and patterns. Marine Ecology Progress Series 230: 195–209.
Hirst, A. G. & C. H. Lucas, 1998. Salinity influences body weight quantification in the scyphomedusa *Aurelia aurita*: important implication for body weight determination in gelatinous zooplankton. Marine Ecology Progress Series 165: 259–269.
Hirst, A. G., J. C. Roff & R. S. Lampitt, 2003. A synthesis of growth rates in marine epipelagic invertebrate zooplankton. Advances in Marine Biology 44: 1–142.
Hosia, A. & U. Båmstedt, 2007. Seasonal changes in the gelatinous zooplankton community and hydromedusa abundances in Korsfjord and Fanafjord, western Norway. Marine Ecology Progress Series 351: 113–127.
Houghton, J. D. R., T. K. Doyle, J. Davenport & G. C. Hays, 2006. Developing a simple, rapid method for identifying and monitoring jellyfish aggregations from the air. Marine Ecology Progress Series 314: 139–170.
Houghton, J. D. R., T. K. Doyle, J. Davenport, M. K. S. Lilley, R. P. Wilson & G. C. Hays, 2007. Stranding events provide indirect insights into the seasonality and persistence of jellyfish medusae (Cnidaria: Scyphozoa). Hydrobiologia 589: 1–13.
ICES, 2000. ICES Zooplankton Methodology Manual. Academic Press, London: 705.
Ishii, H. & F. Tanaka, 2001. Food and feeding of *Aurelia aurita* in Tokyo Bay with an analysis of stomach contents and a measurement of digestion times. Hydrobiologia 451 (Developments in Hydrobiology 155): 311–320.
Ishii, H. & F. Tanaka, 2006. Respiration rates and metabolic demands of *Aurelia aurita* in Tokyo Bay with special reference to large medusae. Plankton and Benthos Research 1: 64–67.
Kaartvedt, S., T. A. Klevjer, T. Torgersen, T. A. Sørnes & A. Røstad, 2007. Diel vertical migration of individual jellyfish (*Periphylla periphylla*). Limnology and Oceanography 52: 975–983.
Kirby, R. R. & J. A. Lindley, 2005. Molecular analysis of continuous plankton recorder samples, an examination of echinoderm larvae in the North Sea. Journal of the Marine Biological Association of the United Kingdom 85: 451–459.
Kohama, T., N. Shinya, N. Okuda, H. Miyasaka & H. Takeoka, 2006, Estimation of trophic level of *Aurelia aurita* using stable isotope ratios in Uwa Sea, Japan. Proceedings of the COE International Symposium 2006 Pioneering Studies of Young Scientists on Chemical Pollution and Environmental Changes. Ehime University, Japan.
Kremer, P., 1976. Population dynamics and ecological energetics of a pulsed zooplankton predator, the ctenophore *Mnemiopsis leidyi*. In Wiley, M. L. (ed.), Estuarine Processes, Vol. 1. Uses, Stresses, and Adaptation to the Estuary. Academic Press, New York:197–215.
Kremer, P., 1977. Respiration and excretion by the ctenophore *Mnemiopsis leidyi*. Marine Biology 44: 43–50.
Kremer, P., 1982. Effect of food availability on the metabolism of the ctenophore *Mnemiopsis mccradyi*. Marine Biology 71: 149–156.
Kremer, P., 1994. Patterns of abundance for *Mnemiopsis* in U.S. coastal waters: a comparative overview. ICES Journal of Marine Science 51: 347–354.
Kremer, P., 2005. Ingestion and elemental budgets for *Linuche unguiculata*, a scyphomedusae with zooxanthellae. Journal of the Marine Biological Association of the United Kingdom 85: 613–625.
Kremer, J. N. & P. Kremer, 1982. A three trophic level estuarine model: synergism of two mechanistic simulations. Ecological Modelling 15: 145–157.
Kremer, P. & M. R. Reeve, 1989. Growth dynamics of a ctenophore (*Mnemiopsis*) in relation to variable food supply. II. Carbon budgets and growth model. Journal of Plankton Research 11: 553–574.
Kremer, P., M. R. Reeve & M. A. Syms, 1986. The nutritional ecology of the ctenophore *Bolinopsis vitrea*: comparisons with *Mnemiopsis mccradyi* from the same region. Journal of Plankton Research 8: 1197–1208.
Larson, R. J., 1986. Water content, organic content, and carbon and nitrogen composition of medusae from the northeast Pacific. Journal of Experimental Marine Biology and Ecology 99: 107–120.
Larson, R. J., 1987. Respiration and carbon turnover rates of medusae from the NE Pacific. Comparative Biochemistry and Physiology A 87: 93–100.
Larson, R. J., 1991. Diet, prey selection and daily ration of *Stomolophus meleagris*, a filter-feeding scyphomedusa from the NE Gulf of Mexico. Estuarine and Coastal Shelf Science 32: 511–525.
Link, J. S. & M. D. Ford, 2006. Widespread and persistent increase of Ctenophora in the continental shelf ecosystem off NE USA. Marine Ecology Progress Series 320: 153–159.
Liu, C.-Y. & Y.-P. Bi, 2006. A method of recapture rate in jellyfish ranching. Fishery Science/Shuichan Kexue 25: 150–151.
Lo, W.-T. & I.-L. Chen, 2008. Population succession and feeding of scyphomedusae, *Aurelia aurita,* in a eutrophic tropical lagoon in Taiwan. Estuarine Coastal and Shelf Science. 76: 227–238.
Lucas, C. H., 1996. Population dynamics of *Aurelia aurita* (Scyphozoa) from an isolated brackish lake, with particular reference to sexual reproduction. Journal of Plankton Research 18: 987–1007.
Lynam, C. P., M. J. Gibbons, B. E. Axelsen, C. A. J. Sparks, J. Coetzee, B. G. Heywood & A. S. Brierley, 2006. Jellyfish overtake fish in a heavily fished ecosystem. Current Biology 16: R492–R493.

Lynam, C. P., S. J. Hay & A. S. Brierley, 2004. Interannual variability in abundance of North Sea jellyfish and links to the North Atlantic Oscillation. Limnology and Oceanography 49: 637–643.

Lynam, C. P., M. R. Heath, S. J. Hay & A. S. Brierley, 2005. Evidence for impacts by jellyfish on North Sea herring recruitment. Marine Ecology Progress Series 298: 157–167.

Mackie, G. O. & C. E. Mills, 1983. Use of the pisces IV submersible for zooplankton studies in coastal waters of British Columbia. Canadian Journal of Fisheries and Aquatic Sciences 40: 763–776.

Mackie, G. O., P. R. Pugh & J. E. Purcell, 1987. Siphonophore biology. Advances in Marine Biology 24: 97–262.

Marrari, M., C. Hu & K. Daly, 2006. Validation of SeaWiFS chlorophyll a concentrations in the Southern Ocean: a revisit. Remote Sensing of Environment 105: 367–375.

Martinussen, M. B. & U. Båmstedt, 1999. Nutritional ecology of gelatinous planktonic predators. Digestion rate in relation to type and amount of prey. Journal of Experimental Marine Ecology and Biology 232: 61–84.

Martinussen, M. B. & U. Båmstedt, 2001. Digestion rate in relation to temperature of two gelatinous planktonic predators. Sarsia 86: 21–35.

Matsakis, S. & R. J. Conover, 1991. Abundance and feeding of medusae and their potential impact as predators on other zooplankton in Bedford Basin (Nova Scotia Canada) during spring. Canadian Journal of Fisheries and Aquatic Sciences 48: 1419–1430.

Mianzan, H. W., 1999. Ctenophora. In Boltovskay, D. (ed.), South Atlantic Zooplankton. Backhuys Publishers, Leiden: 561–573.

Mianzan, H. W. & P. F. S. Cornelius, 1999. Cubomedusae and Scyphomedusae. In Boltovskay, D. (ed.), South Atlantic Zooplankton. Backhuys Publishers, Leiden: 513–559.

Miller, R. J., 1970. Distribution and energetics of an estuarine population of the ctenophore, *Mnemiopsis leidyi*. Ph.D. Thesis, Department of Zoology, North Carolina State University, Raleigh.

Molinero, J. C., F. Ibanez, P. Nival, E. Buecher & S. Souissi, 2005. North Atlantic climate and northwestern Mediterranean plankton variability. Limnology and Oceanography 50: 1213–1220.

Møller, L. F. & H. U. Riisgård, 2007. Respiration in the scyphozoan jellyfish *Aurelia aurita* and two hydromedusae (*Sarsia tubulosa* and *Aequorea vitrina*): effect of size, temperature and growth. Marine Ecology Progress Series 330: 149–154.

Montoya, J. P., S. G. Horrigan & J. J. McCarthy, 1990. Natural abundance of $^{15}$N in particulate nitrogen and zooplankton in the Chesapeake Bay. Marine Ecology Progress Series 65: 35–61.

Nemazie, D. A., J. E. Purcell & P. M. Glibert, 1993. Ammonium excretion by gelatinous zooplankton and their contribution to the ammonium requirements of microplankton in Chesapeake Bay. Marine Biology 116: 451–458.

Oguz, T., 2005a. Black Sea ecosystem response to climatic teleconnections. Oceanography 18: 122–133.

Oguz, T., 2005b. Long-term impacts of anthropogenic forcing on the Black Sea ecosystem. Oceanography 18: 112–121.

Oguz, T., H. W. Ducklow, J. E. Purcell & P. Malanotte-Rizzoli, 2001. Modeling the response of top–down control exerted by gelatinous carnivores in the Black Sea pelagic food web. Journal of Geophysical Research 106: 4543–4564.

Olesen, N. J., J. E. Purcell & D. K. Stoecker, 1996. Feeding and growth by ephyrae of scyphomedusae *Chrysaora quinquecirrha*. Marine Ecology Progress Series 137: 149–159.

Omori, M. & E. Nakano, 2001. Jellyfish fisheries in southeast Asia. Hydrobiologia 451 (Developments in Hydrobiology 155): 19–26.

Omori, M., H. Ishii & A. Fujinaga, 1995. Life history strategy of *Aurelia aurita* (Cnidaria, Scyphomedusae) and its impact on the zooplankton community of Tokyo Bay. ICES Journal of Marine Science 52: 597–603.

Pagès, F., H. E. González & S. R. González, 1996a. Diet of the gelatinous zooplankton in Hardangerfjord (Norway) and potential predatory impact by *Aglantha digitale* (Trachymedusae). Marine Ecology Progress Series 139: 69–77.

Pagès, F., M. G. White & P. G. Rodhouse, 1996b. Abundance of gelatinous carnivores in the nekton community of the Antarctic Polar Frontal Zone in summer 1994. Marine Ecology Progress Series 141: 139–147.

Palomares, M. L. D. & D. Pauly, 2008. The growth of jellyfishes. Hydrobiologia. doi:10.1007/s10750-008-9582-y.

Pauly, D., W. M. Graham, S. Libralato, L. Morissette & M. L. D. Palomares, 2008. Jellyfish in ecosystems, online databases and ecosystem models. Hydrobiologia. doi:10.1007/s10750-008-9583-x.

Pavlova, E. V. & N. I. Minkina, 1993. The respiration rate of the Black Sea invader ctenophore (Ctenophora. Lobata: *Mnemiopsis*) Dokl. RAS 333(5): 682–683.

Pitt, K. A., A.-L. Clement, R. M. Connolly & D. Thibault-Botha, 2008a. Predation by jellyfish on large and emergent zooplankton: implications for benthic–pelagic coupling. Estuarine, Coastal and Shelf Science 76: 827–833.

Pitt, K. A., R. M. Connolly & T. Meziane, 2008b. Stable isotope and fatty acid tracers in energy and nutrient studies of jellyfish: a review. Hydrobiologia. doi:10.1007/s10750-008-9581-z.

Pitt, K. A. & M. J. Kingsford, 2003. Temporal variation in the virgin biomass of the edible jellyfish, *Catostylus mosaicus* (Scyphozoa, Rhizostomeae). Fisheries Research 63: 303–313.

Pitt, K. A., M. J. Kingsford, D. Rissik & K. Koop, 2007. Jellyfish modify the response of planktonic assemblages to nutrient pulses. Marine Ecology Progress Series 351: 1–13.

Pitt, K. A., K. Koop & D. Rissik, 2005. Contrasting contributions to inorganic nutrient recycling by the co-occurring jellyfishes, *Catostylus mosaicus* and *Phyllorhiza punctata* (Scyphosoa, Rhizostomeae). Journal of Experimental Marine Biology and Ecology 315: 71–86.

Pitt, K. A., D. T. Welsh & R. H. Condon, 2008c. Influence of jellyfish blooms on carbon, nitrogen and phosphorus cycling and plankton production. Hydrobiologia. doi:10.1007/s10750-008-9584-9.

Pugh, P. R., 1999. Siphonophorae. In Boltovskay, D. (ed.), South Atlantic Zooplankton. Backhuys Publishers, Leiden: 467–511.

Purcell, J. E., 1988. Quantification of *Mnemiopsis leidyi* (Ctenophora, Lobata) from formalin-preserved plankton samples. Marine Ecology Progress Series 45: 197–200.
Purcell, J. E., 1992. Effects of predation by the scyphomedusan *Chrysaora quinquecirrha* on zooplankton populations in Chesapeake Bay. Marine Ecology Progress Series 87: 65–76.
Purcell, J. E., 1997. Pelagic cnidarians and ctenophores as predators: selective predation, feeding rates and effects on prey populations. Annales de l'Institut océanographique, Paris 73: 125–137.
Purcell, J. E., 2003. Predation on zooplankton by large jellyfish (*Aurelia labiata, Cyanea capillata, Aequorea aequorea*) in Prince William Sound, Alaska. Marine Ecology Progress Series 246: 137–152.
Purcell, J. E. & M. N. Arai, 2001. Interactions of pelagic cnidarians and ctenophores with fishes: a review. Hydrobiologia 451 (Developments in Hydrobiology 155): 27–44.
Purcell, J. E., D. L. Breitburg, M. B. Decker, W. M. Graham, M. J. Youngbluth & K. A. Raskoff, 2001a. Pelagic cnidarians and ctenophores in low dissolved oxygen environments: a review. In Rabalais, N. N. & R. E. Turner (eds), Estuarine Studies 58: Coastal Hypoxia: Consequences for Living Resources and Ecosystems. American Geophysical Union, Washington, DC: 77–100.
Purcell, J. E., E. D. Brown, K. D. E. Stokesbury, L. H. Haldorson & T. C. Shirley, 2000. Aggregations of the jellyfish *Aurelia labiata*: abundance, distribution, association with age-0 walleye pollock, and behaviors promoting aggregation in Prince William Sound, Alaska, USA. Marine Ecology Progress Series 195: 145–158.
Purcell, J. E. & M. B. Decker, 2005. Effects of climate on relative predation by scyphomedusae and ctenophores on copepods in Chesapeake Bay during 1987–2000. Limnology and Oceanography 50: 376–387.
Purcell, J. E. & P. Kremer, 1983. Feeding and metabolism of the siphonophore *Sphaeronectes gracilis*. Journal of Plankton Research 5: 95–106.
Purcell, J. E., T. A. Shiganova, M. B. Decker & E. D. Houde, 2001b. The ctenophore *Mnemiopsis* in native and exotic habitats: U.S. estuaries versus the Black Sea basin. Hydrobiologia 451 (Developments in Hydrobiology 155): 145–176.
Purcell, J. E., S.-I. Uye & W.-T. Lo, 2007. Anthropogenic causes of jellyfish blooms and direct consequences for humans: a review. Marine Ecology Progress Series 350: 153–174.
Purcell, J. E., T. E. Whitledge, K. N. Kosobokova & R. R. Hopcroft. Distribution, abundance, and predation effects of epipelagic ctenophores and jellyfish in the western Arctic Ocean. Deep-Sea Research II (in press).
Rapoza, R., D. Novak & J. H. Costello, 2006. Life-stage dependent, in situ dietary patterns of the lobate ctenophore *Mnemiopsis leidyi* Agassiz 1865. Journal of Plankton Research 27: 951–956.
Raskoff, K. A., 2001. The impact of El Niño events on blooms of mesopelagic hydromedusae. Hydrobiologia 451: 121–129.
Raskoff, K. A., 2003. Collection and culture techniques for gelatinous zooplankton. Biological Bulletin 204: 68–80.
Raskoff, K. A., R. R. Hopcroft, J. E. Purcell & M. J. Youngbluth. Jellies on ice: ROV observations from the Arctic 2005 Hidden Ocean Expedition. Deep-Sea Research II (in press).
Raskoff, K. A., J. E. Purcell & R. R. Hopcroft, 2005. Gelatinous zooplankton of the Arctic Ocean: in situ observations under the ice. Polar Biology 28: 207–217.
Riisgård, H. U., L. Bøttiger, C. V. Madsen & J. E. Purcell, 2007. Invasive ctenophore *Mnemiopsis leidyi* in Limfjorden (Denmark) in late summer 2007—assessment of abundance and predation effects. Aquatic Invasions 2: 395–401.
Rutherford, L. D. Jr. & E. V. Thuesen, 2005. Metabolic performance and survival of medusae in estuarine hypoxia. Marine Ecology Progress Series 294: 189–200.
Ruzicka, J. J., R. D. Broudeur & T. C. Wainwright, 2007. Seasonal food web wodels for the Oregon inner-shelf ecosystem; investigating the role of large jellyfish. CalCOFI Reports 48: 106–128.
Schneider, G., 1989. A comparison of carbon based ammonia excretion rates between gelatinous and non-gelatinous zooplankton: implications and consequences. Marine Biology 106: 219–225.
Seymour, J. E., T. J. Carrette & P. A. Sutherland, 2004. Do box jellyfish sleep at night? Medical Journal of Australia 181: 707.
Shiganova, T. A., et al., 2003. Ctenophore invaders *Mnemiopsis leidyi* (A. Agassiz), *Beroe ovata* Mayer 1912, and their effect on the pelagic ecosystem of the northern Black Sea. Biological Bulletin 2: 225–235.
Siferd, T. D. & R. J. Conover, 1992. Natural history of ctenophores in the resolute passage area of the Canadian High Arctic with special reference to *Mertensia ovum*. Marine Ecology Progress Series 86: 133–144.
Sparks, C., E. Buecher, A. S. Brierley, B. E. Axelsen, H. Boyer & M. J. Gibbons, 2001. Observation on the distribution and relative abundance of the scyphomedusan *Chrysaora hysoscella* (Linné, 1766) and the hydrozoan *Aequorea aequorea* (Forskål, 1775) in the northern Benguela ecosystem. Hydrobiologia 451 (Developments in Hydrobiology 155): 275–286.
SPSS, 1997. SigmaStat® 2.0 for Windows® User's Manual. SPSS Inc., Chicago.
Stoecker, D. K., A. E. Michaels & L. H. Davis, 1987b. Grazing by the jellyfish *Aurelia aurita* on mirozooplankton. Journal of Plankton Research 9: 901–915.
Stoecker, D. K., P. G. Verity, A. E. Michaels, L. H. Davis, 1987a. Feeding by larval and post-larval ctenophores on microzooplankton. Journal of Plankton Research 9: 667–683.
Suchman, C. L., E. A. Daly, J. E. Keister, W. T. Peterson & R. D. Brodeur, 2008. Feeding patterns and predation potential of scyphomedusae in a highly productive upwelling region. Marine Ecology Progress Series 358: 161–172.
Sullivan, B. K., J. R. Garcia & G. Klein-MacPhee, 1994. Prey selection by the scyphomedusa predator *Aurelia aurita*. Marine Biology 121: 335–341.
Sullivan, L. J. & D. J. Gifford, 2004. Diet of the larval ctenophore *Mnemiopsis leidyi* A. Agassiz (Ctenophora, Lobata). Journal of Plankton Research 26: 417–431.
Sullivan, B. K. & M. R. Reeve, 1982. Comparison of estimates of the predatory impact of ctenophores by two independent techniques. Marine Biology 68: 61–65.

Toonen, R. H. & F.-S. Chia, 1993. Limitations of laboratory assessments of coelenterate predation: container effects on the prey selection of the limnomedusa, *Proboscidactyla flavicirrata* (Brandt). Journal of Experimental Marine Biology and Ecology 167: 215–235.

Toyokawa, M., T. Inagaki & M. Terazaki, 1997. Distribution of *Aurelia aurita* (Linnaeus, 1758) in Tokio Bay; observations with echosounder and plankton net. Proceedings of the 6th International Conference on Coelenterate Biology. Natuurhistorisch Museum, Leiden: 483–490.

Uye, S., 2008. Blooms of the giant jellyfish *Nemopilema nomurai*: a threat to the fisheries sustainability of the East Asian Marginal Seas. Plankton and Benthos Research 3 (Suppl.): 125–131.

Uye, S.-I., N. Fujii & H. Takeoka, 2003. Unusual aggregations of the scyphomedusa *Aurelia aurita* in coastal waters along western Shikoku, Japan. Plankton Biology and Ecology 50: 17–21.

Uye, S. & H. Shimauchi, 2005. Population biomass, feeding, respiration and growth rates, and carbon budget of the scyphomedusa *Aurelia aurita* in the Inland Sea of Japan. Journal of Plankton Research 27: 237–248.

West, E. J., D. T. Welsh & K. A. Pitt, 2008. Influence of decomposing jellyfish on the sediment oxygen demand and nutrient dynamics. Hydrobiologia. doi:10.1007/s10750-008-9586-7.

Witt, M. J., A. C. Broderick, D. J. Johns, C. Martin, R. Penrose, M. S. Hoogmoed & B. J. Godley, 2007. Prey landscapes help identify potential foraging habitats for leatherback turtles in the NE Atlantic. Marine Ecology Progress Series 337: 231–243.

Hydrobiologia (2009) 616:51–65
DOI 10.1007/s10750-008-9593-8

JELLYFISH BLOOMS

# Patterns of jellyfish abundance in the North Atlantic

**Mark J. Gibbons · Anthony J. Richardson**

Published online: 25 September 2008


**Abstract** A number of explanations have been advanced to account for the increased frequency and intensity at which jellyfish (pelagic cnidarians and ctenophores) blooms are being observed, most of which have been locally directed. Here, we investigate seasonal and inter-annual patterns in abundance and distribution of jellyfish in the North Atlantic Ocean to determine if there have been any system-wide changes over the period 1946–2005, by analysing records of the presence of coelenterates from the Continuous Plankton Recorder (CPR) survey. Peaks in jellyfish abundance are strongly seasonal in both oceanic and shelf areas: oceanic populations have a mid-year peak that is more closely related to peaks in phyto- and zooplankton, whilst the later peak of shelf populations mirrors changes in SST and reflects processes of advection and aggregation. There have been large amplitude cycles in the abundance of oceanic and shelf jellyfish (although not synchronous) over the last 60 years, with a pronounced synchronous increase in abundance in both areas over the last 10 years. Inter-annual variations in jellyfish abundance in oceanic areas are related to zooplankton abundance and temperature changes, but not to the North Atlantic Oscillation or to a chlorophyll index. The long-term inter-annual abundance of jellyfish on the shelf could not be explained by any environmental variables investigated. As multi-decadal cycles and more recent increase in jellyfish were obvious in both oceanic and shelf areas, we conclude that these are likely to reflect an underlying climatic signal (and bottom-up control) rather than any change in fishing pressure (top-down control). Our results also highlight the role of the CPR data in investigating long-term changes in jellyfish, and suggest that the cnidarians sampled by the CPR are more likely to be holoplanktic hydrozoans and not the much larger meroplanktic scyphozoans as has been suggested previously.

**Keywords** Pelagic cnidaria · Ctenophora · Seasonality · Inter-annual · Climate change · Plankton

Guest editors: K. A. Pitt & J. E. Purcell
Jellyfish Blooms: Causes, Consequences, and Recent Advances

M. J. Gibbons (✉)
Department of Biodiversity and Conservation Biology, University of the Western Cape, Private Bag X17, Bellville 7535, South Africa
e-mail: mgibbons@uwc.ac.za

A. J. Richardson
Climate Adaptation Flagship, CSIRO Marine and Atmospheric Research, Cleveland, QLD 4163, Australia

A. J. Richardson
School of Physical Sciences, University of Queensland, QLD 4072, Australia

A. J. Richardson
Sir Alister Hardy Foundation for Ocean Science, The Laboratory, Citadel Hill, The Hoe, Plymouth PL1 2PB, UK

## Introduction

Jellyfish comprise members of the phyla Cnidaria (classes Cubozoa, Scyphozoa and Hydrozoa) and Ctenophora (classes Tentaculata and Nuda). Amongst the Cnidaria, cubozoans and scyphozoans (except Coronatae) are largely coastal, whilst hydrozoans (medusae and siphonophores) are often a conspicuous component of marine zooplankton assemblages in coastal and oceanic environments (Raymont, 1983). Amongst the Ctenophora, members of the class Nuda are primarily coastal, whilst the more diverse Tentaculata includes orders that are entirely oceanic (e.g. Cestida) or are found in both coastal and oceanic environments: some species of lobate and cydippid ctenophores can be very common in shelf waters (Shiganova et al., 2001; Gibbons et al., 2003). Although some pelagic cnidarians may contain photosynthetic zooxanthellae (Montgomery & Kremer, 1995), and newly released ephyrae and small medusae and ctenophores may consume microplankton (e.g. Stoecker et al., 1987; Boero et al., 2007), all are otherwise largely carnivorous. Most jellyfish in mid and high latitudes display annual "blooms" in abundance that reflect seasonal population processes (Fraser, 1970; Brinckmann-Voss, 1996), and these may be exaggerated by physical aggregation at oceanographic discontinuities (Graham et al., 2001) and advective processes (Schneider, 1987; Greve, 1994). Blooms also occur naturally at periods longer than 1 year, and these have been attributed to climatic events (Goy et al., 1989; Lynam et al., 2004, 2005a). When jellyfish bloom they have a pronounced impact on an ecosystem (Brodeur et al., 2002) and its pelagic fisheries (Möller, 1984; Travis, 1993; Lynam et al., 2005b).

A number of geographically scattered studies have indicated that blooms of jellyfish, especially pelagic cnidarians, are increasing in frequency and persisting longer than usual (Mills, 2001; Purcell, 2005; Link & Ford, 2006; Purcell et al., 2007). However, there is no agreement about the underlying cause of the bloom increases and a number of hypotheses have been advanced, including alien translocations (Purcell et al., 2001), eutrophication (Arai, 2001), an increase in hard substrata for polyp attachment in the case of cnidarians (Parsons & Lalli, 2002), over-fishing (Banse, 1990; Pauly et al., 2002; Link & Ford, 2006) and climate change (Brodeur et al., 1999; Purcell, 2005; Link & Ford, 2006). Perhaps part of the reason for the lack of consensus (if it exists at all, as they are not mutually exclusive processes) reflects the fact that most studies these hypotheses are based on have been conducted in shelf waters (e.g. Brierley et al., 2001; Link & Ford, 2006; Lynam et al., 2006) or in semi-enclosed seas (e.g. Brodeur et al., 1999, 2002). Thus, local explanations have been sought for local phenomena. Having said that, issues of over-fishing and climate change are clearly larger-scale issues and if they are applicable, it might be expected that jellyfish would now be more abundant in open ocean waters too. Unfortunately, however, there are no long-term data on jellyfish from blue-water, and even seasonal studies in these areas are lacking. Just as most studies have been conducted on continental shelves, so too have they tended to focus on scyphozoans (but see references in Mills, 2001), and it is still debatable whether jellyfish as a group are increasing. One of the reasons for this focus on scyphozoans is that they are individually large and conspicuous, and their negative impacts on various economic activities (e.g. Matsueda, 1969) have resulted in extensive recent work.

It is against this background that we determine whether there have been basin-scale changes in the abundance of jellyfish in the North Atlantic Ocean. We use data on the presence of coelenterate tissue from the Continuous Plankton Recorder (CPR) survey, the longest-running, large-scale, marine biological survey in the world (Richardson et al., 2006). From 1946 to 2005, a total of ~200,000 samples have been collected and counted throughout the North Atlantic Ocean, containing more than 30,000 positive occurrences of coelenterates. We investigate whether the observed changes in jellyfish abundance can be explained by changes in the food environment (zooplankton or phytoplankton abundance) or hydroclimate (sea surface temperature, North Atlantic Oscillation). To aid interpretation of the inter-annual patterns in jellyfish abundance, we also examine seasonal patterns of abundance across the region, as these data too are missing for open ocean systems.

## Methods

### Plankton data

The CPR is a robust plankton sampler towed voluntarily and unaccompanied each month behind merchant ships on their normal routes of passage. Although the methods used to analyse jellyfish in CPR samples have remained unchanged since 1946 (Richardson et al., 2006), the CPR is not an ideal jellyfish sampler and several limitations must be borne in mind when using jellyfish data from this device (see also the "Discussion"). Firstly, the CPR is a near-surface (10 m) sampler, so it preferentially collects groups in surface waters, and the abundance of jellyfish throughout the water column is not known. Secondly, jellyfish (especially ctenophores) are extremely delicate and tend to be extensively damaged during capture by the CPR, so with the exception of rigid calycophoran siphonophores they are simply recorded as present or absent, with no species-level information or quantitative abundance estimates for each sample. Jellyfish (except rigid calycophoran siphonophores which are recorded separately) have been recorded by the CPR team as "coelenterates", a term that does not have a universal definition. Historically it has been interpreted to include either cnidarians and ctenophores or just cnidarians (Barnes, 1980). "Coelenterates" in CPR samples have been identified predominantly by their tissue and often (but not always) by the presence of nematocysts (Richardson et al., 2006). This means that siphonophores without substantially rigid bells as well as detached tentacles from those with rigid bells are likely to have been classified as "coelenterates", and we have thus combined data on the presence of total "coelenterates" and the presence of total siphonophores to form a total jellyfish group for analysis. In our interpretation of CPR data we thus differ from Attrill et al. (2007) who considered coelenterate tissue to have originated only from scyphomedusae and hydromedusae (excluding siphonophores); these authors also ignored the presence of ctenophores in coelenterate tissue. From 1946 to 2005, there were 30,297 positive records of jellyfish out of a total of 196,679 samples. CPR data from oceanic regions span 1948–2005 and from the shelf 1946–2005.

Here, we summarise data within 41 CPR standard areas in the North Atlantic (Fig. 1). The positioning and size of these standard areas is not entirely arbitrary; edges of many of the standard areas follow the edge of the continental shelf and the size of the boxes on the shelf are smaller to reflect the more dynamic physical environment, larger biological variability and greater CPR sampling. Annual abundance estimates were calculated for each CPR box according to the method of Colebrook (1975) for years when 8 or more months were sampled. In calculating monthly or annual abundances, we have used the frequency of occurrence of jellyfish on CPR samples, which has the added advantage of accounting for the operational efficiency of the CPR (Hunt & Hosie, 2006).

To investigate potential factors responsible for inter-annual and seasonal variations in jellyfish abundance, both the physical (see next section) and food environments of jellyfish were estimated. We

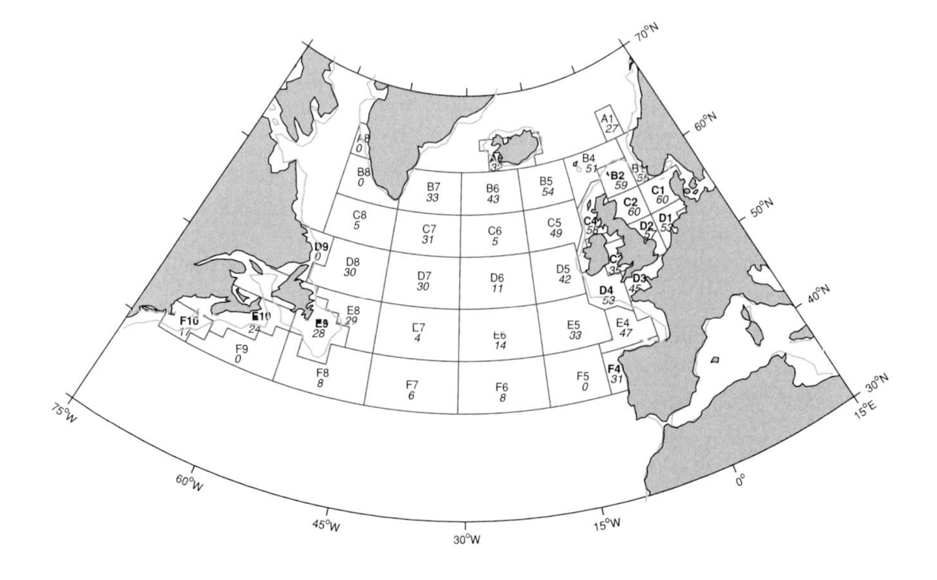


**Fig. 1** Map of the North Atlantic Ocean showing CPR standard areas. In each case, the number of years with more than 8 months of samples is shown. Shelf areas used in our analyses are in bold. The edge of the continental shelf is represented by the 200 m isobath

used two estimates of the potential food environment. The first was total zooplankton, which included chaetognaths, copepods, euphausiids, and larvae of decapods, echinoderms and fish. The other was the Phytoplankton Colour Index (PCI), which is a visual assessment of the greenness of the sampling mesh (silk) in the CPR. This is a good indicator of total phytoplankton biomass (Richardson et al., 2006) and has been included because small jellyfish may consume phytoplankton (e.g. Parsons & Lalli, 2002): it is also generally reflective of primary productivity at that time and place (Raitsos et al., 2005).

We estimated the timing of the seasonal peak throughout the year (the central tendency, $T$) using the month co-ordinate of the centre of gravity of the area below graphs of monthly mean abundances (Colebrook & Robinson, 1965):

$$T = \frac{\sum_{i=1}^{12} m \cdot x_{\mathrm{m}}}{\sum_{i=1}^{12} x_{\mathrm{m}}}$$

where $x_{\mathrm{m}}$ is the mean abundance in month $m$ (January = 1, …, December = 12). This index is sensitive to changes in the timing of the seasonal cycle (Edwards & Richardson, 2004).

The duration of the productive season (in months) was calculated as the standard deviation of the timing, following Colebrook (1979):

$$\sqrt{\frac{\sum_{i=1}^{12} x_{\mathrm{m}}(m - T)^2}{\sum_{i=1}^{12} x_{\mathrm{m}}}}$$

## Environmental data

To assess the importance of hydro-climatic forcing to changes in inter-annual jellyfish abundance we use two indices: viz. Sea Surface Temperature (SST) and the North Atlantic Oscillation (NAO). Sea surface temperature (SST) is an indicator of water mass movement, ocean climate and climate change, and organisms respond physiologically to temperature. Monthly sea surface temperature (SST) data on a 1° by 1° grid for the period 1946–2005 were obtained from the Hadley Centre, UK Met Office (HadISST, Version 1.1). This dataset is a gridded product combining in situ sea surface observations and satellite-derived estimates (Rayner et al., 2003).

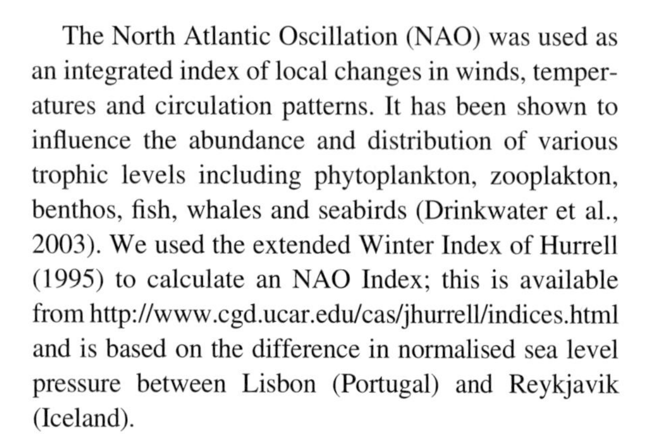

The North Atlantic Oscillation (NAO) was used as an integrated index of local changes in winds, temperatures and circulation patterns. It has been shown to influence the abundance and distribution of various trophic levels including phytoplankton, zooplakton, benthos, fish, whales and seabirds (Drinkwater et al., 2003). We used the extended Winter Index of Hurrell (1995) to calculate an NAO Index; this is available from http://www.cgd.ucar.edu/cas/jhurrell/indices.html and is based on the difference in normalised sea level pressure between Lisbon (Portugal) and Reykjavik (Iceland).

## Statistical analysis

To determine environmental factors (SST, total zooplankton abundance and the PCI) that might influence seasonal changes in jellyfish abundance in the North Atlantic Ocean, we compared the synchrony in the timing of jellyfish with their environment for each habitat (shelf and oceanic, see Fig. 1), using the non-parametric Mann–Whitney *U*-test (Zar, 1984).

We investigated the relative importance of factors influencing inter-annual fluctuations in jellyfish abundance using a model-building approach with total zooplankton, PCI, SST and NAO as predictors. Forward stepwise multiple regression was used rather than more sophisticated non-linear approaches such as generalised additive modelling because an assessment of the data revealed linear relationships were adequate, and because a greater number of degrees of freedom are required for fitting non-linear functions. The response variable, jellyfish abundance, was not transformed as visual plots of normality and homoscedasticity were satisfactory.

To summarise seasonal cycles and inter-annual variation for the CPR standard areas we used a standardised Principal Components Analysis (PCA) based on a correlation matrix. We only report here the first principal component, which represents the major pattern of variation in jellyfish abundance. We performed a separate analysis for CPR standard areas in the open ocean and on the continental shelf. Because of the scarcity of CPR data in the Northwest Atlantic, we only included standard areas east of 43°W in this analysis. Note that we could not apply the central tendency index ($T$) to the seasonal PCA cycles because the index is sensitive to the magnitude of the $y$-values, and these have been standardised in the PCA.

We used the cumulative sums method to detect changes in inter-annual PCA time series (Beaugrand et al., 2003). Because the PCA time series for oceanic and shelf regions were already standardised to a reference value of zero, we successively summed the anomalies forming a cumulative function for each. Consecutive positive anomalies produce a positive slope, whereas successive negative anomalies produce a negative slope.

## Results

### Seasonal patterns

Figure 2 shows the seasonal long-term (1946–2005) mean distribution of jellyfish abundance in the North Atlantic Ocean. Jellyfish in CPR samples are uncommon across the region during winter, but begin to increase along the European continental shelf-break during spring, peaking around the Rockall Bank during summer, and declining slowly during autumn. Abundances are higher in the eastern and south-central sectors of the sampled area than elsewhere. Abundances on the continental shelves of Europe and North America are generally similar to or lower than the open ocean.

The timing of peaks in jellyfish abundance indicates that these occur significantly later in the year over the shelf (mean centroid month = 7.35) than they do in oceanic areas (Fig. 3a; month = 6.89; Mann–Whitney $U = 123$, $P < 0.05$). In oceanic areas, peaks in jellyfish abundance do not lag significantly behind those of either the PCI (month = 6.73; $U = 236$, $P > 0.05$) or total zooplankton abundance (Fig. 3b; month = 6.80; $U = 248$, $P > 0.05$). This suggests a close coupling between the timing of jellyfish occurrence and food environment. By contrast, jellyfish abundances peak significantly ($U = 172$, $P < 0.05$) earlier than peaks in SST (Fig. 3b), suggesting SST has little direct influence on the timing of jellyfish peaks in the open ocean.

Peaks in jellyfish abundance (month = 7.35) over the continental shelf do not lag significantly behind

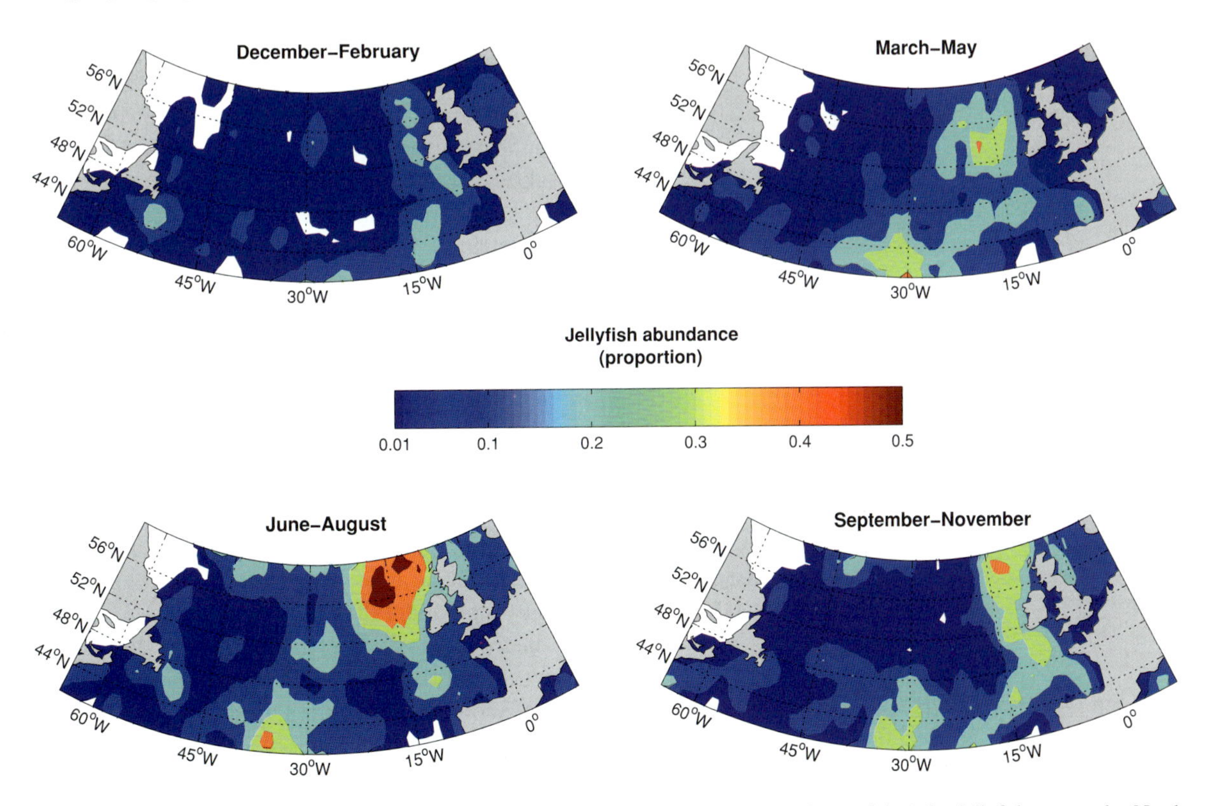


**Fig. 2** Seasonal distribution maps of the frequency of occurrence (proportion of samples positive) for jellyfish across the North Atlantic over the study period. White zones represent regions with poor CPR coverage

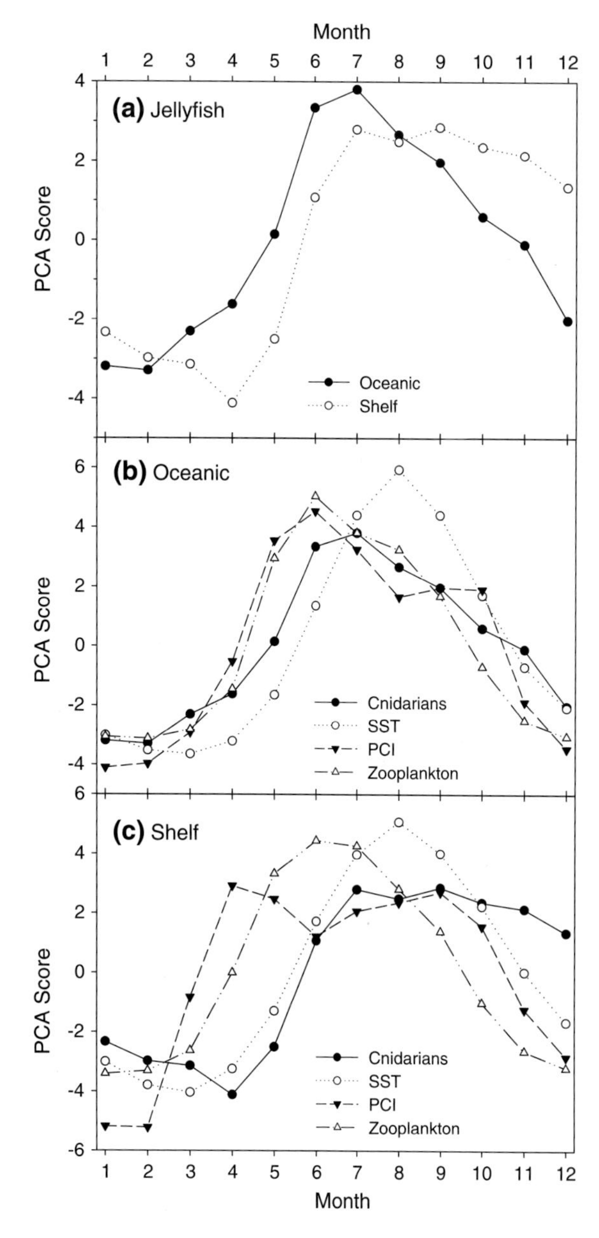


**Fig. 3** 1st principal components of PCAs of seasonal cycles. Comparison of (**a**) jellyfish in oceanic and shelf regions, (**b**) oceanic jellyfish with SST, PCI and zooplankton, and (**c**) shelf jellyfish with SST, PCI and zooplankton. Note that the first PCA shown here describes the main mode of variability in the seasonal cycles but does not give a measure of variability around the monthly values

SST (Fig. 3b; $U = 120, P > 0.05$). This suggests that temperature might have a more direct influence on the abundance of jellyfish in shallow seas. By contrast, both the PCI (month = 6.56) and the total abundance of zooplankton (month = 6.94) peak significantly earlier than jellyfish (Fig. 3b; $U = 36$, $P < 0.001$ and $U = 66$, $P < 0.05$ respectively), which suggests that the food environment does not directly influence seasonality in jellyfish on the shelf.

Jellyfish populations over the shelf had a slightly longer season than those in the open ocean (Fig. 3a, $U = 130$, $P < 0.10$; Fig. 4a). A more pronounced relationship was that jellyfish at higher latitudes have a markedly shorter productive season than those at lower latitudes, so peaks become gradually flatter towards the south (Fig. 4a; $U = 29$, $P < 0.001$). For both oceanic and shelf regions, the length of the productive season for jellyfish is directly correlated with the length of the productive season for different components of the food environment expressed both as PCI (Fig. 4b, $r = 0.64$, $P < 0.001$, $n = 41$) or total zooplankton (Fig. 4c, $r = 0.56$, $P < 0.001$, $n = 41$).

## Inter-annual patterns

Inter-annual variation in jellyfish abundance in oceanic and shelf areas is shown in Fig. 5a. In oceanic areas, the largest peak was in the late 1950s/early 1960s (highest in 1961) and a smaller peak in the late 1970s/early 1980s, and there has been an increase over the last 10 years. A spectral analysis confirmed a significant 20-year cycle. In shelf areas, the largest peak was in the mid-to-late 1980s (highest in 1987), with a more modest increase in more recent years. A spectral analysis documented the strongest signal as a 30-year cycle. There is a relatively weak, marginally significant correlation between jellyfish from oceanic and shelf areas ($r = 0.25$, $P = 0.056$, $n = 56$), with the time series in these regions being more synchronous since 1997 ($r = 0.71$, $P = 0.021$, $n = 10$).

A cumulative sum analysis performed on the 1st principal components of the jellyfish time series highlights the major step changes in oceanic and shelf waters (Fig. 5b). This analysis highlights transitions in the oceanic areas in the early 1960s and on the shelf in the mid-1980s.

In the oceanic area, multiple regression analysis showed that SST and total zooplankton were significantly correlated with the 1st principal component of the inter-annual abundance of jellyfish (Fig. 6; $r = 0.53$, $P < 0.001$, $n = 56$, standardised regression coefficients for total zooplankton = 0.33 and SST = 0.32). Neither the NAO nor the PCI were

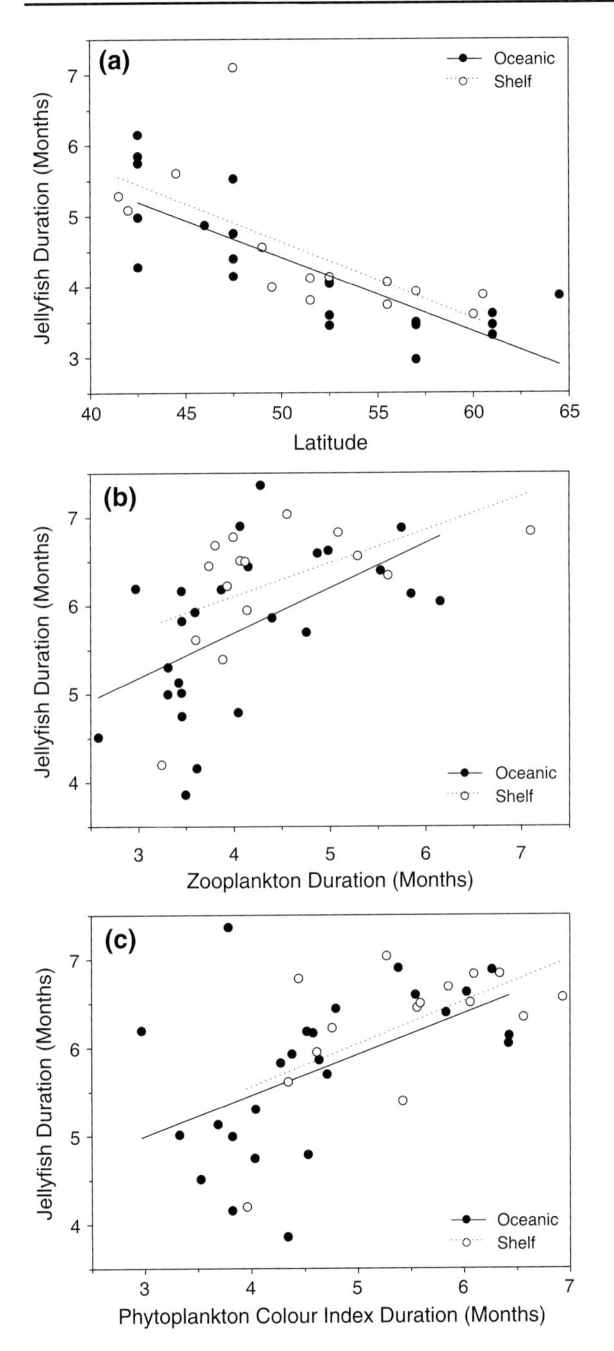


**Fig. 4** Duration of productive season of jellyfish for oceanic and shelf waters regressed against (**a**) latitude, (**b**) zooplankton duration and (**c**) the Phytoplankton Colour Index (PCI) duration

significant. By contrast, in shelf areas only the NAO was just significant ($P = 0.048$), but was non-significant when adjusted for temporal autocorrelation. None of the other environmental variables considered were correlated with jellyfish over the shelf.

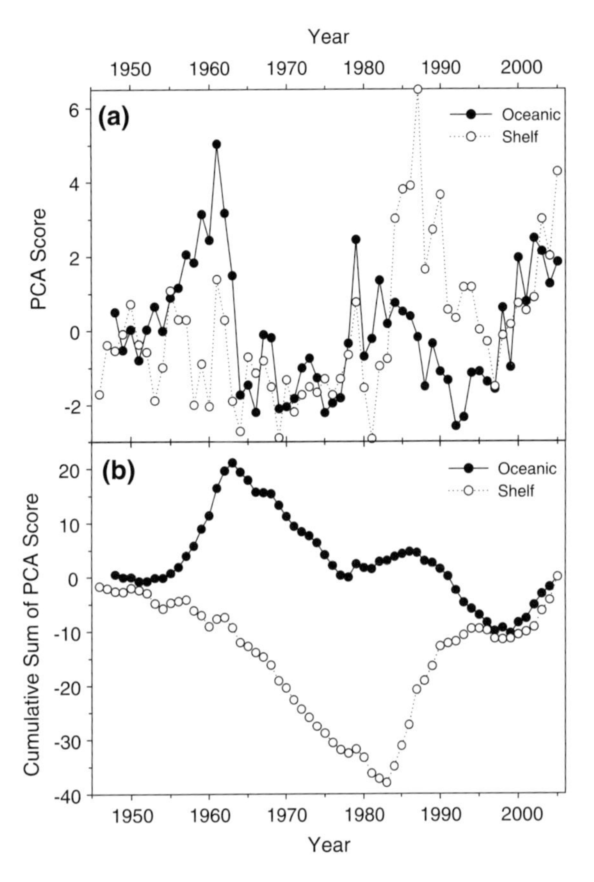


**Fig. 5** (**a**) First principal components of PCAs of inter-annual variation in oceanic and shelf jellyfish from 1946 to 2005, and (**b**) cumulative sums plots of (**a**) highlighting the major step changes in the time series

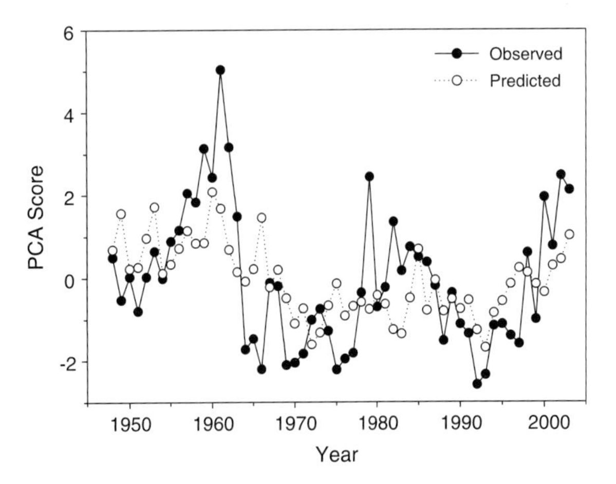


**Fig. 6** The predicted (and observed) time series of jellyfish in oceanic waters of the North Atlantic from a multiple regression model with SST and total zooplankton as predictors

## Discussion

From a 60-year time series of jellyfish based on nearly 200,000 samples in the North Atlantic, we show that there has not been a general increase in jellyfish abundance over this period, but there have been marked cycles of peaks and troughs. We found little coherence in the phases of these cycles between oceanic and shelf populations of jellyfish. However, there has been an increase over the last decade synchronous over both the ocean and the shelf, although present abundances are lower than during times of peak jellyfish abundance since 1946.

### Identity of CPR jellyfish tissue

Knowing the exact identity of the coelenterate material collected by the CPR would substantially enhance interpretation of our results. We need to remember that the CPR is towed in the upper water layers only, and therefore the jellyfish collected are primarily epipelagic but maybe not neustonic. We should also be aware that the aperture of the CPR is only 1.61 $cm^2$, which means that large scyphozoans may be less likely to be sampled than smaller species. An indirect test of the latter assumption is supported by the lack of any significant correlations ($P > 0.05$) between jellyfish from the CPR and abundance data from pelagic trawl data of Lynam et al. (2004, 2005a) on the scyphozoans *Aurelia aurita* (Linné 1758), *Cyanea lamarckii* Péron & Lesueur 1809 and *Cyanea capillata* (Linné 1758), analysed separately or combined, for three areas in the North Sea from 1971 to 1986 (12 separate tests). This finding differs from that of Attrill et al. (2007), who noted a positive relationship (in one of the three regions examined) between CPR "coelenterate" tissue and data for *A. aurita* alone, and this can be attributed to the inclusion of siphonophores in our data. It is also at odds with the conclusions of Attrill et al. (2007) who suggested that jellyfish tissues trapped by the CPR were largely derived from scyphomedusae, based on their single observation. We cannot think of a plausible reason why the CPR might differentially sample *A. aurita* and not collect the two species of *Cyanea* also reported in Lynam et al. (2004, 2005a) because although *A. aurita* reaches a smaller maximum diameter than either of the species of *Cyanea*, the nematocyst-bearing tentacles and oral arms of the latter species can exceed 30 m in length (Russell, 1953) and should be readily sampled.

Other evidence that Attrill et al. (2007) use to support the notion that the CPR predominantly captures scyphozoans is incorrect. They suggested that because the late seasonal peak of CPR "coelenterate" tissue is similar in timing to that of *Aglantha digitale* (O.F. Müller 1776) in the western central North Sea (Nicholas & Frid, 1999) that this provided evidence that the CPR predominantly samples scyphomedusans. In fact, they erroneously attributed *Aglantha digitale* to the Scyphomedusae: it is a small (~40 mm high) holoplanktic hydromedusa (Subclass Trachymedusae, Family Rhopalonematidae).

We therefore caution against accepting the findings of Attrill et al. (2007) that the origin of the jellyfish tissue in the CPR survey are from scyphozoans. We support the observations of Russell (1939, 1953) and Kramp (1959) regarding the distribution of cnidarians, and assume that CPR records made from the open ocean are dominated by holoplanktic Hydrozoa (Siphonophorae, Narcomedusae and Trachymedusae), whilst those from coastal waters include the meroplanktic Hydrozoa (primarily Leptomedusae and Anthomedusae) and neritic ctenophores (Fraser, 1970). We do not consider ctenophores to be likely members of the epipelagic community in oceanic waters because although these organisms may be diverse there, they are rare and mostly confined to the meso- and bathypelagic realm (Mianzan, 1999). That said, definitive proof will only be provided by molecular analysis of CPR coelenterate tissue, as has been done for echinoderms (Kirby & Lindley, 2005).

### Seasonal patterns

The seasonal pattern of distribution and abundance of jellyfish in the North Atlantic (Fig. 2) is essentially similar to that reported by Witt et al. (2007) for both jellyfish and thaliaceans from the CPR survey over the same time period. Although this could be because of the relative rarity of thaliaceans in CPR samples (Richardson et al., 2006), the lack of serious discrepancy may also imply that thaliaceans and jellyfish might display common patterns in the region. This conclusion is supported by the observations of Hunt (1968) who examined thaliacean data from the CPR survey from 1948 to 1965 and noted a similar northward, seasonal progression in their distribution and abundance in the eastern North Atlantic. He considered thaliaceans to be generally oceanic in the region (see also the CPR Atlas, Barnard et al., 2004). Witt et al. (2007) suggested that the aggregation of "gelatinous" zooplankton along the continental shelf break (including the Porcupine Bight and Bank, and the Rockall Bank and Trough) reflects the dynamic nature of the North Atlantic gyre in this area, where divergent and convergent eddies may lead to both elevated productivity and prey aggregation. Such an explanation is not unreasonable, given the relationship between seasonal peaks in jellyfish abundance and their food abundance in the open ocean (see below).

The synchrony between the timing of peaks in jellyfish abundance and those of their ambient food environment in the open ocean is likely to reflect the holopelagic nature of the blue-water fauna. Unfortunately there are no data regarding the seasonality of open water jellyfish from any ocean, but recent work on *Sarsia gemmifera* Forbes 1848 may provide some insight. This species can respond quickly and dramatically to changes in the abundance of its prey field (Stibor & Tokle, 2003). While *S. gemmifera* is an anthomedusa, and is itself unlikely to be found in the open ocean, it can reproduce asexually by producing secondary medusae from buds on the manubrium. Although *S. gemmifera* may be an inappropriate proxy for sexually reproducing holopelagic taxa because its response time to environmental changes is likely to be faster, it should be remembered that the only non-cnidoblast-bearing life-history "stages" of oceanic Hydrozoa are eggs and sperm, and their residence time in the plankton is short.

As far as the CPR is concerned, open ocean jellyfish are likely to be able to respond quickly to changes in the food environment, even when they undergo sexual reproduction, and it should also be remembered that Siphonophorae have prominent asexual life-history stages (Mackie et al., 1987). While peaks in jellyfish abundance occur earlier than peaks in SST in the open ocean, we should realise that warm SST tends to be associated with a stable water column having a nutrient-depleted upper mixed layer and low levels of primary and secondary production (e.g. Mann & Lazier, 1991). The role of temperature in influencing abundance cycles of jellyfish in the open ocean, therefore, is likely to be indirect and to act through its effect on water column stability and the food chain.

The timing of abundance peaks of jellyfish over the continental shelf is more variable, and the productive season is longer, than observed in the open ocean (Fig. 3a). This probably reflects their greater diversity in shallow water, and the greater number of factors controlling their abundance, including advection (Schneider, 1987; Greve, 1994), aggregation (Graham et al., 2001), and life history and population dynamics. The timing of abundance peaks over the shelf is later in the year than in oceanic areas and is linked to SST rather than the food environment. Although a number of environmental cues have been linked to the release of medusae by benthic life-history stages of cnidarians, including lunar cycles (Elmhirts, 1925), salinity (Goy, 1973) and ambient food (Roosen-Runge, 1970; Arai, 1987), temperature has the greatest support in the literature (Carré & Carré, 1990; Bullard & Myers, 2000). Interestingly, however, it is usually the cool and not the warm temperatures as observed here that are associated with medusa release by benthic polyps (e.g. Bullard & Myers, 2000). Most studies of pelagic cnidarians in coastal environments have indicated that Antho- and Leptomedusae peak in abundance from April to July (Watson, 1930; Russell, 1933, 1938; Byrne, 1995; Nicholas & Frid, 1999). Although this date is slightly earlier than observed here (Fig. 3), it is not unreasonable to suppose that they could nevertheless contribute to the initial increases in abundance observed here. The dominant ctenophore in the coastal environments of the European shelf, *Pleurobrachia pileus* (O.F. Müller 1776), similarly displays a peak in abundance during late spring and early summer across much of its distributional range (Greve, 1971; Yip, 1981; van der Veer & Sadée, 1984; Williams & Collins, 1985). While we should remember that in almost all cases, studies of coastal medusae and ctenophores have been conducted over depths shallower than are routinely sampled by vessels deploying the CPR, some of these same studies have also indicated pronounced late season peaks in the abundance of holopelagic taxa such as *Aglantha digitale* (Fig. 7), and there is also evidence that some populations of *Pleurobrachia pileus* may show autumn peaks in abundance (Russell, 1933; Fraser, 1970; Attrill & Thomas, 1996).

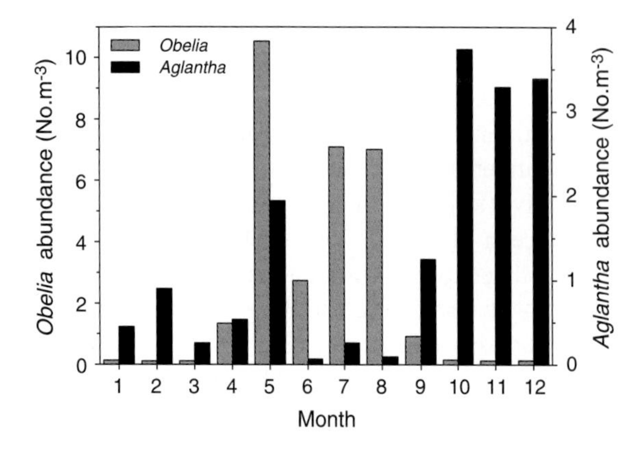


**Fig. 7** Seasonal cycles of *Obelia* (left axis) and *Aglantha digitale* (right axis) in the western central North Sea (Box C2 in Fig. 1). Redrawn from Nicholas & Frid (1999)

The seeding of shelf waters with oceanic taxa (including holopelagic cnidarians) occurs after the spring peak in PCI (Evans, 1972; Iversen et al., 2002), and it is likely to take a while for their populations to build up in the warmer, stratified waters more typical of their oceanic environment. As a consequence, perhaps, later peaks in abundance of jellyfish noted over the shelf reflect processes of advection, aggregation and population dynamics of holoplanktic, not meroplanktic taxa.

That the productive season of jellyfish is substantially shorter in the north than in the south is in agreement with latitudinal trends in the length of annual pelagic production cycles (Longhurst, 1998). Similar observations were also made for phytoplankton from the CPR (Robinson, 1970). The slightly greater length of the productive season in shelf than in oceanic waters has also been reported previously for both phytoplankton and copepods from the CPR (Colebrook & Robinson, 1961; Robinson, 1970). Colebrook & Robinson (1961) attributed these differences to differential impacts of vertical mixing and the timing of stratification in the two areas.

Inter-annual patterns

*Regime shifts*

Data presented here show that jellyfish over the European continental shelf underwent a stepwise change in abundance just before the mid-1980s (Fig. 5). This observation and its timing is coincident with the well-established regime shift in the food web of the North Sea in terms of plankton, fish, marine mammals, and seabirds, which has been attributed to changes in large-scale hydro-meteorological forcing (Reid et al., 2001; Beaugrand, 2004; Weijerman et al., 2005). Step changes in the plankton associated with the regime shift have been noted in diatoms, dinoflagellates, and copepods around 1987 (Edwards et al., 2002, 2006), and the current work extends these changes to jellyfish. It is interesting to note that changes in jellyfish appear to precede those of other planktonic taxa by several years. The regime shift in the North Sea has been accompanied by a biogeographic change in the zooplankton communities; the typical cold-water copepod fauna has been replaced by a warm-water one dominated by smaller, more southerly species (Beaugrand et al., 2001, 2002). We would also expect the regime shift to be associated with an increase in the incidence of pseudo-oceanic holoplanktic taxa, which can dominate shelf communities under warm-water conditions elsewhere (Buecher & Gibbons, 2000).

*Shelf jellyfish*

One of the large-scale hydro-meteorological drivers of the regime shift in the North Sea is thought to be the NAO, particularly the sustained positive phase of recent times (Beaugrand, 2004). We found no link between the NAO and jellyfish abundance estimated by PCA over the European shelf as a whole. Attrill et al. (2007) noted a positive correlation between "coelenterate" CPR abundance from the North Sea and the NAO, which suggests that its role in structuring zooplankton communities may be greater in the relative confines of the North Sea than elsewhere. On close examination, we find significant relationships with the NAO only for 2 out of the 10 CPR standard areas of the European shelf and these were in the western northern and western central North Sea (grid B2, $r = 0.38$, $F = 4.39$, df = 26, $P < 0.05$ and grid C2, $r = 0.49$, $F = 8.00$, df = 25, $P < 0.01$, $n = 57$; significance levels were adjusted for autocorrelation).

*Oceanic jellyfish*

By comparison with the North Sea, there has been far less investigation of long-term changes in the zooplankton of the North Atlantic Ocean beyond the edge of the European shelf. Results of the multiple regression analysis indicate that the food environment (as total zooplankton) has a role to play in fuelling jellyfish abundance. The link between the seasonality of jellyfish and prey in the open ocean has already been noted (Fig. 3b), and bottom-up links between jellyfish and their prey have been commonly reported in the literature for shelf species (e.g. Feigenbaum & Kelly, 1984). The prey field for a carnivore in the open ocean is reduced by comparison with that of the shelf (Raymont, 1983) and if this limits population size, then any increase in prey field will positively act on those species able to take advantage of that increase.

SST was the other environmental variable corresponding to inter-annual changes in jellyfish in the open ocean. Warmer temperatures increase the rate of reproduction for polyps of shelf species of cnidarians (Purcell et al., 1999) and these may be useful proxies for holoplanktic oceanic forms about which very little is known. However, it may be more likely that the warmer temperatures enhance ocean stratification, which favours small phytoplankton cells and longer food chains, which may give an advantage to jellyfish (Parsons & Lalli, 2002).

*Non-climate explanations*

A number of explanations have been advanced to account for the recent increases in jellyfish abundance that have been reported. With our present understanding, three of these can probably be discarded outright when interpreting the oceanic data because they all have a coastal signature: viz. eutrophication, alien translocations and increasing amounts of hard substrata. A number of oceanic fishes feed opportunistically on small jellyfish whilst some larger vertebrates (including turtles and birds) may consume larger species (Arai, 2005). However, few of these predators are obligate jellyfish feeders (Arai, 2005). Given that most high order carnivores are relatively rare at the scale of the ocean basin considered here, it is unlikely that they exert a top-down effect on jellyfish population numbers. Despite the fact that many predator fish populations in the open ocean are currently overexploited (Myers & Worm, 2003)—directly or indirectly—the lack of a long cyclical signal to fishing pressure (FishStat Plus) suggests that fishing itself is unlikely to account for the pattern observed in jellyfish.

By contrast with the open ocean, all the explanations above could potentially be invoked to explain the recent (last 10 years) increases in jellyfish abundance over the shelf waters of north-western Europe. That said, some have greater support than others and none, given the cyclic nature of the patterns observed on the shelf, have as much support as climate. The link between eutrophication and the abundance of pelagic cnidarians has been emphasised (Arai, 2001), and anthropogenic eutrophication in the North Sea has increased dramatically over recent decades (Greve et al., 1996). However, its effects are likely to be relatively localised because 90% of the nutrients in the North Sea enter through Atlantic inflow into the north (North Sea Task Force, 1993).

Alien species of medusae (but not ctenophores, Boersma et al., 2007) to our knowledge have not been widely reported in the shelf waters of north-western Europe. Yet, if these were responsible for the observed increases in abundance they should have been collected during ongoing sampling operations in the region (e.g. those off Northumberland, Helgoland-Roads, and the English Channel), they have not. While there is likely to have been an increase in the amount of anthropogenic hard substrata available for polyp colonisation in the region, this has not been quantified and it is problematic to find a plausible reason to explain any cyclical change in its possible availability.

Many shelf populations of fishes are, or have been, heavily exploited, and it would appear that some populations of jellyfish have expanded in size to take advantage of the vacant ecological space (e.g. Lynam et al., 2006). However, evidence from coastal waters of the eastern North Atlantic is equivocal on the subject. Although fisheries landings from the coastal European waters show no cyclic pattern similar to our findings (FishStat Plus), Lynam et al. (2005b) have shown that the decline in herring spawning stock biomass noted between 1971 and 1986 in the North Sea was accompanied by an increase in the abundance of the scyphozoan *Aurelia aurita*, possibly as a result of competitive release. The rise in biomass of jellyfish may in turn have caused additional decreases in the herring population through negative impacts on recruitment (Lynam et al., 2005b). The relationship is not simple, however, and appears to be mediated through complex climatic interactions.

We show here that jellyfish in the northeast Atlantic show cyclic changes in population sizes and that they are currently increasing in both shelf and oceanic waters. We suggest that the cycles are likely to reflect an underlying climatic signal (and bottom-up control) rather than any change in fishing pressure (top-down control). Our results highlight the role of the CPR survey data in investigating long-term changes in jellyfish: their one principle weakness being an inability to identify the species responsible. Future work needs to include a focus on taxonomic identification of coelenterate tissues contained in the archived samples by employing molecular probes (as Kirby & Lindley, 2005). Aside from making it possible to more meaningfully interrogate CPR data to explore issues of changing species abundance (as Kirby et al., 2007), phenology (as Edwards & Richardson, 2004) and biogeography (as Beaugrand et al., 2002), an improved resolution of the CPR data will allow the definition of prey fields for those species that prey on jellyfish. The CPR data set has recently been used to define prey fields for the wide ranging planktivorous basking shark, *Cetorhinus maximus* (Gunnerus 1765), that feeds on crustacean zooplankton (Sims et al., 2006). The CPR data-set can, therefore, potentially be used to explain the drivers behind spatio-temporal variations in behaviour of planktivores (e.g. Sims et al., 2005). In terms of gelatinous zooplankton, an important obligate jelly-feeder is the leatherback turtle, *Dermochelys coriacea* (Linnaeus 1766), that frequents the North Atlantic in summer and autumn. These turtles target patches of high gelatinous zooplankton abundance feeding on jellyfish such as *Rhizostoma*, *Cyanea* and *Chrysaora* (e.g. Houghton et al., 2006). CPR data have been used previously to define gelatinous zooplankton prey fields for leatherback turtles on the assumption that the CPR predominantly samples scyphozoan jellyfish (Witt et al., 2007). However, the evidence presented here suggests that this might not be the case, re-emphasising the need for more exact identification of the "coelenterate" tissue in the CPR samples to define prey fields more accurately for foraging turtles.

**Acknowledgements** CPR data used in this study were provided by the Sir Alister Hardy Foundation for Ocean Science (SAHFOS) and are freely available to all researchers (http://www.sahfos.ac.uk). The authors are grateful to all past and present members and supporters of the CPR survey, especially the shipping industry that voluntarily tows CPRs on regular routes and funders from UK and elsewhere. MJG is grateful to the Royal Society (London) and the National Research Foundation (South Africa) for funding. Dr Chris Lynam (University of St Andrews) is thanked for providing his scyphozoan data from the North Sea, and Dr Matthew Witt (University of Exeter) is kindly thanked for assistance with preparing Fig. 2. We are particularly grateful to Graeme Hays (University of Wales, Swansea), Kylie Pitt (Griffith University) and an anonymous reviewer for improving earlier versions of the manuscript.

## References

Arai, M. N., 1987. Population ecology of the hydromedusae of Massett Inlet, British Columbia. In Bouillon, J., F. Boero, F. Cicogna & P. F. S. Cornelius (eds), Modern Trends in the Systematics, Ecology and Evolution of Hydroids and Hydromedusae. Clarendon Press, Oxford: 107–116.

Arai, M. N., 2001. Pelagic coelenterates and eutrophication: a review. Hydrobiologia 451: 69–87.

Arai, M. N., 2005. Predation on pelagic coelenterates: a review. Journal of the Marine Biological Association of the United Kingdom 85: 523–536.

Attrill, M. J. & R. M. Thomas, 1996. Long-term distribution patterns of mobile estuarine invertebrates (Ctenophora, Cnidaria, Crustacea: Decapoda) in relation to hydrological parameters. Marine Ecology Progress Series 143: 25–36.

Attrill, M. J., J. Wright & W. Edwards, 2007. Climate-related increases in jellyfish frequency suggest a more gelatinous future for the North Sea. Limnology and Oceanography 52: 480–485.

Banse, K., 1990. Mermaids – their biology, culture and demise. Limnology and Oceanography 35: 148–153.

Barnard, R., S. D. Batten, G. Beaugrand, C. Buckland, D. V. P. Conway, M. Edwards, J. Finlayson, L. W. Gregory, N. C. Halliday, A. W. G. John, D. G. Johns, A. D. Johnson, T. D. Jonas, J. A. Lindley, J. Nyman, P. Pritchard, P. C. Reid, A. J. Richardson, R. E. Saxby, J. Sidey, M. A. Smith, D. P. Stevens, C. M. Taylor, P. R. G. Tranter, A. W. Walne, M. Wootton, C. O. M. Wotton & J. C. Wright, 2004. Continuous plankton records: plankton atlas of the North Atlantic Ocean (1958–1999). II. Biogeographical charts. Marine Ecology Progress Series, Supplement: 11–75.

Barnes, R. D., 1980. Invertebrate Zoology. Saunders College, Philadelphia.

Beaugrand, G., 2004. The North Sea regime shift: evidence, causes, mechanisms and consequences. Progress in Oceanography 60: 245–262.

Beaugrand, G., F. Ibañez & J. A. Lindley, 2001. Geographical distribution and seasonal and diel changes in the diversity of calanoid copepods in the North Atlantic and North Sea. Marine Ecology Progress Series 219: 189–203.

Beaugrand, G., F. Ibañez & J. A. Lindley, 2003. An overview of statistical method applied to the CPR data. Progress in Oceanography 58: 235–262.

Beaugrand, G., P. C. Reid, F. Ibañez, J. A. Lindley & M. Edwards, 2002. Reorganisation of North Atlantic marine copepod biodiversity and climate. Science 296: 1692–1694.

Boero, F., C. Bucci, A. M. R. Colucci, C. Gravili & L. Stabili, 2007. *Obelia* (Cnidaria, Hydrozoa, Campanulariidae): a microphagous, filter-feeding medusa. Marine Ecology 28: 178–183.

Boersma, M., A. M. Malzahn, W. Greve & J. Javidpour, 2007. The first occurrence of the ctenophore *Mnemiopsis leidyi* in the North Sea. Helgololand Marine Research 61: 153–155.

Brierley, A. S., B. E. Axelsen, E. Buecher, C. Sparks, H. Boyer & M. J. Gibbons, 2001. Acoustic observations of jellyfish in the Namibian Benguela. Marine Ecology Progress Series 210: 55–66.

Brinckmann-Voss, A., 1996. Seasonality of hydroids (Hydrozoa, Cnidaria) from an intertidal pool and adjacent subtidal habitats at Race Rocks, off Vancouver Island, Canada. Scientia Marina 66: 89–97.

Brodeur, R. D., C. E. Mills, J. E. Overland, G. E. Walters & J. D. Schumacher, 1999. Evidence for a substantial increase in gelatinous zooplankton in the Bering Sea, with possible links to climate change. Fisheries Oceanography 8: 296–306.

Brodeur, R. D., H. Sugisaki & G. L. Hunt, 2002. Increases in jellyfish biomass in the Bering Sea: implications for the ecosystem. Marine Ecology Progress Series 233: 89–103.

Buecher, E. & M. J. Gibbons, 2000. Interannual variation in the composition of the assemblages of medusae and ctenophores in St Helena Bay, Southern Benguela Ecosystem. Scientia Marina 64: 123–134.

Bullard, L. & A. Myers, 2000. Observations on the seasonal occurrence and abundance of gelatinous zooplankton in Lough Hyne, Co. Cork, south-west Ireland. Biology and Environment – Proceedings of the Royal Irish Academy 100B: 75–83.

Byrne, P., 1995. Seasonal composition of meroplankton in the Dunkellin estuary, Galway Bay. Biology and Environment – Proceedings of the Royal Irish Academy 95B: 35–48.

Carré, D. & C. Carré, 1990. Complex reproductive cycle in *Eucheilota paradoxica* (Hydrozoa: Leptomedusae): medusae, polyps and frustules produced from medusa stage. Marine Biology 104: 303–310.

Colebrook, J. M., 1975. The continuous plankton recorder survey: automatic data processing methods. Bulletin of Marine Ecology 8: 123–142.

Colebrook, J. M., 1979. Continuous plankton records: seasonal cycles of phytoplankton and copepods in the North Atlantic Ocean and the North Sea. Marine Biology 51: 23–32.

Colebrook, J. M. & G. A. Robinson, 1961. The seasonal cycle of the plankton in the North Sea and the North-Eastern Atlantic. Journal du Conseil Interntional pour l'Exloration de lar Mer 26: 156–165.

Colebrook, J. M. & G. A. Robinson, 1965. Continuous plankton records: seasonal cycles of phytoplankton and copepods in the north-eastern Atlantic and the North Sea. Bulletin of Marine Ecology 6: 123–139.

Drinkwater, K. F., A. Belgrano, A. Borja, A. Conversi, M. Edwards, C. H. Greene, G. Ottersen, A. Pershing & H. Walker, 2003. The response of marine ecosystems to climate variability associated with the North Atlantic Oscillation. In Hurrell, J. W., Y. Kushnir, G. Ottersen, & M. Visbeck (eds), The North Atlantic Oscillation: Climatic Significance and Environmental Impact, Geophysical Monograph 134. AGU, Washington, DC: 211–234.

Edwards, M., G. Beaugrand, P. C. Reid, A. A. Rowden & M. B. Jones, 2002. Ocean climate anomalies and the ecology of the North Sea. Marine Ecology Progress Series 239: 1–10.

Edwards, M., D. G. Johns, S. C. Leterme, E. Svendsen & A. J. Richardson, 2006. Regional climate change and harmful algal blooms in the northeast Atlantic. Limnology and Oceanography 51: 820–829.

Edwards, M. & A. J. Richardson, 2004. The impact of climate change on the phenology of the plankton community and trophic mismatch. Nature 430: 881–884.

Elmhirts, R., 1925. Lunar periodicity in *Obelia*. Nature 116: 358–359.

Evans, F., 1972. The permanent zooplankton of Northumberland coastal waters. Proceedings of the University of Newcastle upon Tyne Philosophical Society 2: 25–68.

Feigenbaum, D. L. & M. Kelly, 1984. Changes in the lower Chesapeake Bay food chain in the presence of the sea nettle *Chrysaora quinquecirrha* (Scyphomedusae). Marine Ecology Progress Series 19: 39–47.

FishStat Plus fisheries statistics software on CD-ROM. Food and Agricultural Organization of the United Nations.

Fraser, J. H., 1970. The ecology of the ctenophore *Pleurobrachia pileus* in Scottish waters. Journal du Conseil International pour l'Exploration de la Mer 33: 149–168.

Gibbons, M. J., E. Buecher & D. Thibault-Botha, 2003. Observations on the ecology of *Pleurobrachia pileus* (Ctenophora) in the southern Benguela ecosystem. African Journal of Marine Science 25: 253–261.

Goy, J., 1973. *Gonionemus suvaensis*: structural characters, developmental stages and ecology. Publications of the Seto Marine Biology Laboratory 20: 525–536.

Goy, J., P. Morand & M. Etienne, 1989. Long term fluctuation of *Pelagia noctiluca* (Cnidaria, Scyphomedusa) in the western Mediterranean Sea. Prediction by climatic variables. Deep-Sea Research 36: 269–279.

Graham, W. M., F. Pagès & W. M. Hamner, 2001. A physical context for gelatinous zooplankton aggregations: a review. Hydrobiologia 451: 199–212.

Greve, W., 1971. Ökologische Untersuchungen an *Pleurobrachia pileus*. 1. Freilanduntersuchungen. Helgoländer wiss. Meeresunters 22: 303–325.

Greve, W., 1994. The 1989 German Bight invasion of *Muggiaea atlantica*. ICES Journal of Marine Science 51: 355–358.

Greve, W., F. Reiners & J. Nast, 1996. Biocoenotic changes of the zooplankton in the German Bight: the possible effects of eutrophication and climate. ICES Journal of Marine Science 53: 951–956.

Houghton, J. D. R., T. K. Doyle, M. W. Wilson, J. Davenport & G. C. Hays, 2006. Jellyfish aggregations and leatherback turtle foraging patterns in a temperate coastal environment. Ecology 87: 1967–1972.

Hunt, H. G., 1968. Continuous plankton records: contribution towards a plankton atlas of the North Atlantic and the

North Sea Part XI: the seasonal and annual distributions of Thaliacea. Bulletin of Marine Ecology 6: 225–249. Plates LXVIII–LXXXV.

Hunt, B. P. V. & G. W. Hosie, 2006. Continuous plankton recorder flow rates revisited: clogging, ship speed and flow meter design. Journal of Plankton Research 28: 847–855.

Hurrell, J. W., 1995. Decadal trends in the North Atlantic Oscillation: regional temperatures and precipitation. Science 269: 676–679.

Iversen, S. A., M. D. Skogen & E. Svendsen, 2002. Availability of horse mackerel (*Trachurus trachurus*) in the northeastern North Sea, predicted by the transport of Atlantic water. Fisheries Oceanography 11: 245–250.

Kirby, R. R., G. Beaugrand, J. A. Lindley, A. J. Richardson, M. Edwards & P. C. Reid, 2007. Climate effects and benthic-pelagic coupling in the North Sea. Marine Ecology Progress Series 330: 31–38.

Kirby, R. R. & J. A. Lindley, 2005. Molecular analysis of continuous plankton recorder samples, and examination of echinoderm larvae in the North Sea. Journal of the Marine Biological Association of the United Kingdom 85: 451–459.

Kramp, P. L., 1959. The hydromedusae of the Atlantic Ocean and adjacent waters. Dana Report 46: 1–283.

Link, J. S. & M. D. Ford, 2006. Widespread and persistent increase of Ctenophora in the continental shelf ecosystem off NE USA. Marine Ecology Progress Series 320: 153–159.

Longhurst, A. R., 1998. Ecological Geography of the Sea. Academic Press, San Diego.

Lynam, C. P., M. J. Gibbons, B. A. Axelsen, C. A. J. Sparks, J. Coetzee, B. G. Heywood & A. S. Brierley, 2006. Jellyfish overtake fish in a heavily fished ecosystem. Current Biology 16: 492–493.

Lynam, C. P., S. J. Hay & A. S. Brierley, 2004. Interannual variability in abundance of North Sea jellyfish and links to North Atlantic oscillation. Limnology and Oceanography 49: 637–643.

Lynam, C. P., S. J. Hay & A. S. Brierley, 2005a. Jellyfish abundance and climatic variation: contrasting responses in oceanographically distinct regions of the North Sea, and possible implications for fisheries. Journal of the Marine Biological Association of the United Kingdom 85: 435–450.

Lynam, C. P., M. R. Heath, S. J. Hay & A. S. Brierley, 2005b. Evidence for impacts by jellyfish on North Sea herring recruitment. Marine Ecology Progress Series 298: 157–167.

Mackie, G. O., P. R. Pugh & J. E. Purcell, 1987. Siphonophore biology. Advances in Marine Biology 24: 97–262.

Mann, K. H. & J. R. N. Lazier, 1991. Dynamics of Marine Ecosystems. Blackwell Scientific Publications, Oxford.

Matsueda, N., 1969. Presentation of *Aurelia aurita* at thermal power station. Bulletin of the Marine Biology Station at Asamushi 13: 187–191.

Mianzan, H., 1999. Ctenophora. In Boltovskoy, D. (ed.), South Atlantic Zooplankton. Backhuys Publishers, Leiden: 561–573.

Mills, C. E., 2001. Jellyfish blooms: are populations increasing globally in response to changing ocean conditions. Hydrobiologia 451: 55–68.

Möller, H., 1984. Reduction of a larval herring population by jellyfish predator. Science 224: 621–622.

Montgomery, M. K. & M. M. Kremer, 1995. Transmission of symbiotic dinoflagellates through the sexual cycle of the host scyphozon *Linuche unguiculata*. Marine Biology 124: 147–155.

Myers, R. A. & B. Worm, 2003. Rapid worldwide depletion of predatory fish communities. Nature 423: 280–283.

Nicholas, K. R. & C. L. J. Frid, 1999. Occurrence of hydromedusae in the plankton off Northumberland (western central North Sea) and the role of planktonic predators. Journal of the Marine Biological Association of the United Kingdom 79: 979–992.

North Sea Task Force, 1993. North Sea Quality Status Report 1993. Oslon and Paris Commissions, London/Olsen and Olsen, Fredensborg.

Parsons, T. R. & C. M. Lalli, 2002. Jellyfish population explosions: revisiting a hypothesis of possible causes. La mer 40: 111–121.

Pauly, D., V. Christensen, S. Guenette, T. J. Pitcher, U. R. Sumaila, C. J. Walters, R. Watson & D. Zeller, 2002. Towards sustainability in world fisheries. Nature 418: 689–695.

Purcell, J. E., 2005. Climate effects on formation of jellyfish and ctenophore blooms: a review. Journal of the Marine Biological Association of the United Kingdom 85: 461–476.

Purcell, J. E., T. A. Shiganova, M. B. Decker & E. D. Houde, 2001. The ctenophore *Mnemiopsis* in native and exotic habitats: U.S. estuaries versus the Black Sea basin. Hydrobiologia 451: 145–176.

Purcell, J. E., S.-I. Uye & W.-T. Lo, 2007. Anthropogenic causes of jellyfish blooms and their direct consequences for humans: a review. Marine Ecology Progress Series 350: 153–174.

Purcell, J. E., J. R. White, D. A. Nemazie & D. A. Wright, 1999. Temperature, salinity and food effects on asexual reproduction and abundance of the scyphozoan *Chrysaora quinquecirrha*. Marine Ecology Progress Series 180: 187–196.

Raitsos, D. E., P. C. Reid, S. J. Lavender, M. Edwards & A. J. Richardson, 2005. Extending the SeaWiFS chlorophyll dataset back 50 years in the northeast Atlantic. Geophysical Research Letters 32: L06603.

Raymont, J. E. G., 1983. Plankton and productivity in the oceans, Zooplankton. Pergamon Press, Oxford.

Rayner, N. A., D. E. Parker, E. B. Horton, C. K. Folland, L. V. Alexander, D. P. Rowell, E. C. Kent & A. Kaplan, 2003. Global analyses of sea surface temperature, sea ice, and night marine air temperature since the late nineteenth century. Journal of Geophysical Research 108(D14): 4407.

Reid, P. C., N. P. Holliday & T. J. Smyth, 2001. Pulses in the eastern margin current and warmer water off the north west European shelf linked to North Sea ecosystem changes. Marine Ecology Progress Series 215: 283–287.

Richardson, A. J., A. W. Walne, A. W. G. John, T. D. Jonas, J. A. Lindley, D. W. Sims, D. Stevens & M. Witt, 2006. Using continuous plankton recorder data. Progress in Oceanography 68: 27–74.

Robinson, G. A., 1970. Continuous plankton records: variation in the seasonal cycle of phytoplankton in the North Atlantic. Bulletins of Marine Ecology 6: 333–345.

Roosen-Runge, E. C., 1970. Life cycle of the hydromedusa *Phialidium gregarium* (A Agassiz, 1862) in the laboratory. Biological Bulletin of the Marine Biology Laboratory, Woods Hole 166: 206–215.

Russell, F. S., 1933. The seasonal distribution of meroplankton as shown by catches in the 2-m Stramin ring-trawl in offshore waters off Plymouth. Journal of the Marine Biological Association of the United Kingdom 19: 73–82.

Russell, F. S., 1938. The Plymouth offshore medusa fauna. Journal of the Marine Biological Association of the United Kingdom 22: 411–439.

Russell, F. S., 1939. Hydrographical and biological conditions in the North Sea as indicated by planktonic organisms. Journal du Conseil 14: 171–192.

Russell, F. S., 1953. The Medusae of the British Isles. Cambridge University Press, Cambridge.

Schneider, G., 1987. Role of advection in the distribution and abundance of *Pleurobrachia pileus* in Kiel Bight. Marine Ecology Progress Series 41: 99–102.

Shiganova, T. A., Z. A. Mirzoyan, E. A. Studenikina, S. P. Volovik, I. Siokoi-Frangou, S. Zervoudaki, E. D. Christou, A. Y. Skirta & H. J. Dumont, 2001. Population development of the invader ctenophore *Mnemiopsis leidyi* in the Black Sea and other seas of the Mediterranean basin. Marine Biology 139: 431–445.

Sims, D. W., E. J. Southall, G. A. Tarling & J. D. Metcalfe, 2005. Habitat-specific normal and reverse diel vertical migration in the plankton-feeding basking shark. Journal of Animal Ecology 74: 755–761.

Sims, D. W., M. J. Witt, A. J. Richardson, E. J. Southall & J. D. Metcalfe, 2006. Encounter success of free-ranging marine predator movements across a dynamic prey landscape. Proceedings of the Royal Society B 273: 1195–1201.

Stibor, H. & N. Tokle, 2003. Feeding and asexual reproduction of the jellyfish *Sarsia gemmifera* in response to resource enrichment. Oecologia 135: 202–208.

Stoecker, D. K., A. E. Michaels & L. H. Davis, 1987. Grazing by the jellyfish, *Aurelia aurita*, on microzooplankton. Journal of Plankton Research 9: 901–915.

Travis, J., 1993. Invader threatens Black, Azov Seas. Science 262: 1366–1367.

Van der Veer, H. W. & C. F. M. Sadée, 1984. Seasonal occurrence of the ctenophore *Pleurobrachia pileus* in the western Dutch Wadden Sea. Marine Biology 79: 219–227.

Watson, H. G., 1930. The coelenterate plankton of the Northumbrian coast during the year 1925. Journal of the Marine Biological Association of the United Kingdom 17: 233–239.

Weijerman, M., H. Lindeboom & A. F. Zuur, 2005. Regime shifts in marine ecosystems of the North Sea and Wadden Sea. Marine Ecology Progress Series 298: 21–39.

Williams, R. & N. R. Collins, 1985. Chaetognaths and ctenophores in the holoplankton of the Bristol Channel. Marine Biology 85: 97–107.

Witt, M. J., A. C. Broderick, D. J. Johns, C. Martin, R. Penrose, M. S. Hoogmoed & B. J. Godley, 2007. Prey landscapes help identify potential foraging habitats for leatherback turtles in the northeast Atlantic. Marine Ecology Progress Series 337: 231–244.

Yip, S. Y., 1981. Investigations of the plankton of the west coast of Ireland – VII. A preliminary study of planktonic ctenophores along the west coast of Ireland, with special reference to *Pleurobrachia pileus* Müller, 1776, from Galway Bay. Proceedings of the Royal Irish Academy 81B: 89–109.

Zar, J. H., 1984. Biostatistical Analysis. Prentice-Hall, Englewood Cliffs.

Hydrobiologia (2009) 616:67–85
DOI 10.1007/s10750-008-9583-x

JELLYFISH BLOOMS

# Jellyfish in ecosystems, online databases, and ecosystem models

**Daniel Pauly · William Graham ·**
**Simone Libralato · Lyne Morissette ·**
**M. L. Deng Palomares**

Published online: 26 September 2008


**Abstract** There are indications that pelagic cnidarians and ctenophores ('jellyfish') have increased in abundance throughout the world, or that outbreaks are more frequent, although much uncertainty surrounds the issue, due to the scarcity of reliable baseline data. Numerous hypotheses have been proposed for the individual increases or outbreaks that are better documented, but direct experimental or manipulative studies at the ecosystem scale cannot be used for testing them. Thus, ecological modeling provides the best alternative to understand the role of jellyfish in large fisheries-based ecosystems; indeed, it is an approach consistent with new ecosystem-based fisheries management practices. Here, we provide an overview of online databases available to ecosystem modelers and discuss general aspects and shortcomings of the coverage of jellyfish in these databases. We then provide a summary of how jellyfish have been treated and parameterized by existing ecosystem models (specifically focusing on 'Ecopath with Ecosim' as a standard modeling toolset). Despite overall weaknesses in the parameterization of jellyfish in these models, interesting patterns emerge that suggest some systems, especially smaller and more structured ones, may be particularly vulnerable to long-term jellyfish biomass increase. Since jellyfish also feed on the eggs and larvae of commercially important food fish, outbreaks of jellyfish may ultimately imply a reduction in the fish biomass available to fisheries. On the other hand, jellyfish, which have been traditionally fished for human consumption in East and Southeast Asia, are now seen as a potential resource in other parts of the world, where pilot fisheries have emerged. It is also argued here that reduced predation on the benthic and pelagic stages of jellyfish, both a result of fishing, may be a strong contributing factor as well. For marine biologists specializing on jellyfish, this means that their research might become more applied. This

Guest editors: K. A. Pitt & J. E. Purcell
Jellyfish Blooms: Causes, Consequences, and Recent Advances

D. Pauly (✉) · M. L. Deng Palomares
Sea Around Us Project, Fisheries Centre,
University of British Columbia, 2202 Main Mall,
Vancouver, BC, Canada V6T 1Z4
e-mail: d.pauly@fisheries.ubc.ca

M. L. Deng Palomares
e-mail: m.palomares@fisheries.ubc.ca

W. Graham
Dauphin Island Sea Lab and University of South
Alabama, 101 Bienville Blvd., Dauphin Island,
AL 36528, USA
e-mail: mgraham@disl.org

S. Libralato
Istituto Nazionale di Oceanografia e di Geofisica
Sperimentale—OGS, Borgo Grotta Gigante 42/c,
Sgonico, Trieste 34010, Italy
e-mail: slibralato@ogs.trieste.it

L. Morissette
Institut des Sciences de la Mer de Rimouski, 310, Allée
des Ursulines C.P. 3300, Rimouski, QC, Canada G5L 3A1

implies that they would benefit from adopting some concepts and methods from fisheries biology and ecosystem modeling, and thus from using (and contributing to) online databases, such as SeaLifeBase and FishBase, developed to support such research. This would remedy the situation, documented here, wherein jellyfish are either infrequently included in food web models, typically constructed using the Ecopath with Ecosim software, or included as a single functional group with the characteristic of an 'average' jellyfish. Thus, jellyfish specialists could readily improve the jellyfish-related components of such models, and we show how they could do this. Also, it is suggested that when such improvement is performed, the resulting models can lead to non-intuitive inferences and hence interesting hypotheses on the roles of jellyfish in ecosystems. This is illustrated here through (a) an investigation of whether jellyfish are keystone species and (b) the identification of conditions under which (simulated) jellyfish outbreaks may occur.

**Keywords** Ecopath with Ecosim · FishBase · SeaLifeBase · Keystone species · Fishing down marine food webs · Jellyfish outbreaks

## Introduction

There are a number of examples of what appear to be recent (i.e., past few decades) increases, if not only longer-term climate-driven cycles, of jellyfish populations (reviewed by Mills, 2001; Graham, 2001; Brodeur et al., 2002; Purcell, 2005; Kawahara et al., 2006; Purcell et al., 2007). Since the majority of the world's oceans are under some level of fishing pressure (Pauly et al., 1998a, b; 2002), it is reasonable to say that jellyfish predators, at a minimum, contribute additionally through predation on eggs and larvae of fishes to a multitude of human-induced pressures (Halpern et al., 2008). It only makes sense, therefore, that fisheries, by either contributing directly to increased jellyfish populations or by being impacted by jellyfish (or both), should incorporate jellyfish populations into newly emerging ecosystem-based fisheries management schemes (Ruckelshaus et al., 2008).

Historically, the relationship between jellyfish scientists and fisheries scientists has been obscured. We may attribute this to differences between applied and basic science approaches, and thus where our respective work is ultimately published. It is likely, though, a difference in scientific language, tools, and data requirements contributes as much as simply not reading each other's publications. Therefore, it is of greatest priority that fisheries scientists understand the truly special nature of arguably the most important predators of the sea, the jellyfish, and that jellyfish scientists understand what tools are being used in fisheries research and, consequently, how best to supply good data to, and/or collaborate with, fisheries scientists.

In this contribution, we highlight established toolsets currently in use by fisheries sciences to describe the functioning of the ecosystems upon which fisheries are based, notably the modeling architecture of Ecopath with Ecosim (EwE; www.ecopath.org). More precisely, the three specific aims of this contribution are:

The first is to review the coverage of jellyfishes in two online databases, FishBase and SeaLifeBase. We also seek to review and examine the role of jellyfish in published trophic models of ecosystems, constructed using the 'EwE' software and, in part at least, parameterized with the aid of these databases. We pay special attention to important parameterization features of EwE so that future modeling efforts may correctly incorporate jellyfish.

Our second aim is to provide examples of the kind of hypotheses concerning the ecological role of jellyfish that can be tested using such trophic models. This concerns hypotheses based on the structure (relative biomasses, food web linkages, etc.) of such (Ecopath) models, as well as hypotheses resulting from an examination of their population dynamics (as run by Ecosim).

Our third aim, finally, is to reframe these hypotheses, all eminently testable, around the ubiquitously observed pattern of 'fishing down marine food webs' (Pauly et al., 1998a, b), to explain why jellyfish outbreaks, worldwide, should occur more frequently than hitherto—notwithstanding the fact that the compelling evidence that they do is dubious, owing to the nature of available datasets (Purcell, 2005). We then conclude by outlining what we believe would be fruitful avenues for further research.

## Jellyfish in FishBase

FishBase (www.fishbase.org) is an online database of fish, containing information about all (>30,000) fish species described so far. It provides information on nomenclature and taxonomy, as well as quantitative and qualitative information on biological and ecological aspects of fish (e.g., distribution, food and feeding, growth, and reproduction, all of which can be combined to yield new insights), extracted from the scientific literature and fully referenced (Froese & Pauly, 2000). FishBase is a hugely successful online resource to scientists and laypersons alike, with about 25 millions 'hits' and over 2 million visitors per month.

Initially conceived for the purposes of (single-species) fisheries management, FishBase now increasingly includes data which link fish to their ecosystem, paralleling the move of fisheries science toward ecosystem-based consideration (Pikitch et al., 2006). FishBase thus includes much information on the prey of fishes, including jellyfish, both as reports on individual food items and as components of detailed diet composition studies. Also, FishBase incorporates information on fish predators, including jellyfish. This information, extracted from Arai (1987) and Purcell (1985) and other sources, is summarized in Fig. 1 (top). A total of 124 species of fish are reported as feeding occasionally or predominately on jellyfish. Of these, 11 species appear specialized, and about half belong to the perciform fishes. The size distribution of the fish species feeding on jellyfish is presented in Fig. 2A. As might be seen, the majority of the fish predators fall in the range of 50 to 200 cm maximum length; also, the number of species feeding on jellyfish decreases rapidly as the size increases.

## Jellyfish in SeaLifeBase

SeaLifeBase (www.sealifebase.org) represents an attempt to reproduce the success of FishBase for marine organisms other than fish, thus including jellyfish. Its coverage of jellyfish relies on a taxonomy provided by the Catalogue of Life (www.catalogoflife.org), based, in the case of jellyfish, on the work of Van der Land (2006).

SeaLifeBase has the same basic structure as FishBase, and thus can similarly be used to summarize jellyfish predator/prey information, but for animals other than fish. This is presented here in Fig. 1 (middle) and Fig. 2B. As might be seen, 34 non-fish marine organisms, mainly reptiles, birds, and crustaceans, are reported as feeding, at least occasionally, on jellyfish. The size distribution of these predators evidences, similarly to that observed for fish predators, a decrease in the number of non-fish predator species as the size increases. We do not believe that this pattern will change as the coverage of SeaLifeBase is extended.

## Jellyfish in (Ecopath) ecosystem models

There are various ways to represent ecosystems through formal, quantitative models. The model structure will depend on whether its prime intent is to summarize and validate one's knowledge of an ecosystem, to draw inference on ecosystem properties, to test hypotheses on various aspects of ecosystem structure and functioning, or to run various management scenarios. One of the most commonly used ecosystem modeling approaches is that incorporated in the Ecopath with Ecosim (EwE) software (see www.ecopath.org). Central to this method is the representation of mass-balance trophic flows between functional groups for basic parameterization (i.e., 'Ecopath models'; Christensen & Pauly, 1992). These interactions can then be run in 'Ecosim' and 'Ecospace' for time-dynamic and spatial simulations, respectively (Walters et al., 1997, 1998; Pauly et al., 2000), but this contribution focuses exclusively on Ecopath and Ecosim.

Owing in part to the ease with which such models can be constructed based on previously published data, over 400 EwE models of (mostly marine) ecosystems have been built and documented in media ranging from Masters and PhD theses to reports and articles in the peer-reviewed literature (see www.ecopath.org). Over 100 of these models (all representing marine ecosystems, and most developed by fisheries scientists) were examined in the context of a wider study examining the relationship between the 'quality' of Ecopath models (as defined by the level of detail included and the types of data for model construction) and the resilience exhibited in their time-dynamic behavior (Morissette, 2007). Of these, only 23 explicitly included gelatinous zooplankton (medusae,

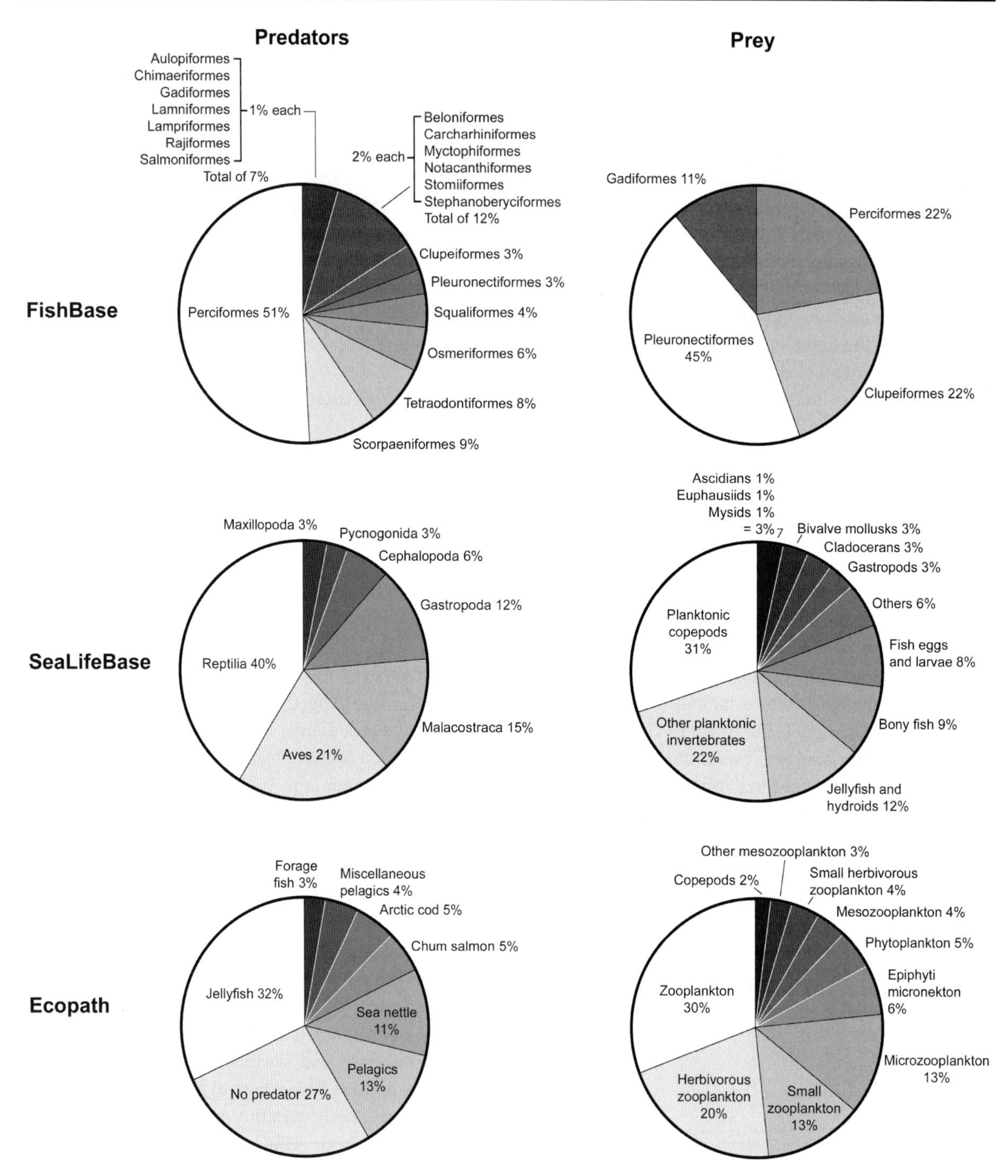


**Fig. 1** Summary of data on the predators and prey of jellyfish available in three online sources. Top left: Orders to which the 124 fish species belong that are reported to prey on gelatinous organisms (from 181 food items records in FishBase); Top right: Orders to which the 11 fish species belong that are preyed on by gelatinous organisms (data from FishBase); Middle left: Non-fish marine organisms (34 species) preying on gelatinous organisms (data from SeaLifeBase); Middle right: Non-fish marine organisms (89 species) consumed by gelatinous organisms (data from SeaLifeBase); Bottom left: Main predators of gelatinous organisms assembled from 21 (Ecopath) ecosystem models; Bottom right: Main prey of gelatinous organisms assembled from the same 21 ecosystem models

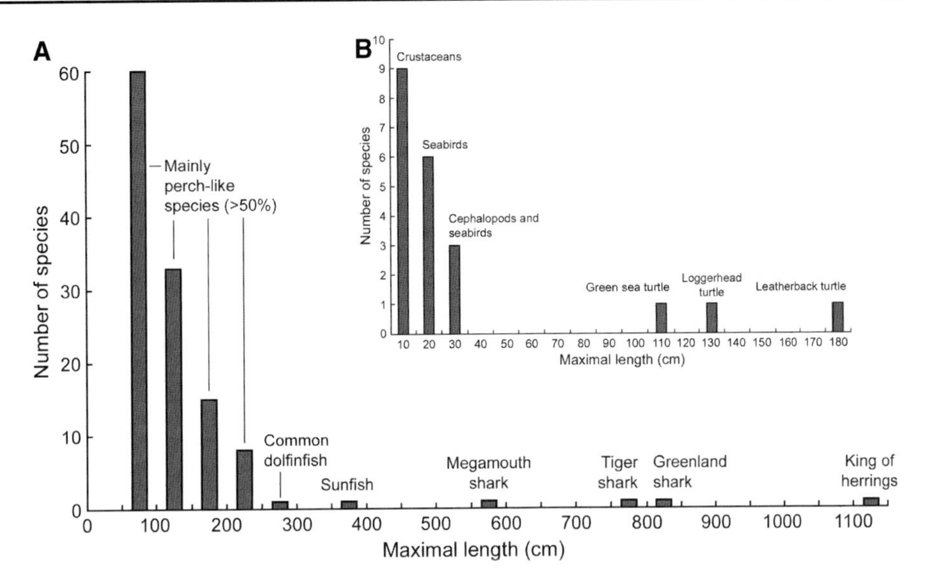


**Fig. 2** Size distribution of (groups of) predators of jellyfish. Size distribution of fish feeding on jellyfish, with the smallest fish feeding mainly on hydroids, and the largest, the King of herrings (*Regalecus glesne* Ascanius, 1772), being documented by only 4 specimens (data from FishBase) (**A**); Size distribution of non-fish organisms feeding on jellyfish, with the smallest consisting mainly of planktonic crustaceans feeding on ephyrae, and the largest consisting of marine turtles (data from SeaLifeBase) (**B**)

ctenophores, or pelagic tunicates) as a distinct functional group in their food webs.

Consistency among models in how detailed they are with respect to 'jellyfish' is an important consideration when attempting to evaluate and compare them. We are mindful that the overwhelming majority of models do not include a jellyfish compartment, and those that do tend to collapse all things considered gelatinous into a single functional 'jellyfish' group. With few exceptions, such as the Chesapeake Bay (Baird & Ulanowicz, 1989; Walters et al., 2005) and the eastern Bering Sea (Trites et al., 1999; NRC, 2003) models, a single jellyfish group poorly represents the true diversity of taxa and trophic interactions involving jellyfish. Hopefully, this can be overcome in the future, in part, by online databases such as SeaLifeBase.

We suspect the low number of models incorporating jellyfish, far from reflecting the true importance of jellyfish in marine ecosystems, in fact highlights the limited knowledge many fisheries scientists or marine biologists have of jellyfish. However, we also believe that the crude representation of 'jellyfish' as a single, undifferentiated functional group in most models that did include gelatinous zooplankton (see Table 1) may also indicate lack of access to or appreciation for the key literature on these taxonomic groups. With this caveat in mind, the main prey and predators of jellyfish included in these 23 models are summarized in Fig. 1 (bottom).

The outcome of a modeling effort is only as good as its initializing parameters. For EwE, there are a suite of common parameters utilized, sometimes available from the above databases, most of the time not. Following is a brief summary of key EwE parameters and how they, typically, have been handled for jellyfish in most existing models. Additionally, we provide brief insight on how these parameters can be individually measured or estimated so that future modeling efforts can handle jellyfish more realistically.

*Diet and Trophic Level.* Within the EwE construct, the diet composition of functional groups (typically species, or groups thereof) defines the position in the food web, and ultimately Trophic Level (TL). To assume that all jellyfish have the same diet and exist at the same trophic level would be very wrong. Indeed, their different morphologies, swimming modes, prey capture, and nematocyst composition yield widely different diets (Purcell & Mills, 1988; Costello & Colin, 1995; Peach & Pitt, 2005). Even within a species, diet may change simply due to an increase in size (Graham & Kroutil, 2001). Finally, it is particularly important to consider multiple jellyfish

**Table 1** Summary of major features of the (Ecopath) ecosystem models used in the analyses of trophic interaction presented here (i.e., Fig. 4); models with an asterisk (*) were also used for Ecosim runs

| Model name | Ecosystem type and location | Years | Living + non-living groups | Gelatinous plankton group (GLP) | GLP ($t \cdot km^{-1}$) | P/B ($year^{-1}$) | Q/B ($year^{-1}$) | EE | Human impacts | References |
|---|---|---|---|---|---|---|---|---|---|---|
| *Chesapeake Bay | Temperate, enclosed coastal area | 1983–1986 | 33 + 1 | Sea nettle; TL = 3.31 | 0.00020 | 5.00 | 2.00 | **0.000** | Highly impacted system | Baird & Ulanowicz (1989) and Walters et al. (2005) |
| *Central North Pacific | Temperate, open ocean | 1990–1998 | 25 living groups; 9 fishing fleets | 'Jellyfish'; TL = 3.0 | 0.00612 | 0.79 | 20.00 | **0.001** | Top predators (tuna) fisheries | After Cox et al. (2002) |
| *Eastern Bering Sea | Temperate shelf and slope down to 500 m | 1955–1960 | 23 + 2; 5 fishing fleets | Mainly *Chrysaeora melanaster*; TL = 3.16 | 0.00016 | 0.88 | 2.00 | **0.018** | Exploited, marine mammals declined | Trites et al. (1999) and NRC (2003) |
| *Georgia Strait | Temperate narrow basin, average depth 156 m | 1950s | 26 + 1; 3 fishing fleets | 'Jellyfish'; TL = 3.02 | 0.01861 | 3.00 | 12.00 | **0.211** | Heavily exploited in the last 90 years | Pauly et al. (1998a, b)Martell et al. (2002) and Walters et al. (2005) |
| Hong Kong waters | Tropical inshore waters | 1990s | 36 + 1 detritus pool; 7 fleets | 'Jellyfish'; TL = 3.0 | 0.00578 | 5.01 | 25.05 | **0.257** | Stocks re-building | Cheung et al. (2002) and Buchary et al. (2002) |
| Lancaster Sound | Shallow area in the Canadian Arctic | Late 1980s | 31 + 1 detritus pool; 1 fishing fleet | *Mertensia ovum*; TL = 2.98 | 0.01190 | 8.82 | 29.41 | **0.806** | Moderate fishing | Mohammed (2001) |
| North Coast of Central Java | Tropical shelf, shallow coastal area | Mid 1970s | 27 + 1 detritus pool; 6 fishing fleets | Mainly Scyphozoans; TL = 3.0 | 0.00045 | 5.01 | 25.05 | **0.528** | Low to moderate fishing; near shore | Buchary (1999) |
| Norwegian-Barents Sea 1950 | Temperate ecosystem; from intertidal area to over 3,000 m depth | 1950s | 30 living groups; 9 fishing fleets | 'Jellyfish'; TL = 3.17 | 0.01762 | 3.00 | 10.00 | **0.343** | Moderate fishing activity; | Dommasnes et al. (2001) |
| Norwegian-Barents Sea 1997 | Temperate ecosystem; from intertidal area to over 3,000 m depth | 1997–2000 | 30 living groups; 9 fishing fleets | 'Jellyfish'; TL = 3.19 | 0.02004 | 3.00 | 10.00 | **0.339** | High fishing activity | Dommasnes et al. (2001) |
| Southeastern United States | Tropical, continental shelf area, from intertidal to 500 m depth | 1995–1998 | 41 + 1 detritus group; 9 fleets | 'Jellyfish'; TL = 2.8 | 0.00028 | 40.00 | 80.00 | 0.950 | Intense trawling activity; highly fished | Okey & Pugliese (2001) |

**Table 1** continued

| Model name | Ecosystem type and location | Years | Living + non-living groups | Gelatinous plankton group (GLP) | GLP ($t \cdot km^{-1}$) | P/B ($year^{-1}$) | Q/B ($year^{-1}$) | EE | Human impacts | References |
|---|---|---|---|---|---|---|---|---|---|---|
| *Gulf of Thailand | Tropical shallow coastal area; 10–50 m depth range | 1970s | 39 + 1 detritus group; 6 fleets | 'Jellyfish'; TL = 3.0 | 0.02198 | 5.00 | **20.00** | **0.000** | Moderate fishing | FAO/FISHCODE (2001) and Walters et al. (2005) |
| US Mid Atlantic Bight | Temperate continental shelf, from upper intertidal to 200 m isobath | 1995–1998 | 55 + 1; 1 fleet | 'Jellyfish'; TL = 3.07 | 0.00018 | 18.25 | 80.00 | 0.900 | Highly exploited area | Okey (2001) |
| Prince William Sound | Cold temperate coastal area in Alaska | 1994–1996 | 46 + 2; 3 fishing fleets | 'Jellyfish'; TL = 2.96 | 0.01409 | 8.82 | 29.41 | **0.006** | Fishing and oil spill impacts | Okey & Pauly (1998) |
| West Florida Shelf | Tropical shelf area from intertidal zone to 200 m depth | | 55 + 4 non-living groups; 11 fisheries | Carnivorous jellyfish; TL = 3.39 | 0.00037 | 40.00 | 80.00 | **0.928** | Problems of nutrient enrichment | Okey et al. (2004) |
| North Central Adriatic Sea | Shelf, 3 miles west (or to 10 m) to 12 miles from the east coast | 1990s | 37 + 3; 5 fisheries | Mainly *Pelagia noctiluca*; TL = 2.66 | 0.01208 | 14.60 | 50.48 | **0.170** | Risk of ecosystem overfishing | Coll et al. (2007) |
| South Catalan Sea | Upper slope from 3 miles of 50 m depth to 400 m depth. | 1994–2000 | 37 + 3; 4 fisheries | 'Jellyfish'; TL = 2.3 | 0.00659 | 26.51 | 56.80 | **0.017** | Large impacts of fisheries | Coll et al. (2006) |

Values in bold were estimated by Ecopath; GLP, gelatinous plankton group; P/B, production/biomass ratio; Q/B, consumption/biomass ratio; EE, ecotrophic efficiency (see text)

compartments when strong trophic dynamics occur among gelatinous zooplankton in the so-called jelly web (Robison, 2004).

Figure 1 (right pie charts) shows that 'jellyfish' as a functional group tend to be carnivorous zooplankton predators, operating at trophic levels between 2.3 and 3.3 (median = 3.0) (Table 1). Most likely, this is an accurate reflection of the TL range based on available diet information in the published literature (reviewed by Arai, 1997). However, diet estimated from jellyfish gut contents biases toward shell or carapace bearing prey and those prey most common in the plankton (e.g., copepods in Fig. 1, lower right pie chart). True energetic ration of jellyfish likely includes substantial fractions of unrecognizable items such as detritus and soft microzooplankton (e.g., naked ciliates) (Colin et al., 2005) or very large and rare prey (Pitt et al., 2008).

*Biomass* (*B*). Jellyfish are usually not included in fisheries stock assessments, and so estimates of biomass are derived either directly from the literature or from estimates based on occasional by-catch information. Either way, estimates of jellyfish biomass are confounded by their tendency to be overdispersed in nature by seasonal blooms or physical aggregation (Graham et al., 2001). New technologies such as acoustics are being successfully applied to jellyfish biomass assessment (Bamstedt et al., 2003; Colombo et al., 2003; Brierley et al., 2005; Lyman et al., 2006), and with routine time-series surveys, these should provide much-needed information to refine EwE simulations.

Even in existing EwE models, biomass estimates and related parameters of jellyfish, such as annual food production relative to biomass (Q/B) and production per unit of biomass (P/B), vary widely (Table 1). This is due, to a large extent, to the biomass of jellyfish not being reported in a fashion consistent with their high water content. It is conventional to report biomass of jellyfish as dry or elemental (e.g., C or N) mass, yet most EwE models are constructed using wet weight (i.e., tonnes $\cdot$ km$^{-2}$). Thus, even our own evaluations must be taken with caution until jellyfish biomass is appropriately, uniformly, and consistently parameterized in EwE models.

*Production* (*P*). Estimates of population production by jellyfish are rare and usually limited to rapid individual growth periods during the early phase of a cohort (e.g., Van der Veer & Oorthuysen, 1985; Olesen et al., 1994; Uye & Shimauchi, 2005). For most coastal medusae, the adult stock is derived annually from the benthic polyp stage, and for ctenophores, seasonal bloom cycles may rely on successful overwintering of a small, marginal population (Costello et al., 2006a, b). In such cases, parameterization of production for EwE may best be accomplished from population energetics (i.e., production equals metabolism plus other expenses such as reproductive output and mucus production; Arai, 1997). On the other hand, production, in EwE models (Table 1), is usually estimated from biomass and turnover rates (P/B), which is equivalent, under equilibrium, to total mortality (Allen, 1971). Palomares & Pauly (2008) suggest that this approach, which requires estimation of individual growth parameters, can be applied to jellyfish, in spite of major differences in life-history characteristics between jellyfish (medusae and ctenophores) and fish.

*Consumption* (*Q*). Prey consumption rates by jellyfish can be estimated directly from gut contents (Purcell, 2003), but this requires knowledge of digestion times that vary widely by prey type and temperature (reviewed by Arai, 1997). Consumption may also be determined in containers or mesocosms by following losses of the ambient prey, though container effects are usually high owing to the very high clearance rates of larger jellyfish (de Lafontaine & Leggett, 1988; Cowan & Houde, 1992, reviewed by Arai, 1997). Ultimately, EwE utilizes biomass-specific consumption (Q/B) and this parameter varies widely among models (from 2 to 80 year$^{-1}$in Table 1). We suspect such differences are again due to inconsistencies in reporting jellyfish biomass as dry or wet mass.

*Ecotrophic efficiency* (*EE*). This parameter of EwE expresses the fraction of the production of jellyfish that ultimately passes on to their predators (or a fishery). Obviously, the value of EE one chooses will depend on one's knowledge of the organisms that prey on jellyfish. As seen in Fig. 1 (pie charts on the left), predation on jellyfish by non-jellyfish consumers occurs (mostly by fish, but also sea turtles, seabird, and a few invertebrates). Indeed, some organisms are obligate jellyfish predators (e.g., ocean sunfish *Mola mola* (Linnaeus, 1758) and leatherback turtles *Dermochelys coriacea* Vandelli, 1761). However, very little is known about predation rates on jellyfish. Thus EE is often set at or near zero (Table 1), assuming jellyfish are largely trophic 'dead ends.' This may be a good approximation in many ecosystems, and seasonal declines in jellyfish are often noted to be coincident with senescence (e.g., Yamamoto et al., 2008; personal observation (WG)). In a few ecosystems (e.g., the Central North Pacific), the top predators rely on throughput of energy from jellyfish to higher trophic level predators in which case EE would have to be set near to 1 (Table 1). Finally, we should mention that the EE is calculated by Ecopath for each functional group for which B, P/B, Q/B, and the diet composition are known.

## Mutual impacts through food webs

In order to further investigate the role of jellyfish in ecosystem models, we selected 5 high quality EwE models, i.e., with highly disaggregated functional groups (Pinnegar et al., 2005), parameterized mostly on the basis of accurate, locally sampled data (Morissette, 2007). These models, each representative of different regions also included EwE simulations that had been fitted to time-series data of the abundance of several of their functional groups (although none was fitted with jellyfish time-series data), which precludes model parameterizations from generating pathological behaviors, e.g., system self-simplification (Christensen & Walters, 2004). All five of these models (highlighted in Table 1) were rechecked by the persons who created them, and they all fitted the available time-series reasonably well.

The mutual impact matrix within Ecopath was used to compare the 'with/without' impact of predation by jellyfish on the ecosystem as a whole (Fig. 3). This method allows the estimation of the relative direct or indirect impact of a (small) change in the biomass of one group on all other functional groups in the ecosystem, given the assumption that their diet composition remains constant (Ulanowicz & Puccia, 1990; Christensen & Pauly, 1992). As can be seen from Fig. 3 (upper panel), jellyfish are mainly impacted by plankton groups (phytoplankton and zooplankton) which are their direct and indirect prey. Figure 3 represents an example for the Gulf of Thailand, but this was true for all studied ecosystems.

The relatively few reported predators (both diversity and number) of jellyfish in ecosystems do not generate an important overall impact on the group. This strongly suggests that jellyfish species are subject to bottom–up control from their forage base rather than top–down controls from predators. One exception supporting strong top–down control

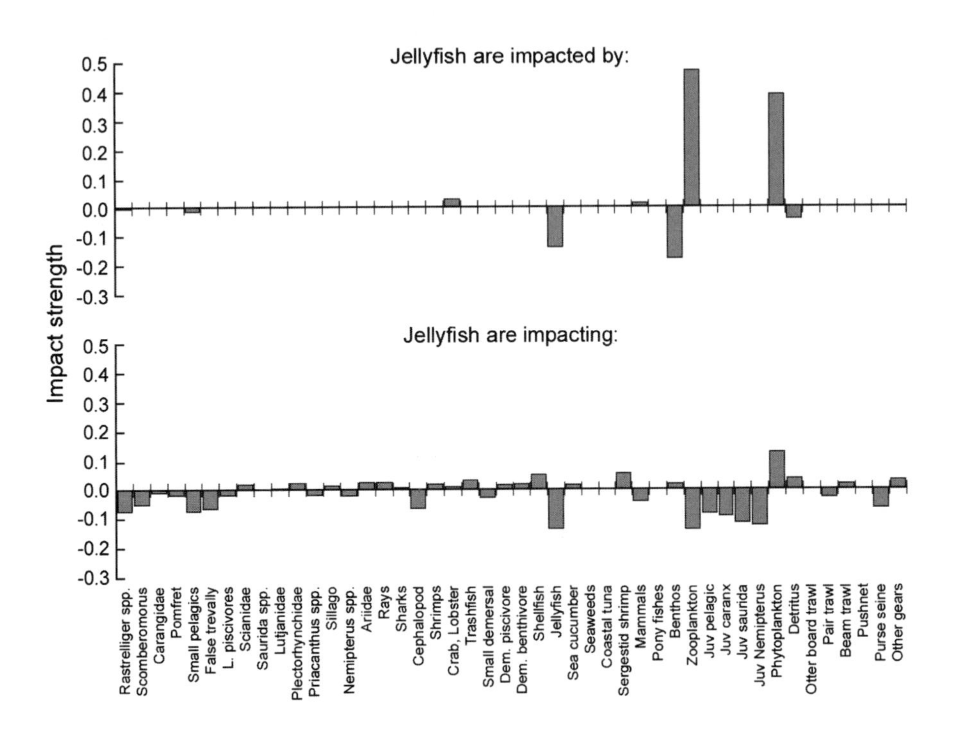


**Fig. 3** Graphic representation of a partial output of the mutual impact matrix of Ecopath, here for the Gulf of Thailand ecosystem model documented in Table 1, and used to evaluate the impacts on jellyfish of a small increase of biomass of the other groups in the system (above), and the impact on the other groups in the system of a small increase of jellyfish in the same system (below)

involves predation by one functional group of jellyfish on another, which of course generates an apparent negative impact of jellyfish on itself. On the other hand, most species of the ecosystem are indirectly impacted by jellyfish (Fig. 3, lower panel). A positive impact of jellyfish predation is seen on phytoplankton (because jellyfish consume herbivorous zooplankton), and on different fish species (for example, Ariidae, rays, shellfish, sergestid shrimp in the Gulf of Thailand) that share the characteristics of being benthivorous species. A negative impact of jellyfish is also seen on different zooplankton groups (their main prey), and indirectly on fish such as Indian mackerels (*Rastrelliger* spp.), Spanish mackerel (*Scomberomorus* spp.), false trevally *Lactarius lactarius* (Bloch & Schneider, 1801), or small pelagics species that are mainly zooplanktivorous.

## Jellyfish keystoneness

The analysis of the role of jellyfish within ecosystems can be broadened by asking whether they can be labeled as keystone species/group. Keystone species, first identified by Paine (1969), are defined by Power et al. (1996) as species that are important to the structure and productivity of the ecosystem in which they occur, but without being abundant. Using the most famous 'keystone' example, in kelp forest ecosystems along the Pacific coast of North America, despite the kelp's structures and the fact that it fuels the entire system, it is the relatively rare sea otter who plays that role, by keeping sea urchins in check, thus preventing the kelp from being grazed and being turned into an urchin 'barren' (Estes & Palmisano, 1974; Power et al., 1996).

One problem with this concept is that assessing whether a given species performs the role of a keystone through field experiments is difficult at best (Paine, 1992), and impossible in the majority of cases. In response to this, Libralato et al. (2006) developed a modeling approach to investigate this concept based on the mutual impact matrix of EwE (see above). Therein the 'keystoneness' of a species in its ecosystem (or more precisely of a functional group representing that species in an ecosystem model) is computed from the ratio of the sum of its direct and indirect impacts on all other groups in the system over its biomass (Libralato et al., 2006). In this sense, the concept of jellyfish keystoneness may be an appropriate approach to understand why certain ecosystems appear to be more vulnerable to jellyfish population increases.

Sixteen ecosystem models were used for this analysis, constituting a subset of the original 42 models included in Libralato et al. (2006) that included jellyfish (see Table 1 for a brief description). The habitat types they cover range from semi-enclosed waters in the case of Chesapeake Bay (Baird & Ulanowicz, 1989) and Prince William Sound (Okey & Pauly, 1998) to large areas of the open ocean such as the Central North Pacific (Cox et al., 2002). The ecosystems represented are also subject to different degrees of human impacts, ranging from heavily fished, e.g., the Hong Kong area (Cheung et al., 2002), to exploitation targeting mainly top predators, e.g., the Central North Pacific (Cox et al., 2002). All are well described, and include a minimum of 24 functional groups. Jellyfish represent a highly variable fraction of the total biomass of these ecosystems, ranging from approximately 1/50 in the Norwegian-Barents Sea (Trites et al., 1999) to less than 1/5,000 in the Eastern Bering Sea (Dommasnes et al., 2001).

Overall, the results show that jellyfishes' keystoneness ranges widely (Fig. 4), being highest in three ecosystem models, i.e., Lancaster Sound, Chesapeake Bay, and the Central North Pacific. Thus, according to the interpretation proposed in Libralato et al. (2006), jellyfishes may be considered

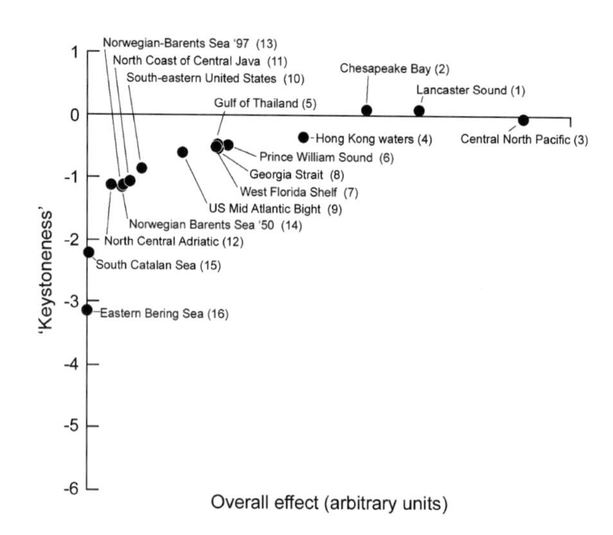


**Fig. 4** Keystoneness of jellyfish in 16 ecosystem models documented in Table 1 against their relative overall effect (see text for interpretation)

keystone groups in these ecosystems. The Central North Pacific may be exceptional since the high overall effect of jellyfish is due to its very large positive impact on sea turtles (leatherback and loggerhead). [In this model, sea turtles are the only predator of jellyfish, and jellyfish constitute 100% of leatherbacks' diet (loggerheads also feeds on epipelagic fish; Cox et al., 2002).] This exclusive predator–prey relationship makes the jellyfish-sea turtle coupling extremely strong, thus explaining the high overall effect. Given the very low turnover rates of sea turtles, this degree of coupling is dubious, yet consistent with the very long-term (equilibrium) solution that is at the base of the methodology proposed by Libralato et al. (2006).

Perhaps better examples of realistically high keystoneness can be seen in the Lancaster Sound and Chesapeake Bay models. In Lancaster Sound, the large effects are attributable to the very high negative impact that one particular species, the ctenophore *Mertensia ovum* (Fabricius, 1780), has on its main prey, i.e., omnivorous and herbivorous zooplankton, ice amphipods, and chaetognaths (Mohammed, 2001). In Chesapeake Bay, overall keystoneness is mainly due to direct positive effects small changes in 'jellyfish' (sea nettle) biomass have on its main predators (fish larvae, blue herring, bay anchovy, menhaden, and shad) and to indirect effects on competitors (zooplankton and ctenophores). In this ecosystem, cascading effects, wherein a predator suppresses a prey, whose own prey then increases, etc., appear to be particularly strong. The positive indirect effects of the sea nettle's predators (bluefish, weak fish, summer flounder, and striped bass) contribute substantially to the sea nettle's overall effect on the food web.

In contrast, jellyfish have low keystoneness in the Eastern Bering and South Catalan Seas, and have negligible values of keystoneness in the remaining ecosystems (Fig. 4). Between the Eastern Bering Sea and the South Catalan Sea models, the relative abundances of jellyfish differ by two orders of magnitude. However, it is still their low overall effect on other groups, not their abundance, which explains the very low keystoneness of jellyfish in these models. In both examples, jellyfish are minor consumers of zooplankton, while cannibalism is the only (Eastern Bering Sea) or the main (South Catalan Sea) form of predation on jellyfish. Such feeding interactions do not count when keystoneness is calculated (see Libralato et al., 2006).

There are also cases where jellyfish have intermediate keystoneness values. Thus, for example, in Hong Kong, jellyfish prey on the zooplankton, but are themselves preyed upon by many pelagic and demersal fishes (Cheung et al., 2002). Although this creates many direct and indirect impacts, they are relatively small, thus yielding a small overall effect.

In summary, we can conclude that very different structural features might lead to jellyfish assuming a keystone role. One of the obvious patterns emerging from Fig. 4 is the physical size of the ecosystem with smaller semi-enclosed systems (Chesapeake Bay and Lancaster Sound) being most sensitive to 'keystone' jellyfish and larger, more open ones being less sensitive (with the exceptional relationship between turtles and jellyfish in the Central North Pacific excluded). Perhaps this is validation of the expected complexity of multichannel food webs in (larger) open systems being less structured and thus less sensitive to trophic cascades (Polis & Strong, 1996). In addition, we suggest three typical ecological configurations where jellyfish keystoneness increases:

- jellyfish are a high-impact predator with a relatively low biomass (Lancaster Sound);
- are a prey of significant importance to some predators in an ecosystem characterized by large indirect effects (Chesapeake Bay); and
- are a part of a mutually exclusive prey–predator pair (Central North Pacific).

## Time-dynamic simulations

Time-dynamic simulations (using the Ecosim component of EwE) were run to complete our preliminary coverage of the behavior of jellyfish in trophic ecosystem models. Specifically, we attempted to identify fishing patterns (on finfishes and invertebrates other than jellyfish) that would generate marked increases in jellyfish biomass within the ecosystem. The simulations were performed on the five representative models identified under the analysis of '*Mutual impacts through food webs*' section, and details of how the models were parameterized are provided in Table 1. Only results from four of these models (Gulf of Thailand, Central North Pacific,

Chesapeake Bay, and Eastern Bering Sea) are presented as they best illustrate the range of the 'keystoneness' aspect discussed earlier as well as the ecological differences (notably the presence or absence of predators on jellyfish).

All of the models were forced, in a similar manner, by a linear increase in zooplankton production over time (i.e., bottom–up control). In the case of the Gulf of Thailand and Central North Pacific models, the models were also forced by an artificially enhanced mortality on the predators of jellyfish (i.e., top–down controls). The Chesapeake Bay and Eastern Bering Sea models have no explicit jellyfish predators identified for those models.

In general, the simulations indicate that strong bottom–up control by increased prey resources may be more important than the top–down controls placed on jellyfish by their predators (Figs. 5–7). In all cases, jellyfish biomass increased over time with increased zooplankton production. Predator mortality simulated by increased fishing mortality on small pelagic fishes in the Gulf of Thailand and by increased sea turtle mortality in the Central North Pacific did not yield marked increases in jellyfish biomass. However, this is all very preliminary because, as mentioned earlier, predation on jellyfish by vertebrates and other invertebrates is poorly understood (Arai, 2005; Purcell & Arai, 2001) and is likely far underestimated in these modeled systems.

Without explicit predators of jellyfish in the Chesapeake Bay and Eastern Bering Sea models, the linearly increasing function of zooplankton representing bottom–up forcing (possibly via eutrophication) causes interesting patterns that seem to reflect the 'keystoneness' of jellyfish. As discussed earlier, the Chesapeake Bay model is highly structured with respect to jellyfish and shows a strong coupling between zooplankton prey and jellyfish as seen in the nearly linear relationship in Fig. 7 (top panel). In contrast, the Eastern Bering Sea model, forced in a similar manner, shows large sinusoidal variations (Fig. 7, bottom panel). The large variations are likely due to the reduced structure in the larger Eastern Bering Sea system (i.e., more explicit redundancies in trophic pathways involving jellyfish). This is an interesting, albeit preliminary, pattern and could be useful to further identify ecosystem attributes (e.g., physical size, food web complexity, etc.) that make them more vulnerable to jellyfish increases.

Likewise, the presence of predators on jellyfish may be important in regulating jellyfish biomass increases in some cases. While the Central North Pacific may be unrealistic in its parameterization of trophic flows involving jellyfish, the presence of jellyfish-eating turtles certainly is a highly coupled

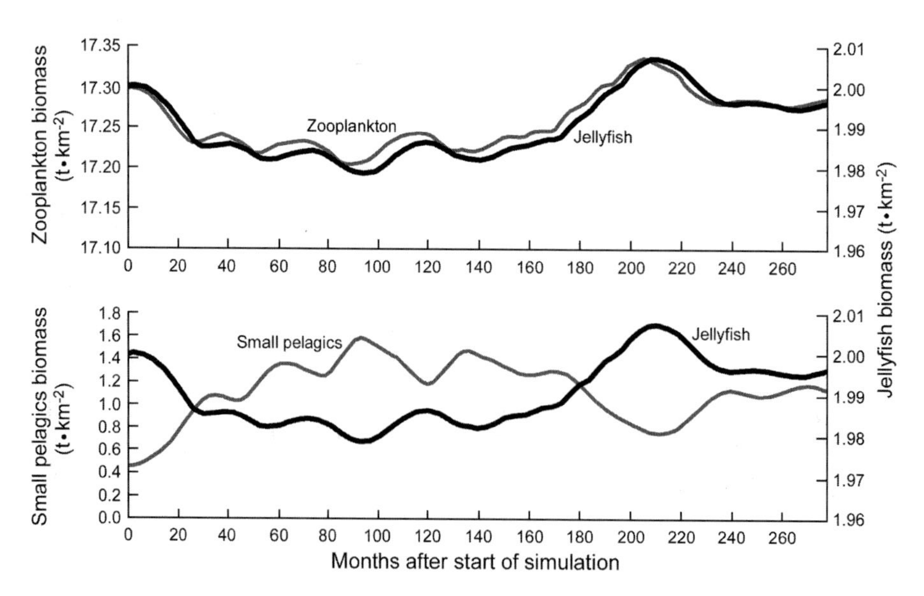


**Fig. 5** Example of simulated biomass trends for jellyfish and their main prey (above) and jellyfish and their main predator (below) in the Gulf of Thailand

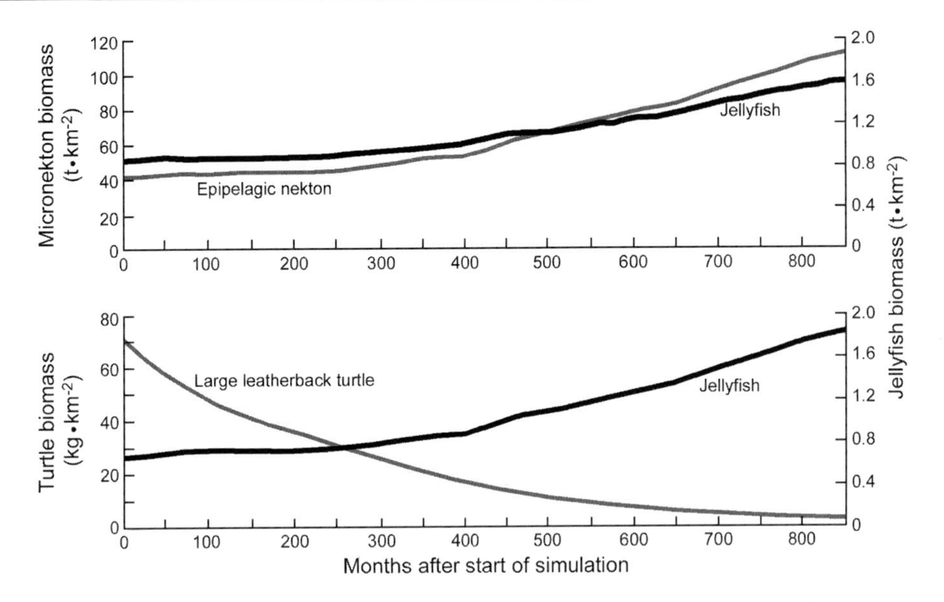


**Fig. 6** Simulation of a jellyfish bloom with ecosystem models: an example for the Central North Pacific (see text)

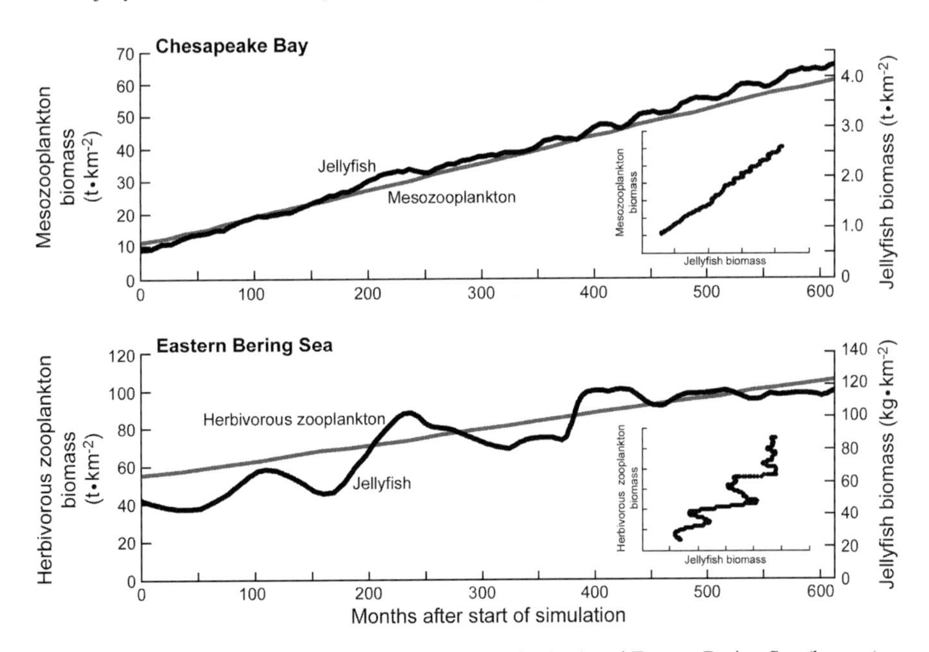


**Fig. 7** Simulating the effect of increasing zooplankton in the Chesapeake (top) and Eastern Bering Sea (bottom) ecosystem models. Note the increase of jellyfish, shown in the insets in relation to the biomass of their prey (see text)

relationship which should be explored in more detail. The Gulf of Thailand model may be the most informative for exploring coupled top–down and bottom–up effects, and could explain how multiple pressures such as combined effects of fishing and eutrophication interact to yield large increases in jellyfish biomass. The obvious next step for such models is to further investigate whether a positive feedback loop exists that further depresses jellyfish-consuming fish through increased ichthyoplankton mortality due to increased jellyfish.

## Fishing down marine food webs and jellyfish outbreaks

The phenomenon wherein fisheries increasingly target smaller fish lower down in the food web, now commonly referred to as 'fishing down marine food webs' (FD; Pauly et al., 1998a, b), is now well documented from a variety of countries and ecosystem types (Pauly & Palomares, 2005). This is one reason why the Convention on Biological Diversity selected the mean trophic level of fisheries catch, renamed 'Marine Trophic Index,' as one of eight indicators for 'immediate testing' by its >180 member countries (Pauly & Watson, 2005).

It is, however, the ecological changes in food web configuration that accompany FD which are relevant here. These may indeed provide a framework within which changes in marine ecosystems brought about by fishing can be assessed, particularly with respect to their possible impacts on jellyfish populations and their dynamics. Essentially, FD consists of the gradual loss of large organisms, species diversity, and structural diversity, i.e., the replacement of recently evolved, derived groups (marine mammals, bony fishes) by more primitive groups (invertebrates, notably jellyfishes, and bacteria). This is best seen when splitting the FD process into three phases, i.e., 1: 'pristine' or 'past'; 2: 'present' or 'exploited'; and 3: 'future' or 'fully degraded', corresponding to the three phases of degradation suggested by Jackson et al. (2001) and Pandolfi et al. (2005). We can then characterize each phase by (i) the main features of the fishes and other nektonic organisms and (ii) the pelagic–benthic coupling and its effect on processes in the water column (see also Fig. 8).

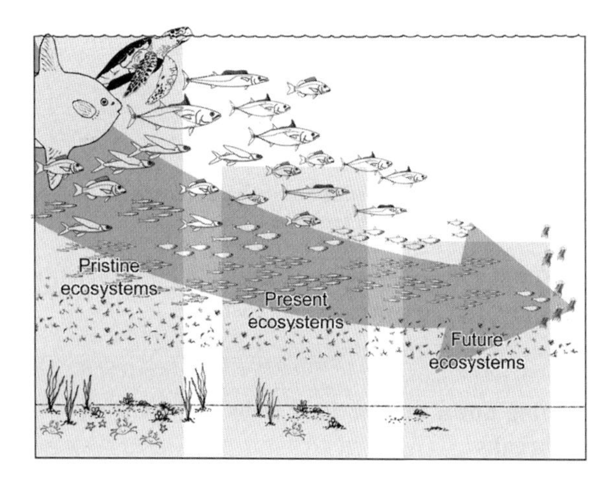


**Fig. 8** Fishing down marine food webs: the three stages (Pristine, Present, Future) offer increasingly better conditions for jellyfish (see text)

The first 'pristine' phase prevailed before humans had a chance to strongly impact ocean ecosystems (Jackson et al., 2001). Few parts of the oceans, notably outlying areas of the North and South Pacific, may still be labeled as pristine. For most of the world's oceans, estimates of pristine abundances must be recovered based on historical accounts and anecdotes (Sáenz-Arroyo et al., 2005; Palomares et al., 2006), or inferred from archeological data (Jackson et al., 2001, Roberts, 2007).

Invariably, a pristine state is characterized by numerous marine mammals and large fish as top predators, the latter group with biomasses often exceeding their present abundance by one or two orders of magnitude. Elevated biomass of top predators also implies an abundance of small prey fishes and invertebrates, though not necessarily of those opportunistic groups (shrimps, squids) that now support increasingly valuable fisheries.

In the pristine environment, benthic life is dominated by an abundant structure-forming and sessile fauna. It is composed of filter feeders and deposit feeders, which suppress phytoplankton and detritus biomass and prevent re-suspension of sediments. As a result, the water column tends to be free of suspended particles and of nutrients leaching from them, i.e., oligotrophic. It is thus hypothesized that jellyfish were less abundant, and perhaps severe bloom cycles less frequent than now, because:

- The pelagic stage of jellyfish (the medusoids) had less food, and far more predators (e.g., leatherback sea turtles) than now; and
- Their sessile polyp stages were exposed to far more benthic predators, and had to contend with far more benthic zooplankton-feeding competitors than now, because of the intact benthic communities then prevailing.

The second 'exploited' phase is where we currently are. It is best characterized by declines, notably in the biomasses of large fishes, sizes and diversity of fishes in fisheries catches, trophic levels of the same (and hence the FD phenomenon), and benthos. Initially, these declines are compensated for by cascade effects (e.g., Myers et al., 2007), manifest in the emergence of new fisheries for squids and other invertebrates, but which eventually decline as well. Benthic life is also modified: biogenic structures, built over centuries by filter and detritus feeders, are increasingly destroyed by bottom trawling, and replaced by small errant benthic animals (Hall, 1999), a situation advantageous for the benthic (polyp) stages of jellyfishes. The water column becomes progressively eutrophied, owing to the increasing scarcity of animals and structures that were cropping phytoplankton and consuming marine snow (detritus), which is now being re-suspended by storms and by trawling itself. For jellyfish, this generally implies improved conditions (Arai 2001). However, jellyfish outbreaks may still require to be triggered by exceptional abiotic conditions (Purcell, 2005).

The third 'fully degraded' phase follows on the continuation of present trends—although in some places, e.g., estuaries such as the Chesapeake Bay, many of the features associated with this third stage have already developed. In the Chesapeake Bay, fishing not only has eliminated virtually all animals above the size of striped bass, the current top predator, but also more importantly has eliminated the key filter feeders, both benthic (oyster; Jackson et al., 2001) and pelagic (menhaden; Franklin, 2007). Oysters, which, until 150 years ago, formed giant reefs, are reported as having been capable of filtering the water of Chesapeake Bay in 3 days (Jackson et al., 2001). Their absence (again, a result of fishing), compounded by the scarcity of menhaden, is the ultimate reason why pollution from effluents can now have such strong effects, and why harmful algae bloom occur. This also applies to other water bodies, estuarine or not, which are rendered less resilient by fishing, and easier for invasive species to overwhelm.

The biological endpoint of ecosystem degradation is the 'dead zone.' It is a zone free of oxygen and multicellular life as a result of excess nutrients, with bacteria instead of benthic animals, processing the resulting abundance of marine snow and other detritus. Such dead zones occur throughout the world, from the northern Gulf of Mexico (Rabalais et al., 2002) to the Benguela system off southwest Africa (Bakun & Weeks, 2004), and their number is growing rapidly (Dybas, 2005).

Jellyfish appear to have a competitive advantage over fish in hypoxic conditions (Sagasti et al., 2001). Such adaptation can lead to the dominance of jellyfish in a highly stressed ecosystem, as reported for example by Lyman et al. (2006) for the Benguela upwelling ecosystem.

## Conclusions

Jellyfish clearly play an extremely important role in the processing of energy in fisheries-based ecosystems. Short of conducting large-scale manipulations of marine ecosystems, the best alternative is ecosystem modeling. In this paper, we have documented that representation of jellyfish in existing EwE models is generally poor in terms of both their simple inclusion in models and their parameterization when they are included. Appropriate parameterization is critical, and we have shown in Table 1 that the parameters of jellyfish in EwE (i.e., their Q/B, P/B, TL and) are unacceptably variable. Some of this may be due to inappropriate use of wet biomass-specific measures, which are highly biased for gelatinous animals. We encourage future modeling studies to not only include jellyfish, but also to carefully consider how the models handle their unique biological nature.

Despite these difficulties, we see interesting and potentially important patterns among various EwE models that have included jellyfish. Specifically, we note the 'keystoneness' values of jellyfish appear to increase as the size of the ecosystem decreases. This is but one example of how ecosystem models can be used to develop and test hypotheses on the functional role of jellyfish in marine ecosystems.

We are now also able to test specific hypotheses derived from 'Fishing Down Marine Food Webs,' an explanation growing in popularity for explaining jellyfish biomass increase in fisheries-based ecosystems. The three phases of fishing down, pristine, exploited, and fully degraded, offering increasingly favorable conditions for jellyfish, are schematic, and they could be further subdivided, and/or defined more rigorously. Still, even in their present, preliminary form, they provide a coherent framework for many of the changes observed in ocean ecosystems, including those that favor jellyfish blooms or outbreaks.

One way to test this framework, and the hypotheses it includes, would be to re-express all models in Table 1 in dry weight or energy units (they currently use wet weight as currency), and rerun all analyses. This would help overcome the likelihood that the high water content of jellyfish has been ignored in some ecosystem models, i.e., that 'biomass' consists overwhelmingly of water. Also, the often assumed diets of jellyfish would have to be verified, as they play, in trophic models, as crucial a role as they play in real ecosystems. Better information on the diet of jellyfish will likely bring to evidence the need for enhancing the resolution currently used for describing these organisms in food web models. Description of jellyfish with a set of appropriate functional groups will certainly help in better understanding their role in food webs and their effects on ecosystems. Once these precautions are taken, however, it should be possible to integrate jellyfish, in spite of their strange biology, into the mainstream of marine biology and fisheries research.

**Acknowledgments** We wish to thank Ms Aque Atanacio for drafting the figures, and Ms Colette Wabnitz for numerous suggestions on an earlier draft. Daniel Pauly wishes to thank Dr. Kylie Pitt for the invitation to present a keynote on which this is partly based. This is a contribution of the *Sea Around Us* Project, initiated and funded by the Pew Charitable Trusts, Philadelphia.

## References

Allen, K. R., 1971. Relation between production and biomass. Journal of the Fisheries Research Board of Canada 28: 1573–1578.

Arai, M. N., 1987. Interactions of fish and pelagic coelenterates. Canadian Journal of Zoology 66: 1913–1927.

Arai, M. N., 1997. A Functional Biology of Scyphozoa. Chapman and Hall, New York: 316.

Arai, M. N., 2001. Pelagic coelenterates and eutrophication: a review. Hydrobiologia 451: 69–87.

Arai, M. N., 2005. Predation on pelagic coelenterates: a review. Journal of the Marine Biological Association of the United Kingdom 85: 523–536.

Baird, D. & R. E. Ulanowicz, 1989. Seasonal dynamics of the Chesapeake Bay ecosystem. Ecological Monographs 59: 329–364.

Bakun, A. & S. J. Weeks, 2004. Greenhouse gas build-up, sardines, submarine eruptions and the possibility of abrupt degradation of intense marine upwelling ecosystems. Ecology Letters 7: 1015–1023.

Bamstedt, U., S. Kaartvedt & M. Youngbluth, 2003. An evaluation of acoustic and video methods to estimate the abundance and vertical distribution of jellyfish. Journal of Plankton Research 25: 1307–1318.

Brierley, A. S., D. C. Boyer, B. E. Axelsen, C. P. Lynam, C. A. J. Sparks, H. J. Boyer & M. J. Gibbons, 2005. Towards the acoustic estimation of jellyfish abundance. Marine Ecology-Progress Series 295: 105–111.

Brodeur, R. D., H. Sugisaki & G. L. Hunt, 2002. Increases in jellyfish biomass in the Bering Sea: implications for the ecosystem. Marine Ecology-Progress Series 233: 89–103.

Buchary, E., 1999. Evaluating the effect of the 1980 trawl ban in the Java Sea, Indonesia: an ecosystem-based approach. M.Sc. thesis. Department of Resource Management and Environmental Studies. The University of British Columbia, Vancouver, Canada.

Buchary, E., T. J. Pitcher, W.-L. Cheung & T. Hutton, 2002. New ecopath models of the Hong Kong marine ecosystem. In Pitcher, T., E. Buchary & P. Trujillo (eds), Spatial Simulations of Hong Kong's Marine Ecosystem. Fisheries Centre Research Reports (This and all other Fisheries Centre research Reports cited therein can be freely downloaded from: http://www.fisheries.ubc.ca/publications/reports/fcrr.php) 10(3): 6–16.

Cheung, W.-L., R. Watson & T. Pitcher, 2002. Policy simulation of fisheries in the Hong Kong marine ecosystems. In Pitcher, T. & K. Cochrane (eds), The Use of Ecosystem Models to Investigate Multispecies Management Strategies for Capture Fisheries. Fisheries Centre Research Reports 10(2): 46–54.

Christensen, V. & D. Pauly, 1992. The ECOPATH II—a software for balancing steady-state ecosystem models and calculating network characteristics. Ecological Modelling 61: 169–185.

Christensen, V. & C. J. Walters, 2004. Ecopath with Ecosim: methods, capabilities and limitations. Ecological Modelling 172: 109–139.

Colin, S. P., J. H. Costello, W. M. Graham & J. Higgins, 2005. Omnivory by the small cosmopolitan hydromedusa *Aglaura hemistoma*. Limnology and Oceanography 50: 1264–1268.

Coll, M., I. Palomera, S. Tudela & F. Sardà, 2006. Trophic flows, ecosystem structure and fishing impact in the South Catalan Sea, Northwestern Mediterranean. Journal of Marine Systems 59: 63–96.

Coll, M., A. Santojanni, E. Arneri, I. Palomera & S. Tudela, 2007. An ecosystem model of the Northern and Central Adriatic Sea: analysis of ecosystem structure and fishing impacts. Journal of Marine Systems 67: 119–154.

Colombo, G. A., H. Mianzan & A. Madirolas, 2003. Acoustic characterization of gelatinous-plankton aggregations: four case studies from the Argentine continental shelf. ICES Journal of Marine Science 60: 650–657.

Costello, J. H. & S. P. Colin, 1995. Flow and feeding by swimming scyphomedusae. Marine Biology 124: 399–406.

Costello, J. H., B. K. Sullivan & D. J. Gifford, 2006a. A physical–biological interaction underlying variable phenological responses to climate change by coastal zooplankton. Journal of Plankton Research 28: 1099–1105.

Costello, J. H., B. K. Sullivan, D. J. Gifford, D. Van Keuren & L. J. Sullivan, 2006b. Seasonal refugia, shoreward thermal amplification, and metapopulation dynamics of the ctenophore *Mnemiopsis leidyi* in Narragansett Bay, Rhode Island. Limnology and Oceanography 51: 1819–1831.

Cowan, J. H. & E. D. Houde, 1992. Size-dependent predation on marine fish larvae by ctenophores, scyphomedusae, and planktivorous fish. Fisheries Oceanography 1: 113–126.

Cox, S. P., T. E. Essington, J. F. Kitchell, S. J. D. Martell, C. J. Walters, C. Boggs & I. Kaplan, 2002. Reconstructing ecosystem dynamics in the central Pacific Ocean, 1952–1998. II. A preliminary assessment of the trophic impacts of fishing and effects on tuna dynamics. Canadian Journal of Fishery and Aquatic Sciences 59: 1736–1747.

de Lafontaine, Y. & W. C. Leggett, 1988. Predation by jellyfish on larval fish: an experimental evaluation employing in situ enclosures. Canadian Journal of Fisheries and Aquatic Sciences 45: 1173–1190.

Dommasnes, A., V. Christensen, B. Ellertsen, C. Kvamme, W. Melle, L. Nottestad, T. Pedersen, S. Tjelmeland & D. Zeller, 2001. An Ecopath model for the Norwegian Sea and Barents Sea. In Guénette, S., V. Christensen & D. Pauly (eds), Fisheries Impacts on North Atlantic Ecosystems: Models and Analyses. Fisheries Centre Research Reports 9(4): 213–239.

Dybas, C. L., 2005. Dead zones spreading in world oceans. BioScience 55(7): 552–557.

Estes, J. A. & J. F. Palmisano, 1974. Sea otters: their role in structuring nearshore communities. Science 185: 1058–1060.

FAO/FISHCODE, 2001. Report of a bio-economic modelling workshop and a policy dialogue meeting on the Thai demersal fisheries in the Gulf of Thailand held at Hua Hin, Thailand, 31 May–9 June 2000, FI: GCP/INT/648/NOR: Field Report F-16 (En), Rome, FAO.

Franklin, H. B., 2007. The Most Important Fish in the Sea: Menhaden and America. Island Press, Washington, DC.

Froese, R. & D. Pauly (eds), 2000. FishBase 2000: Concepts, Design and Data Sources. ICLARM, Los Baños, Philippines [updates in www.fishbase.org].

Graham, W. M., 2001. Numerical increases and distributional shifts of *Chrysaora quinquecirrha* (Desor) and *Aurelia aurita* (Linné) (Cnidaria: Scyphozoa) in the northern Gulf of Mexico. Hydrobiologia 451: 97–111.

Graham, W. M. & R. M. Kroutil, 2001. Size-based prey selectivity and dietary shifts in the jellyfish, *Aurelia aurita*. Journal of Plankton Research 23: 67–74.

Graham, W. M., F. Pages & W. M. Hamner, 2001. A physical context for gelatinous zooplankton aggregations: a review. Hydrobiologia 451: 199–212.

Hall, S. J., 1999. The Effects of Fishing on Marine Ecosystems and Communities. Blackwell Science, Oxford.

Halpern, B. S., S. Walbridge, K. A. Selkoe, C. V. Kappel, F. Micheli, C. D'Agrosa, J. F. Bruno, K. S. Casey, C. Ebert, H. E. Fox, R. Fujita, D. Heinemann, H. S. Lenihan, E. M. P. Madin, M. T. Perry, E. R. Selig, M. Spalding, R. Steneck & R. Watson, 2008. A global map of human impact on marine ecosystems. Science 319: 948–952.

Jackson, J. B. C., M. X. Kirby, W. H. Berger, K. A. Bjorndal, L. W. Botsford, B. J. Bourque, R. Cooke, J. A. Estes, T. P. Hughes, S. Kidwell, C. B. Lange, H. S. Lenihan, J. M. Pandolfi, C. H. Peterson, R. S. Steneck, M. J. Tegner & R. R. Warner, 2001. Historical overfishing and the recent collapse of coastal ecosystems. Science 293: 629–638.

Kawahara, M., S. Uye, K. Ohtsu & H. Izumi, 2006. Unusual population explosion of the giant jellyfish *Nemopilemia nomurai* (Scyphozoa: Rhizostomeae) in East Asian waters. Marine Ecology-Progress Series 307: 161–173.

Libralato, S., V. Christensen & D. Pauly, 2006. A method for identifying keystone species in food web models. Ecological Modelling 195: 153–171.

Lyman, C. P., M. J. Gibbons, B. E. Axelsen, C. A. J. Sparks, J. Coetzee, B. G. Heywood & A. S. Brierley, 2006. Jellyfish overtake fish in a heavily fished ecosystem. Current Biology 16: R492–R493.

Martell, S. J., A. I. Beattie, C. J. Walters, T. Nayar & R. Briese, 2002. Simulating fisheries management strategies in the Gulf of Georgia ecosystem using Ecopath with Ecosim 2002. In Pitcher, T. & K. Cochrane (eds), The Use of Ecosystem Models to Investigate Multispecies Management Strategies for Capture Fisheries. Fisheries Centre Research Reports 10(2): 16–23.

Mills, C. E., 2001. Jellyfish blooms: are populations increasing globally in response to changing ocean conditions? Hydrobiologia 451: 55–68.

Mohammed, E., 2001. A model of the Lancaster Sound Region in the 1980s. In Guénette, S., V. Christensen & D. Pauly (eds), Fisheries Impacts on North Atlantic Ecosystems: Models and Analyses. Fisheries Centre Research Reports 9(4): 99–110.

Morissette, L., 2007. Complexity, cost and quality of ecosystem models and their impact on resilience: a comparative analysis, with emphasis on marine mammals and the Gulf of St. Lawrence. PhD thesis, University of British Columbia, Vancouver, BC.

Myers, R. A., J. K. Baum, T. D. Shepherd, S. P. Powers & C. H. Peterson, 2007. Cascading effects of the loss of apex predatory sharks from a coastal ocean. Science 315: 1846–1850.

NRC, 2003. Decline of the Steller Sea Lion in Alaskan waters: untangling food webs and fishing nets. Committee on the Alaska Groundfish Fishery and Steller Sea Lions, Ocean Studies Board, Polar Research Board, Division on Earth and Life Studies, National Research Council of the National Academies. The National Academies Press, Washington, D.C.

Okey, T. A., 2001. A 'straw-man' Ecopath model of the Middle Atlantic Bight continental shelf, United States. In Guénette, S., V. Christensen & D. Pauly (eds), Fisheries Impacts on North Atlantic Ecosystems: Models and Analyses. Fisheries Centre Research Reports 9(4): 151–166.

Okey, T. A. & D. Pauly (eds), 1998. A trophic mass-balance model of Alaska's Prince William Sound Ecosystem, for the post-spill period 1994–1996, 2nd edn. Fisheries Centre Research Reports 7(4): 144.

Okey, T. A., & R. Pugliese, 2001. A preliminary Ecopath Model of the Atlantic Continental Shelf adjacent to the south-eastern United States. In Guénette, S., V. Christensen & D. Pauly (eds), Fisheries Impacts on North Atlantic Ecosystems: Models and Analyses. Fisheries Centre Research Reports 9(4): 167–181.

Okey, T. A., G. A. Vargo, S. Mackinson, M. Vasconcellos, B. Mahmoudi & C. A. Meyer, 2004. Simulating community effects of sea floor shading by plankton blooms over the West Florida Shelf. Ecological Modelling 172: 339–359.

Olesen, N. J., K. Frandsen & H. U. Riisgård, 1994. Population dynamics, growth and energetics of jellyfish *Aurelia aurita* in a shallow fjord. Marine Ecology Progress Series 105: 9–18.

Paine, R. T., 1969. A note on trophic complexity and species diversity. American Naturalist 103: 91–93.

Paine, R. T., 1992. Food web interaction strength through field measurement of per capita interaction strength. Nature 355: 73–75.

Palomares, M. L. D. & D. Pauly, 2008. The growth of jellyfish. In K. A. Pitt & J. E. Purcell (eds) Jellyfish blooms: causes, consequences and recent advances. Hydrobiologia (this volume). doi:10.1007/s10750-008-9582-y.

Palomares, M. L. D., E. Mohammed & D. Pauly, 2006. On European expeditions as a source of historic abundance data on marine organisms: a case study of the Falkland Islands. Environmental History 11: 835–847.

Pandolfi, J. M., J. B. C. Jackson, N. Baron, R. H. Bradbury, H. M. Guzman, T. P. Hughes, C. V. Kappel, F. Micheli, J. C. Ogden, H. P. Possingham & E. Sala, 2005. Are U.S. coral reefs on the slippery slope to slime? Science 307: 1725–1726.

Pauly, D., V. Christensen, J. Dalsgaard, R. Froese & F. C. Torres Jr., 1998a. Fishing down marine food webs. Science 279: 860–863.

Pauly, D., V. Christensen, S. Guénette, T. J. Pitcher, U. R. Sumaila, C. J. Walters, R. Watson & D. Zeller, 2002. Towards sustainability in world fisheries. Nature 418: 689–695.

Pauly, D., V. Christensen & C. J. Walters, 2000. Ecopath, Ecosim and Ecospace as tools for evaluating ecosystem impact of fisheries. ICES Journal of Marine Science 57: 697–706.

Pauly, D. & M. L. Palomares, 2005. Fishing down marine food webs: it is far more pervasive than we thought. Bulletin of Marine Science 76: 197–211.

Pauly, D., T. Pitcher & D. Preikshot, 1998b. Back to the Future: Reconstructing the Strait of Georgia Ecosystem. Fisheries Centre Research Reports 6(5): 99.

Pauly, D. & R. Watson, 2005. Background and interpretation of the 'Marine Trophic Index' as a measure of biodiversity. Philosophical Transactions of the Royal Society: Biological Sciences 360: 415–423.

Peach, M. B. & K. A. Pitt, 2005. Morphology of the nematocysts of the medusae of two scyphozoans, *Catostylus mosaicus* and *Phyllorhiza punctata* (Rhizostomeae): implications for capture of prey. Invertebrate Biology 124: 98–108.

Pikitch, E. K., C. Santora, E. A. Babcock, A. Bakun, R. Bonfil, D. O. Conover, P. Dayton, P. Doukakis, D. Fluharty, B. Heneman, E. D. Houde, J. Link, P. A. Livingston, M. Mangel, M. K. McAllister, J. Pope & K. J. Sainsbury, 2006. Ecosystem-based fishery management. Science 305: 346–347.

Pinnegar, J. K., J. L. Blanchard, S. Mackinson, R. D. Scott & D. E. Duplisea, 2005. Aggregation and removal of weak-links in food-web models: system stability and recovery from disturbance. Ecological Modelling 184: 229–248.

Pitt, K. A., A.-L. Clement, R. M. Connolly & D. Thibault-Botha, 2008. Predation by jellyfish on large and emergent zooplankton: implications for benthic–pelagic coupling. Estuarine, Coastal and Shelf Science 76: 827–833.

Polis, G. A. & D. R. Strong, 1996. Food web complexity and community dynamics. American Naturalist 147: 813–846.

Power, M. E., D. Tilman, J. A. Estes, B. A. Menge, W. J. Bond, L. S. Mills, G. Daily, J. C. Castilla, J. C. Lubchenco & R. T. Paine, 1996. Challenges in the quest for keystones. Bioscience 46: 609–620.

Purcell, J. E., 1985. Predation on fish eggs and larvae by pelagic cnidarians and ctenophores. Bulletin of Marine Science 37: 739–755.

Purcell, J. E., 2003. Predation on zooplankton by large jellyfish, *Aurelia labiata*, *Cyanea capillata* and *Aequorea aequorea*, in Prince William Sound, Alaska. Marine Ecology Progress Series 246: 137–152.

Purcell, J. E., 2005. Climate effects on formation of jellyfish and ctenophore blooms: a review. Journal of the Marine Biological Association of the United Kingdom 85: 461–476.

Purcell, J. E. & M. N. Arai, 2001. Interactions of pelagic cnidarians and ctenophores with fish: a review. Hydrobiologia 451: 27–44.

Purcell, J. E. & C. E. Mills, 1988. The correlation between nematocyst types and diets in pelagic hydrozoa. In Hessinger, D. A. & H. Lenhoff (eds), The Biology of Nematocysts. Academic Press, New York.

Purcell, J. E., S. Uye & W. T. Lo, 2007. Anthropogenic causes of jellyfish blooms and their direct consequences for humans: a review. Marine Ecology-Progress Series 350: 153–174.

Rabalais, N. N., R. E. Turner & W. J. Wiseman Jr., 2002. Gulf of Mexico hypoxia—a.k.a. 'the Dead Zone'. Annual Review of Ecology and Systematics 33: 235–263.

Roberts, C., 2007. The Unnatural History of the Sea: The Past and Future of Humanity and Fishing. Island Press, Washington, DC.

Robison, B. H., 2004. Deep pelagic biology. Journal of Experimental Marine Biology and Ecology 300: 253–272.

Ruckelshaus, M., T. Klinger, N. Knowlton & D. R. Demaster, 2008. Marine ecosystem-based management in practice: scientific, and governance challenges. Bioscience 58: 53–63.

Sáenz-Arroyo, A., C. M. Roberts, J. Torre, M. Cariño-Olvera & R. R. Enríquez-Andrade, 2005. Rapidly shifting environmental baselines among fishers of the Gulf of California. Proceeding of the Royal Society (B) 272: 1957–1962.

Sagasti, A., L. C. Schaffne & J. E. Duffy, 2001. Effects of periodic hypoxia on mortality, feeding and predation in an estuarine epifaunal community. Journal of Experimental Marine Biology and Ecology 258: 237–283.

Trites, A. W., P. A. Livingston, S. Mackinson, M. C. Vasconcellos, A. M. Springer & D. Pauly, 1999. Ecosystem Change and the Decline of Marine Mammals in the Eastern Bering Sea: Testing the Ecosystem Shift and Commercial Whaling Hypotheses. Fisheries Centre Research Reports 7(1): 106.

Ulanowicz, R. E. & C. J. Puccia, 1990. Mixed trophic impacts in ecosystems. Coenoses 5: 7–16.

Uye, S. & H. Shimauchi, 2005. Population biomass, feeding, respiration and growth rates, and carbon budget of the scyphomedusa *Aurelia aurita* in the Inland Sea of Japan. Journal of Plankton Research 27: 237–248.

Van der Land, J. (ed.), 2006. UNESCO-IOC Register of Marine Organisms (version 7 November 2006). In Bisby, F. A., Y. R. Roskov, M. A. Ruggiero, T. M. Orrell, L. E. Paglinawan, P. W. Brewer, N. Bailly & J. van Hertum (eds), Species 2000 and ITIS Catalogue of Life: 2007 Annual Checklist. CD-ROM; Species 2000: Reading, UK.

Van Der Veer, H. W. & W. Oorthuysen, 1985. Abundance, growth and food demand of the scyphomedusa *Aurelia aurita* in the western Wadden Sea. Netherlands Journal of Sea Research 19: 38–44.

Walters, C. J., V. Christensen, S. J. Martell & J. F. Kitchell, 2005. Possible ecosystem impacts of applying MSY policies from single-species assessment. ICES Journal of Marine Science 62: 558–568.

Walters, C. J., V. Christensen & D. Pauly, 1997. Structuring dynamic models of exploited ecosystems from trophic mass-balance assessments. Reviews in Fish Biology and Fisheries 7: 139–172.

Walters, C. J., D. Pauly & V. Christensen, 1998. Ecospace: prediction of mesoscale spatial patterns in trophic relationships of exploited ecosystems, with emphasis on the impacts of marine protected areas. Ecosystems 2: 539–554.

Yamamoto, J., M. Hirose, T. Ohtani, K. Sugimoto, K. Hirase, N. Shimamoto, T. Shimura, N. Honda, Y. Fujimori & T. Mukai, 2008. Transportation of organic matter to the sea floor by carrion falls of the giant jellyfish *Nemopilema nomurai* in the Sea of Japan. Marine Biology 153: 311–317.

Hydrobiologia (2009) 616:87–97
DOI 10.1007/s10750-008-9594-7



# Quantifying movement of the tropical Australian cubozoan *Chironex fleckeri* using acoustic telemetry

**M. R. Gordon · J. E. Seymour**

Published online: 4 October 2008


**Abstract** Cubomedusae are considered to have superior swimming abilities compared to other pelagic cnidarians, yet many of the theories describing such behaviours are based on anecdotal evidence, sting records or opportunistic sightings, rather than quantitative data. Acoustic telemetry was used to document the movements of adult *Chironex fleckeri* medusae within both coastal and estuarine habitats. The rate at which tagged medusae moved was influenced by an interaction between time period (day or night) and habitat (coastal or estuarine), with rates of travel being relatively similar during the day and night within the coastal habitat, but significantly greater at night than during the day within the estuarine habitat. Medusae in coastal habitats travelled at similar rates throughout all tidal states while estuarine medusae travelled at significantly faster rates towards the middle of the tide than at the low and high ebbs. Such movements occurred with, and independent of, tidally generated currents, but at increased current speeds, medusae tended to travel with the current. Data are also presented that show that large medusae may move from coastal to estuarine habitats.

Guest editors: K. A. Pitt & J. E. Purcell
Jellyfish Blooms: Causes, Consequences, and Recent Advances

M. R. Gordon (✉) · J. E. Seymour
School of Marine and Tropical Biology,
James Cook University, Cairns, QLD 4878,
Australia
e-mail: matthew.gordon@jcu.edu.au

**Keywords** Box jellyfish · Habitat · Tide · Tracking

## Introduction

A range of strategies are used by species of cnidaria to regulate the distance and direction in which they travel, and as a result, some are considered to be more active than others. Hydromedusae and scyphomedusae, for instance, are often considered to be passive in their movements (Shorten et al., 2005). Although they may actively move vertically between currents of differing speed and direction (Rifkin, 1996), movement in a horizontal plane is often an indirect result of their vertical movement. In contrast, the swimming behaviours of cubomedusae appear to surpass all other jellyfish in terms of power and complexity (Rifkin, 1996), enabling them to move independently of currents and local weather conditions (Barnes, 1966; Kinsey, 1986; Hartwick, 1987). *Chironex fleckeri* Southcott is a large cubozoan found in Australia's tropical waters and can attain an unassisted speed of 414 m $h^{-1}$ using a pulse-coast style of jet propulsion (Shorten et al., 2005). Anecdotal reports of *C. fleckeri* medusae swimming at "3 to 4 knots all day" (Kinsey, 1986) suggest that elevated levels of activity can be sustained for extended periods of time. The extent to which speed in this species is influenced by physical and biological factors remains unclear with many currently favoured theories based on qualitative data such as sting records, opportunistic sightings and anecdotal evidence, rather than quantitative data.

Physical factors that may directly or indirectly influence *C. fleckeri* medusae movements include tidal state, water currents (Kinsey, 1986), time of day (Kinsey, 1986, Seymour et al., 2004) and exposure to and changes in local weather conditions (Southcott, 1956; Barnes, 1960, 1965, 1966; Brown, 1973; Yamaguchi, 1982; Kinsey, 1986; Marsh et al., 1986; Hartwick, 1987). Wind speed, for instance, may indirectly influence movement patterns by increasing wave action, as *C. fleckeri* medusae appear to be particularly sensitive to turbulence (Barnes, 1966; Kinsey, 1986). It remains unclear, however, whether the absence of medusae during rough conditions is due to their relocation to less turbulent areas such as sheltered bays (Brown, 1973), estuary systems (Kinsey, 1986) or offshore waters (Barnes, 1966; Kinsey, 1986), or is just sampling bias. The influence of tidal state is also unknown, although anecdotal evidence suggests that medusae are never found at low tide and collections are more successful on a making tide (Kinsey, 1986). Tidal currents can also potentially influence cubomedusae movements in that travel can be directional (with or against the current) or random (Yamaguchi, 1985; Kinsey, 1986). Movement patterns in *C. fleckeri* are further complicated by changes in activity levels, with a resting behaviour reported for medusae in the mid to late afternoon (Kinsey, 1986, Seymour et al., 2004). During this time, medusae lie motionless on the substrate with tentacles either outstretched along the seafloor (Seymour et al., 2004) or retracted to or inside the bell (Kinsey, 1986). As such, any study on movement patterns in *C. fleckeri* needs to incorporate time of day into the study.

Biological factors may also influence movement patterns of *C. fleckeri* medusae. For example, the abundance and distribution of prey, such as fish and prawns, may vary spatially and temporally. Given that large aggregations of prey provide *C. fleckeri* medusae with particularly favourable feeding conditions (Southcott, 1971; Kinsey, 1986; Hartwick, 1991; Rifkin, 1996) and these medusae have the potential to actively locomote to different areas, medusae movements may then be influenced by the abundance of their prey.

If an understanding of the links between physical and biological factors and movement in *C. fleckeri* are sought, a more quantitative approach is required. Acoustic telemetry is widely used to gather data from a diverse range of marine species, although the majority of animals are large in size or have suitable sites for internal or external tag attachment (Egli & Babcock, 2004; Atkinson et al., 2005; Heupel & Simpfendorfer, 2005; Jackson et al., 2005; Kerwath et al., 2005; Heupel et al., 2006; Willis & Hobday, 2007; Wingate & Secor, 2007). Although continued miniaturisation has seen acoustic tags attached to increasingly smaller animals (Jackson et al., 2005), cubomedusae present a number of unique issues that have limited the effectiveness of acoustic telemetry as a tool to study their movement patterns. In particular, cubomedusae are often small in size and their entire body is used in locomotion, making tag attachment difficult. However, a new technique for tag attachment has been developed (Seymour et al., 2004) allowing movement of these animals to be investigated using manual acoustic tracking. The aim of this study was to use acoustic telemetry to quantify the influence of time of day, tidal state, tidal current and habitat on the movements of *C. fleckeri* medusae.

## Materials and methods

### Sample sites

Medusae movements were documented at five sites within three locations in north Queensland, Australia (Fig. 1).

- Location 1—Townsville

  The Australian Institute of Marine Science (AIMS) site was a shallow water embayment into which a permanently flowing estuary drains. Mangroves line the bay which is afforded a degree of protection by a headland and rocky/coral reef to the north.

- Location 2—Mission Beach region

  The Mission Beach region was represented by two sites: Mission Beach and Tully. The

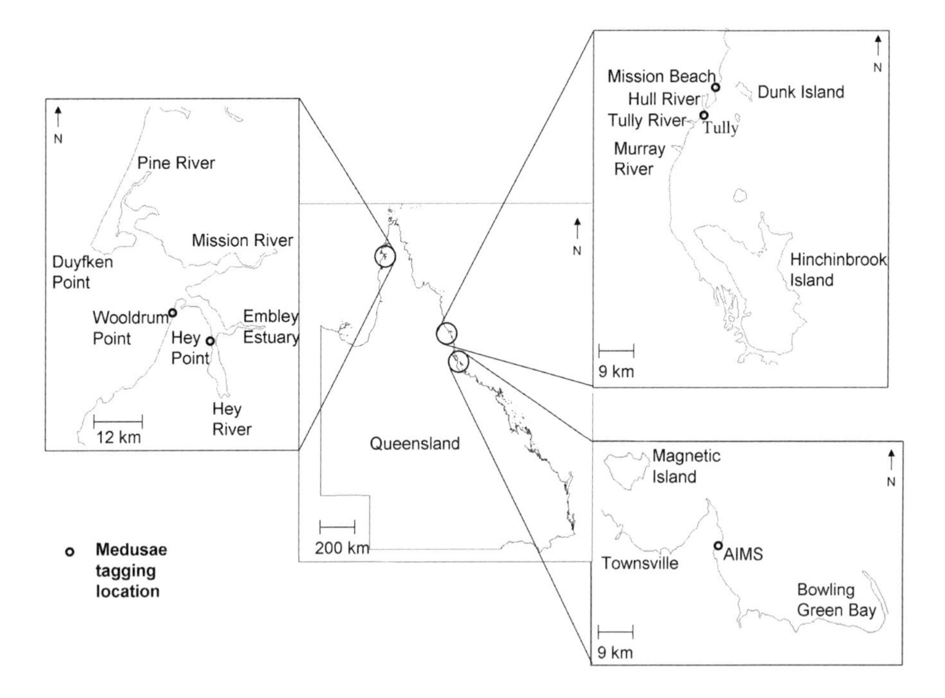


**Fig. 1** Geographic location of AIMS, Mission Beach, Tully, Wooldrum Point and Hey Point sample sites in north Queensland, Australia, showing specific tagging locations

coastline of this region is typified by long stretches of gently sloping sand/mud beachfront, separated occasionally by rocky headlands. Three mangrove-lined estuary systems, the Tully, Hull and Murray rivers, flow directly onto this stretch of coastline. The Mission Beach site was located approximately 8 km north of the Hull River, while the Tully site was located between the Hull and Tully rivers, including the Tully River itself.

- Location 3—Weipa

Weipa was represented by two sites: Wooldrum Point and Hey Point. Located on the eastern boundary of the shallow (7–10 m) embayment of Albatross Bay, Wooldrum Point is an uninterrupted stretch of gently sloping sand/mud beachfront that extends approximately 9 km south of the Embley Estuary mouth. Although protected in the north by Duyfken Point, three mangrove-lined estuary systems, the Mission, Embley and Pine rivers, empty directly into this embayment. The Embley Estuary is a 32 km long system into which the Hey River flows. Both systems are lined with extensive mangrove forests, seagrass beds and intertidal flats, with depth typically increasing rapidly from a shallow (~2 m) narrow (~100 m) intertidal shelf to a maximum depth of 21 m. Hey Point is situated at the intersection of the Hey River and Embley Estuary, at an interface between a mangrove forest and a stretch of muddy intertidal beachfront.

Acoustic tracking

Twelve *C. fleckeri* medusae were manually tracked: two at AIMS, two at Mission Beach, one at Tully, two at Wooldrum Point and five at Hey Point. Medusae selected for tracking had a minimum interpedalia distance (i.e. distance between the midline of two pedalia along the line passing through the rhopalial niche (IPD)) of 80 mm to reduce potential confounding effects due to tag weight. No tag used was more than 0.2% of the medusa's wet weight (where wet weight was calculated using regression equations for a morphometrically similar species, *Chiropsella bronzei* Gershwin (Gordon, 1998). All medusae used had no deformation or damage to the bell and no damaged or missing pedalia. Specimens were collected by hand and their IPD was measured to the nearest mm. An acoustic tag was glued to an external face of the bell with Hystoacryl, a non-toxic surgical glue that is used for closing incisions and sets in seconds. Tags were positioned half way between the top of the bell and the velarium, along

the fold line that forms between the rounded shoulder that gives rise to the pedalia and the more flattened interpedalial face (upon which the rhopalia are located) (Fig. 2). Three types of coded Sonotronics transmitters were used within this study (Pico, IBT 96-2, IBDT 97-2), varying in length and weight from 19 mm and 1 g to 49 mm and 3 g.

The capture and release point of each medusa was determined by a Garmin GPS 12 hand held GPS and the location classified as either estuarine (within an estuary) or coastal (outside an estuary). Individual medusae were actively tracked for between 10 and 38 h (Table 1) using a Sonotronics USR 96 receiver and unidirectional hydrophone. Individuals were relocated between fixes by returning to the last known location and estimating the distance and direction to the new location using signal strength. The tracking vessel was then moved in a straight line along the axis of maximum signal strength until the source of the signal was passed, at which point, a decreasing spiral search pattern allowed the animal's location to be determined. Several visual contacts during daylight hours confirmed an accuracy of approximately 10 m to the animal's true location. The time of successive readings was determined by the Garmin GPS 12 clock, and the location was again classified as either estuarine or coastal. Animals were grouped according to the habitats in which their movement patterns were documented, with three habitat categories identified as:

1) coastal—individuals that remained wholly within the coastal habitat
2) estuarine—individuals that remained wholly within the estuarine habitat
3) combination—individuals that moved between habitats

Details of each medusa's size (IPD in mm), the habitat and site at which it was tracked, the date and time (h) at which it was fitted with an acoustic tag, the total duration over which it was tracked (h) and tag type used are summarised in Table 1.

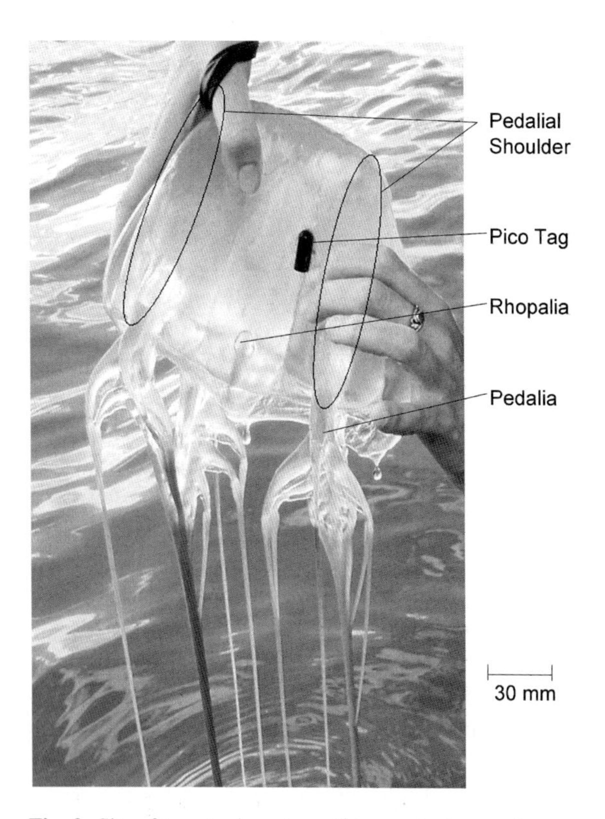


**Fig. 2** Site of tag attachment on *Chironex fleckeri* medusae—half way between the top of the bell and the velarium, along the fold line of the rounded shoulder giving rise to the pedalia, and the interpedalial face

## Calculation of distance estimates

The distance travelled by each medusa was measured as the distance between two successive GPS location readings. This estimate represented the minimum distance travelled as it did not allow for any deviation from travel in a straight line. Since the time between successive readings varied, the rate of travel (m $h^{-1}$) for each estimate was calculated by dividing the minimum distance travelled (m) by the time (h) between readings.

## Environmental factors

To investigate diurnal patterns, rates of travel during the day and night were estimated separately. Given that medusae are more active between 0600 and 1500 h than between 1500 and 0600 h (Seymour et al., 2004), an artificial definition of day (0600–1500 h) and night (1501–0559 h) was used.

The magnitude and duration of individual tides varied between sites (Table 2), with proportion of tidal state rather than absolute tide height used to investigate the effect of tidal state on rates of travel. A complete tidal cycle was defined as two successive high tides separated by a single low tide (Fig. 3A, B). Each complete tidal cycle was divided into its flood

**Table 1** Details of each medusa's size (IPD in mm), the habitat and site at which it was tracked, the date and time (h) at which it was fitted with an acoustic tag, the total duration over which it was tracked (h) and the tag type used

| Animal ID | IPD (mm) | Habitat | Site | Date | Time tagged (h) | Tracking duration (h) | Tag type |
|---|---|---|---|---|---|---|---|
| 1 | 140 | Coastal | AIMS | 13 Jan 2003 | 2100 | 11 | IBDT 97-2 |
| 2 | 145 | Coastal | AIMS | 17 Jan 2003 | 2100 | 12 | IBDT 97-2 |
| 3 | 150 | Coastal | Mission Beach | 9 Apr 2003 | 0930 | 11 | IBDT 97-2 |
| 4 | 175 | Coastal | Mission Beach | 10 Apr 2003 | 0645 | 12 | IBDT 97-2 |
| 7 | 90 | Coastal | Wooldrum Pt | 17 Dec 2006 | 0925 | 38 | Pico |
| 8 | 105 | Coastal | Wooldrum Pt | 17 Dec 2006 | 1344 | 10 | IBT 96-2 |
| 9 | 180 | Combination | Tully | 23 Mar 2005 | 1000 | 26 | IBDT 97-2 |
| 5 | 110 | Estuarine | Hey Pt | 17 Nov 2005 | 1100 | 17 | Pico |
| 6 | 115 | Estuarine | Hey Pt | 17 Nov 2005 | 1115 | 10 | Pico |
| 10 | 96 | Estuarine | Hey Pt | 9 Nov 2007 | 0848 | 15 | Pico |
| 11 | 100 | Estuarine | Hey Pt | 9 Nov 2007 | 0903 | 16 | Pico |
| 12 | 100 | Estuarine | Hey Pt | 9 Nov 2007 | 0914 | 11 | Pico |

**Table 2** Tidal characteristics experienced at each location during tracking of medusae showing minimum and maximum tidal height (m), minimum and maximum tidal amplitude (m) and minimum and maximum tide duration (min)

| Location | Minimum tide height (m) | Maximum tide height (m) | Minimum tidal amplitude (m) | Maximum tidal amplitude (m) | Minimum tide duration (min) | Maximum tide duration (min) |
|---|---|---|---|---|---|---|
| Townsville | 0.58 | 3.48 | 1.40 | 2.90 | 384 | 403 |
| Mission Beach | 1.36 | 2.49 | 0.19 | 2.07 | 237 | 520 |
| Weipa | 0.46 | 2.81 | 0.79 | 2.32 | 364 | 951 |

and ebb components, which were further divided into six equal tidal classes, each representing 1/6 of an individual tide (Fig. 3A). Each tidal class was assigned a tidal proportion class (1–10), according to whether it occurred on a flood or ebb tide (Fig. 3A).

Tidal proportion values for each location reading were calculated as follows:

$$P = \frac{D}{T}$$

where $P$ is the proportion of the individual tide at which a reading was taken, $D$ is the time interval (min) between when a reading was taken and high tide and $T$ is the total duration of the individual tide (min). For instance, a location reading taken 300 min from high tide ($D$) on a tide with a total duration ($T$) of 400 min would have a tidal proportion vale ($P$) of 0.75. The corresponding tidal proportion class for this reading would be 5 if it was taken on an ebb tide but 7 if it was recorded on a flood tide. For ebb tides, $D$ was taken to be the time (min) between when a location reading was taken and the previous high tide while for flood tides, $D$ was calculated as the time (min) between a location reading and the approaching high tide.

Due to the small number of medusae tracked in each habitat (five estuarine, six coastal) it was necessary to use sequential rate of travel (m $h^{-1}$) estimates as replicates to allow statistical analysis to be performed. As such, sequential rate of travel (m $h^{-1}$) estimates for each individual were not truly independent and a repeated measures analysis was conducted to determine this effect of dependency. No significant effect was found ($F = 0.075$, df $= 2 \times 30$, $P = 0.92$) and as such analysis was conducted assuming the replicates were independent. A three-way Analysis of Variance was used to compare rates of travel with respect to habitat (estuarine or coastal), time of day (day or night) and tidal state (tidal proportion class). Least Significant Difference post hoc analysis was carried out on tidal state to determine which states were statistically different. These analyses were

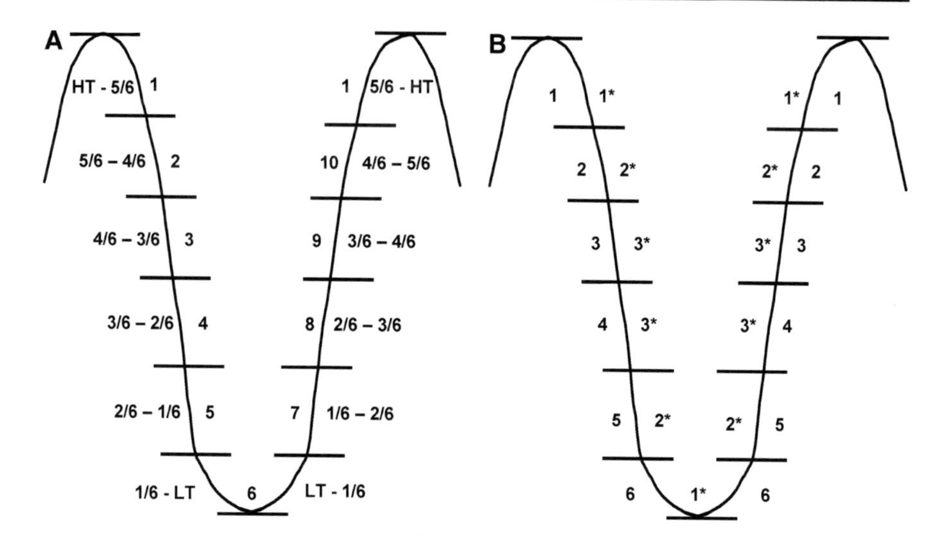


**Fig. 3** One complete tidal cycle showing tidal classes (1/6's) with corresponding tidal proportion classes (1–10) (**A**) and tide portion classes (1–6) with corresponding current numbers (1–3 with *) (**B**)

performed only on medusae that spent their entire time in either the coastal or the estuarine habitat.

To investigate the effect of current speed and direction on medusae movements, a flow index for each rate of travel estimate was calculated. Analysis was limited to data collected from the Weipa sites of Wooldrum Point and Hey Point, as these trials were conducted during the late dry season (as defined by Cyrus & Blaber, 1992) when currents are tidally driven (due to a lack of rainfall runoff). Such parameters have not been defined for east coast sites, especially during the wet season, when trials were conducted at these sites. Individual flood and ebb tides were divided into six equal tide portion classes as shown in Fig. 3B. Current numbers (*C*) were assigned to each tide portion class according to the generalisation that the greatest amount of tide-induced water movement would occur towards the middle of a tide, tapering off towards the top and bottom of the tide. Tide portion classes 1 and 6 were assigned a current number of 1 as they represented the top and bottom of the tide, 3 and 4 were assigned a current number of 3, representing the middle portions of the tide, and the intermediate tidal portions of 2 and 5 were assigned a current number of 2 (Fig. 3B). The flow index (*F*) for each rate of travel estimate was calculated as:

$$F = C \times A$$

where $F$ is the flow index (m), $C$ is the current number and $A$ is the tidal amplitude (m).

Given that each rate of travel estimate arose from two successive location readings, the direction of medusa travel in relation to the direction of tidal current could be established. This was defined as either having been with the current (dependent on) or across/against the current (independent of). Significant relationships between flow index and rates of travel were investigated using linear and curve estimation regression analysis for both dependent and independent travel with respect to current speed and direction.

## Results

No significant interaction between habitat, tidal state and time of day was found, with medusae in both habitats travelling at similar rates throughout the tidal cycle both day and night ($F = 1.836$, df $= 4 \times 83$, $P = 0.130$).

Medusae movement was, however, significantly influenced by an interaction between time of day and habitat ($F = 4.729$, df $= 1 \times 83$, $P = 0.033$). Medusae within the coastal habitat travelled at similar rates during the day ($135 \pm 31$ SE m h$^{-1}$) and night ($88 \pm 22$ SE m h$^{-1}$), but estuarine medusae travelled significantly faster at night ($467 \pm 65$ SE m h$^{-1}$) than during the day ($245 \pm 42$ SE m h$^{-1}$) (Fig. 4).

Rates of travel were also significantly influenced by an interaction between tidal state and habitat ($F = 2.063$, df $= 9 \times 83$, $P = 0.042$). While coastal

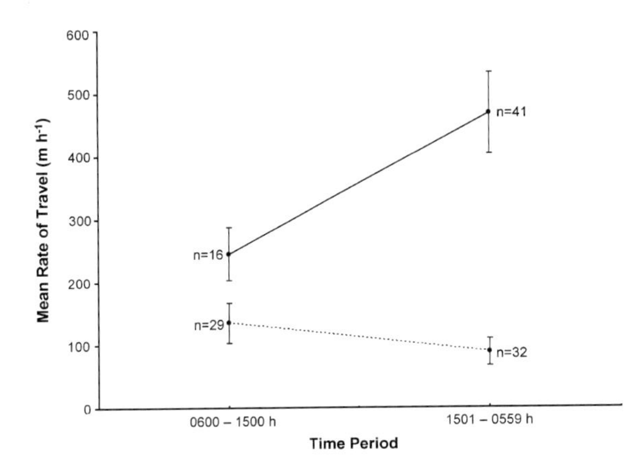


**Fig. 4** Mean rate of travel (m $h^{-1}$) ± 1 SE for medusae within the coastal (. . . . . ) and estuarine (——) habitats during the day (0600–1500 h) and night (1501–0559 h)

individuals travelled at similar rates throughout all tidal states (from 52 ± 26 SE m $h^{-1}$ to 187 ± 53 SE m $h^{-1}$), rates of travel for estuarine medusae varied substantially with tidal state. Rates of travel during slack water at the top of the tide (from 110 ± 53 SE m $h^{-1}$ to 169 ± 27 SE m $h^{-1}$) were similar to those observed within the coastal habitat, but increased throughout the tidal cycle, peaking at 828 ± 306 SE m $h^{-1}$ and 568 ± 241 SE m $h^{-1}$ for the ebb and flood tides, respectively, before decreasing again to 323 ± 77 SE m $h^{-1}$ at low tide (Fig. 5). Rates of travel within each tidal state were relatively consistent between day and night however, with no significant interaction between tidal state and time of day established ($F = 1.096$, df $= 9 \times 83$, $P = 0.375$).

While medusae were found to move both with and independently of the current, they typically travelled with the current at higher flow indexes (Fig. 6). No significant relationship between rate of travel and flow index was established for travel independent of the current ($F = 2.126$, df $= 2 \times 18$, $P = 0.148$); however, a significant relationship did exist between rate of travel and flow index for movements with the current ($F = 11.715$, df $= 2 \times 55$, $P < 0.001$). The rate at which medusae travelled increased with flow index, with this relationship best described by

$$R = 146.023 + (54.478 \times F) + \left(8.381 \times F^2\right)$$

where $R$ is the rate of travel (m $h^{-1}$) and $F$ is flow index (m).

While movement patterns were relatively consistent between the majority of individuals tracked, those of a large adult medusa (180 mm IPD) tagged at the Tully site on 23rd March 2005 differed in a number of key aspects. This was the only individual that used both coastal and estuarine habitats within 24 h, and it did not display the diurnal behaviours observed for other individuals. The medusa was tagged at the mouth of the Hull River at 1000 h and

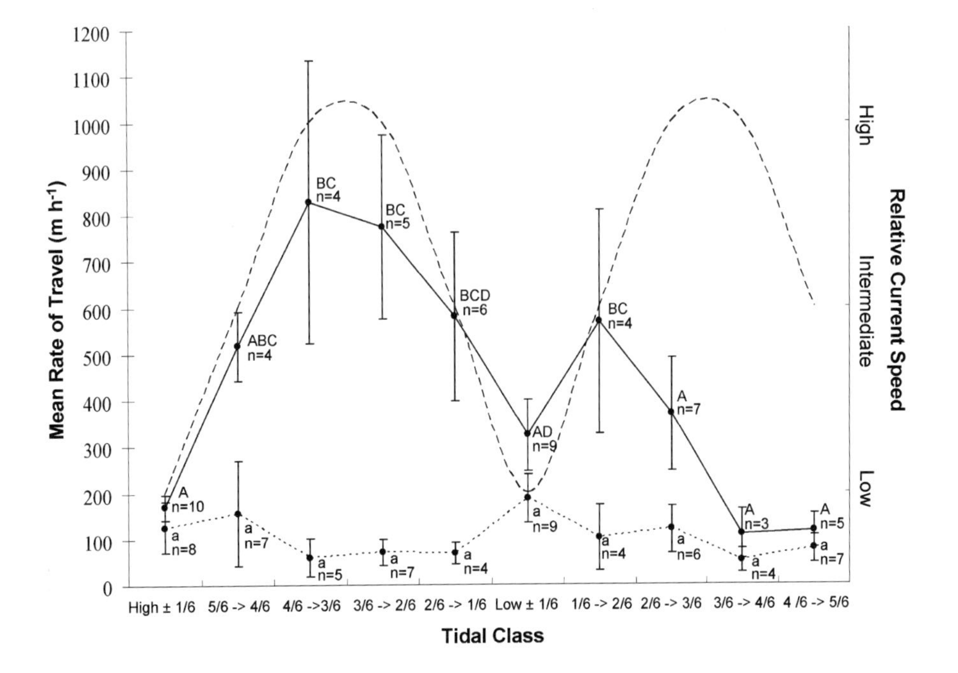


**Fig. 5** Mean rate of travel (m $h^{-1}$) ± 1 SE for medusae within the coastal (. . . . . ) and estuarine (——) habitats with respect to relative current (- - - -) at each tidal class with means followed by the same letter not significantly different at the 95% level

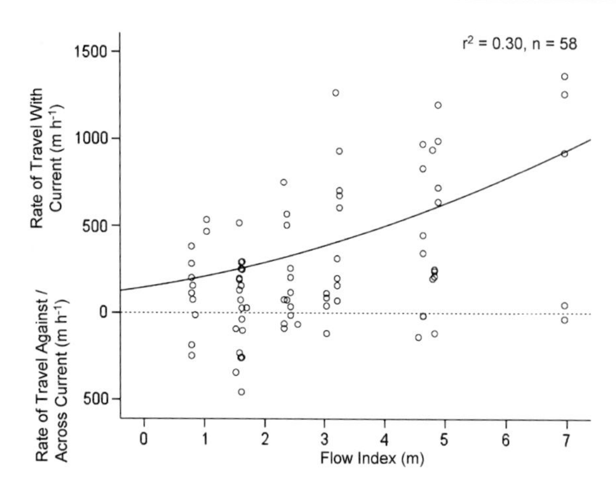


**Fig. 6** Rates of travel (m $h^{-1}$) with and against/across the current flow in relation to flow index (m) showing regression line (——) and zero reference line (- - - -)

moved south at an average speed of 269 ± 81 SE m $h^{-1}$ to be located adjacent to the Tully River mouth at 2100 h. From 2100 h to 0200 h, the medusa returned north along the beachfront at an average speed of 383 ± 105 SE m $h^{-1}$. By 0600 h, it had again returned south to the Tully River mouth at an average speed of 714 ± 186 SE m $h^{-1}$. During the daylight hours of the 24th March (0600 to 1200 h), the medusa moved from the costal to the estuarine habitat, into the Tully River (Fig. 7). Although water current speeds were not collected while this animal was tracked, the first 3 h of movement up the Tully River (average speed 503 ± 85 SE m $h^{-1}$) corresponded with an incoming tide while the second 3 h down the river (average speed 99 ± 15 SE m $h^{-1}$) occurred during an outgoing tide (Fig. 7). While this medusa exhibited a similar trend to coastal medusae whereby rates of travel during the day (average speed 364 ± 110 SE m $h^{-1}$) were similar to those travelled at night (average speed 403 ± 86 SE m $h^{-1}$), the actual rates of travel were of a similar scale to that of estuarine medusae.

## Discussion

Previous research has reported diurnal (Seymour et al. 2004) and afternoon resting behaviours (Kinsey, 1986) for *C. fleckeri*. Medusae are active between 0600 and 1500 h but have extended periods of relative inactivity from 1500 to 0600 h (Seymour et al., 2004). Such diurnal patterns were less pronounced in the current study and were not

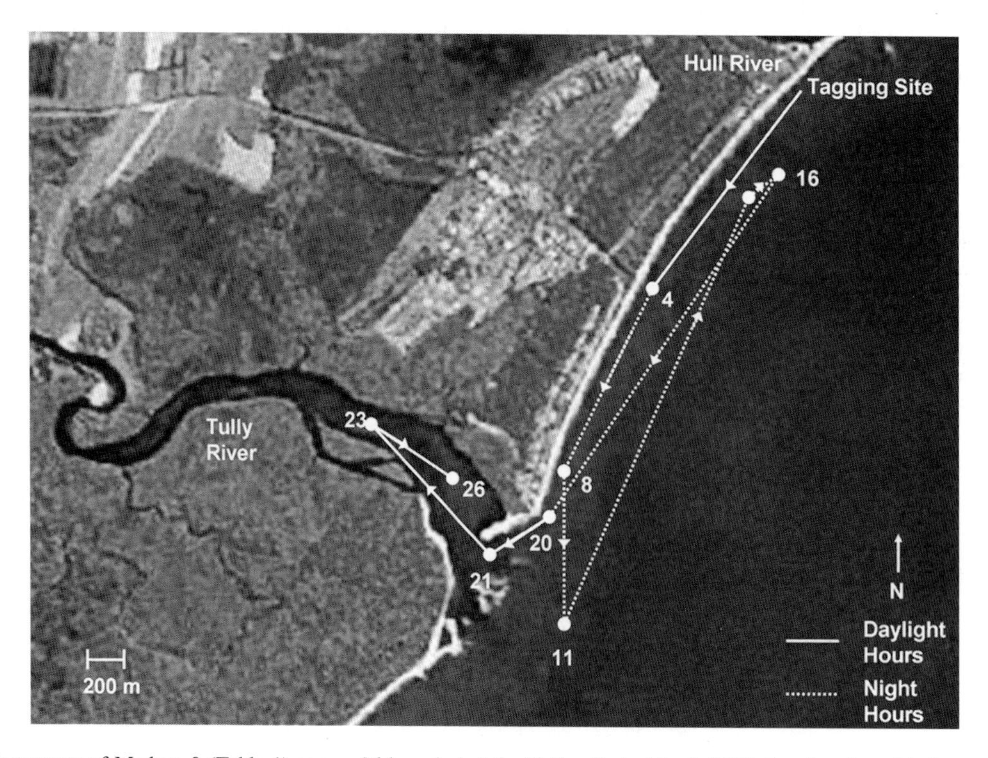


**Fig. 7** Movements of Medusa 9 (Table 1) over a 26-h period at the Tully site in March 2005 showing time (h) since tagging

consistent between habitats. During the day (0600–1500 h), coastal medusae travelled at an average rate of 135 ± 31 SE m $h^{-1}$, which is slower than the 212 m $h^{-1}$ reported by Seymour et al. (2004). A mean rate of travel of 88 ± 22 SE m $h^{-1}$ at night was, however, considerably more than the 10 m $h^{-1}$ previously reported (Seymour et al., 2004). Diurnal patterns were reversed for estuarine medusae that travelled at an average rate of 467 ± 65 SE m $h^{-1}$ at night but only 245 ± 42 SE m $h^{-1}$ during the day (0600–1500 h). The rate travelled at night was also more than 40 times the rate previously established for coastal medusae at night. It remains unclear whether biological or physical mechanisms drove these observations.

The relatively slow rates of travel by medusae during the day, within both the estuarine and coastal habitats, could represent localised hunting and feeding behaviours. When hunting, *C. fleckeri* swim at or near the surface with tentacles outstretched, sampling deeper waters only periodically by suspending bell activity and passively sinking (Barnes, 1966; Kinsey, 1986). Following prey capture, bell activity is typically suspended until prey have been removed from the tentacles (Barnes, 1966), further decreasing the distances medusae are likely to travel while actively feeding. Since prey such as prawns and small fish are characterised by spatial and temporal variability (Blaber et al., 1989, 1995; Vance & Staples, 1992; Xiao & Greenwood, 1993; Omundsen et al., 2000), it may be energetically beneficial for medusae to remain searching within an area of high prey concentration rather than leave the area in search of other prey.

Given that vision appears to play a role in prey location and capture (Kinsey, 1986; Seymour et al., 2004), the ability of medusae to successfully capture prey or avoid underwater obstacles would be greatly reduced at low light intensities (e.g. night). At night, therefore, it may be energetically beneficial for medusae to remain motionless on the substrate. The reduced rates of travel of coastal medusae at night may have arisen from the short periods of activity medusae demonstrate when disturbed from their otherwise motionless, energy conserving state (Seymour et al., 2004). Estuarine medusae, however, may have been unable to sink to the substrate and become inactive due to the relatively strong currents that are likely to persist within estuaries. The greater rates of travel exhibited by estuarine medusae during the night, therefore, may be attributed to medusae moving with the tidal current until sufficient light returned for feeding activities to resume.

Alternatively, tidal state could indirectly influence the rates at which medusae travel by limiting access to either prey or the habitats in which high abundances of prey occur. For instance, greater densities of potential prey occur on the flood rather than the ebb tide within the Embley Estuary (Xiao & Greenwood, 1992), particularly towards the high tide (Vance & Staples, 1992; Vance et al., 1994). In contrast, the majority of fish species found in the nearshore coastal habitat of Weipa are not influenced by tidal state (Blaber et al., 1995). As such, the relatively slow and consistent rates of travel observed for coastal medusae may have arisen from a more reliable access to prey, a situation that does not exist for estuarine individuals, whose movements varied considerably throughout each tidal cycle.

The duration for which abundant prey are accessible to medusae is also likely to vary between habitats. For example, in tropical estuaries, mangrove communities that act as nursery areas (Blaber & Blaber, 1980; Robertson & Duke, 1987, 1990; Staples & Vance, 1987; Blaber et al., 1989, 1995; Vance et al., 1996) and intertidal mudflats adjacent to mangroves which support a greater species composition and biomass of fish (Blaber et al., 1989) would have only been accessible to medusae at higher tidal states. Medusae would have most likely fed, and hence travelled shorter distances, at these times. If medusae were no longer able to feed at intermediate tidal states, however, it may have become energetically efficient to move with the current rather than maintain a position against the current. In contrast, a more uniform topography may have provided coastal medusae with more consistent access to favourable feeding areas, with the more uniform rates of travel arising from their ability to feed throughout a greater proportion of the tidal cycle. This assumes a close association between medusae and their food supply. Given the rapid digestion rates of *C. fleckeri* (Hamner & Doubilet, 1994, Hamner et al., 1995), the high metabolic rates (Gordon, 1998), fast growth rates (Gordon et al., 2004) and rapid digestion rates (Yamaguchi, 1985) of closely related species, this appears to be a reasonable assumption.

Movement patterns of medusae were also influenced by tidally generated currents. At slower flow indexes, medusae moved both with and independently of the current. When flow index (i.e. current) increased, medusae were not only more likely to move with the current, but the rates at which they travelled also increased. Collectively, these data suggest that while medusae possess the ability to move independently of water currents, they do so only until the current reaches a critical velocity. Above this velocity, medusae appeared unable to move against the current or the energetic costs required to move against the current outweighed the benefits. While this critical flow level cannot be quantified here, future studies should focus on this aspect of medusae movement.

The potential for medusae movement patterns to vary between individuals was highlighted by the movements of a large medusa (180 mm IPD) over 26 h at the Tully site in March 2005. This medusa lacked the diurnal behaviours reported elsewhere (Seymour et al., 2004) and the rates at which it travelled were more consistent with those of an estuarine individual, despite remaining within the coastal habitat for all but the final 6 h of tracking. It remains unclear whether such movements were typical of large coastal individuals or occurred because this individual utilised both the estuarine and coastal habitats. Alternatively, medusae are thought to move upstream towards the end of the season to take part in semelparous spawning (Hartwick, 1987, 1991). The observed movements may, therefore, potentially represent an estuary selection process or the first stages of an upstream migration. Further evidence is required to validate such hypotheses.

Overall, this study has quantified the relationships between medusae movements and time of day, tidal state, tidal current and habitat. The movement of a large medusa between the coastal and estuarine habitats has also been documented. These accounts represent a base line to which future studies can compare movement patterns under a range of conditions. It is only when clear links are established between medusae movement patterns and factors such as prey availability, local weather conditions, time of season or medusae size (in particular, large individuals between habitats) that accurate predictions of medusae distribution and occurrence can be made.

**Acknowledgements** This research would not have been possible without financial support from the Lions Foundation of Australia, National Geographic, Mission Beach Tourism Commission, Australian Geographic, Cairns City Council, Cardwell Shire Council, Smart State Queensland, Weipa Houseboats and Comalco. Sincere thanks to Teresa Carrette, Robert Gordon, Anna Kintner, Glenda, Amelia and Benjamin Seymour and Ben Kelly for their assistance in fieldwork.

## References

Atkinson, L. J., S. Mayfield & A. C. Cockcroft, 2005. The potential for using acoustic tracking to monitor the movement of the West Coast rock lobster *Jasus lalandii*. African Journal of Marine Science 27: 401–408.

Barnes, J. H., 1960. Observations on jellyfish stings in north Queensland. Medical Journal of Australia 2: 993–999.

Barnes, J. H., 1965. *Chironex fleckeri* and *Chiropsalmus quadrigatus*: morphological distinctions. North Queensland Naturalist 32: 13–22.

Barnes, J. H., 1966. Studies on three venomous Cubomedusae. In The Cnidaria and Their Evolution: Symposium of the Zoological Society of London. No. 16 Academic Press, London: 307–332.

Blaber, S. J. M. & T. G. Blaber, 1980. Factors affecting the distribution of juvenile estuarine and inshore fish. Journal of Fish Biology 17: 143–162.

Blaber, S. J. M., D. T. Brewer & J. P. Salini, 1989. Species composition and biomass of fishes in different habitats or a tropical Northern Australian estuary: their occurrence in the adjoining sea and estuarine dependence. Estuarine. Coastal and Shelf Science 29: 509–531.

Blaber, S. J. M., D. T. Brewer & J. P. Salini, 1995. Fish communities and the nursery role of the shallow inshore waters of a tropical bay in the Gulf of Carpentaria, Australia. Estuarine, Coastal and Shelf Science 40: 177–193.

Brown, T., 1973. *Chironex fleckeri* – Distribution and Movements Around Magnetic Island, North Queensland. World Life Research Institute, Colton, California. ISBN 095994365X.

Cyrus, D. P. & S. J. M. Blaber, 1992. Turbidity and salinity in a Tropical Northern Australian Estuary and their influence on fish distribution. Estuarine, Coastal and Shelf Science 35: 545–563.

Egli, D. P. & R. C. Babcock, 2004. Ultrasonic tracking reveals multiple behavioural modes of snapper (*Pagrus auratus*) in a temperate no-take marine reserve. Journal of Marine Science 61: 1137–1143.

Gordon, M. R., 1998. Ecophysiology of the Tropical Australian Chirodropid *Chiropsalmus* sp. Honours Thesis, School of Tropical and Marine Biology, James Cook University of North Queensland, Australia.

Gordon, M. R., C. Hatcher & J. E. Seymour, 2004. Growth and age determination of the tropical Australian cubozoan *Chiropsalmus* sp. Hydrobiologia 530/531: 339–345.

Hamner, W. M. & D. Doubilet, 1994. Australian box jellyfish–a killer down under. National Geographic 186: 116–130.

Hamner, W. M., M. S. Jones & P. P. Hamner, 1995. Swimming, feeding, circulation and vision in the Australian box jellyfish, *Chironex fleckeri*. Marine and Freshwater Research 46: 985–990.

Hartwick, R. F., 1987. The box jellyfish. In Covachevich, J., P. Davie & J. Pearn (eds), Toxic Plants and Animals–A Guide for Australia. Queensland Museum Press, Brisbane: 99–105.

Hartwick, R. F., 1991. Distributional ecology and behaviour of the early life stages of the box-jellyfish *Chironex fleckeri*. Hydrobiologia 216/217: 181–188.

Heupel, M. R. & C. A. Simpfendorfer, 2005. Quantitative analysis of aggregation behaviour in juvenile black tip sharks. Marine Biology 147: 1239–1249.

Heupel, M. R., J. M. Semmens & A. J. Hobday, 2006. Automated acoustic tracking of aquatic animals: scales, design and deployment of listening station arrays. Marine and Freshwater Research 57: 1–13.

Jackson, G. D., R. K. O'Dor & Y. Andrade, 2005. First tests of hybrid acoustic/archival tags on squid and cuttlefish. Marine and Freshwater Research 56: 425–430.

Kerwath, S. E., A. Gotz, P. D. Cowley, W. H. H. Sauer & C. Attwood, 2005. A telemetry experiment on spotted grunter *Pomadasys commersonnii* in an African estuary. African Journal of Marine Science 27: 389–394.

Kinsey, B. E., 1986. Barnes on Box Jellyfish. James Cook University of North Queensland (Sir George Fisher Centre for Tropical Marine Studies), Townsville, Australia. ISBN 0864432003.

Marsh, L. M., S. M. Slack-Smith & D. L. Gurry, 1986. Sea Stingers – and other venomous and poisonous marine invertebrates of Western Australia. Western Australia Museum, Perth, Australia.

Omundsen, S. L., M. J. Sheaves & B. W. Molony, 2000. Temporal population dynamics of the swarming shrimp, *Acetes sibogae australis*, in a tropical near-shore system. Marine and Freshwater Research 51: 249–254.

Rifkin, J., 1996. Jellyfish mechanisms. In Williamson, J. J., P. J. Fenner, J. W. Burnett & J. Rifkin (eds), Venomous and Poisonous Marine Animals – A Medical and Biological Handbook. University of New South Wales, Sydney, Australia: 121–173.

Robertson, A. I. & N. C. Duke, 1987. Mangroves as nursery sites: comparisons of the abundance and species composition of fish and crustaceans in mangroves and other nearshore habitats in tropical Australia. Marine Biology 96: 193–205.

Robertson, A. I. & N. C. Duke, 1990. Recruitment, growth and residence time of fishes in a tropical Australian mangrove system. Estuarine, Coastal and Shelf Science 31: 723–743.

Seymour, J. E., T. J. Carrette & P. A. Sutherland, 2004. Do box jellyfish sleep at night? Medical Journal of Australia 181: 706.

Shorten, M., J. Davenport, J. E. Seymour, M. C. Cross, T. J. Carrette, G. Woodward & T. F. Cross, 2005. Kinematic analysis of swimming in Australian box jellyfish, *Chiropsalmus* sp. and *Chironex fleckeri* (Cubozoa, Cnidaria: Chirodropidae). Journal of Zoology 267: 371–380.

Southcott, R. V., 1956. Studies on Australian Cubomedusae, including a new genus apparently harmful to man. Australian Journal of Marine and Freshwater Ecology 7: 254–280.

Southcott, R. V., 1971. The box-jellies or sea-wasps. Australian Natural History 17: 123–128.

Staples, D. J. & D. J. Vance, 1987. Comparative recruitments of the banana prawn, *Penaeus merguiensis*, in five estuaries of South-eastern Gulf of Carpentaria, Australia. Australian Journal of Marine and Freshwater Research 38: 29–45.

Vance, D. J. & D. J. Staples, 1992. Catchability and sampling of three species of juvenile penaeid prawns in the Embley River, Gulf of Carpentaria, Australia. Marine Ecology Progress Series 87: 210–213.

Vance, D. J., D. S. Heales & N. R. Loneragan, 1994. Seasonal, diel and tidal variation in beam-trawl catches of juvenile grooved tiger prawns, *Penaeus semisulcatus* (Decapoda: Penaeidae), in the Embley River, North-eastern Gulf of Carpentaria, Australia. Australian Journal of Marine and Freshwater Research 45: 35–42.

Vance, D. J., M. D. E. Haywood, D. S. Heales, R. A. Kenyon, N. R. Loneragan & P. C. Pendrey, 1996. How far do prawns and fish move into mangroves? Distribution of juvenile banana prawns *Penaeus merguiensis* and fish in a tropical mangrove forest in northern Australia. Marine Ecology Progress Series 131: 115–124.

Willis, J. & A. J. Hobday, 2007. Influence of upwelling on movement of southern bluefin tuna (*Thunnus maccoyii*) in the Great Australian Bight. Marine and Freshwater Research 58: 699–708.

Wingate, R. L. & D. H. Secor, 2007. Intercept telemetry of the Hudson River Striped Bass resident contingent: migration and homing patterns. Transactions on the American Fisheries Society 136: 95–104.

Xiao, Y. & J. G. Greenwood, 1992. Distribution and behaviour of *Acetes sibogae* Hansen (Decapoda, Crustacea) in an estuary in relation to tidal and diel environmental changes. Journal of Plankton Research 14: 393–407.

Xiao, Y. & J. G. Greenwood, 1993. The biology of *Acetes* (Crustacea; Sergestidae). Oceanography and Marine Biology: An Annual Review 31: 259–444.

Yamaguchi, M., 1982. Cubozoans and their life histories. Aquabiology Tokyo 4: 248–254.

Yamaguchi, M., 1985. Occurrence of the Cubozoan medusae *Chiropsalmus quadrigatus* in the Ryukyus. Bulletin on Marine Science 37: 780–781.

Hydrobiologia (2009) 616:99–111
DOI 10.1007/s10750-008-9587-6

JELLYFISH BLOOMS

# Acoustic survey of a jellyfish-dominated ecosystem (Mljet Island, Croatia)

**G. Alvarez Colombo · A. Benović · A. Malej · D. Lučić · T. Makovec · V. Onofri · M. Acha · A. Madirolas · H. Mianzan**

Published online: 15 October 2008


**Abstract** Acoustic techniques have been proposed as a new tool to assess jellyfish populations. However, the presence of mixed echoes from jellyfish and other organisms that share their distribution often prevent accurate estimates of their abundance and distribution being obtained. The isolated population of *Aurelia* inhabiting the Veliko Jezero (Big Lake-BL) of Mljet Island, in the South Adriatic Sea, offered a good opportunity to employ acoustic techniques to assess an entire jellyfish population. During October 2–5, 2006, combined video and acoustic methods were used in BL to determine the vertical distribution of medusae. Two synoptic acoustic surveys were performed during the day and night. In the daylight echograms, medusae were clearly discernible from the acoustic data, and their presence verified by video camera images, as forming a layer of varying density at and below the thermocline (15–30 m). The depth of the jellyfish layer also coincided with the depth of maximum dissolved oxygen concentration. The echointegration of these daylight data enabled quantification the *Aurelia* population, at a frequency of 120 kHz. In the night echograms, the acoustic signals of *Aurelia* were at least partially masked by pelagic and demersal fish, which disaggregated from schools and formed a layer associated with a strong thermocline at 15 m. An average target strength (TS) of −76.4 dB was obtained in situ corresponding to a mean length of 10.8 cm and a mean wet weight of 134 g measured from sampled medusae. These results were combined with echo-integration values to provide an estimate of 4,238,602 individuals and a biomass of 568 tons of *Aurelia* in BL. This study provided a synoptic view of Mljet Lake and illustrated the potential of acoustic surveys of jellyfish populations to contribute to ecosystems studies.

Guest editors: K. A. Pitt & J. E. Purcell
Jellyfish Blooms: Causes, Consequences, and Recent Advances

G. Alvarez Colombo (✉) · M. Acha · A. Madirolas · H. Mianzan
Instituto Nacional de Investigación y Desarrollo Pesquero, Mar del Plata, Argentina
e-mail: acolombo@inidep.edu.ar

A. Benović · D. Lučić · V. Onofri
Institute for Marine and Coastal Research, University of Dubrovnik, Dubrovnik, Croatia

A. Malej · T. Makovec
Marine Biology Station, National Institute of Biology, Piran, Slovenia

M. Acha · H. Mianzan
CONICET Consejo Nacional de Investigaciones Científicas y Técnicas, Buenos Aires, Argentina

M. Acha
Facultad de Ciencias Exactas y Naturales, UNMDP, Mar del Plata, Argentina

**Keywords** Hydroacoustics · Medusae · Scyphozoa · *Aurelia* · In situ target strength · Aggregation structure · Distribution · Biomass assessment

## Introduction

Increases in jellyfish abundance and expansions in geographic range have been reported for many pelagic marine ecosystems worldwide (Mills, 2001; Kideys, 2002; Xian et al., 2005), and could be indicative of climate-induced changes and/or regime shifts (Brodeur et al., 1999; Lynam et al., 2004, 2005; Purcell & Decker, 2005). There are some reasons to suspect that jellyfish-dominated communities will be the end point in ecosystems perturbed by high fishing effort ( Parsons & Lalli, 2002; Pauly et al., 2002), as exemplified in the Benguela upwelling (Lynam et al., 2006). Enhancement of jellyfish populations may be indicative of major fundamental changes in marine ecosystems (Parsons & Lalli, 2002), and therefore, there is a requirement to estimate and to map jellyfish distributions and abundances over time and space.

Traditional surveys using plankton nets have been used to provide semi-quantitative estimates of jellyfish abundance (Brodeur et al., 1999; Purcell, 2003). However, due to the nature and distribution of these organisms, this methodology has shown numerous constraints (Arai, 1988; Mianzan & Cornelius, 1999; Alvarez Colombo et al., 2003). Several alternative methods have been suggested for estimating jellyfish distribution and abundance, including video profiles, scuba diving, and aerial and acoustic surveys (e.g., Båmstedt et al., 2003; Uye et al., 2003; Houghton et al., 2006). Jellyfish were traditionally disregarded as conspicuous sources of sound scattering because their tissues have a high water content which provides a very low-density contrast at the water–body interface (Alvarez Colombo et al., 2003). However, it is clear now that many species of gelatinous plankton are able to produce significant amounts of sound scattering at frequencies routinely employed for fisheries assessments (Toyokawa et al., 1997; Brierley et al., 2001; Mianzan et al., 2001; Alvarez Colombo et al., 2003). Information obtained from laboratory and in situ analyses of individual target strength (TS) (Mutlu, 1996; Monger et al., 1998; Brierley et al., 2004) indicates that even changes in bell shape that occur during swimming can be measured.

Acoustic surveys of jellyfish still pose several problems. Medusae co-occur with other gelatinous (e.g., other medusae, ctenophores, and salps) and non-gelatinous planktonic organisms that may also contribute to the total sound backscattered. This makes interpretation of the echogram difficult, may prevent jellyfish signals from being isolated and, may even mask the presence of medusae. The limited number of field studies attest to the difficulties that jellyfish present for acoustic assessment. During acoustic surveys on the Argentinean Continental Shelf (Alvarez Colombo et al., 2003) and in the Benguela ecosystem (Brierley et al., 2005), only a few net samples were dominated by jellyfish and these were thus used to identify and quantify jellyfish swarms from echograms. Acoustic assessment of medusae, therefore, depends on the acoustic properties of the target species and on the ability to distinguish them from other co-occurring taxa.

The genus *Aurelia* is one of the most widely distributed Scyphozoa. *Aurelia* is able to bloom recurrently and aggregates in semi-enclosed areas such as the Black Sea, the Baltic Sea, and the Seto Inland Sea, as well as other sheltered or enclosed waters such as fjords, bays, and estuaries (e.g., Möller, 1980; Hamner et al., 1994; CIESM, 2001; Uye & Shimauchi, 2005). Sometimes, cooling-water intake ducts of power stations are clogged by swarms of *Aurelia* (e.g., Möller, 1984; Verner, 1984; Rajagopal et al., 1989). Its ecological role is complex. It feeds on different components of the zooplankton, including fish larvae, and impacts on the classical copepod-dominated pelagic food web as well as on the microbial one (Möller, 1980; Båmstedt, 1990; Malej et al., 2007).

An isolated population of *Aurelia* inhabits the Veliko Jezero (BL), an oligotrophic coastal lake on the Croatian island of Mljet in the southern Adriatic Sea. The population persists throughout the year (Benović et al., 2000). Average concentrations of *Aurelia* in BL may attain 10 ind. $m^{-3}$ and up to a few hundred $m^{-3}$ may occur in swarms (Benović et al., 2000). The presence of *Aurelia* in a water body that is almost entirely enclosed provided a natural mesocosm to attempt to quantify the entire population and to establish its impact on the pelagic community.

The goals of this work were to identify the structure of the *Aurelia* aggregations in BL using a combination of video-acoustic methodologies; to analyze their horizontal and vertical distribution patterns; and to quantify the population from a synoptic acoustic survey, showing at the same time an overall picture of the possible interaction of *Aurelia* with other components of the ecosystem.

## Materials and methods

### Study area

Veliko Jezero (Big Lake-BL, to distinguish it from Malo Jezero, a smaller adjacent lake) is a submerged karstic valley situated on the north-western side of Mljet, Croatia, an offshore island in the southern Adriatic Sea that extends in a NW–SE direction (Fig. 1A). Its surface area is 1.45 $km^2$ and it has a maximum depth of 46 m. Recent descriptions of this oligotrophic environment and its biotic components can be found in Benović et al. (2000), and Malej et al. (2007).

### The population

A very important and conspicuous component of the lake is the scyphomedusa *Aurelia*. This species is present in the shallow northern Adriatic and in some semi-enclosed bays along the eastern Adriatic but it is absent in the open Adriatic waters offshore of Mljet Island (Benović, unpublished observations). It has traditionally been termed *Aurelia aurita* (Linnaeus, 1758), although molecular criteria used in recent studies have questioned this designation (Schroth et al., 2002) and the species is currently recognized as *Aurelia* sp. 5 (Dawson et al., 2005). In the present study, therefore, it is referred to as *Aurelia*.

### Sampling strategy

During October 2–5, 2006, a field study was carried out at BL using a 9 m boat. Two acoustic surveys (day and night) that encompassed the entire lake were done, to determine synoptically the distribution of the *Aurelia* population and other biological scatterers like fish (Fig. 1B).

Video profiling and stratified net sampling were also done to corroborate the presence of medusae and to analyze the composition of mesozooplankton at a fixed station (42°46′ N and 17°22′ E) during a 24 h period (Fig. 1B).

### Oceanographic and plankton sampling

An STO Hydrolab Surveyor 3 probe was employed to determine salinity (S), temperature (T), and dissolved oxygen (DO) structure at the fixed station. T, S, and DO data were used to identify strata in the water column and to set the sampling depths of the nets.

Net sampling at a fixed station was used to identify other sources of backscattering. Mesozooplankton was collected during the day by two replicated

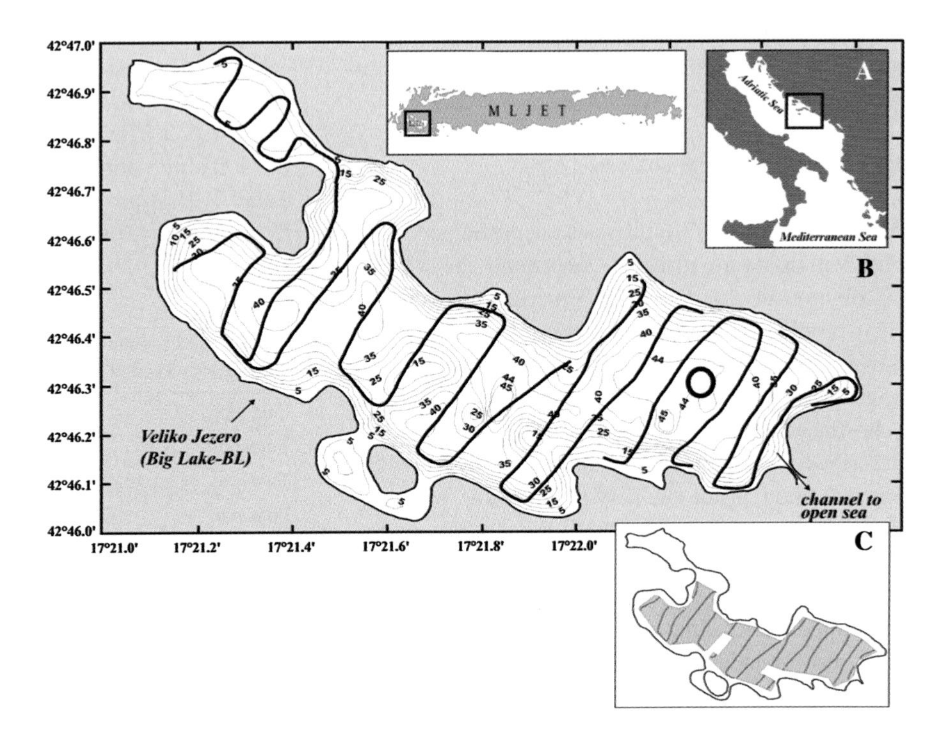


**Fig. 1** Location of Mljet Island, Croatia, in the Adriatic Sea, and the Veliko Jezero Lake (BL) at the western tip of the island (**A**). Bathymetric map of the BL (depths in m), showing the position of the fixed station (open circle) and the vessel track performed during one of the surveys (black line) (**B**). Area selected for echointegration (**C**)

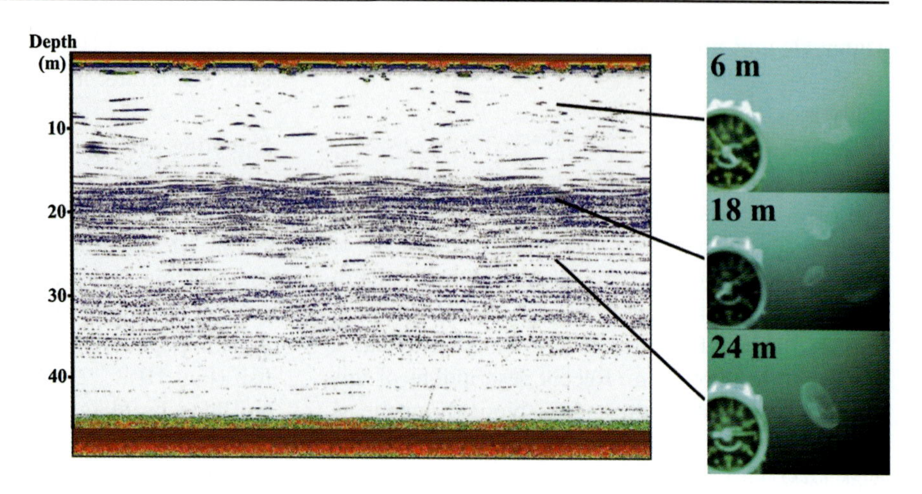


**Fig. 2** Video-acoustic verification of the distribution of *Aurelia* at the fixed station during the day. Images are provided for depths above the thermocline (6 m), at the thermocline (18 m), and below it (24 m)

vertical tows employing a Nansen net equipped with a closing device (125 μm, 54 cm diameter). Two depth strata were sampled: 0–20 m (above the thermocline) and 20–40 m (below the thermocline). Samples were preserved in a 2.5% formaldehyde-seawater solution buffered with $CaCO_3$. Zooplankton was identified, counted, and measured using an Olympus SZX 9 stereomicroscope. Individual *Aurelia* were obtained for morphometric analysis and to determine the size distribution of the population. The jellyfish were randomly collected during the net tows and also by hand by SCUBA divers at the depth the population was most concentrated. Each specimen was placed in a plastic bag to prevent any physical damage to its soft body.

## Video recordings and analysis

At the fixed station, simultaneous acoustic and video observations were made to determine the vertical distribution of *Aurelia*. Eight vertical video profiles (two profiles each during daylight, dusk, night, and dawn) were employed as a ground-truth tool to confirm the presence of medusae identified in the echo-traces. The same methodology was used as described in Malej et al. (2007). Briefly, a video camera fitted inside an acrylic case was manually deployed at very low speed until the sea bottom was visible. Artificial illumination was provided by strobe lights when necessary. Presence of medusae was monitored in real time with a laptop computer along with echosounder observations. Afterward, video recordings were carefully scrutinized in the laboratory using digital video playback and "*frame by frame*" mode to determine the vertical distribution and relative abundance of medusae (Fig. 2).

## Acoustic sampling, data processing, and analysis

Acoustic recordings were obtained by means of a SIMRAD EY500 portable echosounder and a 120 kHz, split-beam transducer (7° beam angle). Angle and power sample data were recorded, providing simultaneous 20 log R and 40 log R time-varied-gain (TVG) functions, for volume backscattering and target strength analysis, respectively. The echosounder was calibrated before sampling with a tungsten carbide calibration sphere following Foote et al. (1987). The main echosounder settings used and the software calibration parameters are presented in Table 1. Post-processing of echodata was done with Echoview v. 4.1 software.

**Table 1** Echosounder parameter settings as employed during the experience, after the system calibration

| | |
|---|---|
| Frequency | 120 kHz |
| Transmitted power | 63 W |
| Two-way beam angle (re 1 Steradian) | –20.6 dB |
| Ping rate | Automatic |
| Transmitted pulse length | 0.3 ms |
| Bandwidth | Automatic |
| Absorption coefficient | 50 dB $km^{-1}$ |
| Sound speed | 1,522 m $s^{-1}$ |
| Sv gain | 26.6 dB |

Target strength measurements

*Aurelia* acoustic target strength (TS) measurements were obtained in situ in accordance with previous acoustic research on jellyfish (see Båmstedt et al., 2003 and Kaartvedt et al., 2007). The TS measurements were made on a low-density layer of individuals, occupying the upper 15 m in the water column. Target identification was corroborated by direct observations with the underwater video camera (Fig. 3A). By using Echoview software, a virtual echogram of single targets was produced from the split-beam transducer data (sample power and sample angle data telegrams). A minimum threshold value of −85 dB was applied to this echogram to avoid inclusion of noise and other very weak unwanted echoes. All echoes located closer than 2 m and further than 15 m from the transducer face were rejected to avoid sound intensity distortions due to the transducer near field and multiple echoes,

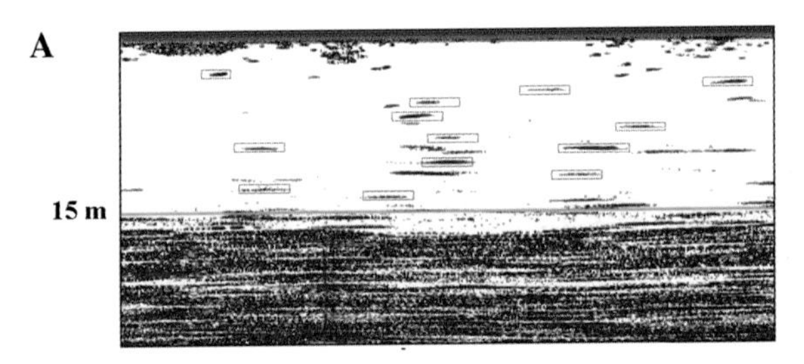


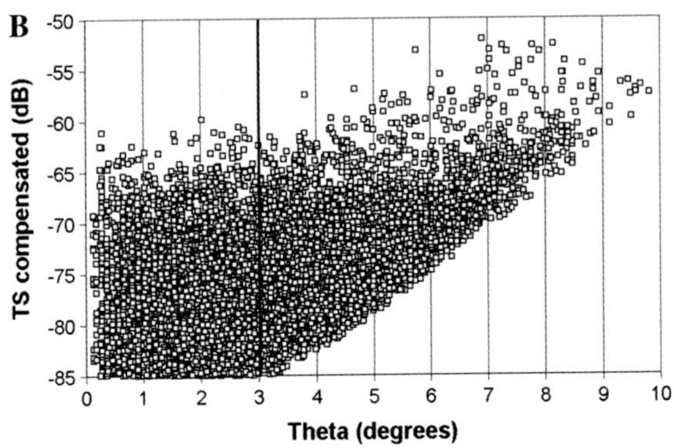


**Fig. 3** 40 log R TVG echogram showing the disaggregated distribution of medusae in the upper 15 m of the water column and examples of individual regions selected for TS analysis (**A**). Based on this echogram, a virtual, single target detection echogram was created, from which single targets were selected after being filtered by the software algorithms. Single targets plotted against the angle from the acoustic axis (theta) (**B**). As shown, only data in the first 3° from the acoustic axis were fully represented at the TS threshold of −85 dB and, therefore, considered for the estimation of the mean TS

respectively. Echotraces consisting of less than four consecutive single echoes were also rejected. Due to the directivity of acoustic transducers, the received echo intensity is maximal at the transducer beam axis and diminishes progressively when the targets are located at increasing distances, i.e., the echo intensity of the same target will decrease with increasing off-axis distance. This can be seen as a progressive reduction of the signal-to-noise ratio with the off-axis angle (*theta*) and results in an increasingly biased distribution of echoes (probability density function-pdf) toward the larger values of TS. This effect is particularly notorious when dealing with weak targets, as is usually the case for medusae. In Fig. 3B, the bias effect introduced by the off-axis angle can be clearly seen as an increase in the lower limit of the TS values as the off-axis angle increases. To avoid a positive bias on the average TS, an angle cut-off value of 3° was established for the off-axis angle according to Zhao (1996) and Ona (1999). Within this range of the off-axis angle (3°), TS appears equally represented for the entire range of values recorded and hence provides an unbiased average TS.

Echointegration surveys directed to *Aurelia* density and biomass estimates

To describe the spatial distribution of *Aurelia*, the total area of BL was surveyed both during the night and during the following morning. Each survey was completed in two hours and their design consisted of parallel transects separated by approximately 120 m, perpendicular to the lake's main axis (Fig. 1B). The relative abundance of *Aurelia* was estimated from the morning survey using the echointegration method (Foote et al., 1991). The echograms were divided into intervals of 10 pings from where the mean area backscattering coefficient $s_A$ ($m^2/nm^2$) was obtained.

*Aurelia* echoes were identified by visual examination of the echograms and from the concurrent information obtained from the video camera and plankton nets. The echointegration layer was set between 15 and 30 m depth and was limited to the central sector of the lake. The surface and bottom layers as well as the near shore waters (<20 m depth) were excluded from the analysis since these sectors contained mostly fish (Fig. 1C). Within the 15–30 m stratum, the abundance and concentration C (ind. $m^{-3}$) were derived from the

echointegration data, using the average backscattering cross section ($\sigma_{bs}$) obtained from in situ TS values.

The following formulae were employed:

$$n_{total} = [[\overline{s_A}/(4\pi \cdot \overline{\sigma}_{bs})]A]$$

$$C = \overline{s_V}/\overline{\sigma_{bs}}$$

where $n_{\text{total}}$ is the total number of medusae; $\overline{s_A}$ is the average acoustic backscattering coefficient per unit area ($m^2/nm^2$); $\overline{\sigma}_{bs}$ is the mean measured backscattering cross-section of *Aurelia* ($m^2$); $A$ is the area of the lake surveyed ($nm^2$); $C$ is medusae concentration; and $\overline{s_v}$ is the average acoustic backscattering coefficient per unit volume ($m^2/m^3$).

Echointegration data showed strong spatial structure, as is typical for most acoustic surveys of aquatic organisms. This promoted the application of Geostatistical analysis, which exploits the spatial structure of the data to evaluate, for instance, the variance of the abundance estimates, and has been applied to the analysis of acoustic survey data for the last 20 years (see Rivoirard et al., 2000 and references therein). The variogram, a common structural tool, was used for the data analysis and the geostatistical variance was obtained by using the software EVA (Petitgas & Prampart, 1993). Finally, the average wet weight obtained from the medusae collected by nets and SCUBA diving was used to estimate the medusae biomass.

## Results

### Oceanographic profiles

Vertical profiles made at the fixed station indicated that the lake was stratified (Fig. 4C). The temperature was 22.6°C at the surface and 9.5°C at the bottom and a distinct thermocline with a vertical gradient of 1.7°C $m^{-1}$ occurred between 15 and 20 m. Salinity values also showed a mild gradient starting at 19 m, with values at the surface and bottom of 37.2 and 39.2 psu, respectively. Similar concentrations of dissolved oxygen were found at surface and bottom (6.9 and 6.6 ml $O_2$ $l^{-1}$). However, higher concentrations occurred at 16 to 30 m and a maximum of

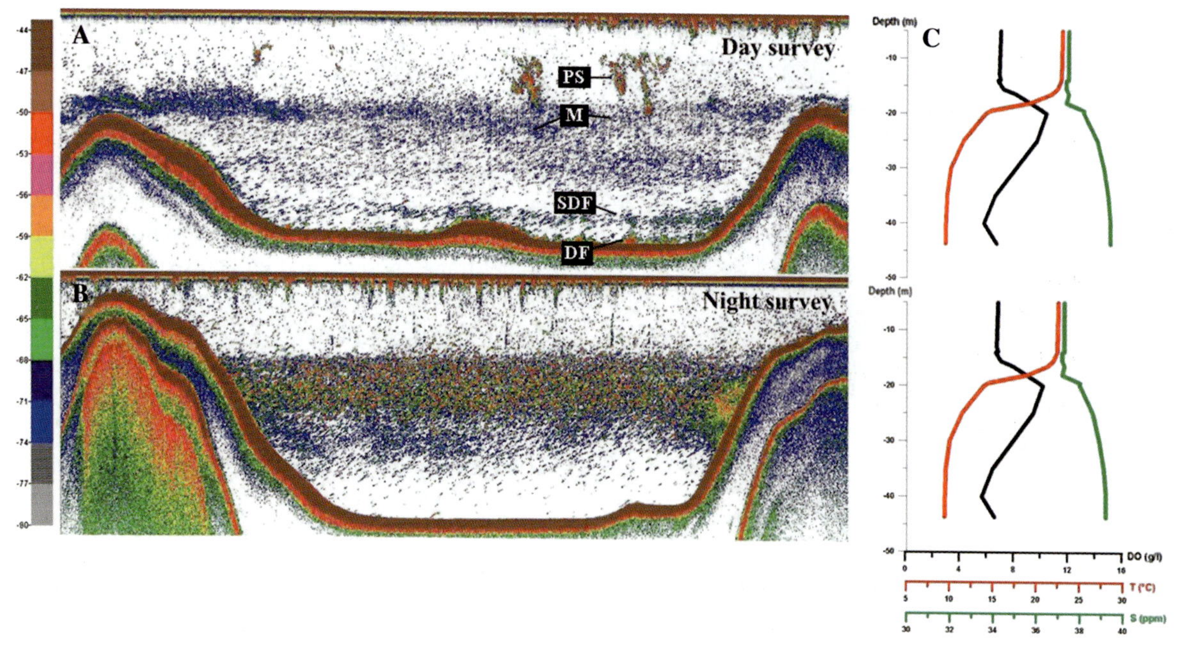


**Fig. 4** Examples of the echograms obtained from survey transects during the day (**A**) and night (**B**) at different sectors of Mljet's Veliko Jezero (BL). At the right (**C**), profiles of dissolved oxygen, temperature, and salinity are paired to the echograms to show the relation between the water column stratification and the distribution of the scatterers. During the day, a layer of medusae (M) can be observed in the 15–30 m layer. Pelagic schools (PS), demersal (DF), and semi-demersal fish layers (SDF) are also observed. During the night, fish aggregations relocated at and below the thermocline depth, masking the presence of the weaker medusae echoes

$8.6 \text{ ml } O_2 \text{ l}^{-1}$ occurred at the lower portion of the thermocline layer (20 m).

Mesozooplankton samples

In BL, copepods dominated the zooplankton fauna with concentrations ranging between 6,639 and 8,102 ind. $m^{-3}$ above thermocline and between 11,686 and 16,102 ind. $m^{-3}$ below it. Adults and copepodites of *Oithona nana* (Giesbrecht, 1892) and *Paracalanus parvus* (Claus, 1863) contributed $\geq 80\%$ of the total copepod abundance. Among other mesozooplankton groups, pteropods of the species *Limacina trochiformis* (d'Orbigny, 1836) were numerous below the thermocline.

*Aurelia* size and weight measurements

The average bell diameter (BD) of *Aurelia* was 10.8 cm (SD = 2.7; $n = 71$) and ranged from 6.3 to 18.5 cm, but most had diameters of 9 to 12 cm (Fig. 5A). The mean wet weight was 134 g (SD = 95.9; $n = 71$) with a range from 65 to 385 g individual$^{-1}$.

In situ target strength measurements

Dispersed individuals observed during the daytime in the upper part of the water column formed the basis for the *Aurelia* in situ target strength measurements. This provided a dataset of echoes recorded from isolated individuals swimming under undisturbed conditions. Hence the measurements were considered to reflect the behavioral component (swimming tilt-angle distribution) of the individual target strength at that moment. After the filtering and thresholding processes, a total of 2,874 single echoes, grouped into 141 medusa echotraces, were selected from this layer (Fig. 3A). The average of number of echoes per trace was 21. The mean TS was −76.4 dB ($\sigma_{bs}$ = 2.28 E−08 $m^2$; SD = 2.63 E−08) (Table 2).

There was close agreement between medusa size and target strength distributions (Fig. 5), with TS values ranging from −83 to −66 dB and BD ranging from 7 to 19 cm. The dominance of the smaller organisms (9–12 cm BD) corresponded to a similar

**Table 2** Acoustic, biological, and physical parameters employed for density and total biomass estimates. Calculations were made in order to estimate densities at the main concentration layer

| Volume of stratum containing medusae | |
|---|---|
| Mean $s_v$ ($m^2/m^3$) | 1.18 E−08 |
| Mean $\sigma_{bs}$ ($m^2$)/(TS = −76.4 dB) | 2.28 E−08 |
| Medusae $m^{-3}$ | 0.52 |
| $s_A$ mean | 5.1 |
| Analyzed area ($nm^2$) | 0.238 |
| *n* medusae | 4,238,602 |
| Mean wet weight (kg) | 0.134 |
| Biomass (tons) | 568 |

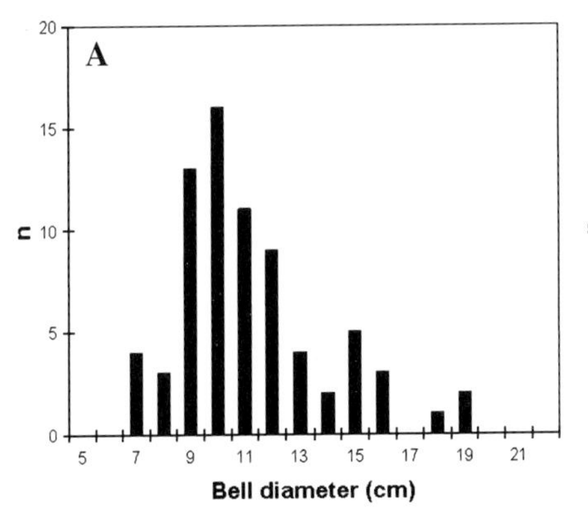


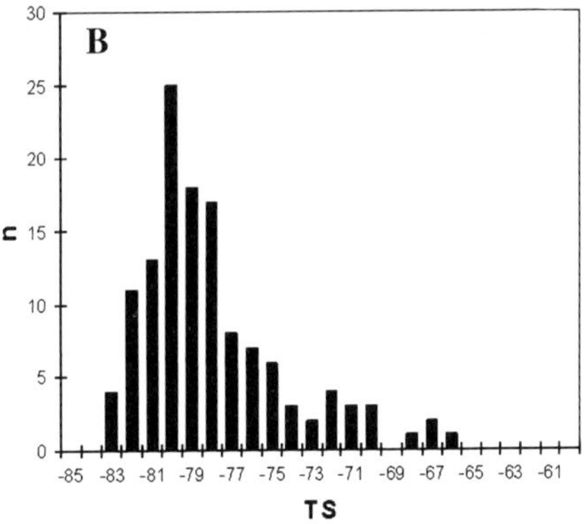


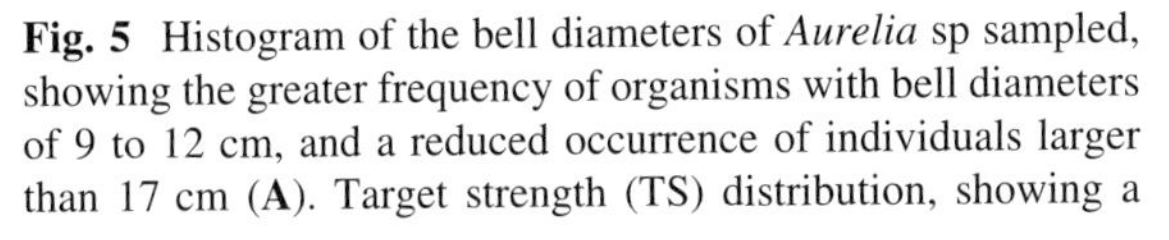

**Fig. 5** Histogram of the bell diameters of *Aurelia* sp sampled, showing the greater frequency of organisms with bell diameters of 9 to 12 cm, and a reduced occurrence of individuals larger than 17 cm (**A**). Target strength (TS) distribution, showing a similar distribution pattern to the *Aurelia* size distribution (**B**). Highest values reached a TS of ~−66 dB at the threshold of −60 dB employed here

proportion of low TS values (−80 to −78 dB). Larger medusae (15–19 cm BD) were collected in small numbers, which coincided with the scarcity of the highest TS values (−72 to −66 dB).

### Day–night surveys comparison

The day and night surveys revealed a very different distribution of organisms in the water column. During daylight, medusae were clearly discernible from the 120 kHz acoustic data, and formed a layer of varying density at and below the thermocline. The depth distribution coincided with the layer of maximum dissolved oxygen concentration (15–30 m). The vertical distribution of medusae was confirmed by observations made using video and while SCUBA diving at the fixed station. In addition, targets associated with isolated individuals of *Aurelia* were detected in the upper 15 m, and extended almost to the surface (6 m depth). Fish were observed aggregated in pelagic, demersal, and semi-demersal schools, often in large concentrations (Fig. 4A).

In contrast, during the night both demersal and pelagic fish disaggregated from schools, migrated toward the thermocline, and masked the presence of medusae. This prevented the use of the echograms to quantify their abundance (Fig. 4B).

### Acoustic estimation of *Aurelia* abundance

More than 4,200,000 individuals and 568 tons of *Aurelia* were estimated to inhabit BL (Table 2). *Aurelia* predominantly occurred in a horizontal layer at 15–30 m depth (Fig. 6A) which extended throughout the lake but there were three areas (two large and one small) where medusae were most concentrated. These areas coincided with the south-east margins of the deepest basins in the lake (>40 m) and comprised 60% of the total biomass (Fig. 6B). The mean concentration of medusae within the horizontal layer was 0.52 ind. $m^{-3}$ (Table 2). Geostatistical analysis indicated good spatial correlation of the echointegration data, as shown in the corresponding empirical variogram (Fig. 7). A theoretical variogram was fitted by choosing a spherical model with a moderate nugget effect (Fig. 7). The estimated variance (derived from the variogram model) was very small: 0.05 (C·V = 4.4%).

## Discussion

The acoustic survey, in conjunction with video observations, was an efficient method for providing, for the first time, a synoptic picture of the distribution and abundance of the population of *Aurelia* in BL. Most of the previous studies of distribution and abundance were based on discrete sampling using plankton nets and SCUBA diving (Benović and Onofri, 1995; Benović et al., 2000; Malej et al., 2007), and the spatial distribution of medusae was not well understood. In this study, it was possible to see and quantify the entire *Aurelia* population of the lake and to obtain a quasi-synoptic picture of the number of aggregations and their structure.

The shape and pattern of swarms of gelatinous organisms is not well known. Acoustic methods, however, enable a synoptic three-dimensional view of the population structure of medusae. Acoustic surveys, combined with hydrological observations, thus improve the understanding of the relationships between the way a population is distributed and its environment (GEOSPACE, 1993). *Aurelia* is known to aggregate frequently as isolated swarms, sometimes of elliptical or round shape in the vertical profile (Toyokawa et al., 1997). In BL, however, a continuous layer of *Aurelia* of varying density was found throughout the entire lake, as reflected by the very low CV obtained. In agreement with Benović et al. (2000), and Malej et al. (2007) most of the *Aurelia* population was located within and slightly below the thermocline and within the dissolved oxygen maximum layer. The association between gelatinous plankton aggregations and physical discontinuities such as thermoclines are well known (Graham et al., 2001) and usually result from a combination of animal behavior and hydrological conditions. Malej et al. (2007) observed that the *Aurelia* in BL undertook daily vertical migrations, with part of the population ascending toward the surface at dusk and descending to deeper waters below 25 m during the night. In the present study, this behavior could have been partially or totally masked in the echograms because of the overlapping of stronger targets, such as fish, during the night. However, the night survey showed the presence of weak targets as deep as 35 m, below the main fish aggregation, possibly indicating the presence of medusae (Fig. 4A, B).

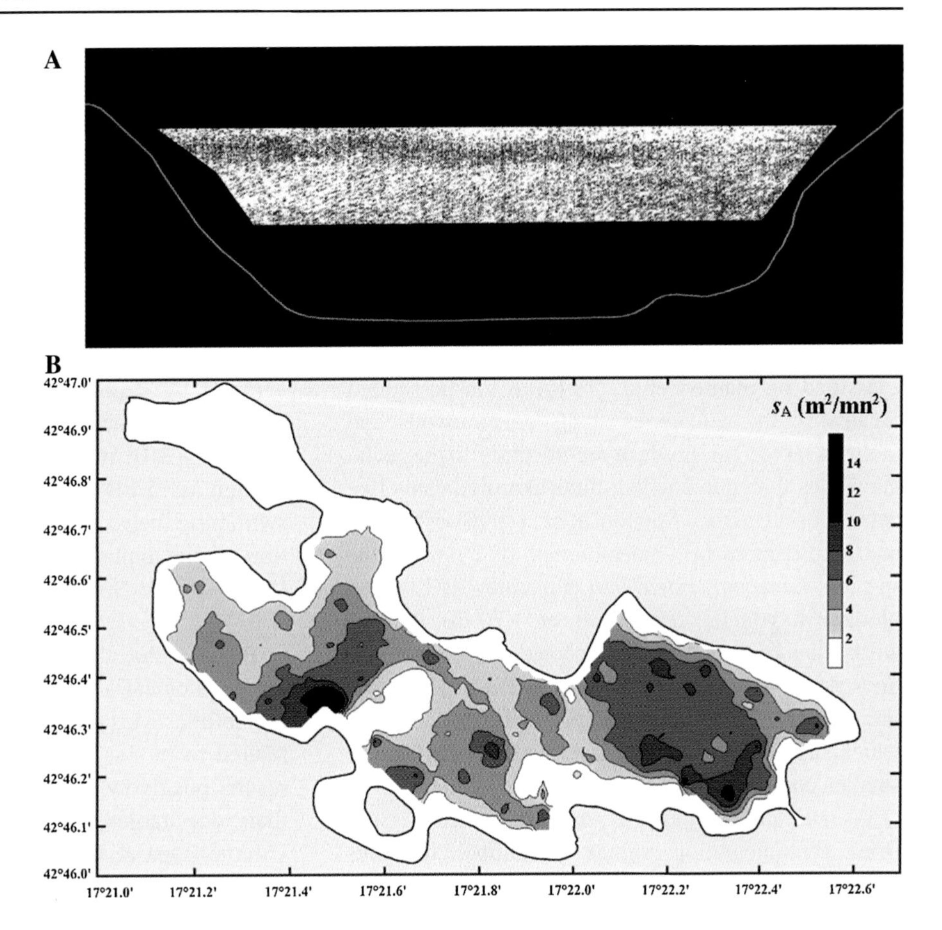


**Fig. 6** Density estimates of *Aurelia* population. Depth stratum (15–30 m) considered for echointegration, showing the main concentration of medusae (**A**). Distribution of the densities obtained in terms of area backscattering strength $s_A$ (m$^2$/nm$^2$) (**B**). White areas represent sectors with depths <20 m, including an island and some reef close it

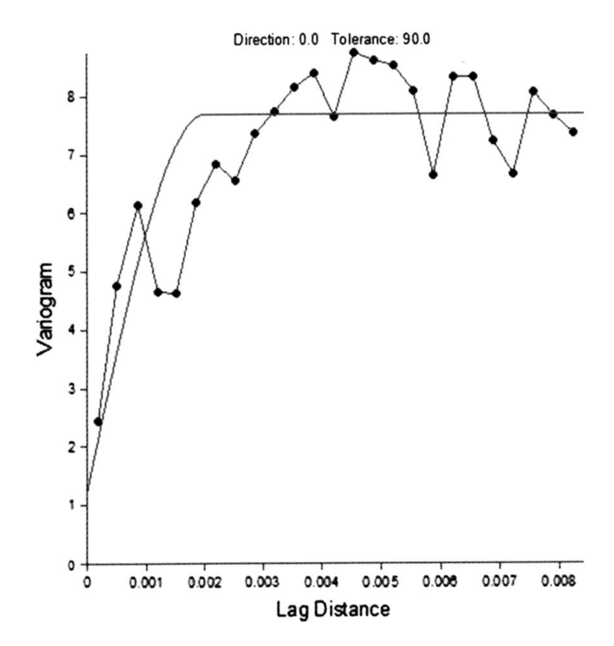


**Fig. 7** Empirical and model fitted variograms of the echointegration data ($s_A$)

There are very few examples of acoustic estimations of total biomass of gelatinous organisms. Lynam et al. (2006) estimated a biomass of almost 12,200,000 tons of *Aequorea forskalea* (Forskål, 1775) in the waters off Namibia, with a mean density of 361.9 tons nm$^{-2}$ for an area of 33,710 nm$^2$. However, the majority of the biomass of *A. forskalea* occurred at few locations where densities were at least an order of magnitude greater than the mean, and similar to the values obtained for this study (2,386 tons nm$^{-2}$). Ishii & Båmstedt (1998) noted that medusae from semi-enclosed/enclosed areas (like BL) have higher population concentrations compared to open areas (like Namibian waters). They attributed this trend to semi-enclosed/enclosed waters retaining planulae, providing suitable depths for settlement of larvae, and sustaining greater densities of polyps. The polyps, in turn, can produce a greater population of ephyrae in the following season.

Some possible sources of error or bias in the abundance estimates must be considered. First, to exclude strong fish echoes, the surface and bottom layers were excluded from the analysis. Those areas, however, possibly contained some medusae which may not have been included in the abundance estimate. On the other hand, however, the presence of some planktonic groups like planktonic mollusks could account for some degree of acoustic backscattering which may have been mistaken for *Aurelia*. Classified by Stanton et al. (1994) as elastic-shelled organisms, the calcareous shells of pteropods and mollusk larvae can produce significantly higher echo intensities than non-shelled, fluid-like organisms like gelatinous plankton. Stanton et al. (1996) estimated that 83 pteropods $m^{-3}$ (mean length of 2 mm) of the species *Limacina retroversa* (Fleming, 1823) can produce a volume echo level of −70 dB. In this work, *Limacina trochiformis* occurred at concentrations of 50 $m^{-3}$ below the thermocline. However, they were much smaller (mean length of 0.33 mm), which suggests they would have made a significantly smaller contribution to the total sound scattering.

As with any other species, the accurate conversion from echointegration values to animal densities depends upon the appropriate estimate of the individual target strength (TS) (MacLennan & Simmonds, 1992). Reliable in situ TS data are not obtained in most studies since the necessary conditions, such as the sufficient dispersion of scatterers in the sampled volume or the absence of non-target organisms, are rarely met (see Ona & Barange, 1999). In some cases, even a multifrequency approach has been necessary to ensure the isolation of single targets (Brierley et al., 2004). Only in very few studies, spatially diluted populations and the paucity of interfering targets made environments ideal for in situ measurements (Båmstedt et al., 2003). A strength of the current study, therefore, was that concurrent in situ measurements of TS were obtained. This was made possible because individuals were sufficiently dispersed and during the day other potential targets, such as fish, did not co-occur with the medusae. The advantage of obtaining in situ TS measurements was that they took into account the actual sizes and tilt-angle distributions of the population being studied.

Reported data about TS measurements seem to be highly variable due to biological factors such as medusae size and shape, body orientation, and behavior. By using frequencies ranging from 38 to 200 kHz, a TS range of −64.4 to −50.1 dB has been reported for *Aurelia* of sizes from 9.5 to 17 cm bell diameter (Mutlu, 1996; Toyokawa et al., 1997). Brierley et al. (2004) statistically inferred a TS of −68 dB from in situ measurements of the comparatively much larger medusa, *Chrysaora hysoscella* (Linnaeus, 1766), with an umbrella diameter of 41 cm. Kaartvedt et al. (2007) found TS values of −75 dB by deploying a 120 kHz transducer close to an aggregation of *Periphylla periphylla* (Péron & Lesueur, 1810) in a Norwegian fjord. TS variations as high as 15 dB were also found to originate in the swimming behavior, mainly the rhythmic contraction of the umbrella, of different medusae species (Mutlu, 1996; Brierley et al., 2004; Kaartvedt et al., 2007).

Besides the above-mentioned biological factors, methodological issues could also account for such variability. Most methodological sources of error related to in situ TS measurements tend to bias the results positively. Multiple targets occur when more than one target occupies the acoustic resolution volume (Ona & Barange, 1999). Under low signal-to-noise conditions, as is the case of weak targets like jellyfish, target position-related bias and also threshold-related bias also will lead to an overestimation of the average TS (Fleischman and Burwen, 2000). In this study, the effect of the off-axis angle on the lower values of TS was evident beyond 3° from the acoustic axis of the transducer (Fig. 3B). A cut-off angle of 3°, therefore, was applied to our measurements and the average TS value obtained (−76.4 dB) resulted in a lower TS than any previously reported measurements for *Aurelia*.

The acoustic survey also provided information about the patterns of distribution of other organisms, such as fish, in BL. Abundant schools of pelagic, demersal, and semi-demersal fishes were observed during day and night. Although fish and medusae were spatially separated during the day, their distributions overlapped during the night. Malej et al. (2007) stated that no obvious predators of *Aurelia* were identified at BL. However, although fish-medusae interactions are not known for this area, a damaged *Aurelia* was observed to be quickly consumed by several fishes within a few minutes (Alvarez Colombo, pers. obs.).

Further research is necessary to assess the impact of the population of *Aurelia* on the trophic ecology of BL. In this sense, the estimate of the total *Aurelia* abundance at BL presented in this work provides a reference value. *Aurelia* is capable of consuming prey of various sizes, ranging from microzooplankton to fish larvae (Möller, 1980; Stoecker et al., 1987; Sullivan et al., 1994; Arai, 2001; Malej et al., 2007). In BL, copepods averaged about 4,000 ind. $m^{-3}$ throughout the year with peaks in September of more than 15,000 ind. $m^{-3}$ (Benović et al., 2000). From enclosure studies, Malej et al. (2007) concluded that *Aurelia* fed on mesozooplankton (mainly *Paracalanus parvus* and *Oithona nana*) and microzooplankton (naked ciliates and copepod nauplii). The estimated *Aurelia* population of more than 4,200,000 of individuals represents the situation during the beginning of autumn when the highest zooplankton values were reported for the lake (Benović et al., 2000). In this study, concentrations of copepods were similar to the maximum concentrations reported by Benović et al. (2000). Small copepods, mainly *Paracalanus parvus* and *Oithona nana*, were the most common species. It is possible that the abundance of *Aurelia* follows the seasonal trends in zooplankton abundances.

This study provided a synoptic view of a Mljet lake and illustrated the potential of acoustic surveys of jellyfish populations to contribute to the understanding of the functioning of particular ecosystems.

**Acknowledgments** This work was supported by the Slovenia–Argentina Cooperation in Science and Technology, Project ES/PA05B02 P1/023, INIDEP, CONICET PIP 5009, IAI CRN 2076, and FONCyT PICT 1553 to HM (Argentina), Research program P1-0273 to AM (Slovenia), and the Ministry of Science Education and Sports of the Republic of Croatia, Research Project No.: 275-0982705-3047 to AB. Special thanks to Zeljko Bace and Marko Zaric for their highly valuable help on field activities at Mljet. This is INIDEP contribution n° 1525.

## References

Alvarez Colombo, G., H. Mianzan & A. Madirolas, 2003. Acoustic characterization of gelatinous plankton aggregations: four study cases from the Argentine continental shelf. ICES Journal of Marine Science 60: 650–657.

Arai, M. N., 1988. Interaction of fish and pelagic coelenterates. Canadian Journal of Zoology 66: 1913–1927.

Arai, M. N., 2001. Pelagic coelenterates and eutrophication: a review. Hydrobiología 451: 69–87.

Båmstedt, U., 1990. Trophodynamics of the scyphomedusae *Aurelia aurita*. Predation rate in relation to abundance, size and type of prey organism. Journal of Plankton Research 12: 215–229.

Båmstedt, U., S. Kaartvedt & M. Youngbluth, 2003. An evaluation of acoustic and video methods to estimate the abundance and vertical distribution of jellyfish. Journal of Plankton Research 25: 1307–1318.

Benović, A., D. Lučić, V. Onofri, M. Peharda, M. Carić, N. Jasprica & S. Bobanović- Ćolić, 2000. Ecological characteristics of the Mljet Island Sea water lakes (Southern Adriatic Sea) with special reference to their resident populations of medusae. Scientia Marina 64: 197–206.

Benović, A., & V. Onofri, 1995. Ecological peculiarity of the Big lake on the Island of Mljet. In Durbešić, P. & A. Benović (eds), Simpozij Prirodne značajke i društvena valorizacija otoka Mljeta, Pomena, Hrvatska, Zagreb 4–10/09/1995: 511–521.

Brierley, A. S., B. E. Axelsen, D. C. Boyer, C. P. Lynam, C. A. Didcock, H. J. Boyer, C. A. J. Sparks, J. E. Purcell & M. J. Gibbons, 2004. Single-target echo detections of jellyfish. ICES Journal of Marine Science 61: 383–393.

Brierley, A., D. Boyer, B. Axelsen, C. Lynam, C. Sparks, H. Boyer & M. Gibbons, 2005. Towards the acoustic estimation of jellyfish abundance. Marine Ecology Progress Series 295: 105–111.

Brierley, A. S., B. E. Axelsen, E. Buecher, C. Sparks, H. Boyer & M. J. Gibbons, 2001. Acoustic observations of jellyfish in the Namibian Benguela. Marine Ecology Progress Series 210: 55–66.

Brodeur, R. D., C. E. Mills, J. E. Overland, G. E. Walters & J. D. Schumacher, 1999. Evidence for a substantial increase in gelatinous zooplankton in the Bering Sea, with possible link to climate change. Fisheries Oceanography 8: 296–306.

CIESM, 2001. Gelatinous zooplankton outbreaks: theory and practice, CIESM Workshop Series 14, 104 pp.

Dawson, M., A. S. Gupta & M. H. England, 2005. Coupled biophysical global ocean model and molecular genetic analyses identify multiple introductions of cryptogenic species. Proceedings of the National Academy of Sciences 102: 11968–11973.

Fleischman, S. J. & D. L. Burwen, 2000. Correcting for position-related bias in estimates of the acoustic backscattering cross-section. Aquatic Living Resources 13: 283–290.

Foote, K. G., H. Petter Knudsen & R. J. Korneliussen, 1991. Post-processing system for echo sounder data. Journal of the Acoustical Society of America 90: 37–47.

Foote, K. G., H. P. Knudsen, G. Vestnes, D. N. MacLennan & E. J. Simmonds, 1987. Calibration of acoustic instruments for fish density estimation: a practical guide. ICES (International Council for the Exploration of the Sea) Cooperative Research Report N° 144.

GEOSPACE Group, 1993. The spacial organization of aquatic populations as observed using hydroacoustic methods (foreword). Aquatic Living Resources 6: 171–174.

Graham, W. M., F. Pages & W. M. Hamner, 2001. A physical context for gelatinous zooplankton aggregations: a review. Hydrobiologia 451: 199–212.

Hamner, W. M., P. P. Hamner & S. W. Strand, 1994. Sun-compass migration by *Aurelia aurita* (Syphozoa):

population retention and reproduction in Saanich Inlet, British Columbia. Marine Biology 119: 347–356.

Houghton, J. D. R., T. K. Doyle, J. Davenport & G. C. Hays, 2006. Developing a simple, rapid method for identifying and monitoring jellyfish aggregations from the air. Marine Ecology Progress Series 314: 159–170.

Ishii, H. & U. Båmstedt, 1998. Food regulation of growth and maturation in a natural population of *Aurelia aurita* (L.). Journal of Plankton Research 20: 805–816.

Kaartvedt, S., T. A. Klevjer, T. Torgersen, T. A. Sornes & A. Rostad, 2007. Diel vertical migration of individual jellyfish (*Periphylla periphylla*). Limnology and Oceanography 52: 975–983.

Kideys, A., 2002. Fall and rise of the Black Sea Ecosystem. Science 297: 1482–1483.

Lynam, C. P., A. S. Brierley & S. J. Hay, 2005. Jellyfish abundance and climate variation: contrasting responses in oceanographically distinct regions of the North Sea, and possible implications for fisheries. Journal of the Marine Biological Association of the United Kingdom 85: 435–450.

Lynam, C. P., M. Gibbons, B. Axelsen, C. A. J. Sparks, J. Coetzee, B. G. Heywood & A. S. Brierley, 2006. Jellyfish overtake fish in a heavily fished ecosystem. Current Biology 16: 492–493.

Lynam, C. P., S. J. Hay & A. S. Brierley, 2004. Interannual variability in abundance of North Sea jellyfish and links to the North Atlantic oscillation. Limnology and Oceanography 49: 637–643.

MacLennan, D. N. & E. J. Simmonds, 1992. Fisheries Acoustics. Chapman and Hall, London.

Malej, A., V. Turk, D. Lučić & A. Benović, 2007. Direct and indirect trophic interactions of *Aurelia* sp. (Scyphozoa) in a stratified marine environment (Mljet Lakes, Adriatic Sea). Marine Biology 151: 827–841.

Mianzan, H. & P. F. S. Cornelius, 1999. Cubomedusae and Scyphomedusae. In Boltovskoy, D. (ed.), South Atlantic zooplankton. Backhuys Publishers, Leiden, The Netherlands: 513–559.

Mianzan, H., M. Pájaro, G. Alvarez Colombo & A. Madirolas, 2001. Feeding on survival-food: gelatinous plankton as a source of food for anchovies. Hydrobiologia 451: 45–53.

Mills, C. E., 2001. Jellyfish blooms: are populations increasing globally in response to changing ocean conditions? Hydrobiologia 451: 55–68.

Möller, H., 1980. Population dynamics of *Aurelia aurita* medusae in Kiel Bight, Germany (FRG). Marine Biology 60: 123–128.

Möller, H., 1984. Some speculations on possibilities of controlling jellyfish blooms. Proceedings Workshop on Jellyfish Blooms in the Mediterranean, UNEP, Athens, Greece, October 31st–November 4th, 1983: 211–215.

Monger, B. C., S. Chinniah-Chandy, E. Meir, S. Billings, C. H. Greene & P. H. Wiebe, 1998. Sound scattering by the gelatinous zooplankter's *Aequorea victoria* and *Pleurobrachia bachei*. Deep-Sea Research II 45: 1255–1271.

Mutlu, E., 1996. Target strength of the common jellyfish (*Aurelia aurita*): a preliminary experimental study with a dual beam acoustic system. ICES Journal of Marine Science 53: 309–311.

Ona, E., 1999. Methodology for target strength measurements. ICES Cooperative Research Report, No. 235 (ISSN 1017-6195), 59 pp.

Ona, E., & M. Barange, 1999. Single-target recognition. In Ona, E. (ed.), Methodology for Target Strength Measurements (with special reference to in situ techniques for fish and micro-nekton). ICES Cooperative Research Report, No. 235: 28–43 pp.

Parsons, T. R. & C. M. Lalli, 2002. Jellyfish population explosions: revisiting a hypothesis of possible causes. La Mer 40: 111–121.

Pauly, D., V. Christensen, S. Guenette, T. J. Pitcher, U. R. Sumaila, C. J. Walters, R. Watson & D. Zeller, 2002. Towards sustainability in world fisheries. Nature 418: 689–695.

Petitgas, P., & A. Prampart, 1993. EVA (*Estimation Variance*). A Geostatistical software on IBM-PC for structure characterization and variance computation. ICES CM 1993/D: 65.

Purcell, J. E., 2003. Predation on zooplankton by large jellyfish (*Aurelia labiata*, *Cyanea capillata*, *Aequorea aequorea*) in Prince William Sound, Alaska. Marine Ecology Progress Series 246: 137–152.

Purcell, J. E. & M. B. Decker, 2005. Effects of climate on relative predation by ctenophores and scyphomedusae on copepods in Chesapeake Bay during 1987–2000. Limnology and Oceanography 50: 376–387.

Rajagopal, S., K. V. K. Nair & J. Azariah, 1989. Some observations on the problem of jellyfish ingress in a power station cooling system at Kalpakkam, east coast of India, Mahasagar. Quarterly Journal in Oceanography, National Institute of Oceanography, Goa, India 22: 151–158.

Rivoirard, J., J. Simmonds, K. G. Foote, P. Fernandes & N. Bez, 2000. Geostatistics for estimating fish abundance. Blackwell Science Ltd., Oxford.

Schroth, W., G. Jarms, B. Streit & B. Schierwater, 2002. Speciation and phylogeography in the cosmopolitan moon jelly, *Aurelia* sp. BioMed Central Evolutionary Biology 2: 1–10.

Stanton, T. K., D. Chu & P. H. Wiebe, 1996. Acoustic scattering characteristics of several zooplankton groups. ICES Journal of Marine Science 53: 289–295.

Stanton, T. K., P. H. Wiebe, D. Chu, M. Benfield, L. Scanlon, L. Martin & R. L. Eastwood, 1994. On acoustic estimates of zooplankton biomass. ICES Journal of Marine Science 51: 505–512.

Stoecker, D. K., A. E. Michaels & L. H. Davies, 1987. Grazing by the jellyfish, *Aurelia aurita*, on microzooplankton. Journal of Plankton Research 9: 901–915.

Sullivan, B. K., J. R. García & G. Klein-MacPhee, 1994. Prey selection by the scyphomedusan predator *Aurelia aurita*. Marine Biology 121: 335–341.

Toyokawa, M., T. Inagaki & M. Terazaki, 1997. Distribution of *Aurelia aurita* (Linnaeus, 1758) in Tokyo Bay, observations with echosounder and plankton net. In Den Hartog, J. C. (ed.), Proceedings of the 6th International Conference on Coelenterate Biology, 1995. National Naturhistorisch Museum, Leiden: 483–490.

Uye, S., N. Fujii & H. Takeoka, 2003. Unusual aggregations of the scyphomedusa *Aurelia aurita* in coastal waters along western Shikoku, Japan. Plankton Biology and Ecology 50: 17–21.

Uye, S. & H. Shimauchi, 2005. Population biomass, feeding, respiration and growth rates, and carbon budget of the syphomedusa *Aurelia aurita* in the Inland Sea of Japan. Journal of Plankton Research 27: 237–248.

Verner, B., 1984. Jellyfish flotation by means of bubble barriers to prevent blockage of cooling water supply and a proposal for a semi-mechanical barrier to protect bathing beaches from jellyfish. Proceedings Workshop jellyfish blooms, Athens, Greece, October 31st–November 4th, 1983: 1–10.

Xian, W., B. Kang & R. Liu, 2005. Jellyfish blooms in the Yangtze Estuary. Science 307: 41c.

Zhao, X., 1996. Target strength of herring (*Clupea harengus* L.) measured by the split-beam tracking method. M.Phil. thesis, Department of Fisheries and Marine Biology, University of Bergen: 103 pp.

Hydrobiologia (2009) 616:113–118
DOI 10.1007/s10750-008-9592-9



# Stock enhancement of the edible jellyfish (*Rhopilema esculentum* Kishinouye) in Liaodong Bay, China: a review

**Jing Dong · Lian-xin Jiang · Ke-fei Tan · Hai-ying Liu · Jennifer E. Purcell · Pei-jun Li · Chang-chen Ye**

Published online: 22 September 2008


**Abstract** Among the edible species, jellyfish *Rhopilema esculentum*, is one of the most abundant and important fishery species in China. The jellyfish fishery is characterized by considerable fluctuations in catch and a very short fishing season. In this article, we first review the research results on the biology of *R. esculentum*, which previously were published in Chinese, as related to the jellyfish enhancement and fishery. Next, we review results from enhancement experiments conducted from 1984 to 2004, with the aims of stabilizing and increasing catch. During 2005 and 2006, stock enhancement of *R. esculentum* was carried out on a large scale for the first time in Liaodong Bay, China, where 414 million juvenile jellyfish (umbrella diameter > 1 cm) were released. We present results of these enhancements, including the survey methods, catch prediction, enhancement assessment, and fishery management. In 2005 and 2006, the recapture rate of released jellyfish was 3.0 and 3.2%, respectively. The fishermen earned ¥ 159 million during the 2 years. The ratio of the input (cost of culturing juvenile jellyfish) to the output (value of the sales) was about 1:18. The high commercial value of *R. esculentum* enhancement in Liaodong Bay makes this a very successful enterprise.

**Keywords** Fisheries · Aquaculture · Management

Guest editors: K. A. Pitt & J. E. Purcell
Jellyfish Blooms: Causes, Consequences, and Recent Advances

J. Dong (✉) · L.-x. Jiang · K.-f. Tan · P.-j. Li · C.-c. Ye
Liaoning Ocean and Fisheries Science Research Institute, Liaoning Key Lab of Marine Fishery Molecular Biology, Heishijiao Street 50, Dalian 116023, China
e-mail: dj660228@mail.dlptt.ln.cn

H.-y. Liu
School of Marine Engineering, Dalian Fisheries University, Dalian 116024, China

J. E. Purcell
Western Washington University, Shannon Point Marine Center, 1900 Shannon Point Rd, Anacortes, WA 98221, USA

## Introduction

Large edible jellyfish, *Rhopilema esculentum* (Kishinouye, 1891) in the order Rhizostomeae are popular food in Chinese cooking (Hsieh et al., 2001). They are also used for medicinal purposes (Omori & Nakano, 2001), such as treatment of high blood pressure and bronchitis. They have been commercially exploited along the coasts of China for more than 1,000 years (Morikawa, 1984; Hsieh et al., 2001; Omori & Nakano, 2001). For these reasons, *R. esculentum* was selected as the species to be cultured and released for commercial harvest (Ye et al., 2006).

Although *R. esculentum* is of great importance as a commodity, only a little was known about the biology and fishery of the edible jellyfish, particularly in Southeast Asia where scientific studies have lagged behind the rapid development of exploitation (Omori & Nakano, 2001). In Liaoning Ocean and Fisheries Science Research Institute, the centre for fishery biological research of *R. esculentum* in China, a series of studies were performed over last 20 years to promote commercial development. The research on *R. esculentum* covers life cycle, reproductive biology, response to environmental factors, and natural ecology. In addition, the technology of artificial breeding, pond culture, and stock enhancement in nature were further developed along the coastal waters of northern China. We published more than 30 papers about *R. esculentum*; however, due to the language barrier and the poor communication with developed countries, almost all articles were published in Chinese.

Liaodong Bay is one of the most important jellyfish fishing grounds in China. The fishery is characterized by considerable fluctuations in annual catch, varying from about 10,000 tons to over 100,000 tons, and very short fishing season. Our earliest enhancement experiment began in 1984 for the purposes of stabilizing and increasing catches. From 1984 to 2004, 11 tentative stock enhancements were conducted (Chen et al., 1994; Wang & Yu, 2002; Li et al., 2004). In this article, the stock enhancement of edible jellyfish, especially the large scale stock enhancement of jellyfish during 2005–2006, in Liaodong Bay is discussed. The survey method, catch prediction, enhancement assessment, and fishery management are described accordingly.

## Biology and culture conditions of *Rhopilema esculentum*

Research on the life cycle of *R. esculentum* has been conducted since the 1970s (Ding & Chen, 1981). Its morphology and structure from the fertilized-eggs to the ephyrae stages were described for the first time. The medusae are dioecious. Fertilization and embryogenesis occur in open seawater. In experimental conditions, actively swimming planulae appear 7 h after fertilization. Most of the planulae metamorphose into scyphistomae with four tentacles in 3–4 days and have 16 tentacles in 15–20 days. During and after this growth, the scyphistomae continuously form podocysts. Strobilation occurs at 18–20°C 2 months later. A strobila generally produces 6–10 ephyrae. The ephyrae grow to about 20 mm in diameter in about 15 days in the laboratory, and may attain 50 mm in diameter within 30 days. In Liaodong Bay, ephyrae grow into mature jellyfish 250–450 mm in diameter in 2–3 months.

The effects of physical factors (i.e., temperature, salinity, light, and food) on different development stages of *R. esculentum* have been studied (Chen & Ding, 1983; Chen et al., 1984; Chen, 1985; Guo et al., 1987; Lu et al., 1989, 1995, 1997; Guo, 1990). (1) Light is a very important factor for planula metamorphosis, podocyst excystment, and strobilation. Suitable weak light can stimulate planula metamorphosis, while dark conditions promote podocyst excystment (Lu et al., 1997), and complete darkness prolongs or prevents strobilation. The survival rate of polyps decreases with increasing light intensity during the low temperature period before strobilation, and the survival rate is high in the dark (Chen et al., 1984). (2) Temperature also has several effects. The podocysts do not excyst below 10°C and the excystment rate increases from between 15 and 30°C (Lu et al., 1997). Increasing temperature from 2–10°C to 22°C in winter induces strobilation in 2 weeks. A scyphistoma produces 7–8 ephyrae, on average (Chen et al., 1983). The optimal growth temperature for ephyrae is 24°C, with a favorable range of 16–28°C (Lu et al., 1995). (3) Salinity also is important. No podocysts are produced when salinity is less than 6‰; the optimum salinity range for podocyst generation is 20–22‰ (Lu et al., 1997). The lower limits for survival are 12‰ salinity for planulae, 10‰ for scyphistomae (optimal range 14–20‰), 8‰ for ephyrae (optimal range 14–20‰), and 8‰ for young medusae (Lu et al., 1989). (4) Food also affected development. Planulae of *R. esculentum* is the most favorable food for its early scyphistomae, while trochophores of shellfish (*Crassostrea gigas* Thunberg) and blastulae of sea urchins (*Hemicentrotus pulcherrimus* Agassiz) are food supplements for early scyphistomae (Guo et al., 1987). Fully-developed scyphistomae, ephyrae, and young medusae can be fed with *Artemia* spp. nauplii and zooplankton (Guo, 1990).

From the above results, we concluded that abundant juvenile jellyfish can be obtained in a very short time by adjusting physical factors (light, temperature, salinity, and food) in jellyfish culture, and the techniques for artificial breeding were developed. In 2005, the total volume of artificial breeding tanks for juvenile *R. esculentum* was about 30,000–40,000 $m^3$ in Liaoning Province. In recent years, over 500 million juvenile jellyfish were cultivated in breeding centers there annually. Consistent and efficient production of juvenile jellyfish enabled the large scale stock enhancement of jellyfish.

## Stock enhancement history of *Rhopilema esculentum*

### Experimental release of cultured jellyfish

The earliest stock enhancement experiment was conducted by Liaoning Ocean and Fisheries Science Research Institute. Tentative stock enhancement of *Rhopilema esculentum* was conducted 11 times (1984–2004) with the aim of stabilizing and increasing catches. During 1984–1986, $2.0 \times 10^5$, $5.0 \times 10^5$, and $2.1 \times 10^5$ ephyrae (umbrella diameter 5–15 mm) were released into Heishijiao Bay (near Dalian, China) in the northern Yellow Sea from June to July each year. Monitoring of the numbers of medusae estimated their densities as 155, 189, and 113 medusae $km^{-2}$ within their assumed dispersal range of 33 $km^2$ between the end of August and the middle of October in 1984–1986. Thus, the recapture rate was estimated to be 1.2–2.5% (Chen et al., 1994).

During 1988–1993, large-scale release experiments of jellyfish (*R. esculentum*) were conducted in the coastal waters near the Dayang River estuary on the northern Yellow Sea. The numbers of ephyrae (umbrella diameter 5–10 mm) released ranged from $4.6 \times 10^6$ to $1.73 \times 10^7$ between May and June each year. Based on the catch in August in the main producing area (north of 39°40′N, 123°15′–123°50′E, with an approximate coverage area of 550 $km^2$), the density ranged from 91 to 682 medusae $km^{-2}$ among years. Thus, the annual recapture rate ranged from 0.07 to 1.02% (Chen et al., 1994).

Release experiments of juvenile *R. esculentum* (umbrella diameter 20 mm) were conducted under the guidance of the Enhancement Management Committee of Haiyangdao Fishing Ground. In 2002, $1.2 \times 10^6$ juveniles were released into Jinpu Bay on the northern Yellow Sea, where the recapture rate was estimated to be 1.2%. In 2004, $5.3 \times 10^6$ juveniles were released in the coastal waters near the Dayang River estuary on the northern Yellow Sea in. The fisheries catch of jellyfish was 79 *t* during the fishing season, with each individual averaging 7 kg wet weight. The recapture rate was about 0.2%.

### Large-scale release of cultured jellyfish

During 2005–2006, the first large-scale release and enhancement of *R. esculentum* was conducted by the Ocean and Fisheries Bureau of Liaoning Province in Liaodong Bay, where 416 million juvenile jellyfish (umbrella diameter > 1 cm) were released. Large numbers of ephyrae, produced by polyps the previous year, were raised in temperature-controlled water to juvenile jellyfish (umbrella diameter > 10 mm) in more than 20 breeding centers. Then, they were transferred into plastic bags with oxygen-saturated sea water and released into natural waters of 3–5 m depth. Young *R. esculentum* jellyfish were released along the northern coast of Liaodong Bay near Huludao, Jinzhou, Panjin, Yingkou, and Wafangdian cities (Fig. 1; Table 1). Jellyfish were released 10 June–25 June (except for Jinzhou in 2005) when they had an umbrella diameter of 10 mm, similar to those in nature. The released juveniles immediately mixed with the natural jellyfish, constituting a mixed stock.

## Assessment of enhancement success and fishery forecast

### Survey outline

In order to determine the survival, growth, and recapture rates of the released jellyfish, 18 survey sites in the juvenile jellyfish (mixed stock) habitat were established and surveyed in 2005 and 2006 (Fig. 1). The same fishing gear (drift-nets) was used in each survey, and the surveys were conducted five times during May to July in both years.

### Mortality after release

Various sources of mortality decreased the numbers of released jellyfish over time. It is very important to understand how to reduce mortality in order to improve the efficiency of the enhancement operation.

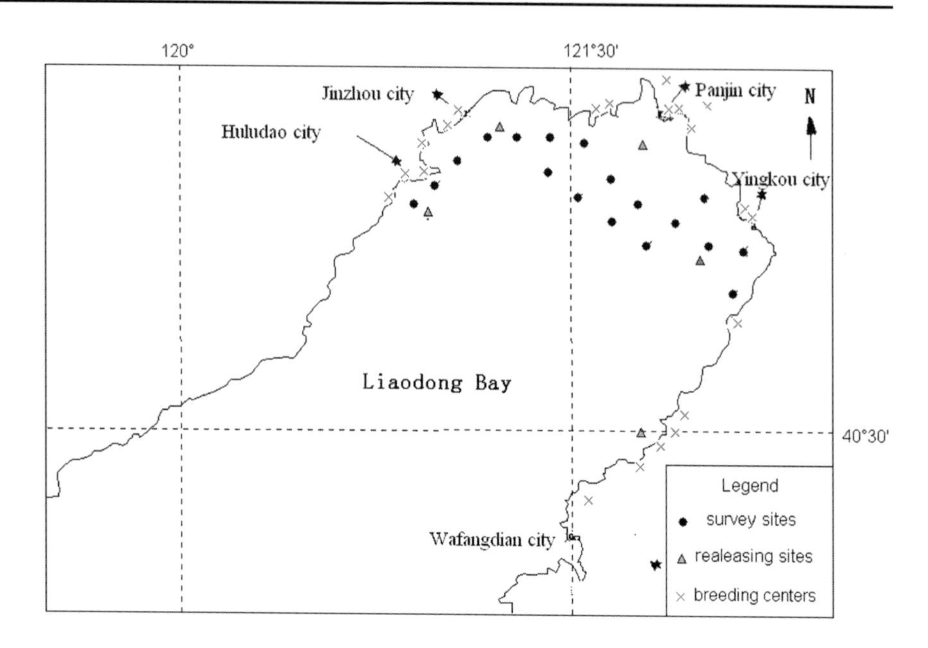


**Fig. 1** Release sites, breeding centers, and survey sites for *Rhopilema esculentum* medusae in Liaodong Bay, China during 2005–2006

**Table 1** Release records of *Rhopilema esculentum* jellyfish in Liaodong Bay, China during 2005–2006

| City | Release sites | Longitude and latitude | Release date | | No. of jellyfish released × $10^7$ | | UD(mm) |
|---|---|---|---|---|---|---|---|
| | | | 2005 | 2006 | 2005 | 2006 | 2005 |
| Jinzhou | Xiaoling River Estuary | 41°51′N 121°13′E | 25–30 May | 12–20 June | 3.50 | 5.62 | 11.5 |
| Huludao | Lianshan Bay | 40°37′N 120°56′E | 16–25 June | 12–20 June | 3.56 | 5.29 | 12.9 |
| Panjin | Shuangtai River Estuary | 40°39′N 121°48′E | 16–25 June | 12–20 June | 3.52 | 7.52 | 12.6 |
| Yingkou | Liao River Estuary | 40°29′N 122°00′E | 16–25 June | 12–20 June | 3.52 | 5.49 | 12.8 |
| Wafangdian | Taiping Bay | 40°00′N 121°46′E | 16–25 June | 12–20 June | 1.56 | 1.88 | 15.0 |
| | Total | | | | 15.66 | 25.8 | 12.7 |

UD, umbrella diameter

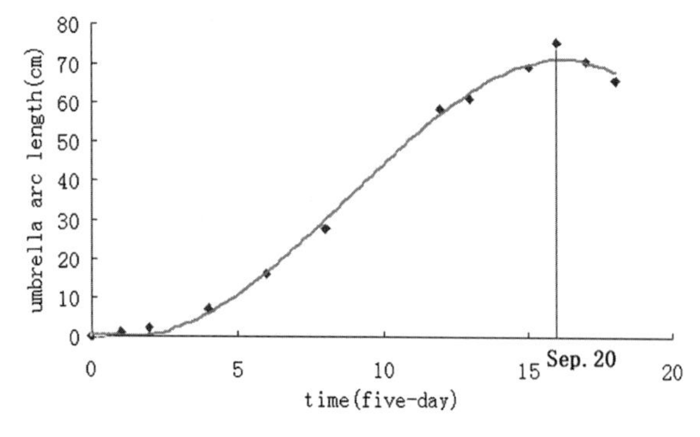


**Fig. 2** The growth curve of umbrella arc length of jellyfish (*Rhopilema esculentum*) in Liaodong Bay, China. More than 1,600 jellyfish samples were measured

Four different causes of mortality are identified. (1) Handling mortality can be easily estimated, and was caused when the jellyfish were transferred from the breeding centers, transported, and released in the natural waters. According to the release data of 2005 and 2006, the average handling mortality was 6.0% of the cultured jellyfish. (2) Abrupt mortality occurred during 2–3 d after the cultured juvenile jellyfish were released into the sea, presumably resulting from the sudden change of physical conditions. In 2006, the abrupt mortality was estimated to be 79% of the released jellyfish. (3) During 40–50 d from the end of June to the beginning of the fishing season, mortality from unknown sources occurs naturally. Based on the relative abundance data in different surveys, the average natural mortality was 55% of the jellyfish surviving after release. (4) Jellyfish numbers decrease from fishing with various kinds of nets before the fishing season opens. The percentage of this unlawful fishing mortality is very difficult to estimate.

### Growth

The growth trajectory of *R. esculentum* jellyfish is important in order to predict the best time for harvest. Since the jellyfish shrank after the first 10-d of September, the growth pattern could not be described by a regular asymptotic growth equation (Li et al., 1988). The growth characteristics of jellyfish in Liaodong Bay were described with a polynomial expression as a function of time, as follows:

$$L_t = 0.2198 + 0.4146t + 0.2203t^2 + 0.03824t^3 - 0.002249t^4$$

$L_t$ designates the arc length of the jellyfish swimming bell. $t$ is time in 5-d units, beginning on June 20 ($t_0$) when the strobilae of jellyfish release ephyrae in the field. The correlation was significant according to the $F$ test statistic ($F = 4{,}126 > F_{0.005}$ $(1, 9) = 13.61$) (Fig. 2).

### Distribution of jellyfish in Liaodong Bay

Distribution of the population refers to the horizontal location and habitat depth of the jellyfish population, which are static characteristics. The distribution of *R. esculentum* in Liaodong Bay are related to their ecological characteristics. Juvenile *R. esculentum* tolerate salinities of 10–20; adult jellyfish tolerate salinities of 12–35, and their optimal salinity is 23–28 (Chen, 1985; Lu et al., 1989). As a relatively independent stock, *R. esculentum* of Liaodong Bay always inhabit both sides of the bay in waters near the 5 m isobath.

### Catch forecast

To forecast the catch, the standard survey method is employed during May to July when the jellyfish occur in high volume. The basic method is to survey the relative abundance with high efficiency fishing nets (drift-nets) during the period of high jellyfish volume. Thirteen sites were established in the main jellyfish fishing areas and one catch made at each site. The relative abundance is the average number of jellyfish caught on one net in one 3-h catch. The yield forecast is obtained based on the model of relative abundance and fishing effort. The forecasting equation (Li et al., 1989) is:

$$Y = -17{,}963 + 98.2x_1 + 4.6x_2$$

where $Y =$ the forecasting output; $x_1 =$ relative abundance; $x_2 =$ fishing effort. This relationship is significant ($R = 0.97$; standard deviation $S = 2{,}898$; $F = 31.57 > F(4, 2)_{0.005} = 26.28$). In recent years, the number of fishing vessels has been constant and had little effect on yield. The accuracy of the forecast index is >90% based on the relative resource and yield.

## Jellyfish fisheries and management

### Jellyfish fisheries

The jellyfish fishing grounds where great numbers of edible *R. esculentum* jellyfish occur are characterized by having a large tidal range, shallow depth, semi-enclosed water mass, freshwater inflow through river systems, reduced salinity, and abundant foods. The jellyfish fishing ground in Liaodong Bay is about 6,860–8,575 km$^2$ and is located in the northern part of 40°30′N with 5–10 m depth. Fishing gear used includes various trawls, set-nets, drift-nets, push-nets, and hand-nets. Drift-nets are the most efficient and are placed across the current flow with a system of floats and sinkers. The length of each net is 30–50 m, with height of 8–12 m. A vessel loads 10–30 nets, which depends on the power of the vessel. The optimal fishing period is 10–20 August. The fishing season is usually ahead of schedule and has lasted only 2–3 d in recent years.

### Management strategy

The management of Chinese Fisheries department includes controlling the jellyfish fishing season, applying for a fishing permit, and preventing illegal fishing. China is actively developing its fishery management plans in an effort to conserve jellyfish resources. In Liaoning Province, the Ocean and Fishery Bureau controls the jellyfish fishing season. Fishermen who engage in jellyfish fishing activities must apply for a fishing permit, in which the time period and type of fishing gear are defined. From the middle of July to the fishing season, Fishery administrative officers patrol the sea to reduce illegal

fishing. Nine radar stations were set up along the coastal waters in Liaoning Province.

## Importance of the Liaoning Bay jellyfish fishery

The recapture rate of released jellyfish in 2005 was 3.2%, equaling 5.02 million jellyfish. The average weight was 2.5 kg, and recapture output was 12,500 tons, which accounted for 13.7% of the total jellyfish harvest in Liaoning Bay. At ¥ 7,000 yuan per ton, the value of the harvest was ¥ 87.5 million yuan. The profit per fishermen was more than ¥ 700 yuan. In 2006, the recapture rate of released jellyfish was 3% (80.7 million jellyfish), the average weight was 1.5 kg, and the recapture yield was 12,100 tons, which accounted for the 36.6% of total jellyfish harvest. At ¥ 6,000 yuan per ton, the fishing output was ¥ 72.6 million yuan. The direct input–output cost ratio was 1:18. The profit per fisherman was more than ¥ 600 yuan.

**Acknowledgments** The authors would like to thank the following people for their contributions to this article: Chu-yang Liu, Bin Wang, and Ming Sun. Especially, we mourn the loss of Prof. Jie-kang Chen and Prof. Geng-wu Ding who contributed their lives to jellyfish research. We are grateful to Dr. Mary Arai who encouraged publication of our research in English. Many thanks to Kylie Pitt for her efforts to enable Jing Dong to present this article at the Second International Jellyfish Bloom Symposium in Australia.

## References

Chen, J.-K., 1985. Cultivation and utilization of *Rhopilema esculentum* Kishinouye. Ocean Press, Beijing (in Chinese).

Chen, J.-K. & G.-W. Ding, 1983. Effect of temperature on the strobilation of jellyfish (*Rhopilema esculentum* Kishinouye, Scyphozoa, Rhizostomeae). Acta Zoologica Sinica 3: 195–205 (in Chinese).

Chen, J.-K., G.-W. Ding & C.-Y. Liu, 1984. Effect of light on the strobilation of jellyfish (*Rhopilema esculentum* Kishinouye). Oceanologia et Limnologia Sinica 4: 310–316 (in Chinese).

Chen, J.-K., N. Lu & C.-Y. Liu, 1994. Resource enhancement experiments in the edible medusae (*Rhopilema esculentum*) in the coastal waters of northern Yellow Sea. Marine Fisheries Research 15: 103–113 (in Chinese).

Ding, G.-W. & J.-K. Chen, 1981. The life history of *Rhopilema esculentum* Kishinouye. Journal of Fisheries of China 5: 93–104 (in Chinese).

Guo, P., 1990. Effect of nutritional condition on the formation and germination of the podocyst of scyphistomae of *Rhopilema esculentum* Kishinouye. Journal of Fisheries of China 3: 206–211 (in Chinese).

Guo, P., C.-Y. Liu & N. Lu, 1987. Food of early larvae of *Rhopilema esculentum* Kishinouye. Fisheries Science 3: 10–13 (in Chinese).

Hsieh, Y.-H. P., F.-M. Leong & J. Rudloe, 2001. Jellyfish as food. Hydrobiologia 451(Developments in Hydrobiology 155): 11–17.

Li, P.-J., H.-Y. Liu & W.-B. Wang, 1989. Study on predicting catch of jellyfish in Liaodong Bay and predicting catch in advance. Fisheries Science 1: 1–4 (in Chinese).

Li, P.-J., K.-F. Tan & C.-C. Ye, 1988. Growth pattern and characteristics of edible jellyfish (*Rhopilema esculentum*) in Liaodong Bay. Journal of Fisheries of China 3: 243–249 (in Chinese).

Li, J.-J., B.-T. Zhang & J.-M. Liu, 2004. Ranching of *Rhopilema esculentum* in coastal Dandong. Fisheries Science 7: 34–35 (in Chinese).

Lu, N., C.-Y. Liu & P. Guo, 1989. Effect of salinity on larva of edible medusae (*Rhopilema esculentum*) at different development phases and a review on the cause of jellyfish resources falling greatly in Liaodong Bay. Acta Ecologica Sinica 4: 304–309 (in Chinese).

Lu, N., J. Shuang & J.-K. Chen, 1995. Effect of temperature and food abundance on the growth of jellyfish (*Rhopilema esculentum* Kishinouye). Oceanologia et Limnologia Sinica 2: 186–190 (in Chinese).

Lu, N., J. Shuang & J.-K. Chen, 1997. Effect of temperature, salinity and light on the podocyst generation of *Rhopilema esculentum* Kishinouye. Fisheries Science 1: 3–8 (in Chinese).

Morikawa, T., 1984. Jellyfish. FAO INFOFISH Marketing Digest 1: 37–39.

Omori, M. & E. Nakano, 2001. Jellyfish fisheries in southeast Asia. Hydrobiologia 451(Developments in Hydrobiology 155): 19–26.

Wang, J.-F. & S.-L. Yu, 2002. Review and proposal on enhancement and releasing of *Rhopilema esculentum* Kishinouye in the coastal waters of Liaoning Province. Modern Fisheries Information 8: 17–18 (in Chinese).

Ye, C.-C., J. Dong & H.-L. Jiang, 2006. Study on selecting releasing species. Fisheries Science 9: 483–484 (in Chinese).

Hydrobiologia (2009) 616:119–132
DOI 10.1007/s10750-008-9581-z

JELLYFISH BLOOMS

# Stable isotope and fatty acid tracers in energy and nutrient studies of jellyfish: a review

K. A. Pitt · R. M. Connolly · T. Meziane

Published online: 22 September 2008


**Abstract** Studies of the trophic ecology of gelatinous zooplankton have predominantly employed gut content analyses and grazing experiments. These approaches record only what is consumed rather than what is assimilated by the jellyfish, only provide evidence of recent feeding, and unless digestion rates of different prey are known, may provide biased estimates of the relative importance of different prey to jellyfish diets. Biochemical tracers, such as stable isotopes and fatty acids, offer several advantages because they differentiate between what is assimilated and what is simply ingested, they provide an analysis of diet that is integrated over time, and may be useful for identifying contributions from sources (e.g., bacteria) that cannot be achieved using gut content approaches. Stable isotope analysis has become more rigorous through recent advances that provide: (1) signature determination of microscopic organisms such as microalgae, (2) analysis of dissolved organic carbon, and (3) improved quantification of relative source contributions. The limitation that natural tracer techniques require different dietary sources to have unique signatures can potentially be overcome using pulse-chase isotope enrichment experiments. Trophic studies of gelatinous zooplankton would benefit by integrating several approaches. For example, gut content analyses may be used to identify potential dietary sources. Stable isotopes could then be used to determine which sources are assimilated and modeling could be used to quantify the contribution of different sources to the diet. Analysis of fatty acid profiles could be used to identify contributions of bacterioplankton to the diet and, potentially, to provide an alternative means of identifying dietary sources in situations where the isotopic signatures of different potential dietary sources overlap. In this review, we outline the application, advantages, and limitations of gut content analyses and stable isotope and fatty acid tracer techniques and discuss the benefits of using an integrated approach toward studies of the trophic ecology of gelatinous zooplankton.

**Keywords** Gelatinous zooplankton · Trophic ecology · Diet · Gut contents

Guest editors: K. A. Pitt & J. E. Purcell
Jellyfish Blooms: Causes, Consequences, and Recent Advances

K. A. Pitt (✉) · R. M. Connolly
Australian Rivers Institute – Coast and Estuaries, and Griffith School of Environment, Griffith University, Gold Coast Campus, Gold Coast, QLD 4222, Australia
e-mail: K.Pitt@griffith.edu.au

T. Meziane
Département Milieux et Peuplements Aquatiques, UMR-CNRS 5178, Biologie des Organismes Marins et Ecosystèmes, MNHN, CP 53, 61 rue Buffon, 75231 Paris cedex 05, France

## Introduction

Changes in the distribution and biomass of some species of jellyfish have increased concerns about their potential impacts on pelagic food webs (Brodeur et al., 2002; Lynam et al., 2006). With few exceptions (Montoya et al., 1990; Malej et al., 1993; Brodeur et al., 2002; Pitt et al., 2008), studies of the diets of gelatinous zooplankton have relied on the analysis of the prey present in the gut of the medusae or grazing experiments. Both of these approaches, however, have limitations for inferring diet and are but two of a suite of tools available for investigating trophic links. All approaches, however, have their own unique set of advantages and limitations that must be considered. The objectives of this paper are to review some of the limitations of gut content analyses and grazing experiments, outline the use of stable isotope and fatty acid tracers, describe the advantages and limitations of tracer techniques and their specific application to pelagic systems, and discuss the benefits of using multiple approaches to elucidate trophic relationships of gelatinous zooplankton.

## Limitations of gut content analyses and grazing experiments

Gut content analyses record the types of prey ingested by a consumer, but they cannot easily discriminate between organisms that are assimilated by the predator and those that are ingested incidentally and either egested or pass through the gut undigested (Fry, 2006; but see Purcell et al., 1991). As most zooplankton are digested within 2–4 h (e.g., Purcell, 1997; Båmstedt & Martinussen, 2000), gut content analyses only provide evidence of recent feeding and extensive spatial and temporal sampling is required to provide a robust analysis. Variation in digestion rates of different prey types also biases estimates of their contributions to the diets of the consumers (Gee, 1989). Digestion rates of different prey have been measured for gelatinous zooplankton, but rates at which individual species of prey are digested vary among individuals, with size of predator and prey, among days, and with feeding intensity and temperature (Heeger & Möller, 1987; Purcell, 1992; Båmstedt & Martinussen, 2000), potentially confounding spatial and temporal comparisons of diet. Gut content analyses of medusae also predominantly focus on mesozooplankton and ichthyoplankton, presumably because they are more visible and retained in the gut for longer than microzooplankton. Studies of the contributions of microplankton are rarer and have been approached using grazing experiments (Stoecker et al., 1987; Sullivan & Gifford, 2004). Grazing experiments have also been used to estimate clearance rates of mesozooplankton (e.g. Fancett & Jenkins, 1988; Hansson et al., 2005). Medusae and zooplankton used in grazing experiments, however, may not behave in captivity as they would in the wild and results need to be interpreted cautiously (Toonen & Chia, 1993). For example, refugia generated by oceanographic features such as stratification may be unavailable to zooplankton in aquaria, which could artificially increase their likelihood of capture. Similarly, confinement may disrupt the flow of water around medusae and reduce their feeding efficiency.

Biochemical tracers, such as stable isotopes and fatty acids, have been used extensively in studies of trophic ecology since the 1970s. For a particular chemical to act as a tracer, its structure must be unaltered or altered in a predictable way as it passes from the dietary source to the consumer. The major advantages biochemical tracers offer over gut content analyses are that they differentiate between what is assimilated and what is simply ingested by the consumer, they provide an analysis of diet integrated over time and may be useful in identifying contributions from sources (e.g., bacteria and detritus) that are not easily determined using gut content approaches. Although tracer techniques have been used extensively to elucidate food webs in terrestrial and estuarine systems, they have been applied less frequently to pelagic systems and relatively rarely to studies involving gelatinous zooplankton (Montoya et al., 1990; Malej et al., 1993; Brodeur et al., 2002; Towanda & Thuesen, 2006; Pitt et al., 2008).

## General principles of isotopic analyses

In recent years, stable isotope techniques have gained wide recognition as a tool to identify and trace energy and nutrient sources in coastal ecosystems (Fry, 2006). Stable isotope analysis of aquatic food webs involves elements that are important in the nutrition of animals and have different naturally occurring isotopes. Their use relies on potential sources, such as different plants or types of prey, having different ratios of the common, light isotope to the heavy, rare isotope. The most commonly used element is carbon ($^{13}C/^{12}C$), which provides the basis for the majority of energetic requirements for pelagic organisms. Nitrogen ($^{15}N/^{14}N$) is also used routinely in aquatic food web studies, and is involved in protein synthesis (West et al., 2006).

Natural variability in the relative abundance of the common and rare isotopes is typically very small, and a special notation is, therefore, used to highlight the differences. Stable isotope ratios are normally reported as parts per thousand deviation from a known international standard, expressed using the delta notation:

$$\delta X = \left[\frac{R_{\text{sample}} - R_{\text{standard}}}{R_{\text{standard}}}\right] \times 10^3 \quad [‰]$$

where R = $^{13}C/^{12}C$ or $^{15}N/^{14}N$ is the ratio of the atom occurrence of the rare to common isotope. The standards are Vienna PDB (equivalent to the original PeeDee Belemnite limestone standard) for $\delta^{13}C$ and atmospheric $N_2$ for $\delta^{15}N$.

The application of stable isotopes to determine energy and nutrient pathways depends on two assumptions:

(1) that stable isotope ratios of potential sources (plants or prey) differ, and
(2) that the ratios are unaltered or altered in a predictable fashion during transfer to higher trophic levels.

In oceanic environments, phytoplankton are the dominant autotrophs that support pelagic food webs. Laboratory studies indicate that isotopic signatures of individual marine phytoplankton species can range widely (e.g., $\delta^{13}C$ −5.5 to −29.7‰; Falkowski, 1991), and that factors such as growth rate and cell size have a strong influence on the isotopic fractionation that occurs during C fixation, and therefore, on their $\delta^{13}C$ (Burkhardt et al., 1999). Thus, individual phytoplankton species can have distinct isotopic signatures that could be used to quantify their contribution to food webs supporting jellyfish. Difficulties in separating individual species in sufficient quantities for analysis from mixed samples collected in the field, however, has generally precluded measurement of species-specific signatures. New methods, such as fluorescence, to separate taxa are now enabling taxon-specific signatures to be obtained (Pel et al., 2003), which will help to elucidate the contribution of different phytoplankton taxa to jellyfish food webs. Pelagic food webs in coastal and estuarine systems may be supported by a greater diversity of autotrophs, including macrophytes such as seagrasses and mangroves. Macrophytes often have distinctive carbon isotope ratios that encompass a wide range of $\delta^{13}C$ units (e.g., Melville & Connolly, 2005; Benstead et al., 2006). These differences result from either different photosynthetic pathways (C3 versus C4 photosynthesis) or whether carbon is obtained from the air or water (Michener & Schell, 1994). The distinctive $\delta^{13}C$ for primary producers in coastal habitats may make it easier to distinguish among potential sources.

Isotopic ratios change slightly from one trophic level to the next in a process known as fractionation. The second assumption, of predictable change, is once again most likely to be met in benthic aquatic systems, where the majority of studies measuring fractionation rates have been done (McCutchan et al., 2003). Average carbon isotope fractionation per trophic level for aquatic animals is 0.4 ± 0.17‰ (McCutchan et al., 2003). Nitrogen isotope ratios generally display a much greater stepwise enrichment between producers and each higher trophic level, but average estimates for aquatic animals have varied from 2.3 ± 0.28‰ (McCutchan et al., 2003) to 3.4 ± 1.1‰ (Minagawa & Wada, 1984). The relatively larger isotopic fractionation of nitrogen has proven useful in assigning relative trophic levels to organisms for which stomach content analysis is difficult.

Although average fractionation rates are frequently applied in isotope studies, the actual range of fractionation rates varies widely among species; standard errors for average fractionation rates calculated across a range of species are ~30% of the mean (Minagawa & Wada, 1984; McCutchan et al., 2003). The degree of fractionation may also vary with food quality (van der Zanden & Rasmussen, 2001). Average fractionation rates should, therefore, be applied cautiously, and carefully controlled experiments to measure fractionation between jellyfish and their prey should be considered as part of isotope studies into jellyfish nutrition. For jellyfish species having symbiotic zooxanthellae, the tight cycling of nutrients between the host and zooxanthellae makes

fractionation particularly difficult to predict, and additional care is needed in the interpretation of isotope results for these species.

## Considerations for sample preparation

### Dealing with salt in jellyfish samples

Samples are dried prior to being analyzed in the mass spectrometer. Dried jellyfish samples, however, contain large amounts of salt. Since salt does not contain the elements usually used for isotopic analyses, it will not affect the ratio of the heavy to light isotopes of interest (i.e., the isotopic signature) in the sample. The amount of salt will, however, influence estimates of absolute amounts of a heavy isotope in the tissues, as a large proportion of the mass of a sample will be composed of salt rather than organic material. This may have implications for measuring assimilation rates in enrichment studies or for comparing assimilation rates between jellyfish and other taxa with much lower salt contents. Salt cannot be easily removed prior to isotopic analysis since rinsing samples with freshwater will lyse cells, resulting in the loss of dissolved organic matter and potentially change the isotopic signature. Knowledge of the relationship between dry weight and ash-free dry weight of the medusae, however, may enable estimates of absolute quantities of a heavy isotope in a sample to be calculated.

### Variation in isotopic signatures among tissues

Isotopic signatures of different tissues within individual organisms may vary (e.g., Lorrain et al., 2002), and this can influence the interpretation of trophic relationships. Variation among tissues can result from different turnover times of elements (Tieszen et al., 1983). For example, carbon in the exoskeletons of mysids and krill is replaced much more rapidly than in muscle tissue (Gorokhova & Hansson, 1999; Schmidt et al., 2003). Tissues that turn over elements rapidly (e.g., gonads and exoskeleton), therefore, may provide information on recent feeding, whereas tissues with longer turnover times (e.g., muscles) may provide information about feeding over longer periods. Differences in isotopic signatures may also occur due to variations in the lipid contents of different tissues (Lorrain et al., 2002). Lipids are more depleted in $^{13}$C than proteins and carbohydrates, and tissues that contain greater proportions of lipids (e.g., gonads and digestive glands) generally have lower $\delta^{13}$C values (De Niro & Epstein, 1977; Lorrain et al., 2002). Some researchers, therefore, advocate the removal of lipids prior to isotopic analysis (e.g., Bodin et al., 2007).

Since the type of tissues selected for analysis can have a strong influence on the interpretation of trophic relationships, it is important to investigate potential variation in isotopic signatures among tissue types before deciding which type of tissue is most appropriate to use. Like most animals, the lipid content of different tissues of jellyfish vary (e.g., Lucas, 1994; Carli et al., 1991), which may influence their $\delta^{13}$C. Variation in isotopic signatures among tissues or areas of the body, however, has been examined for only two jellyfish species. Pitt et al., (2008) found no difference in the isotopic signatures of ectodermal tissue of the umbrella and mesoglea of *Catostylus mosaicus* (Quoy & Gaimard 1824), but Towanda & Thuesen (2006) observed that the mesoglea of *Phacellophora camtschatica* Brandt was greatly enriched in $\delta^{13}$C ($-10.1 \pm 0.9$‰) compared to the whole body ($-25.7 \pm 1.2$‰), gonad ($-27.6 \pm 0.7$‰), and oral arm tissue ($-24.4 \pm 1.1$‰). Pilot studies, therefore, are needed to identify any variation in isotopic signatures among tissues and enable more informed decisions about which types of tissues should be analyzed.

### Removal of inorganic material from potential sources

Gut content analyses indicate that jellyfish ingest large numbers of zooplankton that have inorganic chitinous or calcareous exoskeletons (e.g., mollusc veligers and copepods; Purcell, 2003; Browne & Kingsford, 2005). Inorganic compounds will not normally be assimilated by the jellyfish and isotopic signatures should, therefore, be obtained only for the organic component of potential prey. If the potential prey are large, the exoskeleton can be removed physically, but the simplest way of removing the exoskeletons of small zooplankton is to acidify the sample and redry it prior to analysis. Acidification has no effect on the $\delta^{13}$C of the soft tissues (Bunn et al., 1995; Bosley & Wainright, 1999; Ng et al., 2007), and although some studies indicate that acidification has a negligible influence on $\delta^{15}N$ (Bosley & Wainright, 1999; Ng et al., 2007), another study suggests that acidification depletes $\delta^{15}N$ (Bunn et al., 1995). The effects of acidification on $\delta^{15}N$ should, therefore, be investigated for individual taxa and, if necessary, $\delta^{15}N$ should be analyzed using separate, non-acidified samples.

## Quantitative analysis of contributions of different sources

Recent advances in the analysis of stable isotope data have made isotope studies of food webs more quantitative, more rigorous, and more informative. Early studies compared consumer isotopic ratios with potential source ratios visually, but data are now routinely analyzed using mixing models to quantify the contributions from different sources (Fry, 2006). Mixing equations give unique solutions where the number of potential sources is no more than one greater than the number of elements being used (e.g., two sources for a single element analysis, three sources where, for example, carbon and nitrogen are used). Such equations have been refined so that not only mean values but also the variation around mean values can be used, giving confidence limits around estimates of source contributions (Phillips & Gregg, 2001).

Since jellyfish ingest a diverse suite of taxa, most food web studies involving jellyfish will involve too many potential sources for simple mixing equations to be useful. In these situations, the IsoSource procedure can be applied (Phillips & Gregg, 2003). The IsoSource model calculates all feasible combinations of sources that could explain the consumer isotope value, thereby placing bounds on the dietary contributions of each source. Model output is reported as the distribution of feasible solutions for each source. As an example, consider an IsoSource model that includes a consumer ($\delta^{13}C = -22.0$) and three dietary sources, A) $\delta^{13}C = -26.5$, B) $\delta^{13}C = -24.0$, and C) $\delta^{13}C = -22.0$. The model indicates that the consumer derives no less than 70% and up to 90% of its carbon from Source C and less than 30% from either of the sources A or B (Fig. 1). This methodology, however, sometimes cannot properly delineate source contributions, and a further refinement has been made that better defines potential

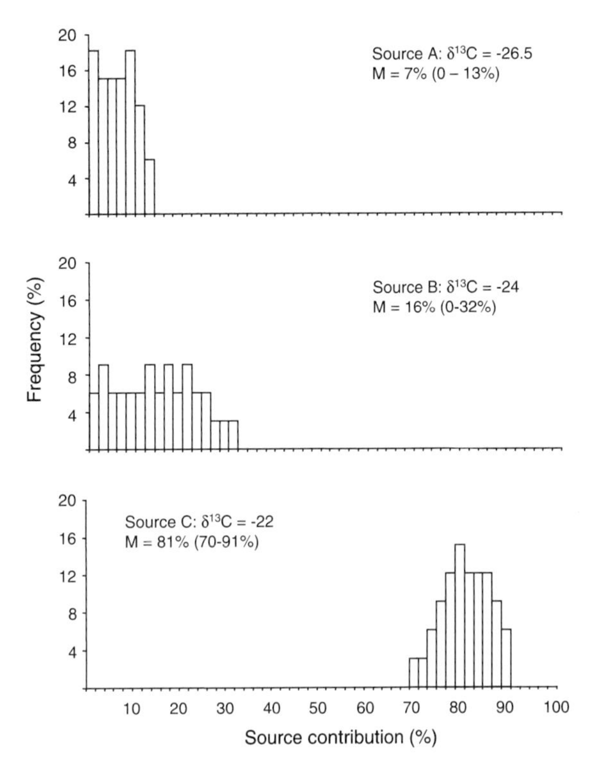


**Fig. 1** Simulation of the distribution of feasible contributions of three sources to the diet of a consumer. M = median (ranges are 1 and 99 percentile values)

contributions by pooling contributions from various groups of sources selected by the researcher (Phillips et al., 2005). Quantitative analysis of isotope data is an active area of research in itself (Fry, 2006), and future developments will no doubt benefit isotope studies of jellyfish nutrition.

## Limitations of isotopic techniques and potential solutions

### Multiple sources with the same signature

Sometimes, multiple sources share the same isotopic signature, preventing the contribution of each source to the diet of the consumer being resolved. In such cases, analysis of an additional element may distinguish among the different sources. For example, analysis of sulfur isotopes has proven useful for distinguishing between sources that share the same carbon and nitrogen signatures (Connolly et al., 2004). Differences in sulfur ratios between different plant and algae types are, on average, much larger than for carbon or nitrogen. Sulfur is especially useful for separating dietary contributions from benthic plants or prey (e.g., seagrass or the animals that live in seagrass) from biota in the water column. Thus, sulfur isotopes may be useful for studies of jellyfish inhabiting estuarine or shallow coastal waters, where emergent demersal zooplankton can form an important part of the diet (Flynn & Gibbons, 2007; Pitt et al., 2008).

Another means of distinguishing between food sources has been developed for situations where analysis of natural isotopic ratios cannot separate the contributions of different food sources. One or more putative sources are spiked artificially with enriched isotopes (e.g., $^{13}$C or $^{15}$N). Any contribution of the artificially enriched source to the diet of consumers is detected as a shift in the isotopic ratios of consumers. The first marine application was the artificial separation of the normally similar nitrogen isotope ratios of seagrass and its epiphytic algae (Winning et al., 1999). Pulse-chase experiments involving the manipulative enrichment of source signatures through the addition of enriched isotopes have since been done in small plots in seagrass (e.g., Mutchler et al., 2004) and on mudflats (e.g., Middelburg et al., 2000), and at a larger scale, in the upper reaches of estuaries (e.g., Gribsholt et al., 2005). There are, however, several difficulties that will need to be overcome for pulse-chase experiments to be applied to jellyfish. For example, since jellyfish prey on a diverse suite of taxa, only a few taxa within the assemblage of zooplankton prey could be labeled at any one time to determine if they are assimilated by the medusae (e.g., two taxa could be tested simultaneously if one taxon was labeled with $^{15}$N and the other with $^{13}$C). For pelagic species, pulse-chase experiments would also have to be undertaken in mesocosms. Such experiments would, therefore, suffer from the artefacts that are associated with all mesocosm experiments. At present, however, mesocosm studies offer the only option for applying this methodology to pelagic jellyfish.

## Variability in isotopic signatures: problems and advantages of shifting baselines

Isotopic signatures of the autotrophs that sustain the food web can vary as a result of changes in the source of nutrients entering a system. This is commonly referred to as a "shifting baseline" and often occurs in coastal environments where terrestrial or anthropogenic nutrients, which have distinct isotopic signatures, enter waterways. For example, nitrogen derived from sewage is typically more enriched in $^{15}$N than natural sources of nitrogen (Heaton, 1986). Autotrophs located close to sewage discharges, therefore, may be more enriched in $^{15}$N than those located elsewhere. Temporal variation in baseline signatures may also occur following heavy rain when terrestrial sources are flushed into coastal waterways or following the commissioning or upgrading of sewage treatment plants (Costanzo et al., 2005). Changes in the isotopic signature of the autotrophs are propagated to higher trophic levels and this can result in substantial temporal and spatial variability in isotopic signatures of consumers and their sources.

The rate at which the isotopic signature of an organism shifts to reflect that of an elemental source varies depending on its turnover time. Organisms turn over isotopes in their tissues during growth and general metabolic maintenance (Fry & Arnold, 1982; Hesslein et al., 1993). Most studies indicate that, for aquatic animals, the majority of carbon and nitrogen turnover results from growth rather than metabolic maintenance (Fry & Arnold, 1982; Hesslein et al., 1993; MacAvoy et al., 2001), although results vary (e.g., Tarboush et al., 2006). Factors that affect growth rates and, to a lesser extent, metabolic rates, such as age (Sakano et al., 2005), temperature (Frazer et al., 1997), and diet (Schmidt et al., 2003), can affect rates at which isotopes are turned over in the tissues. Consequently, rates of turnover may vary within individual organisms, among conspecifics of different ages or life history stages, and among taxa. Variation in turnover times among different components of the food web, or different types of tissues, following a shift in the isotopic baseline can decouple the isotopic relationship between a consumer and its source; this limits the ability of isotopes to reliably identify trophic links (Schmidt et al., 2003). Turnover times of isotopes in gelatinous zooplankton need to be investigated to determine how rapidly they will respond to changes in the baseline signature. Given that medusae grow very rapidly (e.g., Palomares & Pauly, 2008), growth, rather than metabolic maintenance, may be expected to have a greater influence on turnover rates.

Once sources and their consumers have equilibrated to the new isotopic baseline, the shift in the baseline can actually be useful for identifying trophic links. The difference in the isotopic signature between a consumer and its source will generally remain constant, regardless of whether their absolute values change following a shift in the baseline. Consequently, if the isotopic signature of the source and consumer vary in a consistent way through time or among places, i.e., if the signature of the consumer tracks that of its source, it can strengthen conclusions about trophic links (McCutchan & Lewis, 2002; Melville & Connolly, 2003). While two sources may have the same signature at one time or place, they may vary at another. The source can, therefore, be identified as being one of the consumer tracks. A two-dimensional correlation test has recently been developed to measure the strength of the links between sources and a consumer using two elements (e.g., carbon and nitrogen) at the same time (Melville & Connolly, 2003).

Sampling at multiple times and places is essential to determine whether baselines are consistent and to identify whether the different components of the food web have equilibrated to any observed shift in the isotopic baseline. If differences in isotopic signatures between sources and consumers are consistent, even if their actual signatures change, conclusions regarding trophic relationships are more robust. In areas where the isotopic signature of the elemental source changes frequently, variation in turnover times of different components of the food web may reduce the reliability of isotopic approaches.

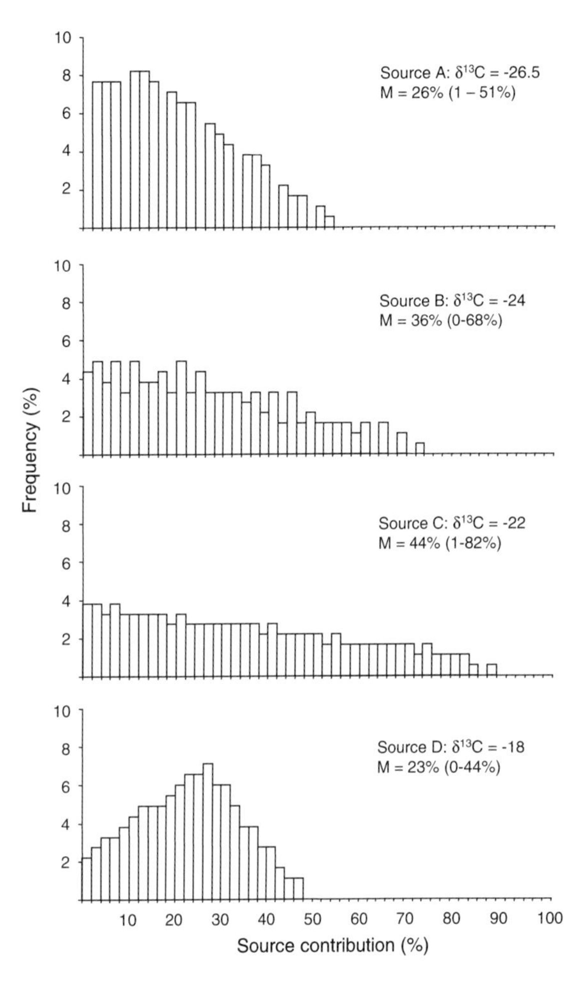


**Fig. 2** Simulation of distribution of feasible contributions of four sources to the diet of a consumer. Details as in Fig. 1

## Missing sources

Interpretation of dietary relationships using isotopic data depends on the relative values of the isotopic signatures of the consumer and its sources. Reliable interpretations of models of isotopic data can only be made if all dietary sources are included in the study. There are two possible outcomes for a model that is missing a source. First, the model may not be resolved. This occurs if, after allowing for appropriate fractionation, the consumer is more enriched or depleted than all sources included in the model. Alternatively, the model may be resolved, but may overemphasise the importance of a source that either makes a minor contribution to the diet or may have no dietary importance. For example, consider if, in the earlier example, a fourth, more enriched source, D, ($\delta^{13}C = -18.0$) was found (e.g., based on gut content analysis) to have been missing from the original model. When revised, the model subsequently indicated that the contribution of Source C to the diet could range anywhere between 0 and 92%, instead of the initial 70–90% (Fig. 2). Models that are missing dietary sources are likely, therefore, to overemphasize the importance of sources that may make only a small contribution to the diet. Unless the model cannot be resolved, the deficiencies of models that fail to include a major source may not be recognised, and therefore, incorrect conclusions may be drawn regarding the dietary importance of different sources. Sources used in models, therefore, need to be carefully considered and justified. Analysis of the contents of the guts may provide an indication of possible dietary sources to include.

Obtaining sufficient material for analysis and "averaging" effects

The greatest technical difficulty in applying natural tracer techniques to planktonic food webs is obtaining sufficient quantities of each taxon for the analysis. For most zooplankton, approximately 5 μg of dried material is required for the analysis of carbon (i.e., after acid washing to remove indigestible carbonates) and 50 μg for the analysis of nitrogen. For larger species of plankton, only tens of individuals may need to be isolated (K. Schmidt, pers. comm.), but for smaller size classes, hundreds of individuals of each species may be required, which is logistically difficult. Many studies have, therefore, either obtained an average signature for bulk samples of zooplankton (e.g., Malej et al., 1993), or fractionated the zooplankton by size class (e.g., obtained by sieving, Rolff, 2000), or used differences in densities to separate zooplankton into coarse taxonomic groups (e.g., copepods and mollusc veligers, Pitt et al., 2008). Some medusae, however, feed selectively on particular types of zooplankton (Purcell, 1997). If the medusae prey on only a subset of species included in the fraction, then the average signature obtained for the fraction may not accurately reflect the actual dietary source. Mixing models are sensitive to small deviations in isotopic signatures and variations of <1‰ may determine whether a model can be resolved. For example, Pitt et al. (2008) examined the contribution of copepods, mollusc veligers, and small shrimp to the diet of the non-zooxanthellate scyphozoan, *Catostylus mosaicus*. Zooplankton was sampled during the day and night to account for possible variation in isotopic signatures associated with emergence of some taxa from the benthos at night. The daytime samples were assumed to comprise species that occurred permanently in the water column, whereas the night samples comprised species that occurred permanently in the water column and those that emerged into the water column from the benthos at night. The copepods sampled at night were 1.5 – 2.8‰ more enriched in $^{13}C$ than the day samples, suggesting that emergent copepods were much more enriched than the diurnal groups (Fig. 3A). The IsoSource model, however, was unable to be resolved using the daytime and night time signatures, because after allowing for fractionation, the medusae were more enriched than all possible sources. Only when a separate signature was subsequently obtained for emergent crustaceans (sampled using emergence traps) could the model be resolved and the likely substantial contribution of emergent species (shrimp and copepods) to the diet be identified (Figs. 3B, 4).

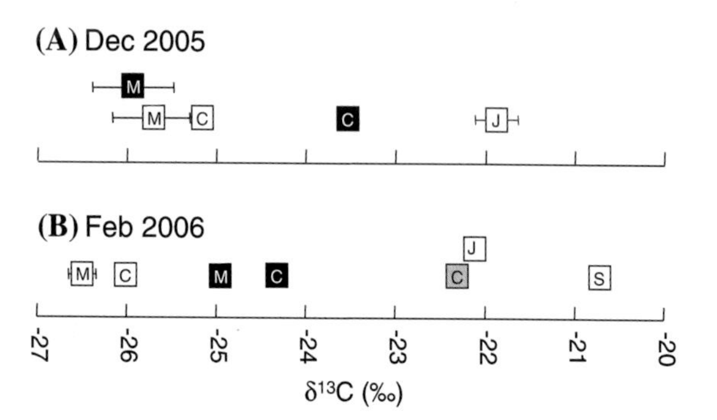


**Fig. 3** (**A**) Mean (± SE) $\delta^{13}C$ values of *Catostylus mosaicus* and its potential diurnal (white) and nocturnal (black) prey in 2005, and (**B**) diurnal, nocturnal, and emergent (grey) prey in 2006. J = *C. mosaicus*, C = copepods, M = mollusc veligers, and S = mysid shrimp. Note that error bars are obscured by symbols in some cases

Recent developments have made it easier to obtain isotopic ratios of inconspicuous sources. The difficulty of isolating enough microalgae for isotope analysis has been overcome in two ways. Hamilton et al. (2005) showed that in many cases, water or sediment samples can be centrifuged in a silica gel to separate microalgae from sediment and detrital matter. Where low algal densities or high detritus loads prevent the use of this method, an alternative is to use compound specific isotopic analysis (Oakes et al., 2005). This approach involves extracting and then obtaining an isotopic signature of a compound, such as phytol, that occurs only in the source of interest. If the isotopic ratio of the compound accurately reflects the isotopic signature of the bulk algal sample, it can be used as a proxy for the algae (Oakes et al., 2005). Although these methods are time-consuming and expensive, they provide useful means for solving otherwise intractable problems. Another related benefit for isotope studies of pelagic systems is the technical advances, making routine isotope analysis of dissolved organic carbon possible (Bouillon et al., 2006).

## General principles of the lipid markers approach

Lipids are key components of cell membranes and exhibit great diversity in structure. Fatty acids (FAs)

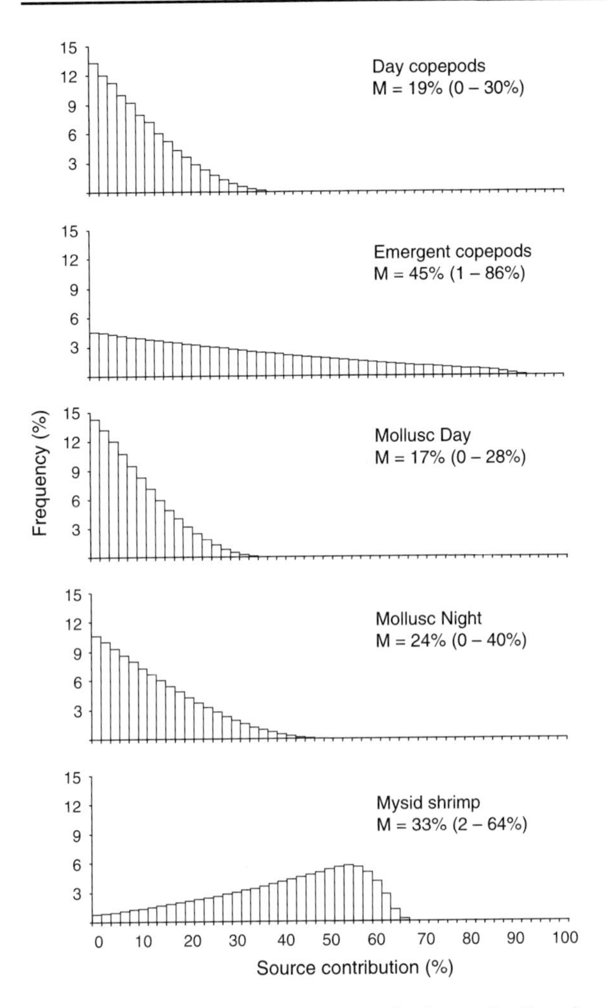


**Fig. 4** Distribution of feasible contributions of diurnal, nocturnal, and emergent sources to the diet of *Catostylus mosaicus*. Details as in Fig. 1

are a particular class of lipids that have a variety of cellular functions. They have a very high energy value and are an important fuel in pelagic ecosystems (Phleger et al., 1998; Falk-Petersen et al., 2002). Their diverse structures enable them to be used as biomarkers of specific organisms (Sargent et al., 1987), because some FAs (and also some sterols, fatty alcohols, and hydrocarbons) occur only in certain taxa, thereby allowing these groups to be distinguished (e.g., Graeve et al., 2002; Meziane et al., 2007). In addition to being a valuable taxonomic tool, the specific fatty acid composition of different animal and plant groups is being increasingly used to map the transfer of the organic matter through aquatic food webs and to understand trophic relationships (Falk-Petersen et al., 2002; Copeman & Parrish, 2003).

The FA trophic marker concept is based on the observation that primary producers are characterized by the presence of certain FAs in their tissues that may be transferred conservatively to, and hence be recognized in, primary consumers (Dalsgaard et al., 2003). In practice, useful FA markers are those that, when transferred throughout the food web, provide knowledge not only about prey–predator relationships but also about the base of the food web. For example, some FAs are only synthesized de novo by plants. These include some polyunsaturated $\omega$3- and $\omega$6-FAs (3 and 6 are the positions of the double bond from the terminal methyl group). Producers other than plants (e.g., fungi and bacteria; Dalsgaard et al., 2003) and sometimes consumers, including jellyfish (Nichols et al., 2003), also have FAs that could be used to trace their transfer through food webs. For example, Nichols et al. (2003) discovered the presence of two rare FAs (24:6$\omega$3 and 24:5$\omega$6) in the tissues of *Aurelia* sp. If these FAs are unique to *Aurelia* sp., then their presence in higher-order consumers will provide evidence of predation on this jellyfish.

The composition of lipids in general has been examined in several cnidarian species. Most attention has focused on the trophic relationship between corals and their symbionts (zooxanthellae) (e.g., Harland et al., 1992; Papina et al., 2003), but more recently, profiles of jellyfish have been investigated (Fukuda & Naganuma, 2001; De Souza et al., 2007). For example, De Souza et al. (2007) used the presence of two diacylglycerols in the tissues of the medusa *Phyllorhiza punctata* von Ledenfeld 1884 as evidence of their translocation from their endosymbiotic zooxanthellae. Also, Fukuda & Naganuma (2001) used temporal variation in the FA compositions of *Aurelia aurita* (Linnaeus) to suggest that its diet may shift between the diatom-based food chain and the detritus-based food chain at different times of the year. FAs and others lipids, such as fatty alcohols, have also been used to elucidate the diet of ctenophores in the Arctic and Antarctic (Ju et al., 2004; Graeve et al., 2008). All three studies analyzed both fatty acid and fatty alcohol compositions and demonstrated that krill and copepods were the major food source of ctenophores in these waters.

## Analysis of fatty acid data

If FAs are unique to particular types of organisms, then the simple presence of these FAs in the tissues of a consumer is sufficient to verify their consumption. In a semi-quantitative way, the prominence of one source over another in the diet of a pelagic consumer can be assessed by comparing the relative abundances of different types of FAs. For example, the ratio of 20:5ω3:22:6ω3 can indicate whether diatoms or dinoflagellates dominate the diet (a high value is indicative of more diatoms and a low value is more typical of a diet dominated by dinoflagellates; Budge & Parrish, 1998).

Individual FAs are useful for tracing the transfer of particular sources in a food web, but additional information can be provided by using multivariate analytical techniques (e.g., ordinations and analyses of similarities) to compare the entire FA profiles of consumers. The advantage of using the entire profile is that it fully utilizes the information that is generated in one species due the changing relative contributions of the numerous FAs present in the tissues. For example, Ju et al., (2004) did a principal components analysis (PCA) on the FA and fatty alcohol profiles of the antarctic ctenophore, *Callianira antarctica* and their potential food sources

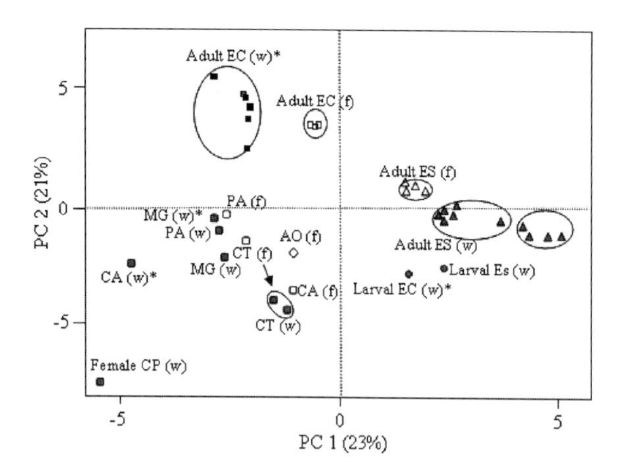


Fig. 5 Two-dimensional PCA plot of the first two principal components based on the combined fatty acid and fatty alcohol profiles for ctenophores and their potential prey in Antarctica. CT = *Callianira antarctica*, ES = *Euphausia superba*, EC = *Euphausia crystallorophias*, PA = *Paraeuchaeta antarctica*, CA = *Calanoides acutus*, AO = *Antarctomysis ohlini*, MG = *Metridia gerachei*, and CP = *Calanus propinquus*. *W* = winter, and *f* = autumn. Redrawn from Ju et al., (2004) with permission

(Fig. 5). PCA enables identification of the variables that contribute the most to the variance. Principle component 1 (PC 1) accounted for 23% of the variation and PC 2 is accounted for 21%; the FA and alcohol profiles of the ctenophores more closely resembled copepods than krill, indicating that copepods make a major contribution and krill a minor contribution to the ctenophore's diet. Multivariate methods are particularly useful when a study includes very large numbers of samples (such as occurs when sampling over multiple spatial and temporal scales), which can make it difficult to readily distinguish the fate of specific markers in the ecosystem (Howell et al., 2003; Meziane et al., 2006). Such analyses can also be used to identify species that occur in similar trophic guilds (Howell et al., 2003).

## Limitations of fatty acid analysis and possible solutions

Like stable isotopes, multiple sources can sometimes share the same FA markers, limiting their utility for tracing material from different producers through food webs (Dalsgaard et al., 2003). Thus, although some FAs are produced in large amounts by some sources, their presence, even in small amounts, in other producers can confound the assignment of food sources to the consumers. One way to overcome this limitation is to have a better understanding of temporal and spatial variations in the FA composition of organic matter at the base of the food web and of the consumer. For example, the absence of a source and its associated FA markers in a consumer at some places or times, but their presence at others, can help elucidate which sources contribute to a consumer's diet (Howell et al., 2003; Meziane et al., 2006).

Multivariate analyses of the entire FA profiles may also help to overcome this problem as they take full advantage of the information that is generated in one species due to the changing relative contributions of the numerous FAs present in the tissues. Finally, if several potential sources have similar FA profiles, the actual or dominant source can be identified using experimental approaches. For example, consumers can be fed a diet with a known FA profile (i.e., one source) and the changes in the FA profile of the consumer can be compared to that of its food source (e.g., Hall et al., 2006). This may allow verification of whether the observed profile of the consumer resulted from the assimilation of one or multiple sources.

Analysis of stable nitrogen isotopes provides information on the number of trophic steps in a food web, but the FA approach is unable to determine whether FAs are transferred directly to consumers or via an intermediary. This is particularly true in benthic systems, where macrofauna can feed on meiofauna, which, in turn, feed on microoganisms living in the sediments (Moens et al., 1999). The FA profile of the macrofauna, therefore, will be similar to the microorganisms, even though they do not feed directly upon them. In this case, knowledge on feeding behavior of the consumers is needed to ascertain a direct trophic link as some organisms are not anatomically equipped to feed on some sources (Meziane et al., 2002). Alternatively, FA analyses may be used in association with analysis of stable nitrogen isotopes to estimate the number of trophic steps. A final limitation of FA markers is that some of these compounds can be metabolized in the tissues of the primary consumer, and therefore, may not readily traced to higher trophic levels. Indeed, polyunsaturated fatty acids (PUFAs) in some invertebrates may elongate and be used to construct other PUFAs that are also potential markers (Ito & Simpson, 1996). Although such examples are uncommon, rigorous application of the FA technique should include not only verification of the presence of specific FA markers in the potential source but their subsequent conservative transfer from the source to the primary and higher order consumers. Whether specific FA are transferred conservatively or, indeed, change structure as they are transferred across trophic levels can be determined using controlled experiments. For example, in a three-step food chain where decaying mangrove leaves were fed to shore crabs (the primary consumers), which, in turn were fed to swimming crabs (the secondary consumers), the FA 18:3$\omega$3 was conservatively transferred from the source to both primary and secondary consumers, whereas 18:2$\omega$6 was not transferred between the primary and secondary consumer (Hall et al., 2006). Indeed, when the secondary consumer was starved, 18:2$\omega$6 actually accumulated in its tissues, indicating that physiological stress during starvation may have caused the synthesis of this FA, which potentially renders it inappropriate to use as a maker of the source (i.e., mangrove leafs) in the diet of higher-order consumers.

## Benefits of using multiple approaches to elucidate trophic relationships in gelatinous zooplankton

All approaches used to study trophic interactions have their own suite of advantages and limitations. To date, most studies of the trophic ecology of gelatinous zooplankton have used gut content analyses or grazing experiments. Integrating multiple approaches, however, is likely to provide more useful and accurate information. For example, analysis of gut contents can provide information about which dietary sources may be useful to include in stable isotope analyses. The stable isotopes will, in turn, provide information about whether the sources are actually assimilated by the jellyfish and modeling can be used to estimate the contributions of the different sources to the diet. Due to the need to collect sufficient quantities of material for analysis, however, stable isotopes are at present difficult to use for examining the potential contributions of microplankton or bacteria to the diets of jellyfish. In such cases, analysis of FA profiles may be helpful as the presence of markers characteristic of bacteria or microplankton in the jellyfish's tissue will provide evidence of their contribution to the diet. FAs may also be useful for identifying sources that may share similar stable isotope signatures, and therefore, are not readily distinguished using isotopic approaches.

A thorough understanding of the advantages and limitations of each approach is required for reliable interpretation of data. In all cases, rigorous temporal and spatial sampling will be required to provide robust analyses. Integrating multiple approaches will provide a more comprehensive and rigorous understanding of the trophic ecology of gelatinous zooplankton.

**Acknowledgments** We thank D. Hall and two anonymous reviewers who provided valuable feedback on this manuscript and R. Harvey for kindly providing Fig. 5.

## References

Båmstedt, U. & M. B. Martinussen, 2000. Estimating digestion rate and the problem of individual variability, exemplified by a scyphozoan jellyfish. Journal of Experimental Marine Biology and Ecology 251: 1–15.

Benstead, J. P., J. B. March, B. Fry, K. C. Ewel & C. M. Pringle, 2006. Testing isosource: stable isotope analysis of a tropical fishery with diverse organic matter sources. Ecology 87: 326–333.

Bodin, N., F. Le Loc'h & C. Hily, 2007. Effect of lipid renoval on carbon and nitrogen stable isotope ratios in crustacean tissues. Journal of Experimental Marine Biology and Ecology 341: 168–175.

Bosley, K. L. & S. C. Wainright, 1999. Effects of preservatives and acidification on the stable isotope ratios ($^{15}N$:$^{14}N$, $^{13}C$:$^{12}C$) of two species of marine animals. Canadian Journal of Fisheries and Aquatic Sciences 56: 2181–2185.

Bouillon, S., M. Korntheuer, W. Baeyens & F. Dehairs, 2006. A new automated setup for stable isotope analysis of dissolved organic carbon. Limnology and Oceanography Methods 4: 216–226.

Brodeur, R. D., H. Sugisaki & G. L. Hunt Jr., 2002. Increases in jellyfish biomass in the Bering Sea: implications for the ecosystem. Marine Ecology Progress Series 233: 89–103.

Browne, J. G. & M. J. Kingsford, 2005. A commensal relationship between the scyphozoan medusae *Catostylus mosaicus* and the copepod *Paramachronchiron maximum*. Marine Biology 146: 1157–1168.

Budge, S. M. & C. C. Parrish, 1998. Lipid biogeochemistry of plankton, settling matter 1 and sediments 2 in Trinity Bay, Newfoundland. II. Fatty acids. Organic Geochemistry 29: 1547–1559.

Bunn, S. E., N. R. Loneragan & M. A. Kempster, 1995. Effects of acid washing on stable isotope ratios of C and N in penaeid shrimp and seagrass: implications for food web studies using multiple stable isotopes. Limnology and Oceanography 40: 622–625.

Burkhardt, S., U. Riebesell & I. Zondervan, 1999. Effects of growth rate, $CO_2$ concentration, and cell size on the stable carbon isotope fractionation in marine phytoplankton. Geochimica et Cosmochimica Acta 63: 3729–3741.

Carli, A., L. Pane & T. Valente, 1991. Lipid and protein content of jellyfish from the Ligurian Sea. First results. In UNEP (United Nations Action Plan), Jellyfish Blooms in the Mediterranean. Proceedings of the II Workshop on Jellyfish in the Mediterranean Sea. Mediterranean Action Plan Technical Reports Series No. 47, UNEP, Athens: 236–240.

Connolly, R. M., M. Guest, A. J. Melville & J. Oakes, 2004. Sulfur stable isotopes separate producers in marine food-web analysis. Oecologia 138: 161–167.

Copeman, L. A. & C. C. Parrish, 2003. Marine lipids in a cold coastal ecosystem: Gilbert Bay, Labrador. Marine Biology 143: 1213–1227.

Costanzo, S. D., J. Udy, B. Longstsaff & A. Jones, 2005. Using nitrogen stable isotope ratios ($\delta^{15}N$) of macroalgae to determine the effectiveness of sewage upgrades: changes in the extent of sewage plumes over four years in Moreton Bay, Australia. Marine Pollution Bulletin 51: 212–217.

Dalsgaard, J., M. St. John, G. Kattner, D. Muller-Navarra & W. Hagen, 2003. Fatty acid trophic markers in the pelagic marine environment. Advances in Marine Biology 46: 225–340.

De Niro, M. J. & S. Epstein, 1977. Mechanism of isotope carbon fractionation associated with lipid síntesis. Science 197: 261–263.

De Souza, L. M., M. Iacomini, P. A. J. Gorin, R. S. Sari, M. A. Haddad & G. L. Sassaki, 2007. Glyco- and sphingo-phosphonolipids from the medusa *Phyllorhiza punctata*: NMR and ESI-MS/MS fingerprints. Chemistry and Physics of Lipids 145: 85–96.

Falkowski, P. G., 1991. Species variability in the fractionation of $^{13}C$ and $^{12}C$ by marine phytoplankton. Journal of Plankton Research 13(supplement): 21–28.

Falk-Petersen, S., T. M. Dahl, C. L. Scott, J. R. Sargent, B. Gulliksen, S. Kwasniewski, H. Hop & R.-M. Millar, 2002. Lipid biomarkers and trophic linkages between ctenophores and copepods in Svalbard waters. Marine Ecology Progress Series 227: 187–194.

Fancett, M. S. & G. P. Jenkins, 1988. Predatory impact of scyphomedusae on ichthyoplankton and other zooplankton in Port Phillip Bay. Journal of Experimental Marine Biology and Ecology 116: 63–77.

Flynn, B. A. & M. J. Gibbons, 2007. A note on the diet and feeding of *Chrysaora hyoscella* in Walvis Bay Lagoon, Namibia, in September 2003. African Journal of Marine Science 29: 303–307.

Frazer, T. K., R. M. Ross, L. B. Quetin & J. P. Montoya, 1997. Turnover of carbon and nitrogen during growth of larval krill, *Euphausia superba* Dana: a stable isotope approach. Journal of Experimental Marine Biology and Ecology 212: 259–275.

Fry, B., 2006. Stable Isotope Ecology. Springer Verlag, New York, USA.

Fry, B. & C. Arnold, 1982. Rapid $^{13}C/^{12}C$ turnover during growth of brown shrimp (*Penaeus aztecus*). Oecologia 54: 200–204.

Fukuda, Y. & T. Naganuma, 2001. Potential dietary effects on the fatty acid composition of the common jellyfish *Aurelia aurita*. Marine Biology 138: 1029–1035.

Gee, J. M., 1989. An ecological and economic review of meiofauna as food for fish. Journal of the Linnean Society 96: 253–261.

Gorokhova, E. & S. Hansson, 1999. An experimental study on variations in stable carbon and nitrogen isotope fractionation during growth of *Mysis mixta* and *Neomysis integer*. Canadian Journal of Fisheries and Aquatic Sciences 56: 2203–2210.

Graeve, M., G. Kattner, C. Wiencke & U. Kartsen, 2002. Fatty acid composition of Arctic and Antarctic macroalgae: indicator of phylogenetic and trophic relationship. Marine Ecology Progress Series 231: 67–74.

Graeve, M., M. Lundberg, M. Böer, G. Kattner, H. Hop & S. Falk-Petersen, 2008. The fate of dietary lipids in the Arctic ctenophore *Mertensia ovum* (Fabricius 1780). Marine Biology 153: 643–651.

Gribsholt, B., H. T. S. Boschker, E. Struyf, M. Andersson, A. Tramper, L. De Brabandere, S. van Damme, N. Brion, P. Meire, F. Dehairs, J. J. Middelburg & C. H. R. Heip, 2005. Nitrogen processing in a tidal freshwater marsh: a whole-ecosystem N-15 labelling study. Limnology and Oceanography 50: 1945–1959.

Hall, D., S. Y. Lee & T. Meziane, 2006. Fatty acids as trophic tracers in an experimental estuarine food chain: tracer transfer. Journal of Experimental Marine Biology and Ecology 336: 42–53.

Hamilton, S. K., S. J. Sippel & S. E. Bunn, 2005. Separation of algae from detritus for stable isotope or ecological stoichiometry studies using density fractionation in colloidal silica. Limnology and Oceanography Methods: 3: 149–157.

Hansson, L. J., O. Moeslund, T. Kiørboe & H. U. Riisgård, 2005. Clearance rates of jellyfish and their potential

predation impact on zooplankton and fish larvae in a neritic ecosystem (Limfjorden, Denmark). Marine Ecology Progress Series 304: 117–131.
Harland, A. D., D. P. Spencer & L. M. Fixter, 1992. Lipid content of some Caribbean corals in relation to depth and light. Marine Biology 113: 357–361.
Heaton, T. H. E., 1986. Isotopic studies of nitrogen pollution in the hydrosphere and atmosphere: a review. Chemical Geology (Isotope Geoscience Section) 59: 87–102.
Heeger, T. & H. Möller, 1987. Ultrastructural observations on prey capture and digestion in the scyphomedusa *Aurelia aurita*. Marine Biology 96: 391–400.
Hesslein, R. H., K. A. Hallard & P. Ramlal, 1993. Replacement of sulfur, carbon, and nitrogen in tissue of growing broad whitefish (*Coregonus nasus*) in response to a change in diet traced by $\delta^{34}$C, $\delta^{13}$C, and $\delta^{15}$N. Canadian Journal of Fisheries and Aquatic Sciences 50: 2071–2076.
Howell, K. L., D. W. Pond, D. S. M. Billett & P. A. Tyler, 2003. Feeding ecology of deep-sea seastars (Echinodermata: Asteroidea): a fatty-acid biomarker approach. Marine Ecology Progress Series 255: 193–206.
Ito, M. K. & K. L. Simpson, 1996. The biosynthesis of $\omega$3 fatty acids from 18:2$\omega$6 in *Artemia* spp. Comparative Biochemistry and Physiology 115B: 67–76.
Ju, S.-J., K. Scolardi, K. L. Daly & H. R. Harvey, 2004. Understanding the trophic role of the Antarctic ctenophore, *Callianira antarctica*, using lipid biomarkers. Polar Biology 27: 782–792.
Lorrain, A., Y.-M. Paulet, L. Chauvaud, N. Savoye, A. Donval & C. Saot, 2002. Differential $\delta^{13}$C and $\delta^{15}$N signatures among scallop tissues: implications for ecology and physiology. Journal of Experimental Marine Biology and Ecology 275: 47–61.
Lucas, C. H., 1994. Biochemical composition of *Aurelia aurita* in relation to age and sexual maturity. Journal of Experimental Marine Biology and Ecology 183: 179–192.
Lynam, C. P., M. J. Gibbons, E. A. Bjørn, C. A. J. Sparks, B. G. Heywood & A. S. Brierley, 2006. Jellyfish overtake fish in a heavily fished ecosystem. Current Biology 16: R492–R493.
MacAvoy, S. E., S. A. Macko & G. C. Garman, 2001. Isotopic turnover in aquatic predators: quantifying the exploitation of migratory prey. Canadian Journal of Fisheries and Aquatic Sciences 58: 923–932.
Malej, A., J. Faganeli & J. Pezdic, 1993. Stable isotope and biochemical fractionation in the marine pelagic food-chain: the jellyfish *Pelagia noctiluca* and net zooplankton. Marine Biology 116: 565–570.
McCutchan, J. H. J. & W. M. J. Lewis, 2002. Relative importance of carbon sources for macroinvertebrates in a rocky mountain stream. Limnology and Oceanography 47: 742–752.
McCutchan, J. H., W. M. Lewis, C. Kendall & C. C. McGrath, 2003. Variation in trophic shift for stable isotope ratios of carbon, nitrogen, and sulfur. Oikos 102: 378–390.
Melville, A. J. & R. M. Connolly, 2003. Spatial analysis of stable isotope data to determine primary sources of nutrition for fish. Oecologia 136: 499–507.
Melville, A. J. & R. M. Connolly, 2005. Food webs supporting fish over subtropical mudflats are based on transported organic matter not in situ microalgae. Marine Biology 148: 363–371.
Meziane, T., F. d'Agata & S. Y. Lee, 2006. Fate of mangrove organic matter along a subtropical estuary: small-scale exportation and contribution to the food of crab communities. Marine Ecology Progress Series 312: 15–27.
Meziane, T., S. Y. Lee, P. L. Mfilinge, P. K. S. Shin, M. H. W. Lam & M. Tsuchiya, 2007. Inter-specific and geographical variations in the fatty acid composition of mangrove leaves: implications for using fatty acids as a taxonomic tool and tracers of organic matter. Marine Biology 150: 1103–1113.
Meziane, T., M. C. Sanabe & M. Tsuchiya, 2002. Role of fiddler crabs of a subtropical intertidal flat on the fate of sedimentary fatty acids. Journal of Experimental Marine Biology and Ecology 270: 191–201.
Michener, R. H. & D. M. Schell, 1994. Stable isotope ratios as tracers in marine aquatic food webs. In Lajtha, K. & R. H. Michener (eds), Stable Isotopes in Ecology and Environmental Science. Oxford Blackwell Scientific Publications, London: 138–157.
Middelburg, J. J., C. Barranguet, H. T. S. Boschker, P. M. J. Herman, T. Moens & C. H. R. Heip, 2000. The fate of intertidal microphytobenthos carbon: an in situ del$^{13}$C-labeling study. Limnology and Oceanography 45: 1224–1234.
Minagawa, M. & E. Wada, 1984. Stepwise enrichment of $^{15}$N along food chains: further evidence and the relation between $\delta^{15}$N and animal age. Geochimica et Cosmochimica Acta 48: 1135–1140.
Moens, T., L. Verbeeck, A. de Maeyer, J. Swings & M. Vincx, 1999. Selective attraction of marine bacterivorous nematodes to their bacterial food. Marine Ecology Progress Series 176: 165–178.
Montoya, J. P., S. G. Horrigan & J. J. McCarthy, 1990. Natural abundance of $^{15}$N in particulate nitrogen and zooplankton in the Chesapeake Bay. Marine Ecology Progress Series 65: 35–61.
Mutchler, T., M. J. Sullivan & B. Fry, 2004. Potential of $^{14}$N isotope enrichment to resolve ambiguities in coastal trophic relationships. Marine Ecology Progress Series 266: 27–33.
Ng, J. S. S., T-C. Wai & G. A. Williams, 2007. The effects of acidification on the stable isotope signatures of marine algae and molluscs. Marine Chemistry 103: 97–102.
Nichols, P. D., K. T. Danaher & J. A. Koslow, 2003. Occurrence of high levels of tetracosahexaenoic acid in the jellyfish *Aurelia* sp. Lipids 38: 1207–1210.
Oakes, J. M., A. T. Revill, R. M. Connolly & S. I. Blackburn, 2005. Measuring carbon isotope ratios of microphytobenthos using compound-specific stable isotope analysis of phytol. Limnology and Oceanography Methods 3: 511–519.
Palomares, M. L. D. & D. Pauly, 2008. The growth of jellyfishes. Hydrobiologia (this volume). doi: 10.1007/s10750-008-9582-y.
Papina, M., T. Meziane & R. Van Woesik, 2003. Symbiotic zooxanthellae provide the host-coral *Montipora digitata* with polyunsaturated fatty acids. Comparative Biochemistry and Physiology Part B 135: 533–537.

Pel, R., H. Hoogveld & V. Floris, 2003. Using the hidden isotopic heterogeneity in phyto- and zooplankton to unmask disparity in trophic carbon transfer. Limnology and Oceanography 48: 2200–2207.

Phillips, D. L. & J. W. Gregg, 2001. Uncertainty in source partitioning using stable isotopes. Oecologia 127: 171–179.

Phillips, D. L. & J. W. Gregg, 2003. Source partitioning using stable isotopes: coping with too many sources. Oecologia 136: 261–269.

Phillips, D. L., S. D. Newsome & J. W. Gregg, 2005. Combining sources in stable isotope mixing models: alternative methods. Oecologia 144: 520–527.

Phleger, C. F., P. D. Nichols & P. Virtue, 1998. Lipids and trophodynamics of Antarctic zooplankton. Comparative Biochemistry and Physiology Part B 120: 311–323.

Pitt, K. A., A. L. Clement, R. M. Connolly & D. Thibault-Botha, 2008. Predation by jellyfish on large and emergent zooplankton: implications for benthic-pelagic coupling. Estuarine, Coastal and Shelf Science 76: 827–833.

Purcell, J. E., 1992. Effects of predation by the scyphomedusan *Chrysaora quinquecirrha* on zooplankton populations in Chesapeake Bay, USA. Marine Ecology Progress Series 87: 65–76.

Purcell, J. E., 1997. Pelagic cnidarians and ctenophores as predators: selective predation, feeding rates and effects on prey populations. Annales de l'Institut océanographique, Paris 73: 125–137.

Purcell, J. E., 2003. Predation on zooplankton by large jellyfish, *Aurelia labiata, Cyanea capillata*, and *Aequorea aequorea*, in Prince William Sound, Alaska. Marine Ecology Progress Series 246: 137–152.

Purcell, J. E., F. P. Cresswell, D. G. Cargo & V. S. Kennedy, 1991. Differential ingestion and digestion of bivalve larvae by the scyphozoan *Chrysaora quinquecirrha* and the Ctenophore *Mnemiopsis leidyi*. Biological Bulletin 180: 103–111.

Quoy, J. R. C. & J. P. Gaimard, 1824. Voyage de l'Uranie. Traité Zool 4: 712.

Rolff, C., 2000. Seasonal variation in $\delta^{13}$C and $\delta^{15}$N of size-fractionated plankton at a coastal station in the northern Baltic proper. Marine Ecology Progress Series 203: 47–65.

Sakano, H., E. Fujiwara, S. Nohara & H. Ueda, 2005. Estimation of nitrogen stable isotope turnover rate of *Oncorhynchus nerka*. Environmental Biology of Fishes 72: 13–18.

Sargent, J. R., R. J. Parkes, I. Muellere-Harvey & R. J. Henderson, 1987. Lipid biomarkers in marine ecology. In Sleigh, M. A. (ed.), Microbes in the Sea. Ellis Horwood Ltd, Chichester: 119–138.

Schmidt, K., A. Atkinson, D. Stubing, J. W. McClelland, J. P. Montoya & M. Voss, 2003. Trophic relationships among Southern Ocean copepods and krill: some uses and limitations of a stable isotope approach. Limnology and Oceanography 48: 277–289.

Stoecker, D. K., A. E. Michaels & L. H. Davis, 1987. Grazing by the jellyfish, *Aurelia aurita*, on microzooplankton. Journal of Plankton Research 9: 901–915.

Tarboush, R. A., S. E. MacAvoy, S. A. Macko & V. Connaughton, 2006. Contribution of catabolic tissue replacement to the turnover of stable isotopes in *Danio rerio*. Canadian Journal of Zoology 84: 1453–1460.

Tieszen, L. L., T. W. Boutton, K. G. Tesdahl & N. A. Slade, 1983. Fractionation and turnover of stable carbon isotopes in animal tissues: Implications for $\delta^{13}$C analysis of diet. Oecologia 57: 32–37.

Toonen, R. J. & R. Chia, 1993. Limitations of laboratory assessments of coelenterate predation: container effects on the prey selection of the limnomedusa, *Proboscidactyla flavicirrata* (Brandt). Journal of Experimental Marine Biology and Ecology 167: 215–235.

Towanda, T. & E. V. Thuesen, 2006. Ectosymbiotic behaviour of *Cancer gracilis* and its trophic relationships with its host *Phacellophora camtschatica* and the parasitoid *Hyperia medusarum*. Marine Ecology Progress Series 315: 221–236.

Van der Zanden, M. J. & J. B. Rasmussen, 2001. Variation in del$^{15}$N and del$^{13}$C trophic fractionation: implications for aquatic food web studies. Limnology and Oceanography 46: 2061–2066.

West, J. B., G. J. Bowen, T. E. Cerling & J. R. Ehleringer, 2006. Stable isotopes as one of nature's ecological recorders. Trends in Ecology and Evolution 21: 408–414.

Winning, M. A., R. M. Connolly, N. R. Loneragan & S. E. Bunn, 1999. $^{15}$N enrichment as a method of separating the isotopic signatures of seagrass and its epiphytes for food web analysis. Marine Ecology Progress Series 189: 289–294.

Hydrobiologia (2009) 616:133–149
DOI 10.1007/s10750-008-9584-9

JELLYFISH BLOOMS

# Influence of jellyfish blooms on carbon, nitrogen and phosphorus cycling and plankton production

**Kylie A. Pitt · David T. Welsh · Robert H. Condon**

Published online: 22 September 2008


**Abstract** Due to their boom and bust population dynamics and the enormous biomasses they can attain, jellyfish and ctenophores can have a large influence on the cycling of carbon (C), nitrogen (N) and phosphorus (P). This review initially summarises the biochemical composition of jellyfish, and compares and contrasts the mechanisms by which non-zooxanthellate and zooxanthellate jellyfish acquire and recycle C, N and P. The potential influence of elemental cycling by populations of jellyfish on phytoplankton and bacterioplankton production is then assessed. Non-zooxanthellate jellyfish acquire C, N and P predominantly through predation on zooplankton with smaller contributions from the uptake of dissolved organic matter. C, N and P are regenerated via excretion of inorganic (predominantly ammonium ($NH_4^+$) and phosphate ($PO_4^{3-}$)) and dissolved organic forms (e.g. dissolved free amino acids and dissolved primary amines). Inorganic nutrients excreted by jellyfish populations provide a small but significant proportion of the N and P required for primary production by phytoplankton. Excretion of dissolved organic matter may also support bacterioplankton production but few data are available. In contrast, zooxanthellate medusae derive most of their C from the translocation of photosynthetic products, exhibit no or minimal net release of N and P, and may actively compete with phytoplankton for dissolved inorganic nutrients. Decomposition of jellyfish blooms could result in a large release of inorganic and organic nutrients and the oxygen demand required to decompose their tissues could lead to localised hypoxic or anoxic conditions.

**Keywords** Cnidaria · Ctenophora · Nutrient recycling · Bacterioplankton · Phytoplankton

Guest editors: K. A. Pitt & J. E. Purcell
Jellyfish Blooms: Causes, Consequences, and Recent Advances

K. A. Pitt (✉) · D. T. Welsh
Australian Rivers Institute – Coast and Estuaries, Griffith University, Gold Coast Campus, QLD 4222, Australia
e-mail: k.pitt@griffith.edu.au

R. H. Condon
Virginia Institute of Marine Science, College of William and Mary, 1208 Greate Rd, Gloucester Point, VA 23062, USA

## Introduction

Jellyfish and ctenophore populations are characterised by large and rapid fluctuations in abundances and often represent a substantial proportion of the pelagic consumer biomass. Due to their sheer abundance and boom and bust population dynamics, jellyfish and ctenophores are likely to influence carbon (C), nitrogen (N) and phosphorus (P) cycling in the ecosystems they inhabit.

Jellyfish and ctenophores acquire C, N and P by assimilating organic compounds from ingested prey, taking up small amounts of dissolved organic material, and some species actively take up dissolved inorganic forms. A proportion of the elements ingested become incorporated into their biomass and undigested material is egested as faeces or released via 'sloppy feeding'. Organic forms of C and N are recycled to the environment as mucus, and both organic and inorganic metabolic products are excreted. During bloom formation, when both individuals and populations are increasing in size, jellyfish and ctenophores act as a net sink for C, N and P; however, they are ephemeral organisms whose populations may decline rapidly. During decomposition, the elements bound within their biomass are regenerated en masse to the water column as dissolved inorganic and organic compounds. Organic C regenerated via mucus production and decomposition may support microbial production, whilst inorganic N and P regenerated by excretion and decomposition may support algal production.

One coronate jellyfish (*Linuche unguiculata* Schwartz) and many rhizostome species (including members of the genera *Cassiopea*, *Mastigias* and *Phyllorhiza*) form symbioses with zooxanthellae. In zooxanthellate medusae, translocation of photosynthetic products from the zooxanthellae is likely to be the major source of C for the host (Balderston & Claus, 1969; Cates, 1975). Also, inorganic excretory products are often translocated from the host to the zooxanthellae instead of being released to the external environment. Consequently, the way in which zooxanthellate jellyfish cycle C, N and P is likely to be very different to non-zooxanthellate taxa.

The aim of this paper is to review the biochemical composition of jellyfish and ctenophores, detail the processes by which they accumulate and release inorganic and organic C, N and P, compare and contrast the influence of zooxanthellate and non-zooxanthellate medusae on C, N and P cycling and examine the extent to which elemental recycling by jellyfish and ctenophore populations could support algal and microbial production.

## Biochemical composition of jellyfish and ctenophores

Due to their large water and salt content, the total organic matter of jellyfish and ctenophores is very small (typically <3% of wet weight: Larson, 1986; Clarke et al., 1992). Proteins are consistently the most abundant organic fraction in both medusae and ctenophores (72 ± 14% of total organic matter (TOM)), followed by lipids (22 ± 12%) and carbohydrates (7 ± 5%; Fig. 1). In jellyfish, proteins serve a variety of functions. Structural proteins occur in the nematocyst capsules (Hessinger & Lenhoff, 1988) and the collagen fibres of the mesoglea. Enzymes (Båmstedt, 1988; Arai, 1997, and references within), toxins (Hessinger & Lenhoff, 1988) and pigments (Blanquet & Phelan, 1987) also largely are composed of proteins. The large protein content in the tissues of medusae and ctenophores is reflected in their low molar C:N (4.5 ± 1.1; Fig. 2). Large quantities of N-rich collagen fibres and inorganic N as $NH_4^+$ in the mesoglea most likely drive the low overall C:N (Arai, 1997).

## Uptake of C, N and P by jellyfish and ctenophores

### Assimilation of C, N and P via predation

Both non-zooxanthellate and zooxanthellate jellyfish acquire C, N and P by ingestion of prey. Predation by jellyfish on zooplankton has been a major focus of

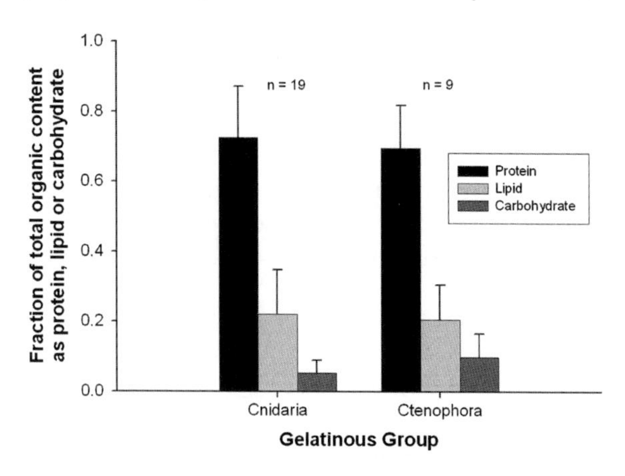


**Fig. 1** Comparison of the average (±SD) protein, lipid and carbohydrate content (expressed as fraction of total organic content) in jellyfish and ctenophores. Values are taken from Ikeda (1972), Percy & Fife (1981), Hoeger (1983), Schneider (1988), Youngbluth et al. (1988), Arai et al. (1989), Clarke et al. (1992), Malej et al. (1993), Lucas (1994), Bailey et al. (1994a, b, 1995), Finenko et al. (2001), Yousefian & Kideys (2003) and Anninsky et al. (2005). $n$ = sample size of data used in analyses

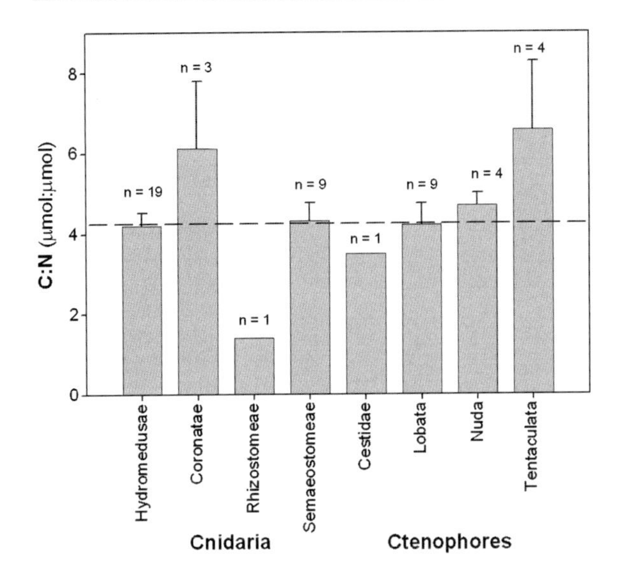


**Fig. 2** Average (±SD) C:N (mol:mol) ratios of various families of jellyfish (Hydromedusae, Coronatae, Rhizostomeae, Semaeostomeae) and ctenophores (Cestidae, Lobata, Nuda, Tentaculata). Dotted line represents average C:N for dataset (4.5 ± 1.1 SD, see text for further detail). C:N values are calculated from Kremer (1977, 2005), Ceccaldi et al. (1978), Reeve et al. (1978), Hoeger (1983), Purcell & Kremer (1983), Shenker (1985), Larson (1986), Lutcavage & Lutz (1986), Kremer et al. (1986), Gorsky et al. (1988), Schneider (1988), Youngbluth et al. (1988), Malej (1989), Clarke et al. (1992), Nemazie et al. (1993), Bailey et al. (1994a, b, 1995), Kasuya et al. (2000), Kinoshita et al. (2000), Yousefian & Kideys (2003) and Uye & Shimauchi (2005). $n$ = sample size of data used in analyses

research and an extensive literature exists on the types of prey captured as well as their clearance and digestion rates. The literature relating to predation on zooplankton has been reviewed previously (e.g. Mills, 1995; Arai, 1997; Purcell, 1997, 2008) and will not be examined further here. Despite the relative wealth of information about predation, there are few estimates of the efficiencies with which ingested C, N and P are assimilated into jellyfish biomass, even though such measurements are critical for calculating elemental budgets. Assimilation generally is defined as the difference between the quantity of an element ingested and the quantity egested (Conover, 1966). Since most jellyfish do not produce distinct faeces, assimilation efficiencies (i.e. (ingestion − egestion)/ingestion × 100) are difficult to quantify. Egested material (i.e. faeces and material lost via 'sloppy feeding') has generally been collected by filtration (e.g. Reeve et al., 1989; Costello, 1991). In siphonophores, however, egested material is produced by the gastrozooids as cohesive particles that can be collected intact (Purcell, 1983).

Assimilation efficiencies of C are typical of carnivores and generally exceed 70% and often exceed 90% (Table 1). The experimental conditions, however, such as the concentration of food supplied and whether food is supplied in a short pulse or continuously appears to greatly influence estimated efficiencies (Table 1) and so care needs to be taken during experiments to ensure experimental conditions reflect natural conditions as closely as possible. The assimilation efficiencies recorded for jellyfish and ctenophores are greater than those measured for doliolids (C assimilation; 20–50%; Katechakis et al., 2004), but are similar to those observed for some copepods (30–80%; Katechakis et al., 2004; generally >70% *Acartia tonsa* Dana; Besiktepe & Dam, 2002). Assimilation efficiencies for N have only been examined in siphonophores and appear to be similar to those for C (>90%; Table 1; Purcell, 1983).

Increasingly, chemical tracers (e.g. radioisotopes such as $^{14}C$) are being used to measure assimilation efficiencies (e.g. Katechakis et al., 2004). These techniques offer some advantages for estimating assimilation efficiencies in jellyfish and ctenophores because food labelled with the tracer (e.g. $^{14}C$-labelled copepods) could be fed to the predators and the amount of tracer ingested could then be compared to that remaining in the body after the evacuation of the gut, thus eliminating the need to collect faeces. Early experiments that measured assimilation efficiencies using radioisotope tracers, however, potentially were confounded because they did not account for recycling of elements through excretion or respiration over the period that assimilation was measured (Conover & Francis, 1973). If experiments are carefully designed to account for excretion and respiration, however, net assimilation can be measured (e.g. He & Wang, 2006). Such approaches have not yet been applied to jellyfish or ctenophores.

## Relative rates of heterotrophy and autotrophy in zooxanthellate jellyfish

Elemental budgets indicate that the C translocated from the zooxanthellae can exceed the C demands of *Mastigias papua* (Lesson) (McCloskey et al., 1994), *Cassiopea xamachana* Bigelow (Verde &

**Table 1** Assimilation efficiencies for jellyfish and ctenophores (ND = no data)

| Taxon | Experimental conditions | C assimilation efficiency (%) (mean ± SD or range) | N assimilation efficiency (%) (mean ± SD) | Source |
|---|---|---|---|---|
| Scyphozoa | | | | |
| *Chrysaora quinquecirrha* | High food (>200 copepods $l^{-1}$) | 88.8 ± 3.8 | ND | Condon & Nees (unpublished data) |
| *Aurelia aurita* | | 36–86 | ND | Anninsky (1988) |
| Hydrozoa | | | | |
| *Cladonema californicum* Anthomedusa | 18°C | 71 ± 4.0 | ND | Costello (1991) |
| *Stephanophyes superba* Chun, Siphonophora | 21–22°C | 92.4 ± 3.8 | 90.7 ± 6.8 | Purcell (1983) |
| *Forskalia* spp. Siphonophora | 21–22°C | 89.5 ± 3.7 | 94.2 ± 3.6 | Purcell (1983) |
| *Diphyes dispar* Siphonophora | 21–22°C | 94.3 | 93.1 | Purcell (1983) |
| *Rosacea* spp. Siphonophora | 21–22°C | 88.2 ± 11.2 | 90.5 ± 7.6 | Purcell (1983) |
| Ctenophora | | | | |
| *Mnemiopsis mccadyi* | 15 min exposure, 100 copepods $l^{-1}$, 26°C | 71.8 ± 12.0 | ND | Reeve et al. (1989) |
| | 3 h exposure to 100 copepods $l^{-1}$, 26°C | 33.3 ± 15.8 | ND | Reeve et al. (1989) |
| *Mnemiopsis leidyi* | Low food (20 – 30 copepods $l^{-1}$) | 67.5 ± 12.0 | ND | Condon & Nees (unpublished data) |
| | High food (>200 copepods $l^{-1}$) | 97.0 ± 1.1 | ND | Condon & Nees (unpublished data) |
| *Pleurobrachia bachei* L. Agassiz | 1 h exposure to: | | ND | Reeve et al. (1978) |
| | 20 copepods $l^{-1}$ | 73 ± 3.7 | | |
| | 100 copepods $l^{-1}$ | 75 ± 5.9 | | |
| | 500 copepods $l^{-1}$ | 52 ± 6.9 | | |
| | 1000 copepods $l^{-1}$ | 60 to −39 | | |

McCloskey, 1998) and *L. unguiculata* (Kremer et al., 1990). Despite this, zooxanthellate scyphomedusae still ingest zooplankton (e.g. Kremer, 2005; Peach & Pitt, 2005). Indeed, when *Cassiopea* sp. is maintained unfed under wavelengths of light appropriate for photosynthesis, the medusae shrink (Pitt & Welsh, unpublished data) indicating that zooplankton contain nutrients that are essential for normal metabolic function. Only Kremer (2005), however, has estimated the contribution of ingested C, N and P to the elemental budgets of an intact symbiosis. Heterotrophy contributed minor amounts of C (1–9%) relative to photosynthesis in *L. unguiculata*. Ingestion of prey was an important source of N, but the relative contribution ranged widely as a function of medusa size. For small medusae, ~10 times more N was acquired through predation than by uptake of dissolved sources, but for large medusae, ingestion and uptake of dissolved forms made similar contributions. For large size classes, ingestion of prey was considered to provide relatively small amounts (<30%) of P.

## Uptake of dissolved organic matter

Free amino acids (FAA) constitute a major component of the dissolved organic matter (DOM) in seawater. Non-zooxanthellate marine invertebrates can assimilate FAA but rates of uptake suggest that DOM generally supplies <10% of their metabolic requirements (Ferguson, 1982). Uptake of FAAs has been demonstrated using $^{14}C$- and $^{15}N$-labelled substrates in several hydrozoans (e.g. *Sarsia* sp. *Liriope exiqua*) and one scyphozoan (*Pelagia cyanella* Péron & Lesueur) medusae (Ferguson, 1988). The experiments were done in the oligotrophic Sargasso Sea and the FAA concentrations used ($1.48 \times 10^{-3}$ µmol $l^{-1}$) were comparable to the small concentrations that occur in the North Atlantic Ocean (Ferguson, 1988). Neritic or estuarine waters, however, generally contain much greater concentrations of DOM (e.g. Minor et al., 2006), and if rates of uptake are proportional to the concentrations available, uptake may be greater in coastal environments. Total DOM concentrations are still small, however, and only about 19% of total DOM in marine environments is labile, and therefore easily utilised and assimilated (Søndergaard & Middelboe, 1995). DOM, therefore, could only ever make a small contribution to the diet. Malej et al. (1993) compared the $\delta^{13}C$ of *Pelagia noctiluca* (Forskål) in the Adriatic Sea with literature-derived values of DOM for neritic systems, and since the medusae were ~2 ppt more enriched in $\delta^{13}C$ than DOM, they concluded that DOM was an insignificant source of C. More rigorous approaches (e.g. isotopic labelling experiments), however, are required to reliably resolve the contribution of DOM to jellyfish diets under a range of conditions.

Like non-zooxanthellate taxa, zooxanthellate medusae also can take up DOM. Quantitative rates of uptake of amino acids have been measured for zooxanthellae isolated from *C. xamachana* (Carroll & Blanquet, 1984) and for the intact symbiosis of *L. unguiculata* (Wilkerson & Kremer, 1992). In *L. unguiculata*, rates of uptake were proportional to the concentration of amino acids supplied and although amino acids were incorporated into both the animal and zooxanthellae fractions, greater concentrations occurred within the animal fraction. There is evidence for corals that amino acids may be transferred between the host and zooxanthellae (Swanson & Hoegh-Guldberg, 1998), so the appearance of the label in the zooxanthellae of *L. unguiculata* also could have occurred via the translocation of dissolved N from the animal. It remains unclear, therefore, as to whether differences in concentrations between the host and zooxanthellae resulted from differences in rates of uptake, a net transfer of amino acids from the zooxanthellae to the host or an internal recycling of N.

## Uptake of dissolved inorganic C, N and P by zooxanthellate jellyfish

Unlike their animal hosts, zooxanthellae from medusae can assimilate dissolved inorganic C (Hofmann & Kremer, 1981), N and P (Wilkerson & Kremer, 1992) from the water column. Thus, zooxanthellate medusae can utilise nutrient pools that are not available to non-zooxanthellate taxa. Uptake has been demonstrated by depletion of nutrient pools (e.g. Pitt et al., 2005) as well as by isotopic tracer experiments (e.g. Hofmann & Kremer, 1981; Wilkerson & Kremer, 1992). Although dissolved $CO_2$ could be taken up by zooxanthellae directly, concentrations of aqueous $CO_2$ are small and, at the normal pH of seawater, inorganic C is most abundant as $HCO_3^-$. $^{14}C$ labelling showed that medusae of *Cassiopea andromedea* (Forskål) took up $HCO_3^-$ from seawater (Hofmann & Kremer, 1981). $HCO_3^-$ then is converted to $CO_2$ using carbonic anhydrase (Weis et al., 1989). Since rates of C fixation during photosynthesis are several times greater than C respiration (e.g. Kremer et al., 1990; Verde & McCloskey, 1998), respiration alone could not meet the $CO_2$ demands for photosynthesis. The majority of DIC, therefore, is likely to be obtained from the water column.

Zooxanthellate medusae take up ammonium in preference to $NO_x$ (nitrite [$NO_2^-$] + nitrate [$NO_3^-$]) (Muscatine & Marian, 1982; Wilkerson & Kremer, 1992; Pitt et al., 2005). Based on depletion experiments, $NO_x$ does not appear to be taken up by *Mastigias* sp. (Muscatine & Marian, 1982), *Phyllorhiza punctata* von Lendenfeld (Pitt et al., 2005) or *L. unguiculata* (Wilkerson & Kremer, 1992). Depletion experiments, however, provide fairly crude information about net uptake/production. Isotopic labelling provides much more sensitivity and information about pathways and fluxes. Wilkerson & Kremer (1992) used $^{15}N$ labelling of $NO_x$ and $NH_4^+$ to show that intact symbioses of *L. unguiculata* took

up $NH_4^+$ approximately 10 times faster and isolated zooxanthellae took up $NH_4^+$ >45 times faster than $NO_3$. When the symbioses were immersed in $^{15}NH_4^+$, the $^{15}N$ appeared in both the zooxanthellae and animal fractions, but concentrations were approximately twice as large in the zooxanthellae. There was no indication of a transfer of DIN between the zooxanthellae and host.

Symbiotic, N-fixing cyanobacteria have recently been discovered in anthozoan cnidarians (Lesser et al., 2004). The cyanobacteria may be an important source of N for corals, whose growth is considered to be N-limited. Whether such symbionts occur in scyphozoans is unknown.

Uptake of inorganic P in zooxanthellate scyphomedusae has been examined only for *L. unguiculata* (Wilkerson & Kremer, 1992) and *P. punctata* (Pitt et al., 2005). *Linuche unguiculata* can deplete $PO_4^{3-}$ from seawater and rates of depletion were similar regardless of whether the medusae were fed. Uptake of $PO_4^{3-}$ continued in darkness, however, if starved and retained in the dark, rates of uptake decreased and net excretion was observed after 4 days. These results contrast with those observed for *P. punctata*. Net excretion of $PO_4^{3-}$ was observed during the day and night even when excess of $PO_4^{3-}$ was available (Pitt et al., 2005). A recent study by West and colleagues (unpublished data) indicates that the duration of retention of N by the zooxanthellate scyphomedusa *Cassiopea* sp. depends upon the availability of particulate and dissolved sources of N. The differences in P uptake/excretion observed for *L. unguiculata* and *P. punctata* may similarly relate to the availability of P. *Linuche unguiculata* inhabits oligotrophic tropical waters that may have limited availability of P, resulting in retention and active uptake of dissolved P. In contrast, P may not be limited in the coastal lagoons and embayments inhabited by *P. punctata*, resulting in net excretion by this species. Further data on the effects of availability of P on rates of uptake/excretion and also N:P ratios of the species is required to explain these observations. Interestingly, *P. punctata* excreted $PO_4^{3-}$ at only 20% of the rate observed in the co-occurring, non-zooxanthellate *Catostylus mosaicus* (Quoy & Gaimard). This suggests that, like $NH_4^+$, some of the inorganic P generated by the host metabolism is translocated to the zooxanthellae in this species.

## Release of C, N and P

Elements assimilated by jellyfish are released via excretion of dissolved inorganic and organic forms, as mucus, and, following death, by decomposition of the tissues. Studies of excretion have mostly focused on inorganic forms and, for most species, the proportions of elements excreted as inorganic and organic forms are unknown. The few studies that have examined excretion of both forms have produced inconsistent results. For example, the ctenophore *Mnemiopsis leidyi* A. Agassiz, releases about 38% of C as dissolved organic C (DOC), 46% of N as dissolved organic N (DON) and 28% of P as dissolved organic P (DOP) (Kremer, 1977). Similarly, the hydromedusa *Cladonema californicum* Hyman was estimated to release approximately 50% of N as DON (Costello, 1991). In both Kremer (1977) and Costello (1991), however, handling of animals may have artificially increased rates of excretion of organic matter (Kremer, pers. comm.). These studies contrast with Shimauchi & Uye (2007) who were unable to detect any organic N or P excretia from the scyphomedusae, *Aurelia aurita* Linnaeus.

## Excretion of inorganic N and P by non-zooxanthellate jellyfish

Ammonium and urea are two common inorganic excretory products (note that urea is an inorganic molecule, despite containing C as the C is oxidised, not reduced). Although urea is commonly excreted by other types of zooplankton (e.g. crustaceans; Miller & Glibert, 1998), urea comprises a negligible proportion (<2%) of the excretia of jellyfish (Kremer, 1975). Ammonium is the dominant form of inorganic N that is excreted and so most studies have focused on this form (e.g. Kremer, 1982; Matsakis, 1992; Nemazie et al., 1993). When standardised to dry weight, rates of excretion of $NH_4^+$ by pelagic cnidarians are <10% of those of zooplankton (Jawed, 1973; Ikeda, 1974; Smith, 1982). When standardised to C biomass, however, rates of excretion are comparable to other zooplankton, indicating that gelatinous and non-gelatinous zooplankton contribute equally to nutrient recycling in the water column on an elemental weight basis (Schneider, 1990). Rates, however, are highly variable, ranging from 110 to 2319 μmol $NH_4^+$ g $C^{-1}$ $d^{-1}$ for ctenophores and 77 to 2639 μmol

$NH_4^+$ g $C^{-1}$ $d^{-1}$ for pelagic cnidarians (Schneider, 1990). Small quantities of $NO_x$ are also released by medusae (Kremer 1975; Pitt et al., 2005; Shimauchi & Uye, 2007). In all cases, $NO_x$ comprised <2% of total DIN that was released. $NO_x$, however, is probably not excreted by the jellyfish but is likely to result from the oxidation of $NH_4^+$ by nitrifying bacteria colonising the jellyfish's surface (Welsh et al., unpublished data), as has been observed for other marine invertebrates (Welsh & Castadelli, 2004; Southwell et al., 2008).

Compared to measures of N excretion, measures of P are rare. P comprises a small component of the particulate organic content of medusae and is, therefore, excreted in small quantities. For example, molar N:P ratios for inorganic excretia range from 6.9 to 11.4 for *A. aurita* (in Schneider, 1989; Shimauchi & Uye, 2007), 7.5 for *P. noctiluca* (in Malej, 1991), 8.7 for *C. mosaicus* (in Pitt et al., 2005) and 7.4 for *M. leidyi* (in Kremer 1975). The N:P ratios of jellyfish excretia are much lower than the 16N:1P required by phytoplankton, as indicated by the canonical Redfield Ratio (Redfield et al., 1963). The larger quantities of P in the excretia, relative to the body (Kremer, 1977), may indicate that surplus P is available and that N is retained preferentially (Malej, 1989; Kremer, pers. comm.).

Rates of excretion in ctenophores and scyphomedusae are greatly influenced by recent feeding history. For example, faster excretion rates are observed in animals that have fed recently compared to those that are starved, and the rate decreases as the duration of starvation increases (Kremer, 1982; Kremer et al., 1986; Malej, 1991; Table 2). Increasing the duration of starvation from 4 to 16 h also reduces rates of excretion in the scyphozoan, *P. noctiluca*, by approximately 67% (Malej, 1991). Kremer et al. (1986) observed similar declines in rates of excretion of the ctenophore *Bolinopsis vitrea* (Agassiz) but over periods of 30 h. In contrast to Malej (1991), Morand et al. (1987) claimed that rates of excretion of $NH_4^+$ (reported as $NH_3$) were constant for *P. noctiluca* that had been starved for periods ranging from a few hours to a few days.

Rates of excretion of inorganic nutrients are also positively correlated with the availability of prey (Table 2). For example, increasing concentrations of copepods from 5 to 500 $l^{-1}$ caused rates of excretion to more than double in the ctenophore *Mnemiopsis mccradyi* Mayer (Kremer, 1982). A similar doubling of excretion rates was observed for the ctenophore *Bolinopsis vitrea* when its food concentration increased from 0 to 100 copepods $l^{-1}$ (Kremer et al., 1986).

Temperature also has a substantial influence on excretion rates (Table 3). For example, rates of excretion of $NH_4^+$ were greater at warmer temperatures in the scyphomedusae *A. aurita* (Shimauchi & Uye, 2007) and *Chrysaora quinquecirrha* (Desor) (Nemazie et al., 1993), the hydrozoan *Clytia* sp. (Matsakis, 1992) and the ctenophore *M. leidyi* (Kremer, 1977; Nemazie et al., 1993). Similar effects of temperature on rates of excretion of $PO_4^{3-}$ were also observed for *A. aurita* (Shimauchi & Uye, 2007) and *M. leidyi* (Kremer, 1977).

Only Matsakis (1992) has examined the interactive effects of food concentration and temperature on rates

**Table 2** Influence of recent feeding history and food concentration on rates of excretion of $NH_4^+$ by jellyfish and ctenophores (dw = dry weight; ind. = individual)

| Species | Temperature (°C) | Feeding history/ food concentration | Specific excretion rate | Source |
|---|---|---|---|---|
| *Pelagia noctiluca* | 19 | Starved for 4 h | 48.4 ± 6.7 μmol ind.$^{-1}$ d$^{-1}$ | Malej (1991) |
| | | Starved for 16 h | 32.1 ± 6.3 μmol ind.$^{-1}$ d$^{-1}$ | |
| *Mnemiopsis mccradyi* | 22 | Freshly collected | 0.7 μg-at h$^{-1}$ g$^{-1}$ dw | Kremer (1982) |
| | | Fed | 0.88 μg-at h$^{-1}$ g$^{-1}$ dw | |
| | | Unfed | 0.52 μg-at h$^{-1}$ g$^{-1}$ dw | |
| | 22 | 0 copepods l$^{-1}$ | 0.42 μg-at h$^{-1}$ g$^{-1}$ dw | Kremer (1982) |
| | | 5 copepods l$^{-1}$ | 0.82 μg-at h$^{-1}$ g$^{-1}$ dw | |
| | | 50 copepods l$^{-1}$ | 1.41 μg-at h$^{-1}$ g$^{-1}$ dw | |
| | | 500 copepods l$^{-1}$ | 1.80 μg-at h$^{-1}$ g$^{-1}$ dw | |
| *Bolinopsis vitrea* | 25 | 0–100 copepods l$^{-1}$ | Twofold increase | Kremer (1986) |

**Table 3** Influence of temperature on excretion rates of jellyfish and ctenophores

| Species | Temperature (°C) | Size range | Specific excretion rate | Relationship between excretion rate, temperature and size (some cases only) | Source |
|---|---|---|---|---|---|
| *Pelagia noctiluca* | 16.5–24 | 10–118 $cm^3$ | | Ln $NH_4^+$ (mg-at $h^{-1}$) = 1.08 ln $V + 0.133T - 5.08$ | Morand et al. (1987) |
| *Pelagia noctiluca* | 12 | 0.06–1.17 g dw | 29.7 ± 10.3 μmol $g^{-1}$ dw $d^{-1}$ | *$NH_4^+$ (μmol g $dw^{-1}$ $d^{-1}$) = $2.55T - 24.24$ ($r^2 = 0.83$) (NB 12°C value was excluded due to poor condition of animals) | Malej (1991) |
| | 16 | 0.17–1.65 g dw | 13.9 ± 4.0 μmol $g^{-1}$ dw $d^{-1}$ | | |
| | 19 | 0.60–1.46 g dw | 28.8 ± 6.6 μmol $g^{-1}$ dw $d^{-1}$ | | |
| | 23 | 0.35–0.19 g dw | 32.4 ± 13.6 μmol $g^{-1}$ dw $d^{-1}$ | | |
| *Aurelia aurita* | 20 | 11–885 g ww | | $NH_4^+$(μmol $ind.^{-1}$ $d^{-1}$) = 0.371 $ww^{1.14}$ | Shimauchi & Uye (2007) |
| | 28 | 78–1330 g ww | | $NH_4^+$(μmol $ind.^{-1}$ $d^{-1}$) = 2.55 $ww^{0.88}$ | |
| *Chrysaora quinquecirrha* | 18–28 | 0.013–2.826 g dw | | log $NH_4^+$ = 0.134 + 0.974 log $W$ + $0.021T$ ($NH_4^+$ measured as μg-at $NH_4^+$ $ind.^{-1}$ $h^{-1}$) | Nemazie et al. (1993) |
| *Mnemiopsis leidyi* | 10.3 | 0.036–0.498 g dw | 11.6 ± 0.6 μg-at $g^{-1}$ dw $d^{-1}$ | *$NH_4^+$ = $1.6304T - 10.116$ ($r^2 = 0.73$) ($NH_4^+$ measured as μg-at g $dw^{-1}$ $d^{-1}$) | Kremer (1977) |
| | 15.8 | 0.071–0.786 g dw | 9.9 ± 0.7 | | |
| | 18.0 | 0.035–0.562 g dw | 20.2 ± 1.7 | | |
| | 20.0 | 0.081–1.078 g dw | 16.8 ± 0.8 | | |
| | 21.8 | 0.038–0.456 g dw | 25.2 ± 1.3 | | |
| | 24.5 | 0.035–0.066 g dw | 35.6 ± 1.6 | | |
| *Mnemiopsis leidyi* | 18–27 | 0.007–0.391 g dw | | log $NH_4^+$ = 1.118 + 0.605 log $W$ + $0.053T$ ($NH_4^+$ measured as μg-at $NH_4^+$ $ind.^{-1}$ $h^{-1}$) | Nemazie et al. (1993) |

$W$ = weight (g dw); ww = wet weight; $V$ = Volume ($cm^3$), $T$ = temperature (°C); other details as per Table 2

*Linear regressions calculated by authors using mean specific excretion rates provided

**Table 4** Interactive effects of temperature and food concentration on rates of excretion of $NH_4^+$ of the hydromedusa *Clytia* spp.

| Temperature (°C) | Food concentration (copepods $l^{-1}$) | Specific excretion rate (μg-at $mg^{-1}$ dw $d^{-1}$) | Relationship |
|---|---|---|---|
| 15 | 7 | 1.9 | $NH_4^+ = 0.0629T + 1.37$ ($r^2 = 0.99$) |
| | 15 | – | |
| | 25 | 2.78 | |
| | 50 | 4.58 | |
| 18 | 7 | 2.64 | $NH_4^+ = 0.0852T + 1.88$ ($r^2 = 0.99$) |
| | 15 | 3.11 | |
| | 25 | 3.80 | |
| | 50 | 6.24 | |
| 21 | 7 | 3.43 | $NH_4^+ = 0.11T + 2.49$ ($r^2 = 0.99$) |
| | 15 | – | |
| | 25 | 4.94 | |
| | 50 | 8.11 | |

Modified from Matsakis (1992); details as per Table 3

of excretion of $NH_4^+$. She observed that rates of excretion increased with increasing food concentration (Table 4). Over the same range of food concentrations, rates of excretion were greater at warmer temperatures until the temperature reached 25°C, at which point feeding was inhibited (Matsakis, 1992).

Since rates of excretion are greatly influenced by environmental conditions, direct comparisons amongst taxa from different ecosystems are problematic. Differences in the temperatures at which measurements were made have been standardised using the $Q_{10}$ law (Ikeda, 1985), but rates still range widely, even within taxa (e.g. Schneider, 1990). Purcell (2008) cautions against adjusting metabolic rates by $Q_{10}$s determined at experimentally manipulated temperatures. Such variability may reflect the differences in the feeding histories of the animals amongst the studies. More detailed information about how the feeding history influences excretion rates and rigorous reporting of the feeding histories of test animals are required to reliably standardise results. Information about the proportions of nutrients excreted as inorganic and organic forms also are required for robust comparisons amongst taxa. These factors may account for some of the large variability in rates of excretion observed amongst and within taxa.

## Excretion of dissolved organic matter

The two primary mechanisms by which jellyfish and ctenophores release DOM are by excretion and mucus production. As proteins constitute the greatest proportion of organic biomass of pelagic coelenterates, products of protein metabolism are likely to constitute the bulk of DON excreted by gelatinous zooplankton. For example, dissolved primary amines (DPA) constitute 21% and 46%, respectively, of the total N and DON excreted by *M. leidyi* (Kremer, 1977), although handling of the ctenophore during the experiment may have increased rates of excretion of organic metabolites (Kremer, pers. comm.). Likewise, dissolved free amino acids (DFAA) are also excreted in large amounts by "jellyfish" (taxa not stated) (15.0 mg N g $dw^{-1}$ $d^{-1}$), with glycine and alanine being the most abundant DFAA species (Webb & Johannes, 1967). The large DPA and DFAA content in the excretory products of jellyfish and ctenophores might be related to rapid turnover of DNA and RNA coupled with fast growth rates (Båmstedt & Skjoldal, 1980). In this case, the metabolism of DNA and RNA could also be a source of DOP. Very few measurements of excretion of DOP, however, have been made. Kremer (1975) reported that 21% of P was excreted in organic form in *M. leidyi*, but Shimauchi & Uye (2007) could not detect organic P in the excretia of *A. aurita*. Only Kremer et al. (1986) have examined factors affecting the proportion of inorganic to organic nutrients excreted. They observed that food concentration and starvation had minimal influence on the $NH_4^+$:TDN (Total Dissolved N) of the ctenophore *Bolinopsis vitrea*.

Glycoproteins are also produced by jellyfish and ctenophores. For example, *M. leidyi* releases modified aminosugar disaccharide metabolites (Cohen &

Forward, 2003). These function as kairomones (molecules that are produced by one organism and that invoke an adventitious change in behaviour or physiology of a second organism; Dicke & Sabelis, 1988) and cause crabs to alter their behaviour to avoid predation. Glycoproteins are also used by zooxanthellate medusae in the formation of pigments that filter injurious solar radiation whilst retaining photosynthetically active wavelengths for zooxanthellae (Blanquet & Phelan, 1987). Release of glycoproteins, either directly or indirectly (e.g. UV breakdown of pigment molecules), would contribute to DOC and DON pools.

Jellyfish are renowned for producing large quantities of mucus that is used in feeding and defence (Heeger & Möller, 1987; Shanks & Graham, 1988; Arai, 1997). The mucus produced by jellyfish is colloidal in nature (Wells, 2002), originating in cells in the epidermis (Heeger & Möller, 1987) and gastrodermis (Arai, 1997). Rates of production of mucus have not been quantified and the biochemical composition of mucus has been examined for only one non-zooxanthellate and one zooxanthellate scyphozoan (Ducklow & Mitchell, 1979). In *A. aurita*, the composition of the mucus resembled the organic composition of the tissues (73% protein, 27% lipid, 5% carbohydrate (total values exceed 100% due to analytical error; H. Ducklow, pers. comm.). In contrast, the mucus of the zooxanthellate medusa *Cassiopea* sp. contained a similar proportion of carbohydrate (2%) but more lipids (38%) and considerably less protein (10%).

Other sources of DOM released from jellyfish include the leaking of digestive enzymes to DOC and DON pools. Possible enzyme species include trypsin and amylase, which have high activities in *A. aurita* medusae (Båmstedt, 1988). The release of damaged cell wall and phospholipid components may also contribute to DOC and DOP pools (Arai, 1997). 'Sloppy feeding' and egestion of undigested prey are other possible mechanisms by which jellyfish could influence the recycling of DOM in marine systems, but no data exist for these processes for jellyfish.

### Recycling of C and N between the host and its symbionts in zooxanthellate jellyfish

A characteristic of symbioses involving zooxanthellae is the tight recycling of inorganic and organic compounds between the zooxanthellae and host. Most studies have been undertaken on anthozoans. Similar physiological mechanisms probably occur in scyphozoans but they mostly remain untested. In anthozoans, zooxanthellae take up dissolved compounds, but some of their C and N requirements are derived from the $CO_2$ and $NH_4^+$ produced by their hosts (Odum & Odum, 1955; Rahav et al., 1989; but see Wang & Douglas (1998) for the alternative 'nitrogen conservation' hypothesis). Inorganic C is fixed into organic forms by the zooxanthellae and recycled back to the host, predominantly as carbohydrates (Muscatine & Cernichiari, 1969; Lewis & Smith, 1971) and some N may also be translocated to the host as amino acids (Swanson & Hoegh-Guldberg, 1998). This tight coupling results in no net release of either $CO_2$ or $NH_4^+$ during periods when medusa–zooxanthellae symbioses are undergoing net photosynthesis. In the dark, a net release of $CO_2$ is observed, but despite the lack of photosynthesis, $NH_4^+$ either continues to be taken up rather than excreted (Muscatine & Marian, 1982; Wilkerson & Kremer, 1992) or $NH_4^+$ is excreted in very small quantities (at rates that are approximately 7% of those of non-zooxanthellate taxa; Pitt et al., 2005).

## Influences of jellyfish and ctenophore blooms on ecosystem level C, N and P cycles

Blooms of jellyfish and ctenophores can attain enormous biomasses and cover extensive areas (Mills, 2001; Brodeur et al., 2002; Hay, 2006). For example, Lynam et al. (2006) reported that blooms of mainly *Aequorea forskalea* (Forskål, 1775) in the Northern Benguela ($\sim$34,000 N $mi^{-2}$) attained mean biomass densities of 361 $\pm$ 22 tonnes N $mi^{-2}$ with the total biomass of the bloom estimated to be 12.2 million tonnes. Equally impressive blooms also occur at smaller scales in coastal embayments and lagoons. For example, in Lake Illawarra, Australia, abundances of *C. mosaicus* increased 30-fold over a period of only 6 weeks and attained $\sim$530 tonnes $km^{-2}$ (Pitt & Kingsford, 2003). During blooms, jellyfish may represent a substantial or even the greatest proportion of the pelagic consumer biomass (Arai, 1997; Pagés et al., 1996; 2001) and the nutrients regenerated by blooms of jellyfish via excretion, mucus production or decomposition may influence plankton production.

## Contribution of recycled inorganic N and P to phytoplankton production

Inorganic nutrients regenerated by medusae supply a small but significant proportion of those required for phytoplankton production. For example, in coastal waters, excretion of $NH_4^+$ by jellyfish blooms has been estimated to supply 8% of the N requirements for phytoplankton in Lake Illawarra, Australia (Pitt et al., 2005), 11% in the Kiel Bight, Western Baltic (Schneider, 1989), 10% in the Inland Sea of Japan (Shimauchi & Uye, 2007) and up to 4% to net microplankton production in Chesapeake Bay, USA (Nemazie et al., 1993). Similarly, a subset of the same studies estimated that $PO_4^{3-}$ excretion could provide 23% in the Kiel Bight and 21.6% in the Inland Sea of Japan of the phytoplankton's P requirements. Although the contribution may appear small, in all cases, except for Chesapeake Bay, the excretory products of jellyfish were estimated to be the second most important source of N and P for primary production, after the sediment (Schneider, 1989; Pitt et al., 2005; Shimauchi & Uye, 2007); therefore, they should be considered in nutrient budgets and models. Indeed, recycled nutrients may be particularly important in estuaries where primary production may be temporarily (e.g. after the spring phytoplankton blooms; Kemp et al., 2005) or permanently (Thingstad et al., 2005) N- or P-limited. Excretion by jellyfish would be expected to be even more important in open oligotrophic waters where allochthonous inputs of nutrients from terrestrial sources are insignificant and where thermal stratification can often greatly limit the transfer of nutrients from the nutrient-rich deep water to the productive surface waters for much of the year. This suggestion is supported by Biggs (1977) in the western North Atlantic, where $NH_4^+$ excretion by gelatinous zooplankton was estimated to provide between 39% and 63% of the N required by phytoplankton. Whilst some doubts exist about the biomass estimates used to calculate these figures (Nemazie et al., 1993), they do demonstrate the potential role that jellyfish and ctenophores could play in supporting phytoplankton production. Indeed, the estimated contributions of regenerated nutrients are likely to be very conservative since $NH_4^+$ may contribute only ~50% of N released by the medusae (Kremer, 1977; Costello, 1991). At least part of the directly excreted DON, in addition to DON that leaches from egested material and occurs in mucus, would be directly available to the phytoplankton (Bronk et al., 2006 and references therein). The remaining DON would subsequently become available to phytoplankton when bacterioplankton mineralise it to $NH_4^+$ (Fenchel et al., 1998). Unfortunately, complete N and C budgets are only available for the hydrozoan *Cladonema californicum* Hyman (Costello, 1991) and the ctenophore *Pleurobrachia* sp. (Reeve et al., 1978). Budgets for a wider range of species need to be done to enable more rigorous assessment of the potential role of jellyfish in elemental cycling and in supporting primary production.

## Contribution of dissolved organic matter to bacterioplankton production

Bacteria are the major consumers of DOM in the oceans (Kirchman, 2000). In most marine systems, the majority of bacterial production is sustained by organic matter derived from primary production or phytoplankton exudates (del Giorgio & Cole, 1998; Kirchman, 2000; Hansell & Carlson, 2002). The bacterial C demand, however, often exceeds rates of primary production (del Giorgio & Peters, 1993; del Giorgio & Cole, 1998) and bulk organic pools contain large concentrations of refractory DOM (Hansell & Carlson, 2002). At such times, the DOM excreted by gelatinous zooplankton may be an important source of C, N and P for bacterioplankton. For example, the excretion of DFAA by "jellyfish" may be a N-rich energy source for bacteria and other marine saprotrophs (Webb & Johannes, 1967). The predominant forms of DOM produced by jellyfish (e.g. DFAAs and DPAs) are labile, suggesting there could be tight coupling between excretion of DOM by jellyfish and bacterioplankton production. Only one study, however, has attempted to correlate abundances of jellyfish and bacterial activity in the field. *Periphylla periphylla* (Péron & Lesueur) undertakes diel vertical migrations in some Norwegian fjords. Riemann et al. (2006) predicted bacterial activity would be stimulated at different depths throughout the diel cycle in correlation with changes in the depth distribution of the medusae. No diel differences in bacterial activity were observed, but the depth at which the greatest biomass of jellyfish occurred coincided with the depth of maximum bacterial activity integrated over 24 h. Care must be taken, however, not to infer causation from correlative data and sampling at multiple places and over periods longer than 24 h are required before robust conclusions regarding the influence of jellyfish on bacterial activity in the field can be drawn.

## Contrasting roles of zooxanthellate and azooxanthellate medusae

The nutrient dynamics of zooxanthellate medusae are dominated by the internal recycling of organic and inorganic compounds between the jellyfish tissues and their algal symbionts (Fig. 3). Consequently, zooanthellate species, in contrast to non-zooanthellate taxa, recycle relatively small quantities of dissolved nutrients to the water column under dark conditions and can be a net sink for inorganic N and P in the light (Muscatine & Marian, 1982; Wilkerson & Kremer, 1992; Pitt et al., 2005; Todd et al., 2006). Similarly, the mucus secreted by at least some zooanthellate species contains much less N than that of non-zooanthellate species (Ducklow & Mitchell, 1979). Zooxanthellate species, therefore, can be considered as a "bottle-neck" in the N and P cycles, as organic nutrients ingested as food are actively retained by the jellyfish and their symbionts, and the majority of accumulated nutrients would become available to other parts of the ecosystem mainly through predation upon the jellyfish or when the medusae die and decompose. Therefore, whilst both non-zooxanthellate and zooxanthallate jellyfish are able to exert top-down pressure on the zooplankton community through predation, only non-zooxanthellate taxa are likely to simultaneously exert substantial 'bottom-up' influences on plankton production via the recycling of nutrients (Fig. 3). Indeed, since zooxanthellate taxa can assimilate dissolved nutrients from the surrounding water, they may even, at times, actively compete with the phytoplankton for dissolved nutrients.

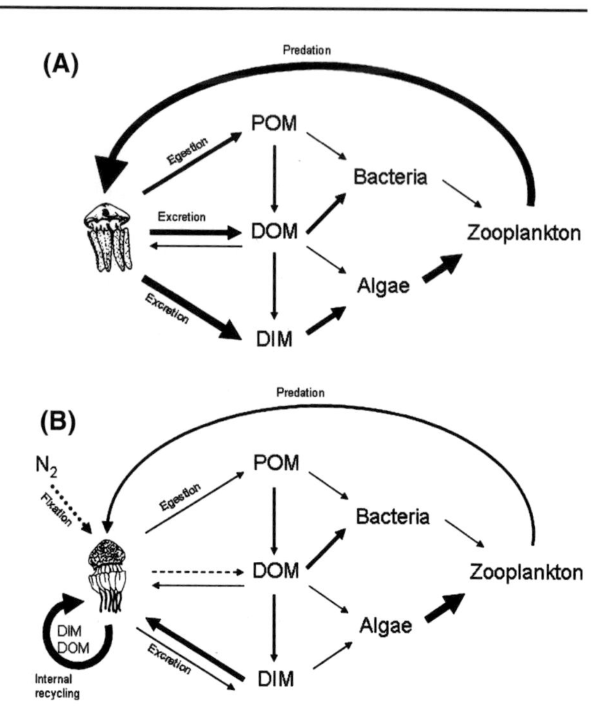


**Fig. 3** Conceptual models of the contributions of non-zooxanthellate (**A**) and zooxanthellate (**B**) medusae to nutrient cycling during growth of jellyfish blooms. The width of the arrows represents their relative contributions. Dotted arrows indicate the contribution has not been confirmed. Microbial feedback loops involving bacteria, phytoplankton and zooplankton are not included. POM = particulate organic matter, DOM = dissolved organic matter and DIM = dissolved inorganic matter

## Decomposition

Jellyfish and ctenophores can form dense populations and, therefore, represent large environmental stocks of nutrients. These stocks will become available and could have significant influences on ecosystem nutrient and oxygen dynamics when the populations die and decompose. Although some authors have proposed that bloom declines and moribund jellyfish biomass could exert significant effects on the environment (e.g. Kingsford et al., 2000; Miyake et al., 2005), decomposition of jellyfish has been examined only twice (Titelman et al., 2006; West et al., 2008). Depending on sinking rates and the depth of the water, jellyfish carcasses may decompose within the water column or on the benthos. Rates of decomposition vary from 4–7 days for *P. periphylla* in the water column (10–12°C) (Titelman et al., 2006) to ~9 days for *C. mosaicus* decomposing on sediments in a coastal lagoon (~25°C; West et al., 2008). The decay of *P. periphylla* involved both direct leaching of DOC from the jellyfish tissues and mineralisation by bacteria, with a large accumulation of $PO_4^{3-}$ and $NH_4^+$ recorded in the water column at the end of the 50-h incubation. Nitrogenous compounds appeared to be hydrolysed faster than C compounds, resulting in an initial increase in the C:N ratio of decaying medusae. Rates of release of total organic carbon were more than an order of magnitude greater ($\sim$0.36 mg C g wet weight$^{-1}$ d$^{-1}$) than that produced by living medusae (*A. aurita* $\sim$0.012 mg C g wet weight$^{-1}$ d$^{-1}$; Hansson & Norman, 1995).

To our knowledge only West et al. (2008) have studied the influence of jellyfish decomposition on sediment biogeochemistry and benthic fluxes. In that study, dead *C. mosaicus* were added to continuous flow mesocosms and benthic fluxes of oxygen and dissolved nutrients were followed for 9 days and compared with control mesocosms. Initially, there were significant peaks of organic nutrient release, but bacterial processes gradually dominated and inorganic nutrient fluxes became more important, with $NH_4^+$ efflux reaching 2 mmol m$^{-2}$ h$^{-1}$, which was close to an order of magnitude greater than the controls.

DOC released during decomposition of gelatinous zooplankton may support bacterioplankton production. Shipboard mesocosm experiments using homogenised *P. periphylla* tissue demonstrated variable results on bacterial communities, with some mesocosms having higher, and some lower, bacterial abundances than those recorded in controls (Titelman et al., 2006). Microscopic observations showed that increases in bacterial abundances in some treatments resulted from growth of specific bacterial phylotypes, indicating that whilst jellyfish tissues stimulated the growth of some bacteria, the growth of others was potentially inhibited. Subsequent experiments found that the degree of inhibition depended on both the concentration of jellyfish homogenate and the types of jellyfish tissues that were used to prepare the homogenates (Titelman et al., 2006).

Despite the paucity of direct experimental data, decomposition of medusae is also likely to have significant effects on oxygen dynamics. The oxygen required to decompose medusae can be calculated from estimates of the carbon biomass. For example, Titelman et al. (2006) observed that 0.36 mg C (g wet weight$^{-1}$ d$^{-1}$), equivalent to 30 µmol C (g wet weight$^{-1}$ d$^{-1}$), was released from *P. periphylla* as DOC. Complete oxidation of the DOC alone, depending on its biochemical composition, would require around 30 µmol of oxygen or approximately 11% of the oxygen present in a litre of oxygen-saturated seawater (Solubility = 276.0 µM at 11°C and 35 salinity). Complete oxidation of the tissues (i.e. C other than that released as DOC), would require considerably more oxygen than this and could result in local hypoxia if the decomposing biomass was large. Whilst the effects of jellyfish decomposition in the water column may be significant, decomposition on the benthos could be more severe. Recently, large accumulations of moribund scyphomedusae, *Crambionella orsini* (Vanhöffen), were observed at depths of 300–3300 m in the Arabian Sea (Billett et al., 2006). The standing stock of C in the decomposing jellyfish was estimated to vary between 1.5 and 78 g C m$^{-2}$, equivalent to 0.125–6.5 mol C m$^{-2}$. Based on C:N ratios of 4.5, the medusae also contained $\sim$0.028–1.44 mol N m$^{-2}$. Complete oxidation of the C alone would require all the oxygen in 367–19,101 l of seawater (solubility of $O_2$ = 340.4 µM at 2°C and 35 salinity), leading to potential anoxia in the boundary layer. Thus, oxygen demands for the mineralisation of C during decomposition of jellyfish could potentially cause or contribute to hypoxic events, particularly in areas with limited water exchange (e.g. coastal embayments) or where stratification inhibits mixing (e.g. some fjords).

Overall, the effects of jellyfish bloom declines and the associated decomposition processes could be similar to those caused by macro-algal blooms, which show similar "boom and bust" population dynamics. Decomposition of these blooms results in massive nutrient recycling and increases in oxygen demand can cause hypoxia, anoxia, and even dystrophy (accumulation of free sulphides in the water column), and decimation of benthic faunal communities (Viaroli et al., 1995; Valiela et al., 1997; Raffealli et al., 1998). Thus, decomposition of jellyfish blooms could have similar impacts, albeit less severe, as algal blooms can achieve significantly higher biomass densities than those of jellyfish.

## Conclusions and future directions for research

This review has highlighted several deficiencies in our understanding about the way in which jellyfish and ctenophores contribute to nutrient cycling. The calculation of elemental budgets is a primary requirement for a complete understanding of nutrient cycling. Whilst elemental budgets have been calculated for several zooxanthellate taxa (e.g. McCloskey et al., 1994; Kremer, 2005), very few budgets have been completed for non-zooxanthellate coelenterates (but see Kremer & Reeve, 1989; Costello, 1991). Moreover, assimilation efficiencies have been estimated for relatively few taxa, probably due to the difficulties in quantifying egestion rates, but are fundamental to the completion of elemental budgets. The use of chemical tracers, such as heavy- and radioisotopes may make assimilation efficiencies easier to quantify. Research on recycling of nutrients has largely focused on excretion of inorganic forms, although these may contribute only ~50% of the total metabolic products (Kremer, 1977; Costello, 1991). Indeed, greater focus on excretion of organic nutrients is required due to their potential importance in supporting bacterioplankton production. The area perhaps most lacking in information, however, is the influence of decomposing medusae on the ecology and nutrient and oxygen dynamics of the water column and benthos.

**Acknowledgements** We thank P. Kremer and an anonymous reviewer who provided valuable feedback on the manuscript. Funding was provided by grant HSF 04-10 from the Hermon Slade Foundation to K. Pitt & D. Welsh.

## References

Anninsky, B. E., 1988. The rate and efficiency of copepod assimilation by scyphoid medusa *Aurelia aurita* L. Ekologiya Morya 28: 58–64.

Anninsky, B. E., G. A. Finenko, G. I. Abolmasova, E. S. Hubareva, L. S. Svetlichny, L. Bat & A. E. Kideys, 2005. Effect of starvation on the biochemical compositions and respiration rates of ctenophores *Mnemiopsis leidyi* and *Beroe ovata* in the Black Sea. Journal of the Marine Biological Association United Kingdom 85: 549–561.

Arai, M. N., 1997. A Functional Biology of Scyphozoa. Chapman & Hall, London.

Arai, M. N., J. A. Ford & J. N. C. Whyte, 1989. Biochemical composition of fed and starved *Aequorea victoria* (Murbach et Shearer, 1902) (Hydromedusa). Journal of Experimental Marine Biology and Ecology 127: 289–299.

Bailey, T. G., J. J. Torres, M. J. Youngbluth & G. P. Owen, 1994a. Effect of decompression on mesopelagic gelatinous zooplankton: a comparison of *in situ* and shipboard measurements of metabolism. Marine Ecology Progress Series 113: 13–27.

Bailey, T. G., M. J. Youngbluth & G. P. Owen, 1994b. Chemical composition and oxygen consumption rates of the ctenophore *Bolinopsis infundibulum* from the Gulf of Maine. Journal of Plankton Research 16: 673–689.

Bailey, T. G., M. J. Youngbluth & G. P. Owen, 1995. Chemical composition and metabolic rates of gelatinous zooplankton from midwater and benthic boundary layer environments off Cape Hatteras, North Carolina, USA. Marine Ecology Progress Series 122: 121–134.

Balderston, W. L. & G. Claus, 1969. A study of the symbiotic relationship between *Symbiodinium microadriaticum* Freudenthal, a zooxanthella, and the upside down jellyfish *Cassiopea* sp. Nova Hedwigia 17: 373–382.

Båmstedt, U., 1988. Interspecific, seasonal and diel variations in zooplankton trypsin and amylase activities in Kosterfjorden, western Sweden. Marine Ecology Progress Series 44: 15–24.

Båmstedt, U. & H. R. Skjoldal, 1980. RNA concentration of zooplankton: relationship with size and growth. Limnology and Oceanography 25: 304–316.

Besiktepe, S. & H. G. Dam, 2002. Coupling of ingestion and defecation as a function of diet in the calanoid copepod *Acartia tonsa*. Marine Ecology Progress Series 229: 151–164.

Biggs, D., 1977. Respiration and ammonium excretion by open ocean gelatinous zooplankton. Limnology and Oceanography 22: 108–117.

Billett, D. S. M., B. J. Bett, C. L. Jacobs, I. P. Rouse & B. D. Wigham, 2006. Mass deposition of jellyfish in the deep Arabian Sea. Limnology and Oceanography 51: 2077–2083.

Blanquet, R. S. & M. A. Phelan, 1987. An unusual blue mesogleal protein from the mangrove jellyfish *Cassiopea xamachana*. Marine Biology 94: 423–430.

Brodeur, R., H. Sugisaki & G. J. Hunt, 2002. Increases in jellyfish biomass in the Bering Sea: implications for the ecosystem. Marine Ecology Progress Series 233: 89–103.

Bronk, D. A., J. H. See, P. Bradley & L. Killberg, 2006. DON as a source of bioavailable nitrogen for phytoplankton. Biogeosciences Discussions 3: 1247–1277.

Carroll, S. & R. S. Blanquet, 1984. Alanine uptake by isolated zooxanthellae of the mangrove jellyfish, *Cassiopea xamachana*. I. Transport mechanisms and utilization. Biological Bulletin 166: 409–418.

Cates, N., 1975. Productivity and organic consumption in *Cassiopea* and *Condylactus*. Journal of Experimental Marine Biology and Ecology 18: 55–59.

Ceccaldi, H. J., A. Kanazawa & S.-I. Teshima, 1978. Chemical composition of some Mediterranean macroplanktonic organisms. 1. Proximate analysis. Tethys 8: 295–298.

Clarke, A., L. J. Holmes & D. J. Gore, 1992. Proximate and elemental composition of gelatinous zooplankton from the Southern Ocean. Journal of Experimental Marine Biology and Ecology 155: 55–68.

Cohen, J. H. & R. B. Forward, 2003. Ctenophore kairomones and modified aminosugar disaccharides alter the shadow response in a larval crab. Journal of Plankton Research 25: 203–213.

Conover, R. J., 1966. Assimilation of organic matter by zooplankton. Limnology and Oceanography 11: 338–345.

Conover, R. J. & V. Francis, 1973. The use of radioactive isotopes to measure the transfer of materials in aquatic food chains. Marine Biology 18: 272–283.

Costello, J., 1991. Complete carbon and nitrogen budgets for the hydromedusa *Cladonema californicum* (Anthomedusa: Cladonemidae). Marine Biology 108: 119–128.

del Giorgio, P. A. & J. J. Cole, 1998. Bacterial growth efficiency in natural aquatic systems. Annual Review of Ecology and Systematics 29: 503–541.

del Giorgio, P. A. & R. H. Peters, 1993. Balance between phytoplankton production and plankton respiration in lakes. Canadian Journal of Fisheries and Aquatic Science 50: 282–289.

Dicke, M. & M. W. Sabelis, 1988. Infochemical terminology: based on cost-benefit analysis rather than compound origins. Functional Ecology 2: 131–139.

Ducklow, H. W. & R. Mitchell, 1979. Composition of mucus released by coral reef coelenterates. Limnology and Oceanography 24: 706–714.

Fenchel, T., G. M. King & T. H. Blackburn, 1998. Bacterial Biogeochemistry: The Ecophysiology of Mineral Cycling. Academic Press, San Diego: 307 pp.

Ferguson, J. C., 1982. A comparative study of the net metabolic benefits derived from the uptake and release of free amino acids by marine invertebrates. Biological Bulletin 162: 1–17.

Ferguson, J. C., 1988. Autoradiographic demonstration of the use of free amino acid by Sargasso Sea zooplankton. Journal of Plankton Research 10: 1225–1238.

Finenko, G. A., B. E. Anninsky, Z. A. Romanova, G. I. Abolmasova & A. E. Kideys, 2001. Chemical composition, respiration and feeding rates of the new alien ctenophore, *Beroe ovata*, in the Black Sea. Hydrobiologia 451: 177–186.

Gorsky, G., S. Dallot, J. Sardou, R. Fenaux, C. Carre & I. Palazzoli, 1988. C and N composition of some northwestern Mediterranean zooplankton and micronekton species. Journal of Experimental Marine Biology and Ecology 124: 133–144.

Hansell, D. A. & C. A. Carlson, 2002. Biogeochemistry of marine dissolved organic matter. Academic Press, San Diego.

Hansson, L. J. & B. Norman, 1995. Release of dissolved organic carbon (DOC) by the scyphozoan jellyfish *Aurelia aurita* and its potential influence on the production of planktonic bacteria. Marine Biology 121: 527–532.

Hay, S., 2006. Marine ecology: gelatinous bells may bring change in marine ecosystems. Current Biology 16: R679–R682.

He, X. & W.-X. Wang, 2006. Releases of ingested phytoplankton carbon by *Daphnia magna*. Freshwater Biology 51: 649–665.

Heeger, T. & H. Möller, 1987. Ultrastructural observations on prey capture and digestion in the scyphomedusa *Aurelia aurita*. Marine Biology 96: 391–400.

Hessinger, D. A. & H. M. Lenhoff, 1988. The Biology of Nematocysts. Academic Press, San Diego.

Hoeger, U., 1983. Biochemical composition of ctenophores. Journal of Experimental Marine Biology and Ecology 72: 251–261.

Hofmann, D. K. & P. Kremer, 1981. Carbon metabolism and strobilation in *Cassiopea andromedea* (Cnidaria: Scyphozoa): significance of endosymbiotic dinoflagellates. Marine Biology 65: 25–33.

Ikeda, T., 1972. Chemical composition and nutrition of zooplankton in the Bering Sea. In Takenouti, A. Y. (ed.), Biological Oceanography of the Northern Pacific Ocean. Idemitsu Shoten, Tokyo: 433–442.

Ikeda, T., 1974. Nutritional ecology of marine zooplankton. Memoirs of the Faculty of Fisheries Hokkaido University 22: 1–97.

Ikeda, T., 1985. Metabolic rates of epipelagic marine zooplankton as a function of body mass and temperature. Marine Biology 85: 1–11.

Jawed, M., 1973. Ammonia excretion by zooplankton and its significance to primary productivity during summer. Marine Biology 23: 115–120.

Kasuya, T., T. Ishimaru & M. Murano, 2000. Metabolic characteristics o the lobate ctenophore *Bolinopsis mikado* (Moser). Plankton Biology and Ecology 47: 114–121.

Katechakis, A., H. Stirbor, U. Sommer & T. Hansen, 2004. Feeding selectivities and food niche separation of *Acartia clausi*, *Penilia avirostris* (Crustacea) and *Doliolum denticulatum* (Thaliacea) in Blanes Bay (Catalan Sea, NW Mediterranean). Journal of Plankton Research 26: 589–603.

Kemp, W. M., W. R. Boynton, J. E. Adolf, D. F. Boesch, W. C. Boicourt, G. Brush, J. C. Cornwell, T. R. Fisher, P. M. Glibert, J. D. Hagy, L. W. Harding, E. D. Houde, D. G. Kimmel, W. D. Miller, R. I. E. Newell, M. R. Roman, E. M. Smith & J. C. Stevenson, 2005. Eutrophication of Chesapeake Bay: historical trends and ecological interactions. Marine Ecology Progress Series 303: 1–29.

Kingsford, M. J., K. A. Pitt & B. M. Gillanders, 2000. Management of jellyfish fisheries with special reference to the order Rhizostomeae. Oceanography and Marine Biology: An Annual Review 38: 85–156.

Kinoshita, J., J. Hiromi & Y. Nakamura, 2000. Feeding of the scyphomedusa *Cyanea nozakii* on mesozooplankton. Plankton Biology and Ecology 47: 43–47.

Kirchman, D. L., 2000. Microbial Ecology of the Oceans. Wiley-Liss, New York.

Kremer, P., 1975. Excretion and body composition of the ctenophore *Mnemiopsis leidyi* (A. Agassiz): comparisons and consequences. In 10th European Symposium on Marine Biology, Ostend, Belgium: 351–362.

Kremer, P., 1977. Respiration and excretion by the ctenophore *Mnemiopsis leidyi*. Marine Biology 44: 43–50.

Kremer, P., 1982. Effect of food availability on the metabolism of the ctenophore *Mnemiopsis mccradyi*. Marine Biology 71: 149–156.

Kremer, P., 2005. Ingestion and elemental budgets for *Linuche unguiculata*, a scyphomedusa with zooxanthellae. Journal of the Marine Biological Association of the United Kingdom 85: 613–625.

Kremer, P., M. F. Canino & R. W. Gilmer, 1986. Metabolism of epipelagic tropical ctenophores. Marine Biology 90: 403–412.

Kremer, P., J. Costello, J. Kremer & M. Canino, 1990. Significance of photosynthetic endosymbionts to the carbon budget of the scyphomedusa *Linuche unguiculata*. Limnology and Oceanography 35: 609–624.

Kremer, P. & M. R. Reeve, 1989. Growth dynamics of a ctenophore (*Mnemiopsis*) in relation to variable food supply. 2. Carbon budgets and growth model. Journal of Plankton Research 11: 553–574.

Larson, R. J., 1986. Water content, organic content, and carbon and nitrogen composition of medusae from the northeast Pacific. Journal of Experimental Marine Biology and Ecology 99: 107–120.

Lesser, M. P., C. H. Mazel, M. Y. Gorbunov & P. G. Falkowski, 2004. Discovery of symbiotic nitrogen-fixing cyanobacteria in corals. Science 305: 997–1000.

Lewis, D. H. & D. C. Smith, 1971. The autotrophic nutrition of symbiotic marine coelenterates with special reference to hermatypic corals. I. Movement of photosynthetic products between the symbionts. Proceedings of the Royal Society of London Series B 178: 111–129.

Lucas, C. H., 1994. Biochemical composition of *Aurelia aurita* in relation to age and sexual maturity. Journal of Experimental Marine Biology and Ecology 183: 179–182.

Lutcavage, M. & P. L. Lutz, 1986. Metabolic rate and food energy requirements of the leatherback sea turtle, *Dermochelys coriacea*. Copeia 1986: 796–798.

Lynam, C. P., M. J. Gibbons, E. A. Bjørn, C. A. J. Sparks, B. G. Heywood & A. S. Brierley, 2006. Jellyfish take over fish in a heavily fished system. Current Biology 16: R492–R493.

Malej, A., 1989. Respiration and excretion rates of *Pelagia noctiluca* (Semaeostomeae, Scyphozoa). In Proceedings of the 21st EMBS. Polish Academy of Sciences, Institute of Oceanology, Gdansk: 107–113.

Malej, A., 1991. Rates of metabolism of jellyfish as related to body weight, chemical composition and temperature UNEP: jellyfish blooms in the Mediterranean. In Proceedings of the II Workshop on Jellyfish in the Mediterranean. UNEP, Athens: 253–259.

Malej, A., J. Faganeli & J. Pezdic, 1993. Stable isotope and biochemical fractionation in the marine pelagic foodchain: the jellyfish *Pelagia noctiluca* and net zooplankton. Marine Biology 116: 565–570.

Matsakis, S., 1992. Ammonia excretion rate of *Clytia* spp. hydromedusae (Cnidaria Thecata): effects of individual dry weight, temperature and food availability. Marine Ecology Progress Series 84: 55–63.

McCloskey, L. R., L. Muscatine & F. P. Wilkerson, 1994. Daily photosynthesis, respiration, and carbon budgets in a tropical marine jellyfish (*Mastigias* sp.). Marine Biology 119: 13–22.

Miller, C. & P. Glibert, 1998. Nitrogen excretion by the calanoid copepod *Acartia tonsa*: results of mesocosm experiments. Journal of Plankton Research 20: 1767–1780.

Mills, C. E., 1995. Medusae, siphonophores, and ctenophores as planktivorous predators in changing global ecosystems. ICES Journal of Marine Science 52: 575–581.

Mills, C. E., 2001. Jellyfish blooms: are populations increasing globally in response to changing ocean conditions? Hydrobiologia 451: 55–68.

Minor, E. C., J.-P. Simjouw & M. R. Mulholland, 2006. Seasonal variation in dissolved organic carbon concentrations and characteristics in a shallow coastal bay. Marine Chemistry 101: 166–179.

Miyake, H., D. J. Lindsay, M. Kitamura & S. Nishida, 2005. Occurrence of the scyphomedusa *Parumbrosa polylobata* Kishinouye, 1910 in Suruga Bay, Japan. Plankton Biology and Ecology 52: 58–66.

Morand, P., C. Carré & D. C. Biggs, 1987. Feeding and metabolism of the jellyfish *Pelagia noctiluca* (scyphomedusae, semaeostomeae). Journal of Plankton Research 9: 651–665.

Muscatine, L. & E. Cernichiari, 1969. Assimilation of photosynthetic products of zooxanthellae by a reef coral. Biological Bulletin 137: 506–523.

Muscatine, L. & R. E. Marian, 1982. Dissolved inorganic nitrogen flux in symbiotic and nonsymbiotic medusae. Limnology and Oceanography 27: 910–917.

Nemazie, D. A., J. E. Purcell & P. M. Glibert, 1993. Ammonium excretion by gelatinous zooplankton and their contribution to the ammonium requirements of microplankton in Chesapeake Bay. Marine Biology 116: 451–458.

Odum, H. T. & E. P. Odum, 1955. Trophic structure and productivity of a windward coral reef community on Eniwetok Atoll. Ecological Monographs 25: 291–319.

Pagés, F., H. E. González & S. R. González, 1996. Diet of the gelatinous zooplankton in Hardangerfjord (Norway) and potential predatory impact by *Aglantha digitale* (Trachymedusae). Marine Ecology Progress Series 139: 69–77.

Pagés, F., H. E. Gonzalez, M. Ramon, M. Sobarzo & J.-M. Gili, 2001. Gelatinous zooplankton assemblages associated with water masses in the Humboldt Current System, and potential predatory impact by *Bassia bassensis* (Siphonophora: Calycophorae). Marine Ecology Progress Series 210: 13–24.

Peach, M. B. & K. A. Pitt, 2005. Morphology of the nematocysts of the medusae of two scyphozoans *Catostylus mosaicus* and *Phyllorhiza punctata* (Rhizostomeae): implications for capture of prey. Invertebrate Biology 124: 98–108.

Percy, J. A. & F. J. Fife, 1981. The biochemical composition and energy content of arctic marine macrozooplankton. Arctic 34: 307–313.

Pitt, K. A. & M. J. Kingsford, 2003. Temporal variation in the virgin biomass of the edible jellyfish, *Catostylus mosaicus* (Scyphozoa, Rhizostomeae). Fisheries Research 63: 303–313.

Pitt, K. A., K. Koop & D. Rissik, 2005. Contrasting contributions to inorganic nutrient recycling by the co-occurring jellyfishes, *Catostylus mosaicus* and *Phyllorhiza punctata* (Scyphozoa, Rhizostomeae). Journal of Experimental Marine Biology and Ecology 315: 71–86.

Purcell, J. E., 1983. Digestion rates and assimilation efficiencies of siphonophores fed zooplankton prey. Marine Biology 73: 257–261.

Purcell, J. E., 1997. Pelagic cnidarians and ctenophores as predators: selective predation, feeding rates and effects on prey populations. Annales de l'Institut océanographique, Paris 73: 125–137.

Purcell, J. E., 2008. Extension of methods for jellyfish and ctenophore trophic ecology to large-scale research. Hydrobiologia (this volume). doi:10.1007/s10750-008-9585-8.

Purcell, J. E. & P. Kremer, 1983. Feeding and metabolism of the siphonophore *Sphaeronectes gracilis*. Journal of Plankton Research 5: 95–106.

Raffealli, D. G., R. A. Raven & L. J. Poole, 1998. Ecological impact of green macroalgal blooms. Oceanography and Marine Biology: An Annual Review 36: 97–125.

Rahav, B. O., Z. Dubinsky, Y. Achituv & P. G. Falkowski, 1989. Ammonium metabolism in the zooxanthellate coral, *Stylophora pistillata*. Proceedings of the Royal Society of London Series B 236. 325–337.

Redfield, A. C., B. H. Ketchum & F. A. Richards, 1963. The influence of organisms on the composition of sea-water. In Hill, M. N. (ed.), The Sea, Vol. 2. Interscience, New York: 26–77.

Reeve, M. R., M. A. Syms & P. Kremer, 1989. Growth dynamics of a ctenophore (*Mnemiopsis*) in relation to variable food supply. I. Carbon biomass, feeding, egg production, growth and assimilation efficiency. Journal of Plankton Research 11: 535–552.

Reeve, M. R., M. A. Walter & T. Ikeda, 1978. Laboratory studies of ingestion and food utilization in lobate and tentaculate ctenophores. Limnology and Oceanography 24: 740–751.

Riemann, L., J. Titelman & U. Båmstedt, 2006. Links between jellyfish and microbes in a jellyfish dominated fjord. Marine Ecology Progress Series 325: 29–42.

Schneider, G., 1988. Chemische zusammensetzung und biomasseparameter der ohrenqualle *Aurelia aurita*. Helgolander Meeresuntersuchungen 42: 319–327.

Schneider, G., 1989. The common jellyfish *Aurelia aurita*: standing stock, excretion and nutrient regeneration in the Kiel Bight, Western Baltic. Marine Biology 100: 507–514.

Schneider, G., 1990. A comparison of carbon based ammonia excretion rates between gelatinous and non-gelatinous zooplankton: implications and consequences. Marine Biology 106: 219–225.

Shanks, A. L. & W. M. Graham, 1988. Chemical defense in a scyphomedusa. Marine Ecology Progress Series 45: 81–86.

Shenker, J. M., 1985. Carbon content of the neritic scyphomedusa *Chrysaora fuscescens*. Journal of Plankton Research 7: 169–173.

Shimauchi, H. & S. Uye, 2007. Excretion and respiration rates of the scyphomedusa *Aurelia aurita* from the Inland Sea of Japan. Journal of Oceanography 63: 27–34.

Smith, K. L. J., 1982. Zooplankton of a bathyal benthic boundary layer: in situ rates of oxygen consumption and ammonium excretion. Limnology and Oceanography 27: 461–471.

Søndergaard, M. & M. Middelboe, 1995. A cross-system analysis of labile dissolved organic carbon. Marine Ecology Progress Series 118: 283–294.

Southwell, M. W., B. N. Popp & C. S. Martens, 2008. Nitrification controls on fluxes and isotopic composition of nitrate from Florida Keys sponges. Marine Chemistry 108: 96–108.

Swanson, R. & O. Hoegh-Guldberg, 1998. Amino acid synthesis in the symbiotic sea anemone *Aiptasia pulchella*. Marine Biology 131: 89–93.

Thingstad, T. F., M. D. Krom, R. F. C. Mantoura, G. A. F. Flaten, S. Groom, B. Herut, N. Kress, C. S. Law, A. Pasternak, P. Pitta, S. Psarra, F. Rassoulzadegan, T. Tanaka, A. Tselepides, P. Wassmann, E. M. S. Woodward, C. W. Riser, G. Zodiatis & T. Zohary, 2005. Nature of phosphorus limitation in the ultraoligotrophic eastern Mediterranean. Science 309: 1068–1071.

Titelman, J., L. Riemann, T. A. Sørnes, T. Nilsen, P. Griekspoor & U. Båmstedt, 2006. Turnover of dead jellyfish: stimulation and retardation of microbial activity. Marine Ecology Progress Series 325: 43–58.

Todd, B. D., D. J. Thornhill & W. K. Fitt, 2006. Patterns of inorganic phosphate uptake in *Cassiopea xamachana*: a bioindicator species. Marine Pollution Bulletin 52: 515–521.

Uye, S. & H. Shimauchi, 2005. Population biomass, feeding, respiration and growth rates, and carbon budget of the scyphomedusa *Aurelia aurita* in the Inland sea of Japan. Journal of Plankton Research 27: 237–248.

Valiela, I., J. McClelland, J. Hauxwell, P. J. Behr, D. Hersh & K. Foreman, 1997. Macroalgal blooms in shallow estuaries: controls and ecophysiological and ecosystem consequences. Limnology and Oceanography 42: 1105–1118.

Verde, E. A. & L. R. McCloskey, 1998. Production, respiration, and photophysiology of the mangrove jellyfish *Cassiopea xamachana* symbiotic with zooxanthellae: effect of jellyfish size and season. Marine Ecology Progress Series 168: 147–162.

Viaroli, P., M. Bartoli, C. Bondavalli & M. Naldi, 1995. Oxygen fluxes and dystrophy in a coastal lagoon colonized by *Ulva rigida* (Sacca di Goro, Po River Delta, Northern Italy). Fresenius Environmental Bulletin 4: 381–386.

Wang, J.-T. & A. E. Douglas, 1998. Nitrogen recycling or nitrogen conservation in an alga-invertebrate symbiosis. Journal of Experimental Biology 201: 2445–2453.

Webb, K. L. & R. E. Johannes, 1967. Studies of the release of dissolved free amino acids by marine zooplankton. Limnology and Oceanography 12: 376–382.

Weis, V. M., G. J. Smith & L. Muscatine, 1989. A "$CO_2$ supply" mechanism in zooxanthellate cnidarians: role of carbonic anhydrase. Marine Biology 100: 195–202.

Wells, M. L., 2002. Marine colloids and trace metals. In Hansell, D. A. & C. A. Carlson (eds), Biogeochemistry of Marine Dissolved Organic Matter. Academic Press, San Diego: 367–404.

Welsh, D. T. & G. Castadelli, 2004. Bacterial nitrification activity directly associated with isolated benthic marine animals. Marine Biology 144: 1029–1037.

Wilkerson, F. P. & P. Kremer, 1992. DIN, DON and $PO_4$ flux by a medusa with algal symbionts. Marine Ecology Progress Series 90: 237–250.

West, E. J., D. T. Welsh & K. A. Pitt, 2008. Influence of decomposing jellyfish on sediment oxygen demand and nutrient dynamics. Hydrobiologia (this volume). doi: 10.1007/s10750-008-9586-7.

Youngbluth, M. J., P. Kremer, T. G. Bailey & C. A. Jacoby, 1988. Chemical composition, metabolic rates and feeding behavior of the midwater ctenophore *Bathocyroe fosteri*. Marine Biology 98: 87–94.

Yousefian, M. & A. E. Kideys, 2003. Biochemical composition of *Mnemiopsis leidyi* in the southern Caspian Sea. Fish Physiology and Biochemistry 29: 127–131.

Hydrobiologia (2009) 616:151–160
DOI 10.1007/s10750-008-9586-7



# Influence of decomposing jellyfish on the sediment oxygen demand and nutrient dynamics

**Elizabeth Jane West · David Thomas Welsh · Kylie Anne Pitt**

Published online: 22 September 2008


**Abstract** Jellyfish populations can grow rapidly to attain large biomasses and therefore can represent significant stocks of carbon and nitrogen in the ecosystem. Blooms are also generally short-lived, lasting for just weeks or months, after which time the population can decline rapidly, sink to the bottom and decompose. The influence of decomposing jellyfish (*Catostylus mosaicus*, Scyphozoa) on benthic dissolved oxygen and nutrient fluxes was examined in a mesocosm experiment at Smiths Lake, a coastal lagoon in New South Wales, Australia. Sediment (10 l) was placed in each of 10 mesocosms (50 × 40 cm, 30 cm deep and ~60 l volume) which were supplied a continuous flow of fresh lagoon water. One jellyfish (1.6 kg wet weight or ~25 g C $m^{-2}$) was added to each of five mesocosms, with the remaining five mesocosms serving as controls. Exchanges of dissolved oxygen, organic and inorganic nutrients between the benthos and water column were measured 14 times over a period of nine days. The addition of dead jellyfish tissue to the mesocosm sediments initially resulted in an efflux of phosphate, dissolved organic nitrogen and dissolved organic phosphorus to the water column. Dissolved organic nitrogen and dissolved organic phosphorus effluxed at rates more than 8 and 25 times greater than those measured in control mesocosms, respectively. This was probably due to the intracellular nutrients leaching from the jellyfish tissues. As decomposition proceeded, a large quantity of ammonium was released to the water column and sediment oxygen demand increased, indicating bacterial decomposition was dominant. Overall the addition of dead jellyfish caused a 454% increase in ammonium efflux and 209% increase in sediment oxygen demand over the 9-day experiment relative to the controls. The decomposition of large numbers of jellyfish after major bloom events could be a significant source of nutrients and, depending on the system, could have a major impact on primary production. Moreover, depending on the degree of mixing in the water column, decaying jellyfish may also contribute to bottom water hypoxia.

**Keywords** Decomposition · Decay · Organic matter · *Catostylus mosaicus* · Scyphozoa · Flux

Guest editors: K. A. Pitt & J. E. Purcell
Jellyfish Blooms: Causes, Consequences, and Recent Advances

E. J. West (✉) · D. T. Welsh · K. A. Pitt
Australian Rivers Institute, Griffith School of Environment, Griffith University, Gold Coast Campus, Gold Coast, QLD 4222, Australia
e-mail: e.west@griffith.edu.au

## Introduction

The appearance and disappearance of large numbers of medusae is a common characteristic of jellyfish populations (Graham, 2001; Mills, 2001). Blooms of jellyfish can attain very large biomasses and are generally short-lived, lasting for just weeks or months, after which time the population may crash rapidly (Mills, 2001). Jellyfish blooms rapidly assimilate carbon and nutrients from their planktonic prey and the nutrients assimilated in the large living biomass of jellyfish blooms can represent a significant stock of carbon and nutrients within the ecosystem (Pitt et al., this volume). Jellyfish tissue mainly consists of labile components, such as lipids, carbohydrates, and proteins, and has a low C:N ratio (Larson, 1986; Gorsky et al., 1988; Arai et al., 1989; Clarke et al., 1992). Therefore, microbial decomposition of jellyfish biomass could be rapid and result in a large and concentrated release of nutrients to the water column. Consequently, the collapse and decomposition of large jellyfish blooms could have significant impacts on ecosystem oxygen and nutrient dynamics (Hay, 2006).

In deep waters (e.g., fjords), a substantial proportion of decomposition may occur in the water column. For example, approximately 95% of the jellyfish, *Periphylla periphylla* (Péron & Lesuer, 1809), decomposed within five days when suspended in the water column in Raunefjorden, Norway (Titelman et al., 2006). During decomposition the jellyfish released large amounts of organic carbon to the water column, and while growth of some phylogenetic groups of bacteria was enhanced, growth of other groups was inhibited (Titelman et al., 2006). The rapid decay of jellyfish corpses and large release of nutrients in the water column represents an important trophic link in the ecosystem (Titelman et al., 2006). However, some moribund jellyfish would also accumulate on the surface of the sediment especially in shallow waters (Arai, 1997; Kingsford et al., 2000). For example, carcases of the giant jellyfish, *Nemopilema nomurai* (Kishinouye, 1922), sink to the seafloor upon death because the dead animals have a greater density than live animals and the surrounding seawater (Yamamoto et al., 2008). On the sediment surface, the decomposition of organic matter may enhance benthic microbial communities, causing a release of inorganic nutrients and increase in sediment oxygen demand, which then could influence nutrient and oxygen concentrations in the water column (Blackburn & Blackburn, 1993).

The presence of dead jellyfish on the seafloor was recorded as early as 1880 (Billett et al., 2006), and is evident in fossil records since the Cambrian (Hagadorn et al., 2002). Billett et al. (2006) reported a large decaying bloom of jellyfish on the seafloor off the coast of Oman, where jellyfish were observed tumbling down the continental shelf and accumulating over wide areas at depths between 350 and 3300 m. In the southwest Sea of Japan carcasses of *N. nomurai*, which can grow up to 2 m diameter and 200 kg, have been reported at densities of 0.2 to 5.1 individuals per $km^2$ (Yamamoto et al., 2008). Large numbers of jellyfish corpses are likely to be an important source of organic matter to the benthos and may play an important role in the transfer of organic matter from the pelagic zone to the seafloor and associated benthic communities (Billett et al., 2006; Yamamoto et al., 2008). There have been no previous studies that measure the influence of decomposing jellyfish on the sediment surface.

*Catostylus mosaicus* (Quoy & Gaimard, 1824) is a scyphozoan jellyfish that is very common in the estuaries of eastern Australia. Populations of *C. mosaicus* frequently form spectacular blooms that can exceed 500 ton/$km^2$ (Pitt & Kingsford, 2003a). Growth rates of *C. mosaicus* are extremely rapid and small medusae can grow up to 4.81 mm bell diameter per day (Pitt & Kingsford, 2003b). During rapid growth, *C. mosaicus* obtains nutrients from planktonic prey and a large proportion of the available nutrients in the system may potentially be incorporated into the jellyfish biomass. The entire life cycle of *C. mosaicus* may potentially be completed within 2 to 3 months, and abundances can fluctuate enormously over periods of weeks and months (Pitt & Kingsford, 2003b). When a population of *C. mosaicus* collapses, the large quantity of nutrients released during decomposition may have major effects on nutrient and oxygen dynamics. The aim of this study was to investigate the changes in sediment oxygen demand and nutrient dynamics associated with the benthic decomposition of the jellyfish, *C. mosaicus*.

## Materials and methods

### Experimental protocol

The influence of decomposing jellyfish on sediment and water column oxygen and nutrient fluxes was investigated in February and March 2006 at Smiths Lake (152°52′ E, 32°39′ S), a largely unmodified, intermittently closed and open coastal lagoon on the east coast of Australia. Approximately 10 l of evenly-mixed sediment was collected from Smiths Lake at a depth of ~0.5 m and placed into each of 10 black plastic tubs, 50 × 40 cm, 30 cm deep and 60 l capacity, hereafter called mesocosms. The mesocosms were placed in Smiths Lake at 1.5 m depth from February 10 until March 3 to allow natural sediment chemistry and redox profiles to re-establish. They were then carefully removed from the lake and placed into a large pool (305 cm diameter and 75 cm deep) located on the shore. Natural water was continuously pumped (2800 RPM pump; ONGA, Australia) from ~20 m from the shoreline of Smiths Lake (~1.5 m water depth) into a 200 l overflowing drum. Water was gravity fed from the drum to 10 pipes with taps that flowed into individual mesocosms at a rate of ~3 l $min^{-1}$. Water overflowed from the mesocosms into the pool, which maintained a constant natural temperature in the mesocosms. The water level of the pool was lower than the rim of the mesocosms which prevented any mixing of water between mesocosms. Mesocosms were therefore considered independent. Physical parameters in the mesocosms remained relatively constant during the 9-day experiment, and similar to those of Smiths Lake (pH ~8; salinity ~20 ppt; temperature ~30°C; dissolved oxygen ~94% saturation). Five *C. mosaicus* (wet weight 1.6 ± 0.0 kg and ~5 g C each) that had previously been caught with a dip net from Smiths Lake and sacrificed by freezing, were thawed and placed on the sediment surface of five randomly selected mesocosms. No additions were made to the other five mesocosms, which served as controls. The mesocosms were maintained in the dark using black plastic sheeting so that the biogeochemical processes could be measured in the absence of photosynthetic production. Water in the mesocosms was continuously stirred either by the inflow of new water, or during incubations, using small electronic water pumps placed inside each mesocosm.

Determination of sediment-water column oxygen and nutrient fluxes

In total, 14 flux incubations were done over the 9 days that the jellyfish took to visibly decompose. One flux incubation of all mesocosms was done before the addition of jellyfish to mesocosms and the timing of the other 13 was designed to ensure that both short-term and long-term changes in oxygen and nutrient fluxes would be measured. On the first 3 days of the experiment, 2 to 3 incubations were done per day; from day 4 onward, 1 incubation was done each day. During each incubation, water flow was simultaneously stopped in all mesocosms and floating lids were placed over each mesocosm to isolate them from the atmosphere. Small electronic water pumps inside each mesocosm were switched on to mix the water at a rate that did not disturb the sediments. The incubation times were varied to ensure that the oxygen exceeded 80% saturation throughout the incubation. The approximate incubation times required were predicted based on measurements made in previous incubations. Exact incubation times were recorded and ranged between 2 and 4 h. Water samples taken from each mesocosm before and immediately after incubations were analyzed for dissolved oxygen (DO), ammonium ($NH_4^+$), nitrite and nitrate ($NO_x$), phosphate ($PO_4^{3-}$), dissolved organic nitrogen (DON), dissolved organic phosphorus (DOP), total dissolved nitrogen (TDN), and total dissolved phosphorus (TDP) concentrations.

Water analyses

To measure DO, one water sample (12 ml) was collected from each mesocosm (~10 cm water depth) with a 30 ml sterile syringe (Termo, USA). Syringes had 10 cm of tubing placed on their nozzle to minimize contact with the air. Samples were carefully transferred into gas-tight 12 ml borosilicate glass vials (Labco, UK) containing a small glass ball. Water samples were immediately fixed with Winkler reagents A and B (APHA, 1999), and secured with a lid with small rubber septum (Labco, UK) to ensure that the samples were airtight. Care was taken to ensure that no bubbles were present in the samples, which were mixed using the small glass balls. Samples were stored at 4°C and analyzed within 24 h. DO concentrations were measured by the Winkler method with azide modification (APHA, 1999).

For inorganic and organic nutrient analyses, two 10 ml water samples were collected from each mesocosm (~10 cm water depth) with a 30 ml sterile syringe (Termo, USA), filtered (0.45 μm) into sterile plastic vials (Sarstedt, Australia) and frozen (−20°C). For all laboratory and field analyses, gloves were worn and glassware and plasticware were cleaned by soaking in 10% (v/v) HCl (>24 h) and rinsing with de-ionized water (Milli-Q; 18 MΩ cm) prior to use. The concentrations of inorganic nutrients, $NH_4^+$, $NO_x$, and $PO_4^{3-}$, were measured using an Easychem Plus colorimetric analyzer (Systea Analytical technologies S.r.l., Anagni, Italy). Low nutrient seawater (GF filtered) was used for the preparation of standards for an 8-point calibration and for quality controls. Certified reference materials (CRMs) from the Queensland Health Scientific Services were used to verify samples. Recovery rates between 90 and 102% were achieved.

TDN and TDP were determined following oxidation. Water samples, blanks, standards, and CRMs were autoclaved (first digestion 45 min at 121°C; second digestion 15 min at 121°C) with 2:1 ratio of sample to digestion solution (20 g $K_2S_2O_8$ and 5.5 g NaOH in 1 l Milli-Q water). This digestion oxidized all organic and inorganic nitrogen and phosphorus to $NO_x$ and $PO_4^{3-}$, respectively. TDN and TDP concentrations were measured as $NO_x$ and $PO_4^{3-}$, as described above. Concentrations of DON were calculated as the concentration of TDN minus the sum of $NH_4^+$ and $NO_x$. Similarly, concentrations of DOP were calculated as the concentration of TDP minus $PO_4^{3-}$.

Flux calculations

For each measured compound, the net flux between the sediment and water column during each incubation was calculated as:

$$\text{Flux} = \frac{(DC \times V)}{(A \times T)}$$

where *DC* is the difference in concentration (μmol) between the water samples taken before and after the incubation, *V* is the mesocosm volume (l), *A* is the sediment surface area ($m^2$), and *T* is the incubation time (h). If the flux was positive, it indicated that the compound effluxed from the sediment (or sediment and jellyfish) to the water; if negative, it indicated the net consumption of the compound by the sediment (or sediment and jellyfish).

Sediment profiles

At the end of the experiment, sediment samples were taken from each mesocosm to analyze depth profiles

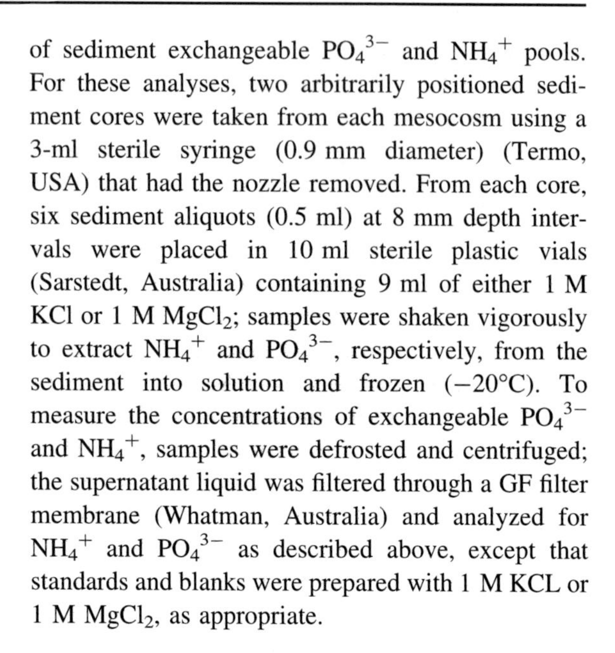

of sediment exchangeable $PO_4^{3-}$ and $NH_4^+$ pools. For these analyses, two arbitrarily positioned sediment cores were taken from each mesocosm using a 3-ml sterile syringe (0.9 mm diameter) (Termo, USA) that had the nozzle removed. From each core, six sediment aliquots (0.5 ml) at 8 mm depth intervals were placed in 10 ml sterile plastic vials (Sarstedt, Australia) containing 9 ml of either 1 M KCl or 1 M $MgCl_2$; samples were shaken vigorously to extract $NH_4^+$ and $PO_4^{3-}$, respectively, from the sediment into solution and frozen (−20°C). To measure the concentrations of exchangeable $PO_4^{3-}$ and $NH_4^+$, samples were defrosted and centrifuged; the supernatant liquid was filtered through a GF filter membrane (Whatman, Australia) and analyzed for $NH_4^+$ and $PO_4^{3-}$ as described above, except that standards and blanks were prepared with 1 M KCL or 1 M $MgCl_2$, as appropriate.

Statistical analyses

One-way Analyses of Variance (ANOVAs) were used to test for differences in fluxes of each measured compound between jellyfish and control mesocosms separately for each of the 14 incubations. One-way ANOVAs were also used to test for differences in sediment exchangeable $PO_4^{3-}$ and $NH_4^+$ pools between control and jellyfish mesocosms, separately for each sediment depth, as depths were not independent. The assumption of homoscedasticity was tested with Levene's test and data were transformed using ln(x) transformations when Levene's test was significant. If the data had negative values, the smallest number was added to each value before transformation to make the values positive (Quinn & Keough, 2002). Levene's test showed that all data transformations removed heterogeneity; therefore, analyses were done on transformed data.

## Results

The decomposition of *C. mosaicus* medusae was rapid; after 9 days, the jellyfish mesocosms were visually similar to the controls. During decomposition, the jellyfish mesocosms contained black stained, sulfide-rich sediment around the jellyfish, had an obvious white layer of sulfur oxidizing bacteria, and had a pungent smell of hydrogen sulfide.

## Sediment oxygen demand

Before jellyfish were added to the mesocosms, the sediment oxygen demand (SOD) in the jellyfish and control mesocosms were similar with a mean (±SE) of 2177 ± 272 μmol $m^{-2}$ $h^{-1}$ (Fig. 1). The mean SOD in control mesocosm incubations remained low for the entire experiment, fluctuating between 1132 ± 788 and 2590 ± 718 μmol $m^{-2}$ $h^{-1}$ (Fig. 1). Conversely, SOD in the jellyfish treatment increased considerably and was significantly greater than the controls at 15 h ($F_{1,9} = 5.5$; $P = 0.04$), 30 h ($F_{1,9} = 15.8$; $P < 0.1$), 42 h ($F_{1,9} = 5.71$; $P = 0.04$), 73 h ($F_{1,9} = 5.8$; $P = 0.04$), 85 h ($F_{1,9} = 11.1$; $P = 0.01$), and 96 h ($F_{1,9} = 8.7$; $P = 0.02$). The average SOD in the jellyfish mesocosm reached a maximum of 6161 ± 3422 μmol $m^{-2}$ $h^{-1}$ at 85 h, which was nearly 9 times greater than the control at that time. SOD in the jellyfish treatment remained high over the following 5 days, with a mean of 4418 ± 737 μmol $m^{-2}$ $h^{-1}$. Variability in the jellyfish incubations during this time was large, however, and no significant differences between the jellyfish and control mesocosms were detected. Overall, the addition of the jellyfish represented an average 209% increase in SOD compared with the controls when integrated over 9 days.

## Nutrient fluxes

The mean $NH_4^+$ (±SE) fluxes in the jellyfish and control treatments were similar at Time 0, prior to the addition of the jellyfish (13 ± 48 μmol $m^{-2}$ $h^{-1}$; Fig. 2A). The control fluxes of $NH_4^+$ remained low throughout the experiment and ranged between −70 ± 32 and 588 ± 62 μmol $m^{-2}$ $h^{-1}$ (Fig. 2A). In the jellyfish mesocosms, the efflux of $NH_4^+$ increased and reached a maximum of 2044 ± 519 μmol $m^{-2}$ $h^{-1}$ at 73 h, which was nearly 9 times greater than the control mesocosms (Fig. 2A). Jellyfish mesocosm incubations showed significantly greater efflux than controls at 42 h ($F_{1,9} = 12.0$; $P = 0.01$), 60 h ($F_{1,9} = 7.8$; $P = 0.02$), 73 h ($F_{1,9} = 12.1$; $P = 0.01$), 114 h ($F_{1,9} = 8.34$; $P = 0.02$), and 141 h ($F_{1,9} = 6.0$; $P = 0.04$) after the jellyfish were added. Thereafter, fluxes slowed and were similar to the controls. Overall, efflux of $NH_4^+$ was 454% higher in the jellyfish mesocosms compared with the controls when integrated over the entire 9 days. Mean $NO_x$ fluxes were relatively low and variable for both the jellyfish and control mesocosms for all incubations, fluctuating between −235 ± 109 and 144 ±

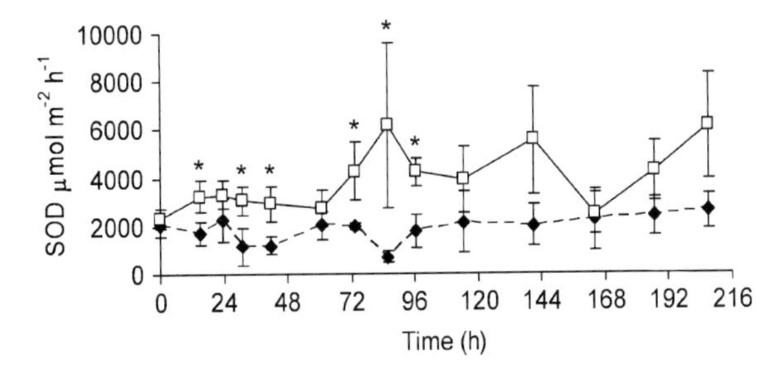


**Fig. 1** Sediment oxygen demand (SOD) during incubations of *Catostylus mosaicus* jellyfish (□) and control (◆) mesocosms before (Time 0) and after jellyfish were added. * indicates significant differences between jellyfish and control mesocosms ($P < 0.05$). Each point (mean ± SE) represents 5 replicates

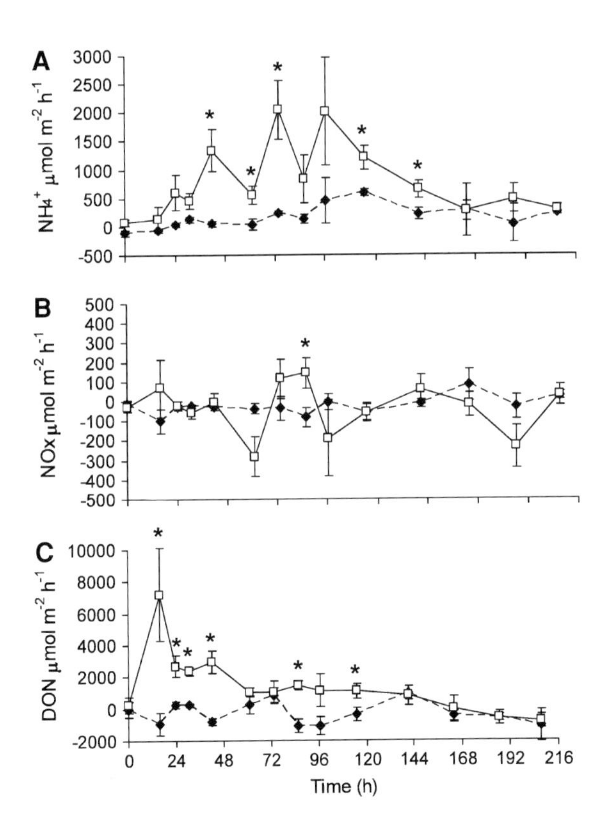


**Fig. 2** Water-column nutrient fluxes during incubations of *Catostylus mosaicus* jellyfish (□) and control (◆) mesocosms before (Time 0) and after jellyfish were added. (**A**) ammonium $NH_4^+$, (**B**) nitrite and nitrate $NO_x$ and (**C**) dissolved organic nitrogen DON. * indicates significant differences between jellyfish and control mesocosms ($P < 0.05$). Each point (mean ± SE) represents 5 replicates

78 µmol $m^{-2}$ $h^{-1}$ for the entire experiment (Fig. 2B); however, $NO_x$ efflux was significantly greater than the controls 85 h ($F_{1,9} = 6.1$; $P = 0.04$) after the jellyfish were added to the mesocosms.

Before the addition of the jellyfish, the mean fluxes of DON in the control and jellyfish mesocosms were similar, with an average of 125 ± 338 µmol $m^{-2}$ $h^{-1}$ (Fig. 2C). The flux of the control treatments remained low, averaging 403 ± 163 µmol $m^{-2}$ $h^{-1}$ during the 9 days (Fig. 2C). Following the addition of jellyfish to the mesocosms there was a rapid efflux of DON, which reached a maximum of 7182 ± 2934 µmol $m^{-2}$ $h^{-1}$, 8-fold higher than the controls (Fig. 2C). DON efflux was significantly different than the controls at 15 h ($F_{1,9} = 2.7$; $P = 0.03$), 24 h ($F_{1,9} = 11.65$; $P = 0.01$), 30 h ($F_{1,9} = 145.3$; $P < 0.1$), 42 h ($F_{1,9} = 27.1$; $P < 0.1$), 85 h ($F_{1,9} = 23.5$; $P < 0.1$), and 114 h ($F_{1,9} = 5.6$; $P = 0.04$) after the jellyfish were added to the mesocosms. Thereafter, DON effluxes in the jellyfish mesocosms gradually declined. When integrated over the entire experiment, DON effluxes were 316% greater in the jellyfish mesocosms compared with the controls. For the first 60 h of the experiment, the efflux of TDN in the jellyfish mesocosms was mainly DON (75–92%); for the remaining incubations, efflux was mostly DIN (84–100%), with the exception of the incubation at 73 h, in which TDN was comprised equally of organic and inorganic components.

Before the addition of jellyfish, mean fluxes of $PO_4^{3-}$ for the jellyfish and control mesocosms were very low and similar, with a mean of 2 ± 1 µmol $m^{-2}$ $h^{-1}$ (Fig. 3A). The $PO_4^{3-}$ flux in the jellyfish treatment was significantly greater than in the control treatment 15 h after addition of the jellyfish ($F_{1,9} = 37.6$; $P < 0.1$). During this incubation, the jellyfish mesocosm reached a maximum of 287 ± 47 µmol $m^{-2}$ $h^{-1}$ (Fig. 3A), which was 448 times greater than the controls. For the remainder of the experiment, the fluxes of $PO_4^{3-}$ in the jellyfish and control treatments were low and similar (Fig. 3A). DOP showed similar trends, with jellyfish mesocosms remaining low and similar to the controls (average −2 ± 2 µmol $m^{-2}$ $h^{-1}$), except for a rapid efflux in the jellyfish mesocosms at 15 h. This was significantly different than the control mesocosms ($F_{1,9} = 31.1$; $P < 0.1$). At 15 h, DOP in the jellyfish mesocosms reached a maximum of 725 ± 124 µmol $m^{-2}$ $h^{-1}$, which was more than 25 times higher than the mean

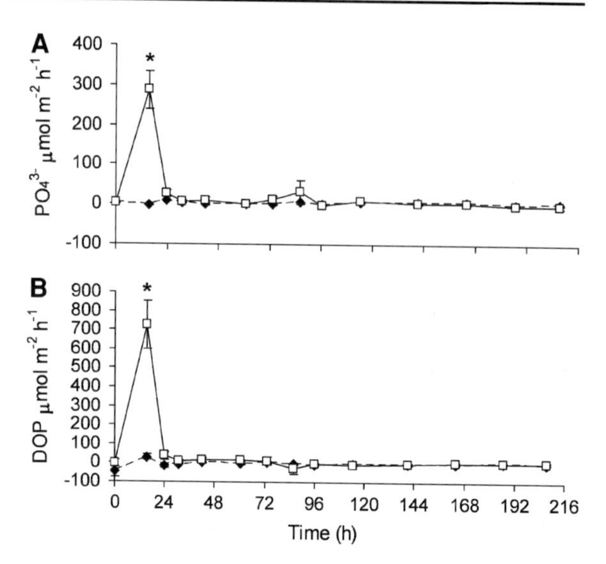


**Fig. 3** Water-column nutrient fluxes during incubations of *Catostylus mosaicus* jellyfish (□) and control (◆) mesocosms before (Time 0) and after jellyfish were added. (**A**) phosphate $PO_4^{3-}$ and (**B**) dissolved organic phosphorus (DOP). * indicates significant differences between jellyfish and control mesocosms ($P < 0.05$). Each point (mean ± SE) represents 5 replicates

control flux (Fig. 3B). TDP flux at 15 h in the jellyfish mesocosms was composed mainly of DOP (>70%).

## Sediment profiles

Depth profiles of exchangeable $NH_4^+$ were similar in the jellyfish and control treatments at all sediment depths (0.0–4.8 cm) at the end of the experiment. Exchangeable $PO_4^{3-}$ concentrations were significantly different ($F_{1,9} = 8.22$; $P = 0.02$) at sediment depth 4.0–4.8 cm, with the jellyfish mesocosms averaging 5 times greater than the controls (Fig. 4). All other sediment depths were not significantly different (Fig. 4).

## Discussion

Key factors regulating benthic nutrient regeneration are the quality, quantity, and spatial distribution of organic matter deposited on the sediment (Blackburn & Henriksen, 1983; Blackburn & Blackburn, 1993). Jellyfish tissue has a simple matrix, a high water content (>90%), and low carbon to nitrogen ratio (Shenker, 1985; Larson, 1986; Gorsky et al., 1988; Clarke et al., 1992; Shushkina et al., 2000), therefore

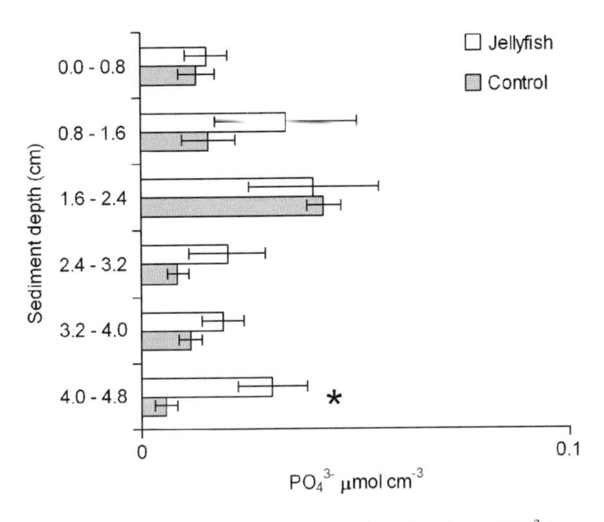


**Fig. 4** Concentration of exchangeable phosphate $PO_4^{3-}$ at different sediment depths (0.8 cm intervals) in *Catostylus mosaicus* jellyfish and control mesocosms at the end of the 9 day experiment. * indicates significant differences between jellyfish and control mesocosms ($P < 0.05$). Each point (mean ± SE) represents 5 replicates

it consists of high quality labile organic matter and is likely to rapidly decompose, causing large increases in oxygen demand and nutrient regeneration rates (Blackburn & Henriksen, 1983; Blackburn & Blackburn, 1993). In this study, the decomposition of dead, *C. mosaicus* jellyfish conformed to predictions for a high surface input of labile organic matter, because decay was rapid (9 days), organically bound nutrients were quickly recycled to the water column, and SOD increased, suggesting that microbial processes were enhanced.

The decomposition of dead jellyfish tissue initially showed a large efflux of organic nutrients. DOP was more than 25-times higher in the jellyfish mesocosms compared with the controls 15 h after the jellyfish was added. DON efflux during this time was 8-fold larger in the jellyfish mesocosms compared with the controls, and while the controls were a moderate sink for DON, the jellyfish mesocosms reversed the direction of fluxes and were a large source. From estimates of C:N ratio of *C. mosaicus* (West, unpublished data) and literature values of P (Schneider, 1990; Arai, 1997), we estimate that 5 g of C, 1.5 g of N and 0.1 g of P were in the jellyfish biomass that was added to the mesocosms. Therefore the initial efflux during the first 24 h represents ~18% of the total nitrogen and ~35% of the total phosphorus in the original jellyfish tissue. This is likely to be a result of soluble compounds rapidly leaching from dead jellyfish tissues. These results are consistent with the rapid leaching of dissolved organic carbon (DOC) observed from dead *P. periphylla* tissue in the water column (Titelman et al., 2006). Although freezing the jellyfish to kill them may have resulted in cell damage and an overestimation of the rate of leaching, these organic nutrients are likely to represent the intracellular pool that would diffuse from the cells, albeit at a slower rate than if they died naturally.

Following the initial leaching period, there was a gradual efflux of DIN from the jellyfish biomass. This efflux lasted for ~5 days and overall 455% more $NH_4^+$ effluxed from the jellyfish than the control mesocosms. The efflux of $NH_4^+$ is likely to be a result of microbial mineralization of the jellyfish tissue, which is further supported by the consumption of oxygen in the jellyfish mesocosms indicating the aerobic respiration by micro-organisms. SOD in the jellyfish mesocosms integrated over all incubations showed a 209% increase when compared with the controls. While DIP was not released to the water column after the first day of mineralization, the increased exchangeable $PO_4^{3-}$ in the sediments suggests that some of the phosphorus may have accumulated in the sediments. Exchangeable $PO_4^{3-}$ in the sediments is indirectly linked to rates of sulfide production and reoxidation (Azzoni et al., 2001). Thus, the black iron sulfide of the top sediment layers was reduced and $PO_4^{3-}$ diffused and accumulated in deeper sediments (4.0–4.8 cm), which were likely to have more iron hydroxides available. There was also no large change in the concentration of $NO_x$ in the jellyfish mesocosms compared with the controls. This is largely as expected because chemoautotrophic organisms, which are responsible for nitrification, tend to be out competed by heterotrophs for available oxygen (Herbert, 1999). Thus, whilst $NH_4^+$ was in excess in the jellyfish treatment it probably could not be extensively utilized by the nitrifiers due to the low availability of oxygen on the surface of the sediment.

Ecological stoichiometry considers the elemental composition in biological transformations, where an elemental imbalance or mismatch of consumers (e.g., microbes) and resources (e.g., organic matter) will determine ecological interactions (Sterner & Elser, 2002). For example, the decomposition of organic matter will be fastest when the elemental composition (or C:N:P) of the organic matter is similar to that of the microbes because the consumer obtains elements from the resource in the same proportions required for their own growth and reproduction; however, as organic matter decomposes, the elemental stoichiometry changes (Frost et al., 2002). In this study, the C:N of jellyfish tissue when it was first introduced to the mesocosms was ~3.9 (West, unpublished data). This ratio is similar to that of heterotrophic bacteria (~4; Sterner & Elser, 2002), which, therefore, can easily utilize the jellyfish tissue as a resource. As labile components are utilized, the residual jellyfish biomass becomes more refractile and therefore the C:N ratio increases and differs from that of the heterotrophic bacteria. Thus the continual high SOD and low effluxes of N and P in the jellyfish mesocosms may be due to preferential microbial retention of these nutrients in their biomass, as the C:N:P of the residual jellyfish tissue increased during the decomposition process.

The extent to which decomposing jellyfish will influence nutrient recycling in an estuary will largely rely on the biomass of the population. Billett et al. (2006) observed large aggregations of dead jellyfish on the seabed off the coast of Oman that ranged from a few individual jellyfish to a continuous layer of rotting jellyfish slime. These jellyfish were estimated to have a standing stock of carbon that spatially varied between 1.5 and 78 g C per $m^2$. In some areas, the downward flux of organic carbon during this time exceeded the annual average estimates by more than an order of magnitude (Billett et al., 2006). In the present study, we report the large increase in SOD and release of organic and inorganic nutrients from a jellyfish that covered ~20% of the sediment surface and was equivalent to ~25 g C per $m^2$, which is well within the range reported by Billett et al. (2006). In some areas, Billett et al. (2006) reported patches that were several metres in diameter and covered 17–100% of the sediment surface. Decomposing blooms of this scale are likely to affect nutrient concentrations in the water column and oxygen demand, which could have cascading ecosystem effects.

The recycled inorganic nutrients, which can be a limiting factor in many marine systems (Fenchel et al., 1998), are likely to enhance primary production in the overlying water column. Apparent limited top–down control of jellyfish populations often leads to the erroneous assumption that they are dead ends or sinks in marine trophic food webs (Arai, 1988, 2005). However, jellyfish have recently been identified as an important component of pelagic food webs providing inorganic nutrients to primary producers during excretion (e.g., Pitt et al., 2005) and as a source of prey for many animals including fish, turtles, and birds (Arai, 2005). Further, the large release of nutrients during decomposition of jellyfish tissue, shown in the present study, may stimulate primary production and is also likely to represent an important trophic link to pelagic environments.

The large SOD of decomposing jellyfish may result in changes in the overall ecosystem function. Oxygen depletion as a result of the breakdown of labile organic matter in coastal areas is a global concern, particularly in enclosed or stratified water bodies (Keister et al., 2000). Oxygen depletion can affect community dynamics by causing direct mortality; reducing growth and reproduction rates; limiting abundances and distributions; and altering interactions among organisms (Keister et al., 2000). Flushing and mixing regimes are likely to determine the severity of oxygen depletion (Keister et al., 2000). Impacts of blooms are likely to be particularly severe in closed or intermittently-closed waterways with limited flushing with the ocean. For example, Smiths Lake, where this study was done, opens to the ocean every 1.5 years, on average, and therefore has a long water residence time. Additionally, stratification of water bodies can result from density differences in marine and fresh water or temperature differences, which prevent vertical mixing and can isolate the bottom waters and give rise to hypoxic or anoxic conditions. One example of an anoxic event from the aerobic decomposition of animal tissues was reported in Mariager Fjord, Denmark. There, the rapid decomposition of a large population of mussels, *Mytilus edulis* (Linnaeus, 1758), consumed oxygen, which subsequently caused anoxia in the water column (Lomstein et al., 2006).

In the present study, experiments were done in isolated tubs that did not incorporate the role of macro-fauna in decomposition. Benthic macro-infauna can alter sediment particle size and pore size, and can physically stir sediments, which may alter biogeochemical processes (Welsh, 2003). Predation of jellyfish corpses would also direct nutrients to predators rather than to the water column or sediments. Yamamoto et al. (2008) caught four species of benthic scavengers in crab traps baited with dead jellyfish, *N. nomurai*, and measured an average 40% of the carcass mass was reduced after 23 h. Predation on dead jellyfish, however, is likely to be determined by factors, such as the number of corpses and types and abundances of scavengers. Further, sulfidic black sediments and reduced oxygen concentrations resulting from decaying jellyfish may kill some macro-fauna. Further studies on predation rates on jellyfish corpses and effects on benthic macro-faunal communities are required.

## Conclusion

Decomposition of *C. mosaicus* resulted in a rapid leaching of organic nutrients from the dead tissues. Following this, microbial mineralization dominated, which consumed oxygen and released dissolved inorganic nutrients. Although this release of nutrients may serve as a trophic link to the pelagic primary producers, it may also lead to environmental problems associated with low oxygen concentrations, depending on the size of the bloom and the degree of mixing in the system.

**Acknowledgements** We thank A. L. Clement, L. Pettifer, and J. P. van de Merwe for field support, and M. Jordon for advice on nutrient analyses. Thanks also to D. Hair and I. Suthers (University of New South Wales) for the use of aquarium equipment and Smiths Lake field station. Funding was provided by the Hermon Slade Foundation.

## References

APHA, 1999. Standard Methods for Examination of Water and Wastewater, 20th ed. American Public Health Association, Washington, DC.

Arai, M. N., 1988. Interactions of fish and pelagic coelenterates. Canadian Journal of Zoology 66: 1913–1927.

Arai, M. N., 1997. A Functional Biology of Scyphozoa. Chapman & Hall, London.

Arai, M. N., 2005. Predation on pelagic coelenterates: a review. Journal of the Marine Biological Association of the United Kingdom 85: 523–536.

Arai, M. N., J. A. Ford & J. N. C. Whyte, 1989. Biochemical composition of fed and starved *Aequorea victoria* (Murbach et Shearer, 1902) (Hydromedusa). Journal of Experimental Marine Biology and Ecology 127: 289–299.

Azzoni, R., G. Giordani, M. Bartoli, D. T. Welsh & P. Viaroli, 2001. Iron, sulphur and phosphorus cycling in the rhizosphere sediments of a eutrophic *Ruppia cirrhosa* meadow of the (Valle Smarlacca, Italy). Journal of Sea Research 45: 15–26.

Billett, D. S. M., B. J. Bett, C. L. Jacobs, I. P. Rouse & B. D. Wigham, 2006. Mass deposition of jellyfish in the deep Arabian Sea. Limnology and Oceanography 51: 2077–2083.

Blackburn, T. H. & N. D. Blackburn, 1993. Rates of microbial processes in sediments. Philosophical Transactions: Physical Sciences and Engineering 344: 49–58.

Blackburn, T. H. & K. Henriksen, 1983. Nitrogen cycling in different types of sediments from Danish waters. Limnology and Oceanography 28: 477–493.

Clarke, A., L. J. Holmes & D. J. Gore, 1992. Proximate and elemental composition of gelatinous zooplankton from the Southern Ocean. Journal of Experimental Marine Biology and Ecology 155: 55–98.

Fenchel, T., G. M. King & T. H. Blackburn, 1998. Bacterial Biogeochemistry: The Ecophysiology of Mineral Cycling. Academic Press, San Diego.

Frost, P. C., R. S. Stelzer, G. A. Lamberti & J. J. Elser, 2002. Ecological stoichiometry of trophic interactions in the benthos: understanding the role of C:N:P ratios in lentic and lotic habitats. Journal of the North American Benthological Society 2: 515–528.

Gorsky, G., S. Dallot, J. Sardou, R. Fenaux, C. Carre & I. Palazzoli, 1988. C and N composition of some northwestern Mediterranean zooplankton and micronekton species. Journal of Experimental Marine Biology and Ecology 124: 133–144.

Graham, W. M., 2001. Numerical increases and distributional shifts of *Chrysaora quinquecirrha* (Desor) and *Aurelia aurita* (Linne) (Cnidaria: Scyphozoa) in the northern Gulf of Mexico. Hydrobiologia 451: 97–111.

Hagadorn, J. W., R. H. Dott Jr. & D. Damrow, 2002. Stranded on a late Cambrian shoreline: medusae from central Wisconsin. Geology 30: 147–150.

Hay, S., 2006. Marine ecology: gelatinous bells may ring change in marine ecosystems. Current Biology 16: R679–R682.

Herbert, R. A., 1999. Nitrogen cycling in coastal marine ecosystems. FEMS Microbiology Reviews 23: 563–590.

Keister, J. E., E. D. Houde & D. L. Breitburg, 2000. Effects of bottom-layer hypoxia on abundances and depth distributions of organisms in Patuxent River, Chesapeake Bay. Marine Ecology Progress Series 205: 43–59.

Kingsford, M. J., K. A. Pitt & B. M. Gillanders, 2000. Management of jellyfish fisheries with special reference to the order Rhizostomeae. Oceanography and Marine Biology: An Annual Review 38: 85–156.

Larson, R. J., 1986. Water content, organic content, and carbon and nitrogen composition of medusae from northeast Pacific. Journal of Experimental Marine Biology and Ecology 99: 107–120.

Lomstein, B. A., L. B. Guldberg & J. Hansen, 2006. Decomposition of *Mytilus edulis*: the effect on sediment nitrogen and carbon cycling. Journal of Experimental Marine Biology and Ecology 329: 251–264.

Mills, C. E., 2001. Jellyfish blooms: are populations increasing globally in response to changing ocean conditions? Hydrobiologia 451: 55–68.

Pitt, K. A. & M. J. Kingsford, 2003a. Temporal variation in the virgin biomass of the edible jellyfish, *Catostylus mosaicus* (Scyphozoa, Rhizostomeae). Fisheries Research 63: 303–313.

Pitt, K. A. & M. J. Kingsford, 2003b. Temporal and spatial variation in recruitment and growth of medusae of the jellyfish, *Catostylus mosaicus* (Scyphozoa: Rhizostomeae). Marine and Freshwater Research 54: 117–125.

Pitt, K. A., K. Koop & D. Rissik, 2005. Contrasting contributions to inorganic nutrient recycling by the co-occurring jellyfishes, *Catostylus mosaicus* and *Phyllorhiza punctata* (Scyphozoa, Rhizostomeae). Journal of Experimental Marine Biology and Ecology 315: 71–86.

Quinn, G. P. & M. J. Keough, 2002. Experimental Design and Data Analysis for Biologists. Cambridge University Press, Cambridge.

Schneider, G., 1990. Phosphorus content of marine zooplankton dry material and some consequences; a short review. Plankton Newsletter 12: 41–44.

Shenker, J. M., 1985. Carbon content of the neritic scyphomedusa *Chrysaora fuscescens*. Journal of Plankton Research 7: 169–173.

Shushkina, E. A., E. I. Musaeva, L. L. Anokhina & T. A. Lukasheva, 2000. Role of gelatinous macroplankton: medusas *Aurelia* and ctenophores *Mnemiopsis* and *Beroe* in the planktonic communities of the Black Sea. Okeanologiya 40: 859–866.

Sterner, R. W. & J. J. Elser, 2002. Ecological Stoichiometry: The Biology of Elements from Molecules to the Biosphere. Princeton University Press, Princeton.

Titelman, J., L. Riemann, T. A. Sornes, T. Nilsen, P. Griekspoor & U. Båmstedt, 2006. Turnover of dead jellyfish: stimulation and retardation of microbial activity. Marine Ecology Progress Series 325: 43–58.

Welsh, D. T., 2003. It's a dirty job but someone has to do it: the role of marine benthic macrofauna in organic matter turnover and nutrient recycling to the water column. Chemistry and Ecology 19: 321–342.

Yamamoto, J., M. Hirose, T. Ohtani, K. Sugimoto, K. Hirase, N. Shimamoto, T. Shimura, N. Honda, Y. Fujimori & T. Mukai, 2008. Transportation of organic matter to the sea floor by carrion falls of the giant jellyfish *Nemopilema nomurai* in the Sea of Japan. Marine Biology 153: 311–317.

Hydrobiologia (2009) 616:161–191
DOI 10.1007/s10750-008-9620-9

JELLYFISH BLOOMS

# A review and synthesis on the systematics and evolution of jellyfish blooms: advantageous aggregations and adaptive assemblages

**William M. Hamner · Michael N Dawson**

Published online: 11 November 2008


**Abstract** Pelagic gelatinous invertebrates in many diverse phyla aggregate, bloom, or swarm. Although typically portrayed as annoying to humans, such accumulations probably are evolutionary adaptations to the environments of pelagic gelatinous zooplankton. We explore this proposition by systematic analysis completed in three steps. First, using the current morphological taxonomic framework for Scyphozoa, we summarize relevant information on species that aggregate, bloom, and swarm and on those species that do not. Second, we establish a molecular phylogenetic framework for assessing evolutionary relationships among classes and many orders of Medusozoa and among most families of Scyphozoa (particularly Discomedusae). Third, we interpret the phylogenetic distribution of taxa and of characteristics of jellyfish that aggregate, bloom, or swarm, in terms of species diversity—a proxy for evolutionary success. We found that: (1) Medusae that occur en masse are not randomly distributed within the Phylum Cnidaria but instead they are found primarily within the Scyphozoa which have a metagenic life history. (2) Midwater and deep-sea medusae rarely bloom or swarm. (3) Epibenthic medusae do not swarm. (4) Large carnivores that feed on large prey do not bloom strongly. (5) Large medusae that feed exclusively on small prey both bloom and swarm. (6) *Pelagia*, the only holoplanktonic, epipelagic scyphomedusan, both blooms and swarms, demonstrating that a metagenic life cycle is not required for blooming or swarming at sea. (7) Environmental change (overfishing, species introductions, and eutrophication) may induce or inhibit blooms. (8) Taxa that bloom or swarm are often more diverse than taxa that do not. (9) Speciation in scyphozoans can occur rapidly. (10) Morphological stasis in holozooplankton masks genetic variability. (11) Selection for convergent evolution in the sea is strong because mass occurrence has evolved multiple times in independent evolutionary lineages under similar circumstances. Thus, attributes possessed by many taxa that occur en masse appear to be evolutionarily advantageous, i.e., adaptations.

**Keywords** Behavior · Ecology · Evolution · Life-history · Oceanography · Phylogeny · Taxonomy · Scyphozoa

**Electronic supplementary material** The online version of this article (doi:10.1007/s10750-008-9620-9) contains supplementary material, which is available to authorized users.

Guest editors: K. A. Pitt & J. E. Purcell
Jellyfish Blooms: Causes, Consequences, and Recent Advances

W. M. Hamner (✉)
Department of Ecology and Evolutionary Biology,
University of California, Los Angeles, CA 90095-1606,
USA
e-mail: whamner@ucla.edu

M. N Dawson
School of Natural Sciences, University of California,
Merced, CA 95344, USA

> "In view of their size and abundance the Scyphomedusae must play a large part in the economy of the seas" (Russell, 1970: Preface, p. X)

## Introduction

Jellyfish blooms are natural population phenomena that occur seasonally, often predictably, and yet seemingly abruptly each spring and summer. Blooms are one aspect of a metagenic life cycle of alternating asexual and sexual reproduction but are also enhanced by human activities through species introductions, commercial fishing activity, or coastal eutrophication (Arai, 2001; Mills, 2001; Purcell et al., 2007). Jellyfish also often exhibit apparent blooms, temporary increases in their local population density due to local re-distribution or re-dispersion in concentration of a stable population by physical and chemical factors and by directed swimming behavior of the medusae themselves that result in dense swarms (Graham et al., 2001). Both types of blooms, true (demographic) blooms and apparent blooms, are often described in negative terms because large jellyfish in large numbers can compete with commercially valuable fisheries for prey, consume larvae that otherwise would have been commercially valuable, break trawlers' nets, clog intake cooling pipes for coastal power plants, and sting and sometimes kill tourists (Arai, 1997, p. xv; Hay, 2006; Purcell et al., 2007).

Although proximate causes are often known, and mass occurrences often result in feeding and reproduction, activities necessary for successful completion of their complex life cycles, nothing is known of the evolution of accumulations or therefore of the aspects of aggregations, blooms, and swarms that are subject to natural selection. Thus, our focus here is to explore the evolutionary advantages of jellyfish assemblages and other adapted attributes of aggregating medusae, rather than to review the anthropogenic aspects of adventitious accumulations of jellyfish (Purcell et al., 2007).

Blooms and swarms characterize many evolutionarily distinct phyla of pelagic, gelatinous zooplankton, from cnidarians to chordates, and similar traits favoring formation of blooms have apparently evolved independently via convergence (Bone, 2005). However, it is an open question whether blooming and swarming are attributes of all or a few taxa *within a given phylum*. Answering this question, of how taxa that bloom and how traits that enable blooming are distributed across well-resolved phylogenetic trees, is key to understanding the evolutionary reasons that jellyfish accumulate to varying degrees. If species that bloom possess many similar phenotypic traits (e.g., of morphology, behavior, physiology) and also occur predictably in a subset of environments, and if these are clustered in multiple independent clades on distinct branches of the phyletic tree, then strong natural selection causing *convergence* is the predominant evolutionary pattern if, and only if, their most recent common ancestors did not bloom. In contrast, if closely related taxa evolved in the same way, maintaining similarity to each other but both diverging from a common ancestor which did not bloom, then this is a different phylogenetic pattern referred to as *parallel evolution*. Other lineages may demonstrate *stasis*, such that all descendent taxa within a clade share the same ancestral state, in which case extant taxa all bloom because their most recent common ancestor bloomed. A fourth, quite different but plausible, pattern—*random* species bloom, with appearance of this trait in multiple orders, families, and genera within each phylum and being found in many habitats—would result if traits enabling blooms were evolutionarily labile, originating and decaying easily and often, producing some ancestors that bloom while others did not, and derived some taxa that bloom while others did not.

The significance of these distinctive evolutionary patterns becomes clear when considering the ecology of adaptive radiations (Schluter, 2000). Schluter listed four criteria as necessary to detect adaptive radiations: common ancestry, phenotype–environment correlations, rapid speciation, and trait performance (Schluter, 2000, pp. 10–14). The first three of these criteria can only be assessed with reference to phylogeny; all four criteria are now widely employed to designate the occurrence of adaptive radiations (see Lukoschek & Keogh, 2006). We suggest that these issues can be enjoined best if addressed sequentially, and we apply Schluter's scheme in this review and in our companion paper (Dawson & Hamner, this volume). First, and the topic of this review, must be a broad understanding of common ancestry (Moore & Willmer, 1997). Unless we know the general shape of the phylogenetic tree and the route of inheritance, we cannot distinguish when or how novel traits arose or their impact, if any, on diversity. Second, identifying phenotype–environment correlations suggests possible causal agents of natural selection. Third, recognition of rapid speciation (Carroll et al., 2007; Hendry et al., 2007), in which the origins of branches of a phylogenetic tree are unusually closely clustered in time and the number of taxa arising is large, suggests that something unusual happened at a specific time in the history of that lineage, thereby stimulating speciation. Fourth, the topic of our companion paper (Dawson & Hamner, this volume), we attempt to determine which particular traits might be selected for blooming and swarming, thus providing the final category of evidence necessary to identify adaptive radiation.

## Telos and design of our review

We wish to determine whether jellyfish with a proclivity to aggregate, bloom, or swarm are a random or non-random sample with respect to their phylogeny, biology, and ecology. If non-random, did relevant traits arise and spread quickly in highly speciated genera? We limit ourselves to considering predominantly the scyphozoan jellyfish within the cnidarian clade Medusozoa, which includes cubomedusae, hydromedusae, and scyphomedusae, for which there is a rich taxonomic history from Linnaeus to recent years (e.g., Linnaeus, 1758; Collins et al., 2006; Daly et al., 2007). We will not review non-cnidarian pelagic invertebrates discussed at this or the prior symposium on jellyfish blooms (Graham et al., 2001; Mills, 2001; Purcell et al., 2001).

We have divided our review into three sections. In Part I, we summarize the current understanding of species diversity and species that occur en masse, commenting incidentally on some organismal and environmental characteristics that strike us as important. First, (A) we review the taxonomic literature describing the Medusozoa (primarily Kramp, 1961; Mianzan & Cornelius, 1999) to compile lists of jellyfish and their characters that are found in blooms. Although these taxonomic publications are dated, they still provide the most concise sources of information regarding the specific attributes of defined taxa, such as their morphology, distribution, and life history. We note that taxa that bloom appear to be clustered taxonomically and geographically and possess a subset of characters. Then, (B) we describe new, and synthesize recent, molecular systematic studies to arrange the traditional taxa phylogenetically. This permits us to understand common ancestry, inheritance, and origination, and to pose questions about the evolutionary ecology of jellyfish blooms for the first time for many taxa. We necessarily pay particular attention to the discovery of new and cryptic species, and we ask whether the advent of molecular phylogenetics is likely to change the conclusions reached using the extant taxonomic literature in Part IA. (The answer is probably not much.) In Part II, we examine whether aggregating, blooming, or swarming has promoted speciation and rapid radiation. This leads us to ask why there is such a profusion of cryptic species among the Medusozoa and how have jellyfish diversified into new geographic portions of the sea despite remaining "morphologically austere … lacking features that can be used to distinguish species effectively" (Bickford et al., 2007). We hypothesize that morphologically cryptic cladogenesis masks the accrual of genetic and physiological variability in large populations that allows for rapid adaptation to new environments, even if the predominant patterns of evolution have been intense convergence in an epipelagic ocean devoid of cover and complexity (Hamner et al., 1975; Hamner, 1995; Dawson & Hamner, 2005; Colin et al., 2006).

## Terminology

The various terms used to describe jellyfish accumulations are not interchangeable, even though the terms are somewhat imprecise and there may be overlap in usage. The vocabulary can be complex to master because collective nouns for aggregations of organisms have etymological roots in medieval "terms of venery," words for aggregations of English animals that were chased and hunted for food and sport [Latin, *venari*, to hunt game]. The peculiarities of this vocabulary are illustrated by the proper term for a jellyfish swarm: a "smack" (Lipton, 1991). The ambiguity, which confounds understanding by conflating causes, led Graham et al. (2001) to distinguish 'true' (demographic) from 'apparent' (non-demographic) blooms. *True blooms* are in part a consequence of seasonal life cycles, and consequently, all metagenic medusae (hydromedusae, cubomedusae, and scyphomedusae) as well as some holoplanktonic cnidarians, ctenophorans, and tunicates likely have the potential to bloom (generally in spring and summer in temperate regions); thus normal and/or abnormal seasonal abundance is directly attributable to population increase due to reproduction and growth, sometimes enhanced by anthropogenic activity. *Apparent blooms*, in contrast, are local increases in abundance of animals associated with temporary or transient physical or chemical phenomena (such as aggregation at fronts or local advection to a new location), or longer-term accumulation of large numbers in enclosed habitats; apparent blooms may, but do not necessarily, reflect true blooms that occurred elsewhere. Additionally, Mills (2001) distinguished '*natural*' from '*unnatural*' blooms, when "...parts of the ocean become increasingly disturbed and overfished... [with] energy...switched over to production of pelagic Cnidaria or Ctenophora"; further, she noted that "...what we interpret as a jellyfish bloom may reflect our expectations about an ecosystem," highlighting that careful objective analyses are necessary to understand aggregations, blooms, and swarms.

During review of the literature for the current study, we found it necessary to further refine the lexicon applicable to jellyfish blooms, as follows. An *accumulation* is an increase in the number of corporeal objects, relative to a different time or place; both animate and inanimate objects accumulate. An *aggregation* is an accumulation of individuals that consequently likely interact (e.g., winnow, reproduce, compete); short-term concentrations, possibly, but not necessarily, within a natural or apparent bloom, due to passive drifting in currents, active behavior of individual medusae, or interaction of the two, in the absence of either increased population growth or mortality. Although non-living objects such as rocks and shells accumulate on beaches where they thereafter sometimes are winnowed by waves into size-specific or object-specific assemblages, more often the term "aggregation" implies that corporeal attributes of living organisms interact to produce species-specific aggregations. *Swarms* are dense re-aggregations of animals due to behavior, but animals within swarms are not spaced and oriented in polarized arrays as are fish in schools or birds in flocks (see Parrish & Hamner, 1997). After strobilation, aggregations of tiny ephyrae are dispersed into the surrounding water column. Later, following metamorphosis and substantial growth, juvenile medusae become large enough to swim horizontally against prevailing currents and can then re-aggregate into swarms (e.g., for reproduction).

In some cases, we encountered instances when it was unclear whether we were dealing with an aggregation, a true bloom, or a swarm. In other cases, we found the need to refer collectively to aggregations, true blooms, and swarms. In such cases, we applied the term *mass occurrence*, which is similar to 'accumulation' (above) but necessarily very large in magnitude.

## Part I. Inventory and description of jellyfish biodiversity

Not all medusozoans produce medusae. Those with medusae are known commonly as cubomedusae (equivalent to Class Cubozoa), hydromedusae (a paraphyletic or polyphyletic group within Class Hydrozoa), and scyphomedusae (equivalent to Class Scyphozoa). Within these groups, most medusae are neritic, metagenic annuals that have a preponderance of new growth (more individuals and increased biomass) in the spring and summer, like many temperate-zone plants. Similar dynamics are apparent in subtropical and tropical scyphomedusae, perhaps related to seasonal monsoons (e.g., Daryanabard & Dawson, 2008; M. N. Dawson, L. Martin, S. Patris, unpubl. data). The general life history of jellyfishes is common knowledge, but we review it briefly here as it is important for understanding aggregations, blooms, and swarms.

The medusa is but one part of a bipartite life cycle that alternates between an asexual, benthic polyp and a sexual, pelagic medusa. Medusae typically are produced asexually in abundance and grow rapidly in spring and summer, although the timing may vary (e.g., Russell, 1970). To our knowledge, there has been only one published study describing the distributions of different life history stages in a single population (Hamner et al., 1994) but it describes a possibly general mechanism allowing newly liberated medusae, which must usually disperse in most coastal bodies of water due to tidal flow and turbulence (e.g., Albert, 2007; but see e.g., Pitt & Kingsford, 2000), to gather again when larger, re-aggregating into reproductive swarms (Hamner et al., 1994). When swarming, the jellyfish spawn and release planulae which settle and metamorphose into a new generation of benthic polyps. Meanwhile, abruptly, the entire population of medusae usually senesces and dies (see Billett et al., 2006). There are interesting and odd exceptions to this general pattern, some of which we will discuss later (Hamner et al., 1994; Albert, 2005; Martin et al., 2005; Sørnes et al., 2007), but most cnidarian jellyfish form annual bipartite life cycles. This life cycle is shared by diverse (but not all) cubozoans, hydrozoans (e.g., Trachymedusae and Narcomedusae are essentially oceanic holoplankton), and scyphozoans from subpolar to tropical latitudes, yet there is great disparity with which blooms have been reported in these groups. In the following sections, we inventory and describe the biodiversity of jellyfish using traditional taxonomic approaches and using recent molecular phylogenetic and statistical morphologic analyses.

(A) Species diversity, mass occurrences, and trait variation described taxonomically

*Cubozoa*

There are very few references in the world literature to mass occurrence of species within Class Cubozoa although two surface aggregations of *Alatina* (*Carybdea*) *alata* (Reynaud) were sampled at different locations in the Indo-Pacific during the Albatross Expedition in 1899 (Mayer, 1910). More recently, a species identified as *A. alata* also has affected water sports along Waikiki Beach, Oahu (B. Holland pers. comm.). *Chironex fleckeri* Southcott accumulates in mangrove openings at several locations along the northeastern Australian coast, and *Chiropsella* (*Chiropsalmus*) *bronzie* Gershwin occurs en masse off Port Douglas, Queensland, Australia (J. Seymour, personal communication). The strong swimming and obvious behavioral repertoire of cubomedusae suggests these aggregations are swarms. However, since publications are few and the systematics and natural history of cubomedusae are currently under revision by a variety of investigators (see contributions on cubomedusae in this Symposium), it is premature to generalize on the taxonomic (or phylogenetic) distribution of mass occurrences or their causes in cubomedusae. Moreover, if traditional estimates of species richness and geographic distributions are good metrics of true richness and distribution, then cubomedusae, with ca. 40 species in shallow, coastal, tropical, and subtropical waters, may contribute only a small number of blooms and swarms globally and few phylogenetic contrasts that might inform us of their evolution. Alternatively, as we argue later in Dawson & Hamner (this volume), if cubopolyps exhibit complete metamorphosis of each polyp into a single cubomedusan, then perhaps this taxon does not have the capacity to form "extraordinary blooms." We do not suggest that particular species or mass occurrences are not important, but simply that our inability to describe cubomedusan blooms and swarms is not fatal to our attempts to describe scyphozoan blooms (a point that is supported by phylogenetic analyses, see Part II).

*Hydrozoa*

Considering the diversity of hydromedusae (ca. 850 species; Bouillon & Boero, 2000), there are surprisingly few records of species that bloom (Purcell, 1989; Purcell et al., 1999; Mills, 2001). Yet true seasonal blooms of hydromedusae must occur in at least the metagenic hydrozoans simply as a consequence of alternating benthic polypoid and planktonic medusoid stages. There is therefore some concern that the small number of mass occurrences recorded for hydromedusae may be an artifact of under-reporting due to their small size, even though the small, hydrozoan-sized, scyphozoans *Pelagia* and *Linuche* are among those best known to occur en masse. Preliminary consideration of historical texts, historical and modern practitioners, and recent publications indicate there is no major bias against reporting of hydromedusae and that unnatural blooms and swarms of Hydrozoa are indeed rare (see Electronic Supplementary Material, Appendix A).

The conditions under which mass occurrences of hydromedusae occur are therefore interesting and perhaps informative of factors supporting a tendency to aggregate, bloom, and/or swarm. Mass occurrences of *Aequorea* spp. have been reported multiple times from the northeastern Pacific, western North Atlantic, and eastern South Atlantic (e.g., Purcell, 1989; Mills, 2001; Sparks et al., 2001; Purcell, 2005; Møller & Riisgård, 2007). *Aequorea* is the largest hydromedusan. Mass occurrences are also known for small leptomedusan *Obelia* spp., anthomedusae *Sarsia* spp.

and *Rathkea octopunctata* (M. Sars) (see Møller & Riisgård, 2007), and the trachymedusan *Aglantha digitale* (O. F. Müller) (see Pertsova et al., 2006; Takahashi & Ikeda, 2006 and earlier papers cited therein). The holoplanktonic *Aglantha digitale* can reproduce asexually very rapidly, going through several generations in 1 year (Russell, 1970, p. 451). Surface accumulations of another trachymedusa, *Liriope tetraphylla* (Chamisso & Eysenhardt) (b.d. 10–30 mm), have been described in Hiroshima Bay off Japan at concentrations up to 3,000 individuals $m^{-2}$ (Ueno & Mitsutani, 1994); this pan-oceanic holoplanktonic medusa has also been reported en masse in shallow water along South America's central Atlantic coast during austral summer (Mianzan et al., 2000) and also common off southern France and around India (Buecher et al., 1997; Santhakumari & Nair, 1999). Accumulations by invasive populations of the hydromedusa *Craspedacusta* have been reported for freshwater lakes (Pennak, 1956). These examples provide us with valuable points of reference with regard to size, environment, life history, and invasion to which we will return in Part II. Although small species of hydromedusae may increase in abundance seasonally, may aggregate vertically or horizontally and actively or passively (Arai, 1992; Graham et al., 2001), and may swim great distances—the narcomedusa *Solmissus albescens* (Gegenbaur) (b.d. 25–30 mm) undergoes a nocturnal migration from 400 to 700 m to the upper 100 m (Mills & Goy, 1988)—most of the time, most hydromedusae do not experience the necessary confluence of biological and environmental characteristics to aggregate, bloom excessively, or swarm (see discussion of swimming and Reynolds number in Dawson and Hamner, this volume).

*Scyphozoa*

In contrast, reports of scyphomedusae (which comprises only circa 200 species) occurring en masse are commonplace. Thus, Lynam et al. (2004) reported evidence of blooms of three scyphozoans but of no hydrozoans. Eleven species in the Order Rhizostomeae are reliably sufficiently abundant to be regularly fished commercially in the Indo-Pacific (Omori & Nakano, 2001), and others have caused problems with coastal power generation (e.g., Daryanabard and Dawson, 2008). Members of the order Semaeostomeae are sometimes troublesome for fisheries and coastal industries, and they are also a cause of injuries and death due to their extensive and well-armed tentacles (Arai, 1997, p. xv; Mills, 2001; Purcell et al., 2007). It is therefore necessary to review scyphomedusae in more detail to elucidate the biodiversity and relationships of blooming species within this Class and thereby circumscribe the suite of organismal and environmental characteristics and their interactions that facilitate most mass occurrences.

*Order Coronatae*

Coronates are not well known for species that occur en masse (Table 1). Most coronate medusae inhabit the deep sea and never or rarely occur in epipelagic waters. This may lead to concern that the small number of coronates classified as aggregating, blooming, or swarming simply reflects under-reporting due to the difficulties of sampling the deep sea. However, surveys of the deep sea have been sufficiently extensive to demonstrate that they are manyfold less abundant than neritic species that are classed as aggregators, bloomers, or swarmers (see Electronic Supplementary Material, Appendix B). Moreover, conclusions based on deep-sea surveys are supported by inference from life histories about species' inabilities to occur en masse. Thus, two independent lines of evidence strongly suggest deep-sea medusae predominantly do not occur en masse. Deep-sea species are usually widely dispersed and holoplanktonic with direct development (Jarms et al., 1999), which eliminates the need for benthic habitat for polyps, and with year-round continuous production of relatively few gametes (Larson, 1986), consistent with small overall population densities and low biomass of prey in the deep sea (Herring, 2002). Life history, population density, distribution, and resources therefore make deep-sea coronates unlikely participants in pestiferous swarms. Concomitantly, Larson (1986) reported densities of *Atolla chuni* (Vanhöffen), *Atolla wyvillei* (Haeckel), and *Periphylla periphylla* (Péron & Lesueur), the three most abundant coronates in the Southern Ocean, as on average between 7 and 21 medusae per 2 h, 3 m Isaacs-Kidd midwater trawl.

Three contrasting, exceptional, situations illustrate the generality that coronates do not occur en masse. *Periphylla periphylla* is a mesopelagic species except at high latitude where it occurs as shallow as 50 m

**Table 1** Approximate diversity and reports of mass occurrences of medusae in order Coronatae Vanhöffen, 1888

| Family (total 6) | Genus (total 12) | Species (total 47) | Aggregate, bloom, or swarm? (References) |
|---|---|---|---|
| Atollidae [DW] | *Atolla* | 6 | No data—inference: do not commonly occur en masse |
| Atorellidae | *Atorella* | 5 | No data—inference: do not commonly occur en masse |
| Linuchidae | *Linantha* | 1 | No data—inference: do not commonly occur en masse |
| | *Linuche* | 3 | Natural breeding aggregations (Mayer, 1910; Kremer et al., 1990; Larson, 1992) |
| Nausithoidae | *Nausithoe* | 18 | No swarms or blooms (L. E. Martin & M. N Dawson, personal observations) |
| | *Palephyra* | 3 | No data—inference: do not commonly occur en masse |
| | *Thecoscyphus* | 1 | No data—inference: do not commonly occur en masse |
| Paraphyllinidae [DW] | *Paraphyllina* | 3 | No data—inference: do not commonly occur en masse |
| Periphyllidae [DW] | *Nauphantopsis* | 1 | No data—inference: do not commonly occur en masse |
| | *Pericolpa* | 3 | No data—inference: do not commonly occur en masse |
| | *Periphylla* | 1 | Very high density achieved only in fjords (Youngbluth & Båmstedt, 2001; Jarms et al., 2002; Båmstedt et al., 2003; Titelman et al., 2006; Sørnes et al., 2007) |
| | *Periphyllopsis* | 2 | No data—inference: do not commonly occur en masse |

Putative aggregating, blooming, or swarming species are noted (with sources). We consider the lack of any data on some species in literature spanning ca. 150 years indicative that those taxa do not aggregate, bloom, or swarm with any frequency, although such inference remains to be substantiated by targeted research. Exclusively deepwater families are indicated by superscript [DW] (following Jarms et al., 2002). Taxa, organized alphabetically, are based largely on Kramp (1961) and Mianzan & Cornelius (1999); compiled primarily by M.N Dawson, G. Jarms, A.C. Morandini, and F. Pàges. Source: *The Scyphozoan* website, 2006–2007

(but usually deeper) in some Norwegian fjords (Youngbluth & Båmstedt, 2001; Båmstedt et al., 2003). In Lurefjorden, medusae reach densities over one to three orders of magnitude greater than known elsewhere (Youngbluth & Båmstedt, 2001; Sørnes et al., 2007). Thus, the high densities appear to be a function of being restricted to these specific deep-ocean-like fjords (Sørnes et al., 2007). By analogy, other members of order Coronatae that evolved in, and currently inhabit, the deep sea are unlikely to occur en masse. We believe this conclusion probably can be applied with confidence to members of the families Atollidae, Periphyllidae, and Paraphyllinidae which are composed exclusively of deep water species (Jarms et al., 2002).

By contrast, do shallow water coronates necessarily occur en masse? *Linuche* is an exclusively shallow water tropical genus because it is zooxanthellate. Photosymbiotic dinoflagellates provide a substantial proportion of its energetic requirements, and a diurnal light cycle is necessary for these jellyfish to spawn at dawn (Kremer et al., 1990). *Linuche unguiculata* Swartz regularly aggregates at the surface "to breed" causing "hundreds of swarms" (Conklin, 1908 in Mayer, 1910, p. 559) as does *Linuche aquila* Haeckel (Mayer, 1910, p. 560). Thus, these annual breeding aggregations, which are maintained by wind-driven Langmuir circulation cells and circular swimming (Larson, 1992), are a normal part of the reproductive biology of this genus. *Nausithoe* also has shallow water tropical representatives, of which at least one species has been observed in a small fjord-like cove in Palau (L. E. Martin & M. N Dawson, personal observations). The population of this small (few mm diameter) medusa, however, is at sufficiently low density that sampling with nets is problematic, and populations of other medusae (the rhizostome *Mastigias* and cubomedusan *Chiropsoides quadrugatus* (Haeckel)) in the same cove are much more impressive due to the much greater biomass of large individuals (L. E. Martin & M. N Dawson, pers. obs.). These observations, although few, suggest that, like most other coronates, *Nausithoe* does not bloom.

*Order Semaeostomeae*

Large populations of many semaeostome scyphomedusae occur around the globe, but they are best known from the temperate northern hemisphere. Consequently, although Semaeostomeae is not the most speciose order (ca. 60 traditionally recognized species [plus 11 new cryptic species of *Aurelia*]; Table 2), reports of semaeostomes predominate in the literature, particularly in discussions of the impacts of climate change (e.g., Purcell, 2005). Whether studies of semaeostomes offer insights into other jellyfish depends on whether key ecological traits are shared, to which our study may speak.

*Family Cyaneidae* Cyaneidae includes three genera of large medusae ($\leq$2 m bell diameter): *Cyanea*, *Drymonema*, and *Desmonema* (Table 2). All are capable of capturing and ingesting many relatively large prey (e.g., *Aurelia*, 0.15–0.30 m b.d.), a process that probably involves extracellular and/or extra-coelenteronic digestion (Hansson, 1997, 2006; Hansson & Kiorboe, 2006), and on which their own population dynamics may be dependent. Several species have been called swarming species in the literature, although we consider evidence of true swarming to be weak.

*Cyanea capillata* Eschscholtz variety 'arctica' reaches bell diameters of 2.3 m and is the "largest of all known medusae." Mayer (1910, p. 597) saw "swarms of them about 20 miles off Barnegat Bay, New Jersey, early in August." Although impressive, such sightings are rare, suggesting this was an unusual accumulation of medusae along an oceanic front rather than a true bloom (Graham et al., 2001). *Cyanea capillata* occurs in portions of the central Baltic Sea in densities of 1–3 individuals 100 m$^{-3}$ in midsummer, but these populations are advected into the area from the west and thus are not true, local, blooms (Barz et al., 2006). Purcell (2005) discussed changes in the density of *Cyanea capillata*, *C. lamarckii* Péron & Lesueur, and *Aurelia aurita* (Linné) in the North Sea in relation to changes in temperature during the North Atlantic Oscillation, noting that increased *Cyanea* and *Aurelia* populations were both associated with cold conditions and with Arctic boreal plankton. Although such increases could be due to local blooms—*Cyanea* feed voraciously on *Aurelia* (Hansson, 1997) which in turn feed on boreal zooplankton—Purcell (2005) suggested advection of populations of medusae from the Arctic was the more likely cause. Purcell (2003) also stated that "*Cyanea capillata*…does not form aggregations."

The clearest report to date of *Cyanea* occurring in situ and en masse comes from a highly artificial system in the Yangtze River Estuary (Xian et al., 2005; N.B. the medusa was incorrectly identified as *Sanderia malayensis* Goette, a pelagiid, by Xian et al., 2005). Overfishing of *Rhopilema esculenta* Kishinouye contributed to an increase in *Cyanea capillata*, from 0.4% of total samplings in 1998 to 85.5% in 2003, corresponding with the time that the Three Gorges Dam began to store water, leading to the estuary becoming warmer, saltier, and having more zooplankton. *Cyanea* constituted 98.4% of the fisheries catches in 2005 (Xian et al., 2005).

During a perceived "population explosion" in 1999 (Williams et al., 2001), densities of *Drymonema dalmatinium* Haeckel around Puerto Rico reached 12 medusae 25 m$^{-2}$, averaging ca. 1 per 100 m$^{2}$ over large (kilometer-scale) distances. The largest *D. dalmatinium* were reported as 1.5–2 m b.d. Like *Cyanea* in the North Sea above, this *D. dalmatinium* event was associated with "an outbreak" of *Aurelia*, which was possibly 20 times more numerous than *D. dalmatinium*, and which provided the food that sustained these very large predators (Larson, 1987). Whether these were both in situ blooms as opposed to accumulations, however, is unclear.

*Family Pelagiidae* Pelagiidae includes species in three families of large (*Chrysaora*) or medium-to-small (*Pelagia*, *Sanderia*) medusae. A long history and large number of reports describe "great swarms in…bays and estuaries" (Mayer, 1910, p. 570). Among the most notable has been *Pelagia noctiluca* (Forskål) in the northern and central Adriatic Sea during the 1980s, which stimulated a 3-year regional investigation of jellyfish blooms coordinated by the United Nations (e.g., see Rottini Sandrini & Avian, 1983, 1989, 1991; Zavodnik, 1987; Malej, 1989). *P. noctiluca* is undoubtedly a swarming species (Malej, 1989; Arai, 1997, 2001; Mills, 2001; Purcell, 2005), with typically sudden and often irregular appearance, persistent swarms of immense size maintained for days to weeks at densities that locally and routinely exceed 100 animals m$^{-3}$ (Zavodnik, 1987; Malej, 1989). Dense temporary aggregations, often close to shore and occasionally containing up to 600 medusae m$^{-3}$, are caused by the interaction of wind, tides, topography, and behavior, apparently including reproductive behavior (Zavodnik, 1987). Interestingly, *P. noctiluca* exhibits direct pelagic development from planula to ephyra

**Table 2** Approximate diversity and reports of mass occurrences of medusae in order Semaeostomeae L. Agassiz 1862

| Family (3) | Genus (19) | Species (71) | Aggregate, bloom, or swarm? (References) |
|---|---|---|---|
| Cyaneidae | *Cyanea* | 14 | Aggregates rarely, does not swarm (Mayer, 1910; Purcell, 2005; Barz et al., 2006) Occurs en masse where introduced into disturbed enclosed environment (Xian et al., 2005) |
| | *Desmonema* | 4 | No data—inference: do not commonly occur en masse |
| | *Drymonema* | 2 | Does not swarm; may aggregate or bloom (Larson, 1987; Williams et al., 2001) |
| Pelagiidae | *Chrysaora* | 14 | Aggregates, blooms, and swarms (Kramp, 1961; Martin et al., 1997; Mianzan & Cornelius, 1999; Masilamoni et al., 2000; Graham, 2001; Sparks et al., 2001; Brodeur et al., 2002; Lynam et al., 2005, 2006; Purcell, 2005; Purcell et al., 2007; Decker et al., 2007 |
| | *Pelagia* | 2 | Aggregates, blooms, and swarms (Mayer, 1910; Rottini Sandrini & Avian, 1983, 1989, 1991; Larson, 1987; Morand et al., 1987; Zavodnik, 1987; Goy et al., 1989; Malej, 1989; Arai, 1997, 2001; Mills, 2001; Purcell, 2005) |
| | *Sanderia* | 1 | No data—inference: do not commonly occur en masse |
| Ulmaridae | *Aurelia* | 17 | Aggregates, blooms, and swarms (Mayer, 1910; Bigelow, 1926; Yasuda, 1968, 1969, 1971; Russell, 1970; Hamner & Jenssen, 1974; Möller, 1980; Schneider, 1988, 1989; Hamner et al., 1994; Lucas, 1996, 2001; Miyake et al., 1997; Dawson & Martin, 2001; Graham, 2001; Purcell, 2003; Albert, 2005; Purcell, 2005; Barz et al., 2006) |
| | *Aurosa*[a] | 1 | No data—inference: do not commonly occur en masse |
| | *Deepstaria*[DW] | 2 | No data—inference: do not commonly occur en masse |
| | *Discomedusa*[a] | 3 | No data—inference: do not commonly occur en masse |
| | *Diplulmaris*[a] | 2 | No data—inference: do not commonly occur en masse |
| | *Floresca*[a] | 1 | No data—inference: do not commonly occur en masse |
| | *Parumbrosa*[a] | 1 | No data—inference: do not commonly occur en masse |
| | *Phacellophora* | 1 | Does not swarm |
| | *Poralia*[DW] | 1 | Rare aggregation (Smith, 1982; Hamner, pers. obs.) |
| | *Stellamedusa*[DW] | 1 | No data—inference: do not commonly occur en masse |
| | *Sthenonia* | 1 | No data—inference: do not commonly occur en masse |
| | *Stygiomedusa*[DW] | 1 | No data—inference: do not commonly occur en masse |
| | *Tiburonia*[DW] | 1 | No data—inference: do not commonly occur en masse |
| | *Ulmaris*[a] | 2 | No data—inference: do not commonly occur en masse |

The species diversity of *Aurelia* is based on recent estimates from molecular analyses that support similar estimates based on morphological differences during the early-twentieth century. Putative aggregating, blooming, or swarming species are noted (with sources). Genera that are primarily or solely deep sea species are indicated by superscript [DW]. Taxa, organized alphabetically, are based largely on Kramp (1961) and Mianzan & Cornelius (1999); compiled primarily by M.N Dawson, G. Jarms, A.C. Morandini, and F. Pàges. Source: *The Scyphozoan* website, 2006–2007. [a] Taxonomic status 'valid' but record credibility 'unverified' per ITIS Report (http://www.itis.gov/servlet/SingleRpt/SingleRpt?search_topic=TSN&search_value=51714)

without an intermediate benthic scyphistoma (Rottini Sandrini & Avian, 1983). This has the potential for rapid and positive feedback on the population dynamics of the pelagic life cycle. Indeed, *P. noctiluca* eats a wide diversity of meso- and macro-zooplankton, and grows fast enough to match its metabolic needs to the density of the zooplankton and micronekton in the entire water column (Larson, 1987; Morand et al., 1987; Malej, 1989). Moreover, in large persistent swarms, finding mates and spawning successfully may not require complex behavioral mechanisms or navigational skills for re-aggregation, which may limit other species.

An apparent 12-year cycle in abundance of *Pelagia noctiluca* in the Mediterranean Sea is possibly related to weather cycles of temperature and rain (Goy et al., 1989). Although a link between coastal degradation and jellyfish blooms is commonly stated, and although *P. noctiluca* appears to grow more rapidly in eutrophic environments, populations of *P. noctiluca* have not been sustained in the Adriatic since 1986 "in spite of continued eutrophication" (Arai, 2001).

Mass occurrences of medusae in the genus *Chrysaora* are perhaps even more prevalent (Table 2). *Chrysaora* species can be dominant carnivores within large bays and estuaries (e.g., Chesapeake Bay), as the genus is apparently adapted to living in low salinity waters (Purcell, 2005; Decker et al., 2007). However, the genus is also a component of almost every continental shelf plankton community that has been investigated, in the Bering Sea (Brodeur et al., 2002) and the Gulf of Mexico (Graham, 2001), along the coast of California (Graham & Largier, 1997), the entire coastline of Brazil (Morandini et al., 2004) and Argentina (Mianzan & Cornelius, 1999), the west coast of South Africa (Sparks et al., 2001), the east coast of India (Masilamoni et al., 2000), and widely throughout the Indo-Pacific (Mayer, 1910). It is seemingly strange that huge populations of *Chrysaora* species can be maintained worldwide, year to year, with abundances increasing locally along open continental shelves in response to increases in food due to eutrophication (Arai, 2001) or decreases in competitors due to overfishing (Mills, 2001; Brodeur et al., 2002; Lynam et al., 2006). This is quite a different scenario from that of *Chrysaora* being physically entrained within embayments or estuaries and establishing a local population of polyps that respond to seasonal and annual changes in food, salinity, and temperature (Purcell, 2005), and quite different from apparent blooms arising from oceanographic events such as the occasional appearance of the large deep-red-to-black (and therefore likely midwater or deep sea) *Chrysaora achlyos* Martin, Gershwin, Burnett, Cargo & Bloom in epipelagic waters off Southern California (Martin et al., 1997). Continental shelf populations can be maintained only if directional swimming behavior of the medusae counteracts the dispersive effects of currents above the shelf that run directionally and parallel to shore, preventing medusae from advecting away and permitting persistent populations of pelagic and benthic stages of the life cycle. For example, in northeastern Monterey Bay, the region of an upwelling shadow (Graham & Largier, 1997), enormous numbers of *C. fuscescens* Brandt, observed via an ROV, were swimming directionally toward the front at depth and at night (Morris, 2006). In the absence of celestial cues, this suggests that medusae oriented toward the front via shear forces present in the stratified water column (e.g., Rakow & Graham, 2006). Medusae south of the front were swimming northward while medusae north of the front were swimming southward (Morris, 2006), thus maintaining the approximate geographic position of the population.

Such behavior by *Chrysaora* may facilitate reproduction by maintaining sufficient densities of reproductive adults in the vicinity of habitat suitable for polyps, and may also facilitate feeding. Like *Pelagia*, *Chrysaora* feeds on diverse prey, from dinoflagellates (Flynn & Gibbons, this symposium) through large mesozooplankton to juvenile fish and squid (pollock in the Bering Sea, *Loligo opalescens* Berry in Monterey; Hamner, personal observations). This ability to feed on the diverse prey that co-occur at high concentrations along upwelling coastal fronts is highly advantageous. *Chrysaora* appears to include aggregating, blooming, and swarming species.

*Family Ulmaridae* Ulmaridae is the most diverse family of semaeostomes in terms of genera, species, habitat, distribution, and morphology (e.g., Table 2). *Aurelia* is the only genus that obviously and recurrently occurs en masse, which it does globally between ca. 70°N to 55°S, in full or dilute sea water, high or reduced oxygen, and with exceptionally plastic phenology (e.g., Mayer, 1910; Russell, 1970; Lucas, 2001). Its nearshore distribution in temperate regions of Europe, North America, and Japan has contributed to its being the best studied genus of scyphomedusae. The attributes of *Aurelia* that contribute to its cosmopolitan distribution, abundance, and exceptional range of habitats and ecologies have been reviewed by Lucas (1996, 2001), Arai (1997), Miyake et al. (1997), Dawson & Martin (2001), and others. These attributes include the ability to utilize diverse small sources of energy and nutrition (from dissolved organic matter, through micro-zooplankton, to larval fish), a variety of vertical and horizontal migration patterns that are clearly adapted to local conditions (e.g., Hamner et al., 1994), and the ability to persist in enclosed ecosystems in small populations (Lucas, 1996; Dawson & Martin, 2001; Albert, 2005, 2007) and also in large populations in coastal bays (e.g., Monterey Bay) and the open ocean (e.g., between Papua New Guinea and Palau). Its moderate size (up to 0.4 m b.d.), and its ability to prosper in peripheral habitats (e.g., Hamner et al., 1982), recruit rapidly (Lucas, 2001), and navigate directionally are among the important aspects of its biology with regard to aggregations, blooms, and swarms.

*Phacellophora camtschatica* Brandt is a large (up to 0.6 m b.d.; Mayer, 1910, p. 613) predator, much like medusae in Family Cyaneidae. It feeds voraciously on *Aurelia labiata* Chamisso & Eysenhardt and ctenophores in Saanich Inlet, Vancouver Island, Canada (Strand & Hamner, 1988). It has wide, curtain-like oral arms, often holding 20 or more partially digested *Aurelia* and tens of thousands of ctenophores at the same time, presumably breaking prey tissues down and removing water prior to final stomach ingestion, as observed for *Cyanea* feeding on *Aurelia* (Hansson, 1997). Like species in Family Cyaneidae, *P. camtschatica* does not form blooms. In the Antarctic, the confamilial *Diplulmaris antarctica* Maas eats medusae and ctenophores (Larson, 1986), but little is known of its population dynamics.

In addition, there are at least seven ulmarid genera of midwater or deep-sea species. Like many coronates, these ulmarids live in cold water with relatively rare prey, low population densities, and with continuous production year-round of relatively few gametes and direct development (Larson, 1986), attributes not conducive to mass occurrence. However, Smith (1982; cited in Larson, 1986) found *Poralia rufescens* Vanhöffen at densities of 10–250 medusae 1,000 $m^{-3}$ in a zone 1–50 m above the bottom. WMH (viewing from the research submarine *Alvin*) also has seen *Poralia* at densities of 20 medusae 100 $m^{-2}$ just above the bottom of the Monterey Canyon. These occurrences likely reflect the interaction of local environment and perhaps organismal traits that enable accumulation or aggregation, as has also been reported for some deepwater hydromedusae (Gili et al., 1999), although such interactions do not appear to be commonplace for *Poralia* or other deepwater ulmarids.

### *Order Rhizostomeae*

Rhizostomeae is the most species-rich order of scyphomedusae (Table 3). Although distributed circumglobally between about 50°N and 40°S, it is predominantly a tropical and subtropical Indo-Pacific taxon (Vanhöffen, 1888; Kramp, 1970). It is composed of 10 families about which relatively little is known; indeed, too little is known to comment generally on the species within Lychnorhizidae. However, a minority of species are known comparatively well because they have notable life histories (e.g., *Cassiopea*), habitats (e.g., *Mastigias*), or human uses (e.g., a dozen fished species; Fig. 1).

*Family Cassiopeidae* Medusae in the monogeneric Cassiopeidae never form swarms, although the epibenthic medusae often accumulate in small, isolated, densely packed populations (see Arai, 2001). *Cassiopea* species are found in quiet, shallow waters resting on sandy mud bottoms, often in or near mangrove forests, where they are not dislodged by waves or currents but where there is sufficient direct sun to sustain their symbiotic zooxanthellae. Patches thus form because, in addition to the sheltered habitat, *Cassiopea* medusae do not disperse via the water column, the planulae and ephyrae swim actively but probably do not disperse far prior to metamorphosis, and thus re-aggregation is not required for reproduction. Increases in the abundance and distribution of *Cassiopea* have been related to anthropogenic species introductions (Holland et al., 2004).

*Families Cepheidae, Catostylidae, Lobonematidae, Rhizostomatidae, Stomolophidae* Eleven species of rhizostome jellyfish in families *Cepheidae*, *Catostylidae*, *Lobonematidae*, *Rhizostomatidae*, and *Stomolophidae* have been fished commercially in Southeast Asia for over 1,700 years (Omori & Nakano, 2001; Fig. 2), evidence that these species must occur locally en masse during a particular period of each, or at least most, years. *Stomolophus*, for example, apparently always has been exceptionally abundant in the Pacific and Atlantic, "often occur[ing] in vast swarms, occupying an area which is sometimes over 100 miles in length" (Mayer, 1910, p. 711; Larson, 1976, 1986). However, recent years have seen "unprecedented mass occurrences of rhizostomes sometimes disturb net fishing, while on other occasions they suddenly disappear from fishing grounds" (Omori & Nakano, 2001). In the Gulf of Mexico, *Stomolophus* is now also fished commercially (Hsieh et al., 2001), but it is unclear whether this new fishery exploits resources that have always been present or that result from recent ecosystem change. For comparison, it may be instructive that *Nemopilema nomurai* (Kishinouye) (a medusa up to 2 m b.d. in Family Rhizostomatidae) was first reported off Japan in 1920, presumably prior to massive land

**Table 3** Approximate diversity and blooming characteristics of medusae in order Rhizostomeae Cuvier 1799

| Family (8) | Genus (25) | Species (92) | Aggregate, bloom, or swarm? (References) |
|---|---|---|---|
| Cassiopeidae | *Cassiopea* | 10 | Invasive and natural accumulations (Arai, 2001; Holland et al., 2004). |
| Catostylidae | *Acromitoides* | 2 | No data—inference: do not commonly occur en masse |
| | *Acromitus* | 5 | Blooms in harbors, shrimp ponds (W.-T. Lo [Taiwan], Sreepada [India], personal communications) |
| | *Catostylus* | 10 | Aggregates, blooms, putatively swarms (Kingsford et al., 2000; Pitt, 2000; Pitt & Kingsford, 2000, 2003; Omori & Nakano, 2001) |
| | *Crambione* | 3 | Aggregates, blooms, putatively swarms (Omori & Nakano, 2001) |
| | *Crambionella* | 3 | Aggregates, natural blooms, putatively swarms (Omori & Nakano, 2001; Billett et al., 2006; Daryanabard & Dawson, 2008) |
| | *Leptobrachia* | 1 | No data—inference: do not commonly occur en masse |
| Cepheidae | *Cephea* | 7 | Aggregates, blooms (Omori & Nakano, 2001) |
| | *Cotylorhiza* | 4 | Accumulates, does not bloom, putatively swarms (Kiortsis, 1964; Kikinger, 1992) |
| | *Netrostoma* | 4 | Aggregates (Stephens and Calder, 2006) |
| | *Polyrhiza* | 1 | No data—inference: do not commonly occur en masse |
| Lobonematidae | *Lobonema* | 2 | Aggregates, blooms, putatively swarms (Omori & Nakano, 2001) |
| | *Lobonemoides* | 3 | Aggregates, blooms, putatively swarms (Omori & Nakano, 2001) |
| Lychnorhizidae | *Anomalorhiza* | 1 | No data—inference: do not commonly occur en masse |
| | *Lychnorhiza* | 3 | No data—inference: do not commonly occur en masse |
| | *Pseudorhiza* | 2 | No data—inference: do not commonly occur en masse |
| Mastigiidae | *Mastigias* | 9 | Aggregates, swarms, does not bloom (Hamner, 1982; Hamner et al., 1982; Martin et al., 2005) |
| | *Mastigietta* | 1 | No data—inference: do not commonly occur en masse |
| | *Phyllorhiza* | 4 | Invasive blooms, putatively swarms (Kramp, 1965; Mills, 2001; Graham et al., 2003; Hewitt et al., 2004; Johnson et al., 2005) |
| Rhizostomatidae | *Eupilema* | 2 | No data—inference: do not commonly occur en masse |
| | *Nemopilema* | 1 | Aggregates, blooms, putatively swarms (Omori & Nakano, 2001; Kawahara et al., 2006) |
| | *Rhizostoma* | 3 | Aggregates, blooms, putatively swarms (Zaitsev & Mamaev, 1997; Omori & Nakano, 2001; DalyYahia et al., 2003; Houghton et al., 2006) |
| | *Rhopilema* | 5 | Aggregates, natural and invasive blooms, putatively swarms (Calder, 1973; Lotan et al., 1992, 1994; Mills, 2001; Omori & Nakano, 2001) |
| Stomolophidae | *Stomolophus* | 3 | Aggregates, blooms, putatively swarms (Hsieh et al., 2001; Omori & Nakano, 2001) |
| Thysanostomatidae | *Thysanostoma* | 3 | No data—inference: do not commonly occur en masse |
| Versurigidae | *Versuriga* | 1 | No data—inference: do not commonly occur en masse |

Putative aggregating, blooming, or swarming species are noted (with sources). Taxa, organized alphabetically, are based largely on Kramp (1961) and Mianzan & Cornelius (1999); compiled primarily by M.N Dawson, G. Jarms, A.C. Morandini, and F. Pàges. Source: *The Scyphozoan* website, 2006–2007

development along the southeastern Asian coast. Mass occurrences of *N. nomurai* were subsequently reported off Japan in 1920, 1958, and 1995 (Kawahara et al., 2006). Most recently, "ecosystem change and annual hydrographical conditions…in and near the main habitat of *Nemopilema nomurai*" may be the cause of sometimes exceptionally strong blooms in the Sea of Japan, far outside its normal distributional range, almost every year since 2002 (Kawahara et al., 2006). In addition to blooming, we know that most of these species can become aggregated (e.g., Shanks & Graham, 1987), but in general it is difficult to evaluate whether most of these fished species also swarm because "almost nothing is known about the

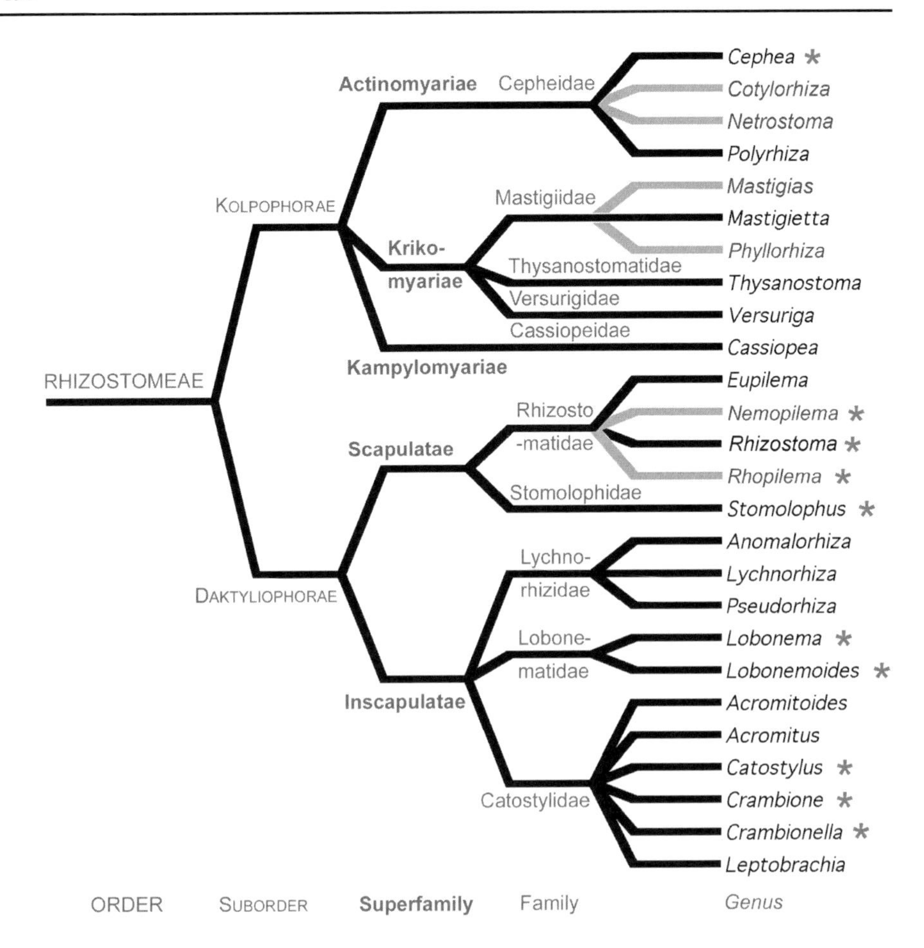


**Fig. 1** Classification of the order Rhizostomeae according to Kramp (1961) by suborder, superfamily, family and genus. Genera shown with asterisks are those fished commercially in Southeast Asia (Omori & Nakano, 2001). Those genera in gray font contain species that, according to searches using "[genus] swarm*" on Google, web of science, and zoological record, have been reported to swarm: *Cotylorhiza* (Kiortsis, 1965), *Netrostoma* (Lester and Calder, 2006), *Mastigias* (Hamner and Hauri, 1981), *Phyllorhiza* (Graham et al., 2003), *Nemopilema* (Kawahara et al., 2006), and *Rhopilema* (Lotan and Ben-Hillel, 1994). Blooming and swarming status is reviewed in the text and in Table 3

biology and ecology of edible jellyfish in Southeast Asia" (Omori & Nakano, 2001).

To illustrate this point, consider species in Catostylidae, a family of moderately large generally strongly swimming medusae. *Catostylus mosaicus* Quoy & Gaimard is commonplace in estuaries and enclosed lagoons along the eastern and southeastern states of Australia (e.g., Pitt & Kingsford, 2000; Dawson, 2005a), occurring in sufficiently high densities for commercial exploitation (Kingsford et al., 2000). Differences in strobilation and/or accumulation of *C. mosaicus* contribute to cohorts of medusae that differ in abundance and timing, suggesting blooms (Pitt & Kingsford, 2000, 2003), but it is only speculated, so far in the absence of data on estuarine circulation or physical isolation of lagoons, that the medusae may be retained by directional swimming behavior (Pitt & Kingsford, 2000; Hamner, pers. obs.). We conclude that *C. mosaicus* likely forms aggregations and blooms and perhaps swarms. *Catostylus townsendi* Mayer may also occur with reasonable frequency and abundance in coastal regions of Taiwan (W.-T. Lo, pers. comm.). *Acromitus flagellatus* (Maas) has occurred at problematically high densities in shrimp farms in southwestern India (R.A. Sreepada, pers. comm.). The apparent association between high population abundances and enclosed or semi-enclosed water bodies, however, suggests that particular hydrographic regimes may be required for mass occurrences of species of Catostylidae and that, intrinsically, they may not be swarming species.

*Rhopilema nomadica* Galil, Spanier, & Ferguson (Rhizostomatidae) has become extremely abundant in the Mediterranean after being introduced from the Red Sea in 1976 (Lotan et al., 1992, 1994; reviewed by Mills, 2001). Further expansion of *R. nomadica* into temperate seas may be limited by temperature sensitivity of the polyps (Lotan et al., 1994), or perhaps because congeneric species already exist there. Species of *Rhizostoma octopus* (Macri) native to cool temperate European waters also appear to be

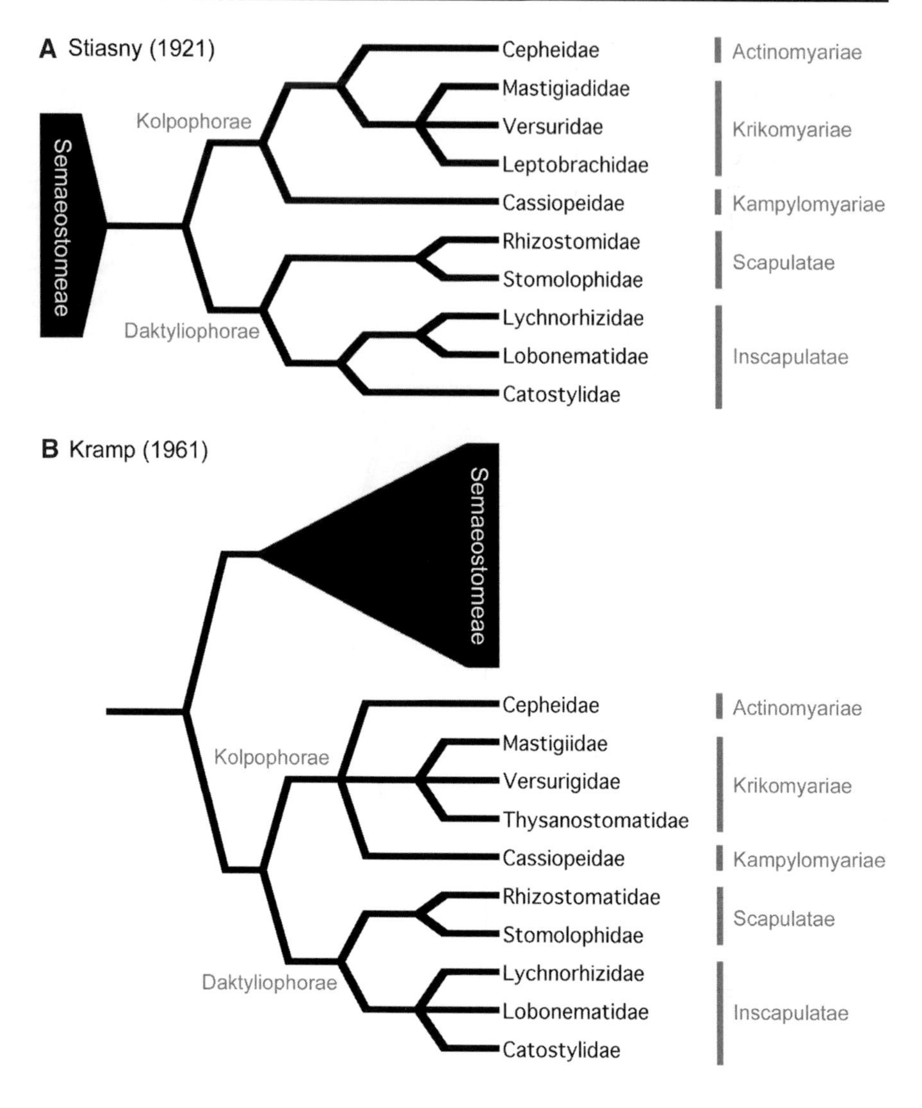


**Fig. 2** Comparison of morphology-based phylogenetic and taxonomic hypotheses of relationships among Rhizostomeae. **A** Phylogenetic relationships posited by Stiasny (1921), in which rhizostomes are derived semaeostomes. **B** Taxonomic relationships according to Kramp (1961), in which semaeostomes are a sister-taxon to the rhizostomes

exceptionally abundant in recent years in the north Atlantic (Houghton et al., 2006), although this species has been observed "in vast swarms" from the cold waters of the Atlantic coast of Europe and the Baltic for over 100 years (Mayer, 1910; Russell, 1970). We can interpret these reports as evidence that *Rhopilema* blooms and does (and/or can be caused to) aggregate but, again, whether it swarms is unclear.

*Family Mastigiidae* Populations in only two of the five genera of Mastigiidae are relatively well known. If representative of the family, these illustrate that mastigiids do not bloom or swarm but may accumulate in specific circumstances, which is illustrative for our study of aggregations and assemblages.

*Mastigias* medusae (hundreds of thousands to millions of ramets) and polyps both occur in exceptionally high abundance year-round, year to year, in fully enclosed marine lakes in Palau (Hamner & Hauri, 1981; Hamner et al., 1982; but see Martin et al., 2005) and eastern Borneo (Tomascik et al., 1997). In contrast, in Palau, populations of the same species of *Mastigias* that are in small semi-enclosed fjord-like coves consist of only tens to a couple of hundreds of medusae that are not consistently present (Dawson & Hamner, 2003; Dawson & Hamner pers. obs.). It is rare to see even a single *Mastigias* medusa in the open lagoon, although they do occur in coastal-oceanic waters (Hamner & Hauri, 1981). Thus, *Mastigias* does not bloom exceptionally, but it does aggregate and swarm. The extraordinary abundance, patchiness, and dense aggregations of medusae in lakes (Hamner & Hauri, 1981; Dawson & Hamner, 2003) appear in large part to be a function of their peculiar physical environment—there is no advective loss of ephyrae, resources (zooplankton, sunlight, nutrients, benthic habitat for polyps) are abundant, lakes are not affected by overfishing or other localized anthropogenic change—and locally adapted life histories, physiologies, and behaviors (Dawson & Hamner, 2005).

In contrast, populations of *Phyllorhiza* clearly aggregate, probably bloom, and may swarm. *Phyllorhiza punctata* von Lendenfeld is "very common" in many of the same estuaries and bays in eastern Australia as is *Catostylus* (Kramp, 1965). Historical documentation describing the fauna of Port Jackson, Sydney (von Lendenfeld, 1887), suggests that it is native to eastern Australia, although this is not beyond question because ship traffic beginning in the early 1800s connected tropical Indo-West Pacific nations, where *P. punctata* also occurs, to Australia (Hewitt et al., 2004). In recent years, *P. punctata* is known to have invaded at least the Mediterranean Sea (Mills, 2001) and to have been introduced, by oceanic currents passing the Yucatan region, into the Gulf of Mexico (Graham et al., 2003; Bolton & Graham, 2004; Johnson et al., 2005). Since 2000, huge numbers of this large jellyfish have formed apparent blooms on the US Gulf of Mexico coast during summer. These apparent blooms may result from a true bloom at, or by seasonal advection from, an unknown source region. They may also reflect some behavior of the medusae that facilitates aggregation in opposition to the natural dispersive tendency of moving water (if they are not simply in an essentially laminar mass transport system).

(B) Molecular phylogenetics, species richness, and phenotypic diversity

Phylogenetic analyses of DNA sequences tend to be less susceptible to convergence than are analyses of morphological similarity in phenotypically simple organisms. Molecular phylogenetics can provide better estimates of diversity and evolutionary relationships than homoplastic morphologies, first by using many additional characters, which increases the probability that groups will be differentiated with high statistical support, and second by allowing phylogenetic reconstruction independent of morphological characters, thereby promoting analyses of phenotypic evolution. Phylogenetic analyses of DNA sequence data are now available for all classes, orders, and a small fraction of families and species within Medusozoa (Collins, 2002; Dawson, 2004; Collins et al., 2006; Daryanabard & Dawson, 2008). Phylogenetic analyses of DNA sequence data generally confirm that each higher taxon recognized on morphological criteria for most of the twentieth century corresponds one-to-one with molecular clades, although the relationships among the taxa may differ (Figs. 2 cf. Fig. 3). For the families from which multiple genera have been analyzed and for the genera from which multiple species have been analyzed, in general there are few lower taxa that have been attributed incorrectly to their concomitant higher taxon. Recent exceptions are the genus *Dactylometra* and the species *Pelagia colorata*, which are now both considered members of the confamilial genus *Chrysaora* (Gershwin & Collins, 2002). In contrast, late twentieth century estimates of species diversity based on morphological comparisons may underestimate 'phylogenetic' species diversity by an order of magnitude (Dawson, 2004). Thus, we have reasonable confidence that traditional taxonomy provides robust information on the attribution of phenotypes to a particular class, order, family or genus, but that patterns of common ancestry inferred from taxonomy need to be revised, and patterns of species diversity must be re-assessed.

*Phylogeny and common ancestry*

*Material and methods (for phylogeny and common ancestry)* Tissue from the bell margin, gonad, gut, or oral arms of medusae was preserved in DMSO + NaCl solution or >70% ethanol (Dawson et al., 1998) and stored in a freezer after shipping. Total DNA was extracted using a CTAB extraction protocol (Dawson et al., 1998; Dawson & Jacobs, 2001). PCRs used *Taq* Polymerase (Perkin Elmer) and Perkin Elmer 2720 thermocyclers. Amplification of nuclear 18S rDNA used primers 18SA and 18SB (Medlin et al., 1988; Collins, 2002) and began with an 8 min denaturation at 94°C, 2 min annealing at 48°C, 2 min extension at 72°C, 4 min denaturation at 94°C, 2 min annealing at 49°C, 3 min extension at 72°C, followed by 38 cycles, each consisting of

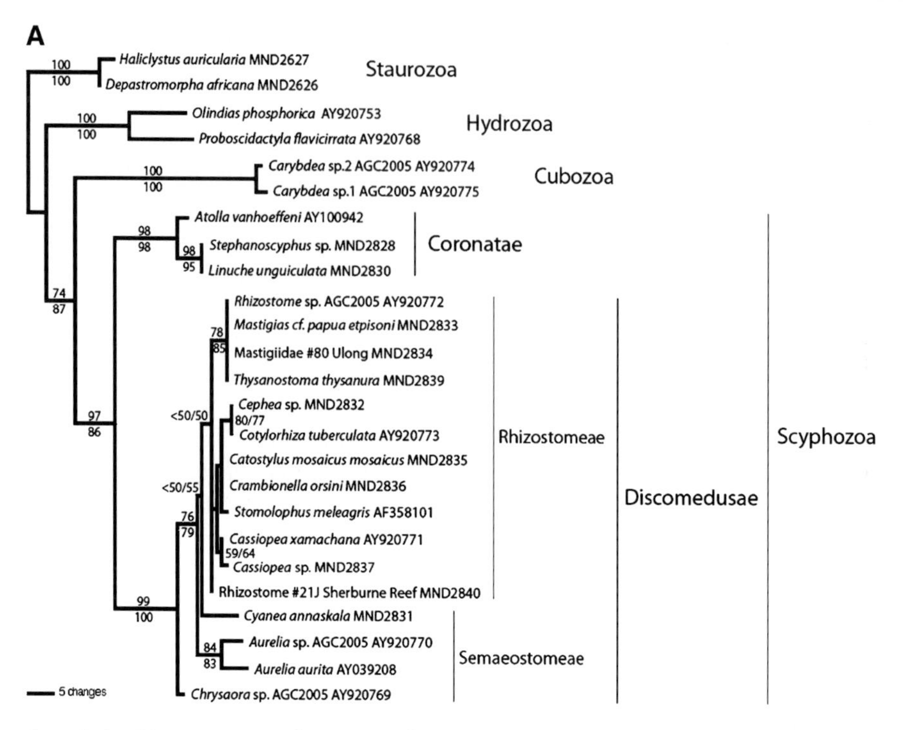


**Fig. 3** Phylogenetic relationships among medusozoans. **A** Class-level relationships; bootstrap support values below branches are from unweighted maximum parsimony while those above branches are from maximum likelihood analyses. **B** Phylogenetic relationships among scyphomedusae; bootstrap support values below branches are from unweighted maximum parsimony while those above branches are from maximum likelihood analyses. **C** Phylogenetic relationships among rhizostomes, redrawn from Daryanabard & Dawson (2008). Above the branches, ML bootstrap values precede MP bootstrap values

30 s at 94°C, 45 s at 50°C, and 180 s at 72°C. Amplification of nuclear 28S rDNA used primers LSUD1F and LSUD4Ra (Matsumoto et al., 2003) and began with an 8 min denaturation at 94°C, 2 min annealing at 53°C, 2 min extension at 72°C, 4 min denaturation at 94°C, 2 min annealing at 54°C, 1.5 min extension at 72°C, followed by 35 cycles, each consisting of 30 s at 94°C, 30 s at 55°C, and 70 s at 72°C. PCRs terminated with a 10 min extension step (72°C) then refrigeration (4°C). Amplification of mitochondrial cytochrome *c* oxidase subunit I (COI) used primers LCOjf and HCO2198 (Folmer et al., 1994; Dawson, 2004) and began with an 8 min denaturation at 94°C, 2 min annealing at 48°C, 2 min extension at 72°C, 4 min denaturation at 94°C, 2 min annealing at 49°C, 1.5 min extension at 72°C, followed by 35 cycles, each consisting of 45 s at 94°C, 45 s at 50°C, and 60 s at 72°C. PCRs terminated with a 10 min extension step (72°C) then refrigeration (4°C). PCR products were cleaned for sequencing by incubating 47 μl product with 4 μl exonuclease and 4 μl shrimp alkaline phosphotase (USB Corporation, Cleveland) at 37°C for 15 min, 80°C for 15 min, then 4°C. The concentration of DNA in the cleaned PCRs was quantified by comparison against a DNA standard ladder (100 bp DNA ladder Plus; Applied Biosystems, Foster City) on a 1.4% agarose in 0.5x TBE minigel run at 130 V for 25 min. Concentration was adjusted to 26 ng/μl for COI and 16 ng/μl for NCR and then sequenced using 3 μM PCR primers and ABI BigDye Terminator v3.1 Cycle Sequencing chemistry on an ABI 3730 Capillary Electrophoresis Genetic Analyzer (Applied Biosystems, Foster City) in the UC Davis DNA sequencing facility.

Sequences were assembled into contigs for each specimen in Sequencher v4.7 (Gene Codes Corporation, Ann Arbor) and edited by eye to remove all primer sequences and ambiguous terminal portions and correct (or delete or insert) incorrect base (or indel) calls. All sequences from all specimens were then assembled into a single contig, and all

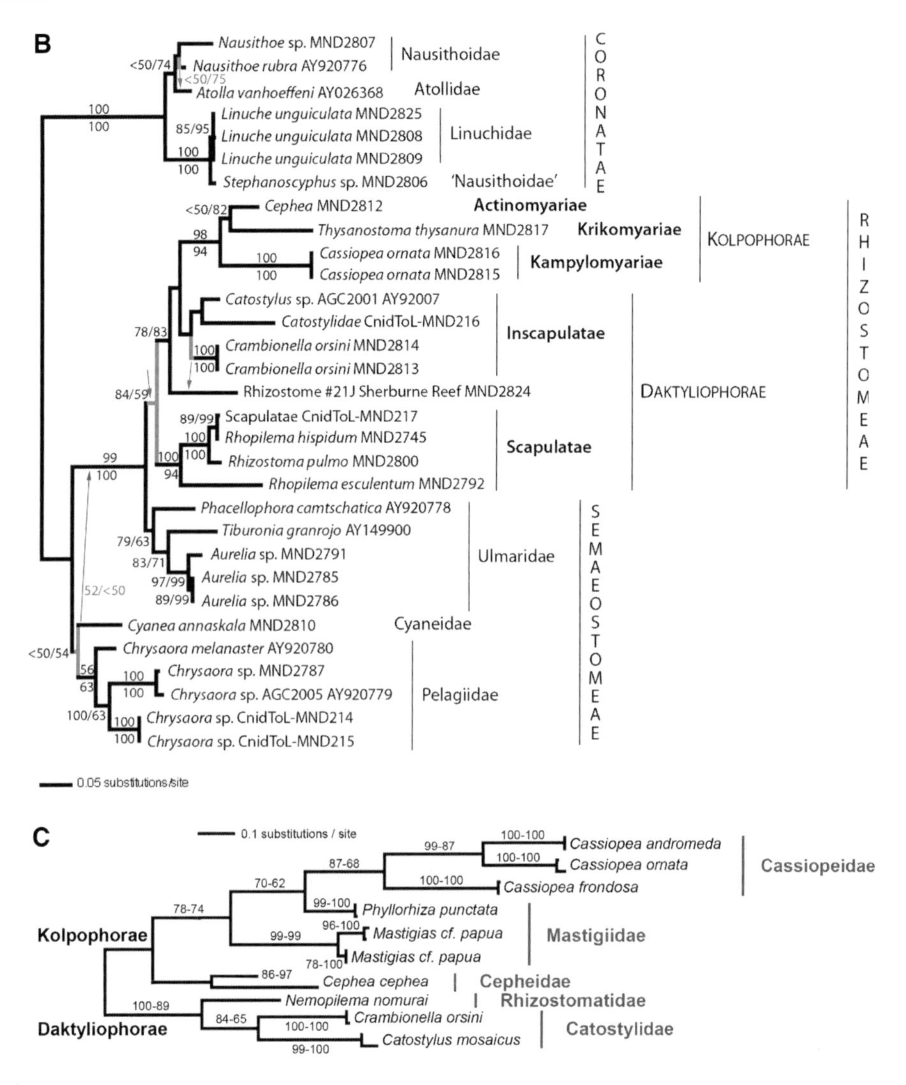


**Fig. 3** continued

autapomorphies and synapomorphic positions checked, and corrected if necessary in the original specimen contigs. The corrected specimen contigs were then exported as FASTA-concatenated format and aligned using CLUSTALX (gap-opening penalty 5 or 10; gap-extension penalty 1, 2, or 5; Jeanmougin et al., 1998). The aligned datamatrices, minus gapped, missing, or ambiguous positions, were 911 nucleotide positions long got 18S rDNA and 605 nucleotide positions long got 28S rDNA. Nuclear 5.8S rDNA alignments were used from Dawson (2004). Datamatrices are available from the authors on request.

Unweighted maximum parsimony (UMP) and maximum likelihood (ML) analyses of the aligned 18S rDNA and 28S rDNA datasets were used to generate gene trees in PAUP* 4.0b10 (Swofford, 2002). In UMP analyses, 1,000 heuristic searches each began with a random tree and then was permuted using Tree-Bisection-Reconnection (TBR). In ML analyses, the GTR + I + G substitution model was selected using MODELTEST v.3.7 (Posada & Crandall, 1998). These same substitution models were employed in bootstrap analyses using UMP (10,000 replicates, each using 10 heuristic searches) and ML (100 replicates, each using two heuristic searches).

A phylogenetic hypothesis for Rhizostomeae based on combined analyses of the 18S rDNA, 5.8S rDNA, partial 28S rDNA, and partial COI mtDNA (aligned by amino acid positions) was generated using UMP, weighted Maximum Parsimony (WMP), and ML analyses in PAUP* 4.0b10 (Swofford, 2002). Genera included in the analyses were *Cassiopea* (two spp.), *Catostylus*, *Cephea*, *Crambionella*, *Cotylorhiza*, *Mastigias*, *Phyllorhiza*, *Rhizostoma*, *Rhopilema* (two spp.), *Stomolophus*, and *Thysanostoma*. All markers were not available for all taxa, so we consider this tree to be preliminary, albeit closely matched to prior morphological hypotheses. UMP and WMP analyses employed TBR search algorithm on no more than 5,000 trees $\geq$5 to 8 steps longer than the shortest tree resulting from 10,000 random starting trees. Gaps were coded as a fifth state. In WMP analyses, markers were weighted in proportion to the percentage of nucleotides that were constant: 18S weight = 15, 5.8S weight = 5, 28S weight = 2, COI weight = 1. ML analyses used the GTR + I + G model (Base = 0.2496 0.2098 0.2684, Nst = 6, Rmat = 0.6367 3.0623 2.4586 0.8973 9.0725, Rates = gamma, Shape = 0.6371, Pinvar = 0.5991) selected using MODELTEST v. 3.7 (Posada & Crandall, 1998).

Trees were drawn by PAUP* 4.0b10 and annotated in Adobe Photoshop CD3 v.10.0.1 or exported from PAUP* 4.0b10 in Newick format and drawn using FigTree v.1.1.1 (A. Rambaut, 2007, http://www.tree.bio.ed.ac.uk/software/figtree/) and MacClade v. 4.08 (Maddison & Maddison, 1989). To generate an hypothesis for the evolution of blooming and swarming in Rhizostomeae, the rhizostome molecular phylogeny was used as a backbone constraint tree into which were inserted new nodes and branches according to relevant subtrees in the morphologic analyses of Stiasny (1921) and Kramp (1961).

*Results (on phylogeny and common ancestry)* The 18S rDNA gene tree (Fig. 3A) shows high support for Class Scyphozoa being composed of two reciprocally monophyletic clades: Coronatae and Discomedusae. Information on relationships within these clades is lacking due to limited taxonomic sampling of Coronatae and poorly supported internal nodes in Discomedusae. The 28S rDNA gene tree (Fig. 3B) resolves relationships within Discomedusae, providing some support (albeit not always strong) for monophyly of the 'semaeostome' families Pelagiidae and Ulmaridae; Cyaneidae is represented by only one species preventing any inference about its monophyly. The interfamilial relationships of semaeostomes, particularly Cyaneidae and Pelagiidae, are not well resolved but family Ulmaridae is clearly sister taxon to rhizostomes. The majority of evidence supports a monophyletic Rhizostomeae containing Daktyliophorae (composed of superfamilies Scapulatae and Inscapulatae) paraphyletic with respect to Kolpophorae (composed of superfamilies Actinomyariae, Krikomyariae, and Kampylomyariae). Monophyly of Kolpophorae is very strongly supported. These phylogenetic relationships within rhizostomeae are robust to taxonomically broader sampling an inclusion of additional molecular markers (Fig. 4A) and therefore provide a robust preliminary framework for interpreting phenotypic evolution of the rhizostome medusae.

*Discussion (of phylogeny and common ancestry)* The new analyses of nuclear 18S and 28S ribosomal DNA sequence data (Fig. 3A, B) and prior analyses (e.g., Dawson, 2004; Daryanabard & Dawson, 2008, see Fig. 3C) yield few surprises considering prior morphologic analyses. Their major contribution to the phylogenetic framework is to provide finer resolution for the families within Discomedusae than available previously. In the absence of rate calibrations using the fossil record (but see Cartwright et al., 2007) or major geological events (but see Dawson, 2005a), branch lengths assumed generally indicative of relative divergence times also provide tantalizing suggestions of different rates of evolution in major clades of scyphozoans.

*Class-level comparisons* Phylogenetic analyses of DNA sequence data clearly support morphological hypotheses that Cubozoa, Hydrozoa, and Scyphozoa are distinct classes but they are ambiguous regarding sister taxa relationships among these classes (e.g., Dawson, 2004; Collins et al., 2006; Fig. 3A). Weak statistical support for Cubozoa–Scyphozoa and against Cubozoa–Hydrozoa or Hydrozoa–Scyphozoa is most safely interpreted at this time as lack of resolution due to insufficient suitable data, although this could also indicate events that happened rapidly in quick succession. We do know from traditional taxonomic descriptions that staurozoans certainly cannot, hydrozoans may not, and cubozoans might aggregate, bloom, or swarm, albeit none to the same extent as scyphomedusae. Thus, it is unlikely that the most recent common ancestor of these three classes aggregated, bloomed, or swarmed, and therefore,

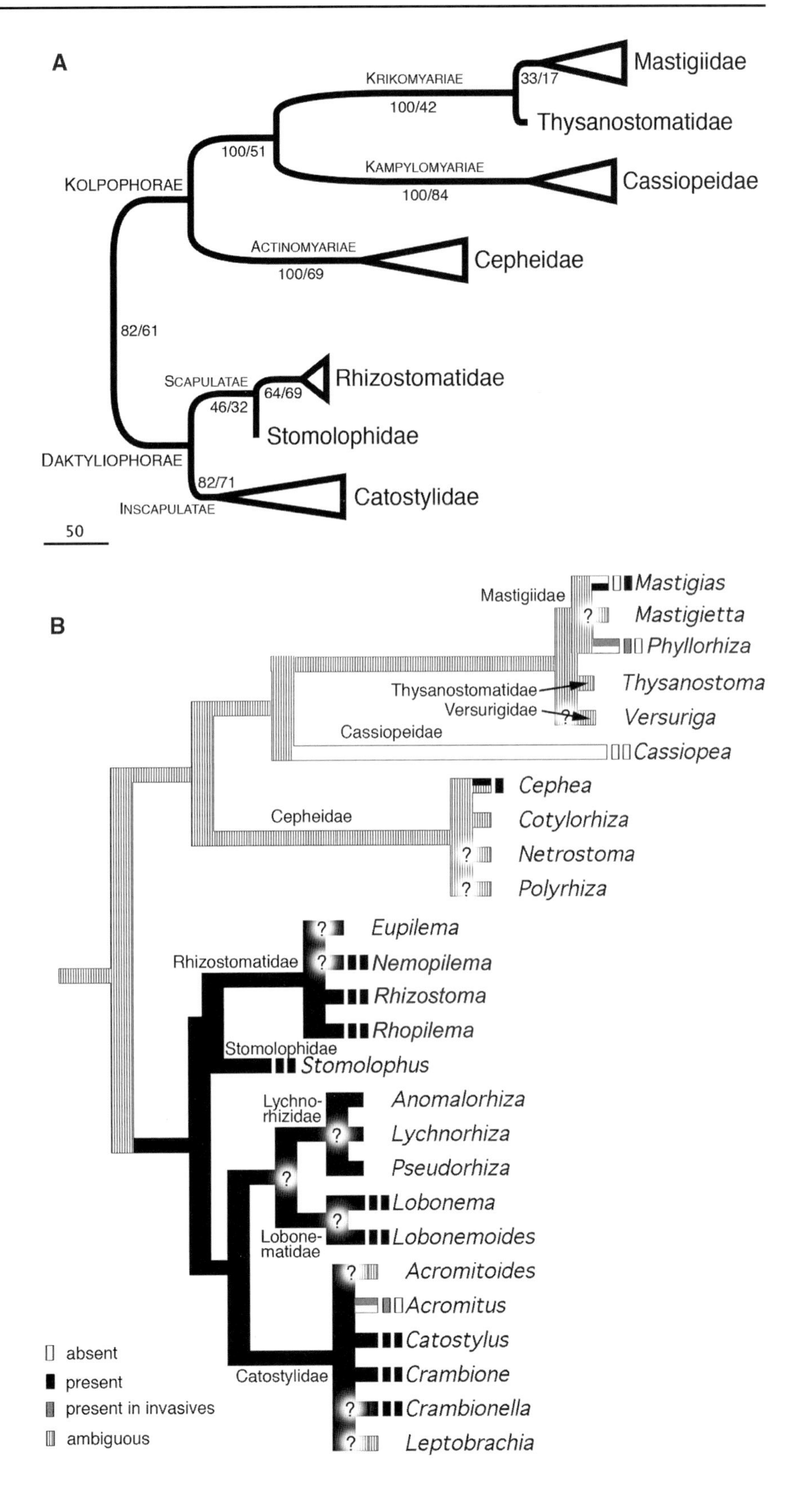


**Fig. 4** **A** Phylogenetic hypothesis for Rhizostomeae. The tree shown is one of the 33 shortest trees recovered by both unweighted (U) and weighted (W) maximum parsimony (MP) analyses (U-MP tree length = 1,443 steps, U-MP consistency index = 0.7859, W-MP consistency index = 0.8165). Values adjacent to branches show first the percentage of those 33 trees in which the branch occurred, followed by the UMP bootstrap value. Alternative groupings in a maximum likelihood tree (-Ln likelihood = 8955.79139) included *Mastigias* + *Thysanostoma* as opposed to *Mastigias* + *Phyllorhiza* and *Catostylidae* paraphyletic with respect to other Daktyliophorae. **B** Hypothesized evolution of states of 'blooming' and 'swarming' traits in rhizostomes. Boxes at the terminal branches, left of the taxa, show how a genus was coded as blooming (left box) or swarming (right box), according to the key (bottom left); states of genera for which there is no information are coded as 'missing data' (taxa with one or zero blocks). Branch shading shows phylogenetically inferred routes of inheritance of blooming and swarming states; "ambiguous" indicates where parsimony inference of the ancestral character state is uncertain; for the few taxa that have different states for blooming and swarming, the upper half of the branch is shaded for blooming, and the lower half of the branch for swarming

cubomedusae, hydromedusae, and scyphomedusae that do occur en masse likely do so because the traits that enable mass occurrences have arisen independently in each of these taxa.

*Orders within Class Scyphozoa* Phylogenetic analyses confirm that the lineage Coronatae is sister taxon to a clade composed of Semaeostomeae and Rhizostomeae, as proposed by Mayer (1910) and Stiasny (1921; cf. Kramp, 1961; Fig. 2). The molecular analyses most inclusive of Scyphozoa also predominantly favor placing order Rhizostomeae as a derived lineage within order Semaeostomeae (combined as the resurrected order or sub-class Discomedusae; Dawson, 2004; Marques & Collins, 2004; Collins et al., 2006; Fig. 3A). Unfortunately, most molecular phylogenetic analyses provide only a hint of relationships among semaeostome families, due to sometimes conflicting reconstructions from different datasets and often only moderately strong bootstrap support (perhaps indicating a period of rapid diversification). However, some consistent patterns do emerge. Pelagiidae commonly branches basally in Discomedusae, and there is little evidence that strongly contradicts the reconstruction of Cyaneidae as basal in a clade also containing Ulmaridae and Rhizostomeae (Fig. 3B). Further, the Ulmaridae and Rhizostomeae seem likely to be reciprocally monophyletic sister taxa on the basis of molecular analyses and distinctive morphology of rhizostomes, and deepwater semaeostomes probably are derived ulmarids (Fig. 3B).

*Family comparisons within Rhizostomeae* Phylogenetic analyses among Rhizostomeae generally confirm the phyletic classification of Stiasny (1921), although basal relationships suffer from relatively low bootstrap values due to somewhat short branches. The only modification suggested is that Kampylomyariae and Krikomyariae are more closely related to each other than either is to Actinomyariae (Fig. 3C). Consequently, the long-standing taxonomy of rhizostomes has been consistent in its inference that almost all aggregating, blooming, or swarming species are members of the same lineage, Daktyliophorae. Mass occurrence is relatively common, however, only within one Family: Rhizostomatidae (Table 3). A variety of species in the families Catostylidae, Cepheidae, Lobonematidae, and Stomolophidae might be considered to be putative swarmers. The two known aggregating and/or blooming species outside Daktyliophorae are both members of Mastigiidae and are both found in unusual situations: enclosed habitat (*Mastigias*) or invasive (*Phyllorhiza punctata*; Graham et al., 2003; Johnson et al., 2005).

*Species comparisons* Based on morphological similarity and on the assumption that similarity and dissimilarity reflect natural units such as species, it has been estimated for many years that the number of morphospecies of many holoplanktonic species, including meroplanktonic scyphomedusae, is surprisingly low. Analysis of DNA between morphologically seemingly similar individuals, however, has occasionally demonstrated the widespread presence of cryptic, sibling and/or sister species, i.e., "two or more distinct species that are erroneously classified (and hidden) under one species name." These exist primarily due to non-visual mating systems and morphological stasis (Bickford et al., 2007), as exist in scyphomedusae.

Cryptic species and/or cryptic subspecies have been discovered in all of the genera of scyphomedusae thus far examined by allozyme and/or DNA sequence analyses (Greenberg et al., 1996; Dawson & Jacobs, 2001; Schroth et al., 2002; Dawson, 2004; Holland et al., 2004). Probably the most impressive case of cryptic species in scyphozoans to date is that of *Aurelia*. *Aurelia* is the best studied genus within the scyphomedusae (Arai, 1997), with hundreds of biological, ecological, and taxonomic publications over the past two centuries culminating in the conclusion that there were only three morphospecies worldwide (*Aurelia aurita*, *A. limbata* (Brandt), and *A. labiata*) of which one (*A. aurita*) was a circumglobal, nearly ubiquitous, ecological generalist. However, the most recent DNA-based phylogenetic analyses of this genus demonstrate that there are at least 14 species of *Aurelia* (Dawson et al., 2005), almost certainly an underestimate of the global species count for this genus, given that large areas of the globe remain unsampled. The molecular phylogenetic analyses of diversity may simply be redressing overly zealous 'lumping' during the mid-twentieth century (Kramp, 1961, 1965, 1968), before which as many as 20 species or varieties of *Aurelia* were recognized (Mayer, 1910).

Similar stories can be told about almost every other genus studied to date using these techniques. Three described and three undescribed species are now recognized in *Cassiopea*, and considering older taxonomic publications and the extent of modern sampling, there are probably more species to be found (Holland et al., 2004). Preliminary analyses of *Catostylus mosaicus* illustrate two cryptic subspecies, concordant with distinct morphogroups suggested by von Lendenfeld (1884) and Dawson (2005b), and there must be also additional cryptic species in Papua New Guinea and the Philippines (Kramp's 1961 list of occurrences). *Cyanea capillata* is not a circumglobal species, but instead is a complex of at least three (and probably many more) cryptic species (Dawson, 2005c). Given these figures, we believe that Dawson's (2004) estimate that "a survey of [genetic variation] within morphospecies indicates that scyphozoan species diversity may be approximately twice recent estimates based on morphological analyses" is excessively conservative, returning estimates of species diversity to at least, and probably much higher than, the estimates at the start of the twentieth century. However, it is pertinent to emphasize here that the expected increase in accepted species diversity is unlikely to destabilize the higher taxonomy of semaeostomes which has remained almost constant for 100 years and appears supported by molecular data.

Indeed, in all cases in which distinct genetic units have been re-analyzed using morphometric and statistical analyses, it has been possible to distinguish between previously seemingly similar species and to clarify the provenance of morphological differences that had been described correctly but attributed incorrectly among taxa (Greenberg et al., 1996; Dawson 2003, 2005b, c; Gershwin, 2001). In an explicit analytical framework, these morphological variations can be used to propose detailed phylogenetic relationships and patterns of evolution as done by Marques & Collins (2004) for the Medusozoa and by Gershwin & Collins (2002) for Pelagiidae. Other phenotypic features, such as adaptation to different temperatures (Dawson & Martin, 2001; Schroth et al., 2002), are also applicable in a comparative phylogenetic framework, although more rarely used, and these also have merit for comparisons between populations within species (e.g., Bolton & Graham, 2004; Dawson, 2005d, e). The potential for new morphological or other phenotypic characters to resolve phylogenies, though, may be limited. Greenberg et al. (1996) found that only two of 12 morphological characters (manubrium length and canal structure) distinguished reliably two distinct genetic varieties of *Aurelia*, and other studies have found fewer (Dawson, 2003, 2005b, 2005c), which suggests there is a dearth of good characters even for sophisticated morphometrics. Nonetheless, detailed documentation of morphological variation that is not useful in a traditional taxonomic framework is still useful in understanding patterns of evolution (Dawson & Hamner, this volume).

### *Conclusions from Part I*

1. *Taxa that occur en masse are not randomly distributed within the Phylum Cnidaria.*
   Within the Scyphozoa only a relatively small number of semaeostome and rhizostome taxa are found in aggregations, blooms, or swarms. By inference, traits or suites of traits that promote co-occurrence are not evolutionarily labile and may be of selective benefit.
2. *Scyphozoan taxa that occur en masse are not randomly distributed among marine habitats.*
   Some particular environments provide stronger selective benefits for accumulations of medusae.
3. *Environmental attributes and organismal traits interact to produce mass occurrences.*
   Although some traits and some environments seem to favor mass occurrences, usually an interaction of the two is required to yield a mass occurrence. For example, irrespective of other organismal traits or patterns of occurrence:

   a. *midwater and deep-sea medusae rarely aggregate and do not bloom or swarm* because patterns of resource availability, life histories, and longevity all favor maintenance of populations without large fluctuations. This generalization is pertinent to most coronates and to most ulmarid semaeostomes. Exceptions may include apparent blooms of *Chrysaora achlyos* and seabed aggregations of *Poralia rufescens*.
   b. *epibenthic medusae do not swarm*, although the epibenthic Cassiopeidae medusae and benthic stauromedusae may occur at high densities.
   c. *large carnivores that feed on large prey rarely occur* en masse. Many species in order

Cubozoa and the large semaeostome medusae in family Cyaneidae and genus *Phacellophora* (F. Ulmaridae) feed at relatively high or intermediate levels in the food chain (e.g., on fish or gelatinous prey such as *Aurelia* and ctenophores). A given body of water cannot support enough prey to feed many of these large predators at any one time. By contrast, *Nemopilema nomurai* is an exceptionally large medusa that blooms, but it feeds primarily on small zooplankton.

d. *the only holoplanktonic, epipelagic scyphomedusan* Pelagia noctiluca *both blooms and swarms*, perhaps because growth rates are rapid and generation times can be short.

4. *Environmental change can modify whether species form blooms.*

   a. *Overfishing* promotes blooms of *Chrysaora* (Lynam et al., 2006) and *Cyanea* (Xian et al., 2005).
   b. *Introductions* of species into novel environments have been followed by blooms of *Aurelia*, *Phyllorhiza*, and *Rhopilema*. The reasons are not well documented for jellyfish, but possibly include lack of competition or an absence of predators or parasites (e.g., Mitchell & Power, 2003). Species introductions emphasize the importance of interactions between environment and phenotype because some introduced populations have not invaded nearby habitats (e.g., *Cassiopea* from New Guinea in Hawaii (Holland et al., 2004), *Aurelia* from Japan in Spinnaker Bay, Long Beach, and San Francisco Bay, California (Greenberg et al., 1996; Gershwin, 2001; Dawson et al., 2005).
   c. *Environmental change can inhibit blooms of blooming species and stimulate others* (Mills, 2001).

5. *Molecular genetic and statistical morphological analyses reveal additional variation* not recognized through conventional taxonomic approaches at both population and species levels. We need more information about the suites of evolved traits of particular species to better understand the causes and importance of blooms and swarms.

## Part II. Diversification and rapid speciation

Following Schluter's (2000) logic, if occurrence en masse is evolutionarily advantageous then taxa possessing attributes associated with aggregations, blooms, and/or swarms might rapidly radiate. Unfortunately, the phylogeny of Scyphozoa currently is not well suited to answering this prediction, for two reasons. First, the phylogeny of Scyphozoa is still incomplete, and we do not yet have an adequate understanding of true species diversity within any clade, although existing taxonomy may provide a reasonable estimate (see Part IB) if cryptic species are relatively evenly distributed taxonomically. Second, for jellyfish there are few opportunities to verify the timing of evolutionary events due to the paucity of fossils or to make molecular comparisons of taxa that speciated allopatrically in association with major tectonic events. Consequently, we take two different approaches here. (1) We compare diversity of sister taxa which must be the same age because they originated at the same cladogenetic event; this is the most powerful approach for identifying correlates of net diversification (Barraclough et al., 1998). Where the phylogeny is not well resolved, we relax the criterion of sister taxa and assume approximately similar age on the basis that the lack of resolution in the tree may (but need not) reflect relatively similar divergence times. (2) We argue for the plausibility of rapid speciation in scyphozoans on the basis of patterns seen in non-cnidarian taxa, relative ages of scyphozoan clades, and a single well-documented example of rapid speciation in Scyphozoa.

### (1) Diversification

Considering the scyphomedusae, the two highest level sister taxa are Coronatae and Discomedusae. Of these, only Discomedusae is notable for species that form unusual blooms and swarms, and is also the more diverse ($\sim$155 species versus $\sim$50); this relationship, like most others below, would not change even if all recently reported 'phylogenetic' species of *Aurelia* were excluded. The shallow water coronates are small medusae, whereas big coronates all inhabit deep water and do not have the organismal or environmental resources required for blooming.

Considering Discomedusae, there are four clades—Cyaneidae, Pelagiidae, Ulmaridae, and Rhizostomeae—whose relationships are not well resolved (Fig. 3), although it seems likely that Cyaneidae or Pelagiidae is basal and Rhizostomeae is certainly derived. Of the semaeostome families, Cyaneidae, Pelagiidae, and Ulmaridae are comparably diverse although Ulmaridae and Pelagiidae are the only two we consider to include strongly blooming or swarming species. In the Ulmaridae (*Aurelia*, *Phacellophora*, and diverse genera of deep-sea medusae), all of the genera, as far as we know, are of approximately the same age, yet the only blooming genus is *Aurelia* and it is by far the most speciose, whether based on traditional taxonomic or phylogenetic estimates of species diversity. The remaining comparison utilizing semaeostomes, i.e., all ancestral Semaeostomeae vs. the diversity of derived Rhizostomeae, shows the latter, which includes the most well-known blooming and swarming species, is most diverse (71 versus 92 [~67 vs. 92 if 'phylogenetic' species are excluded]). Within Rhizostomeae, the blooming Daktyliophorae is only slight more diverse (49 species) than the non-blooming Kolpophorae (44 species).

Comparing genera within families, the pattern is mixed. First, within the Coronatae, which are not typically blooming or swarming species, there is no evidence that mass occurrence leads to diversification. There are only three species in the genus *Linuche* in comparison to a mean of four species (range 1–18) in genera that do not bloom. In contrast, blooming or swarming genera are indisputably more diverse than non-accumulating genera in the semaeostome Pelagiidae (15 *Chrysaora* spp. cf. 1 *Sanderia* and 1 *Pelagia*) and Ulmaridae (17 *Aurelia* spp. cf. mean 1.3 spp. in non-accumulating genera [range 1–3]). In Rhizostomeae, though, the distinction is not so clear. In the four families in which there are blooming and non-blooming genera, the blooming genera are on average more species-rich than the non-blooming genera, but all blooming genera are not always more species-rich than all non-blooming genera (mean [range]: Catostylidae, 5.3 [3–10] cf 2.7 [1–5]; Cepheidae 7.0 [na] cf. 3.0 [1–4]; Mastigiidae 4.0 [na] cf. 3.5 [1–9]; Rhizostomatidae 3.0 [1–5] cf. 2.0 [na]).

Consequently, it appears that for semaeostomes, mass occurrence (and the traits that support it) may represent an adaptive zone (and key innovations)—evolution of a phenotype that allows a lineage to exploit a new way of life and be particularly successful—but that this is not generally the case for coronates or rhizostomes which require a particular suite of characters to come into alignment, except perhaps for *Catostylus*, *Cephea*, and *Rhopilema*. One reason that there may be a difference in the association between blooming and high diversity in semaeostome genera, but not in rhizostome genera, may be that photosymbiosis provided an alternative successful adaptive zone in rhizostomes, confounding our comparisons within rhizostomes (see Dawson and Hamner, this volume).

*Radiation despite morphological conservation*

Numerical comparisons between taxa also raise the issue of what processes allowed evolution of such a large number of species in some genera but so little morphological variation. This question is best investigated by reference to *Aurelia*, in which the evidence for radiation of species in association with evolution of the blooming phenotype is strongest.

For the past 200 years, *Aurelia* has been a widely reported swarming genus (e.g., Mayer, 1910; Russell, 1970; Mills, 2001, Purcell, 2005). It has been the subject of taxonomic study predating 1758 (Linnaeus, 1758), and its biology has been extensively investigated and reviewed (Yasuda, 1968, 1969, 1971; Russell, 1970; Lucas & Williams, 1994; Lucas, 1996, 2001; Miyake et al., 1997; Dawson & Martin, 2001). By classifying all non-boreal moon jellyfish as *Aurelia aurita*, Kramp (1968) made the explicit statement that *Aurelia* possessed little morphological variation in any obvious geographic, ecological, or other manner. The paucity of morphological variation has been confirmed by detailed morphometric analyses (Greenberg et al., 1996; Dawson, 2003). Nonetheless, *Aurelia* occurs in an incredible diversity of locations worldwide; it has a flexible life history, an exceptional range of acceptable prey, and an ability to starve for long periods of time; and it can behaviorally migrate to topographically unique spawning sites using rhopaliar ocelli to sense directional sunlight (Hamner et al., 1994), displays various patterns of diel vertical migration, and can survive via asexual polyp propagation for many years. Also, genetically exceptionally variable (Dawson & Martin, 2001), it has produced many morphologically cryptic species around the world, and has invaded many habitats. The densest population blooms of *Aurelia* often occur in fully or semi-enclosed, often very small, embayments (Lucas & Williams, 1994; Ishii & Båmstedt, 1998; Albert, 2005, 2007), yet it also occurs widely dispersed on continental shelves (Möller, 1980; Graham, 2001). It is now clear from preliminary molecular investigations that an abundance of cryptic species, generally geographically distinct, characterizes this genus (Dawson & Jacobs, 2001; Dawson & Martin, 2001; Schroth et al., 2002; Dawson et al., 2005). Thus, *Aurelia* has diversified via accumulation of variation at multiple levels that is not reflected to a similar degree in morphology.

Conservation, or stasis, of morphology despite diversification of physiology, behavior, etc, may result from several processes, including canalization (lack of genetic variation) of loci related to morphology but not other phenotypes, developmental plasticity that counteracts directional selection, and stabilizing selection. Several lines of evidence suggest that the most likely processes are a mix of plasticity and stabilizing selection. Evidence against canalization comes from morphological and genetic variation within and between species (Gershwin, 2001; Dawson, 2003), population genetic theory which predicts many neutral or nearly neutral polymorphisms in large populations, and evidence that the evolution of adaptive polymorphism is facilitated by overlapping generations (Ellner & Hairston, 1994; Sasaki & Ellner, 1997; Sasaki & De Jong, 1999) such as occurr in scyphomedusae. Evidence for phenotypic plasticity comes from reports of acclimation (e.g., Arai, 1997, p. 191), perhaps counter-gradient variation (Dawson & Martin, 2001), and theory which suggests that asexual reproduction should favor evolution of epigenetic mechanisms that modify an organism's response to long-term temporal change or spatial heterogeneity despite constant genomic content (Jablonka & Lamb, 2005). Evidence for stabilizing selection comes from the many symplesiomorphic morphological character states conserved within many genera in similar environments, compared against the evolutionary loss of traits for accumulation in lineages that have invaded environments that disfavor aggregation. For example, the deep-sea ulmarids have often lost the high fecundity, annual metagenic life cycles, transparency, and U.V. screening pigments that characterize epipelagic gelatinous zooplankton from which they are derived. The deep water medusae possibly also mostly lost the behavior that facilitates reproductive aggregations or swarms because life in the deep sea is diffuse and biomass depleted with depth.

These phylogenetic inferences are important because explanations of morphological similarity as the result of conservation of characters (or morphological stasis) is quite different from the phenomenon of convergent evolution, which has been the primary reason discussed previously to explain similarity of form in diverse epipelagic phyla (e.g., Hamner, 1995; Bone, 2005; Colin et al., 2006). Also, it is apparent that closely related modern species may possess similar phenotypes due to parallel evolution, in which directional selection and/or evolution along 'genetic lines of least resistance' (Schluter, 1996) modifies ancestral phenotypes in a similar manner in two or more daughter lineages.

(2) Rapid radiation

It has been widely argued for many years that rates of evolution in the sea are much slower than on land or in fresh water. The net diversification interval (NDI), the time between originations of lineages, is typically longer for marine animals such as bryozoans, bivalves, corals, echinoids, gastropods, and teredinids (NDI: 6.7–16.4 million years [My]) than it is for terrestrial animals (0.3 to 14.2 My) and freshwater animals (0.004–7.0 My; reviewed in Table 12.1 of Coyne & Orr, 2004). Similarly, the time for speciation, the time for one species to split into two daughter species, is expected to be longer in marine systems than in terrestrial or freshwater systems, presumably due primarily to much larger panmictic gene pools in marine systems. However, this does not mean rapid speciation is not possible in the sea (Norris, 2000). NDI for trilobites is 0.8–1.6 million years and for plankton is 0.1 My (Coyne & Orr, 2004, Table 12.1).

Although the current phylogenetic analysis of Scyphozoa is insufficient for estimating rates or magnitudes of diversification, there is some indication that evolution of mass occurrence, or correlated traits, has coincided with periods of rapid evolution. If short branch lengths and lack of statistical resolution can be interpreted as evidence of rapid evolution, then Fig. 7B suggests periods of rapid evolution when Discomedusae and Rhizostomeae, the two major accumulating clades, diversified. This inference also gains circumstantial support from the observation that Rhizostomeae is the most-derived, i.e., youngest, order within scyphomedusae but also the most diverse.

However, there is one example that does provide incontrovertible evidence of the potential of at least some scyphozoans for rapid evolutionary radiation. In the marine lakes in Palau, the rhizostome jellyfish *Mastigias papua* (Lesson) is recognizably different and exhibits different behaviors from lake to lake (Hamner & Hauri, 1981; Hamner, 1982; Hamner et al., 1982; Hamner & Hamner, 1998), presumably because of rapid genetic differentiation within isolated populations (Hamner & Hauri, 1981). This has now been supported by DNA sequence comparison with geographically isolated ancestral populations that still inhabit the main Palau lagoon (Dawson & Hamner, 2005; Dawson and Hamner, this volume). The rates of evolution of *Mastigias* medusae in the marine lakes has been exceptionally rapid, with incipient species forming in as little as 0.006 My. It might be argued that marine lakes are unusually rare environments and do not provide a reliable gauge for other marine systems. This may be true, but marine lakes in Palau, Indonesia, Vietnam, and elsewhere globally, as well as embayments, estuaries, lagoons, and oceanic islands, provide somewhat isolated habitats for almost every taxon of coastal marine vertebrate and invertebrate. If rapid speciation is possible within the genus *Mastigias*, then rapid speciation must be possible for at least some other jellyfish, perhaps most, assuming they have sufficient genetic variation and appropriate opportunity.

### *Conclusions from Part II*

1. *Traits that promote blooming and/or swarming occur in clades that are more speciose*, in many cases, than in clades composed of species that do not bloom or swarm. Blooming and swarming are, by inference, adaptations favored by natural selection.
2. *Selection on morphology in the sea is strong.* Successful ancestral characteristics are commonly retained by daughter taxa within Scyphozoa. Convergence is rife among diverse epipelagic invertebrates as well. Strong stabilizing or strong directional selection toward common solutions and limited morphological variation among taxa frequently characterize jellyfishes.
3. *Morphological stasis often masks genetic, behavioral, and/or physiological diversification and speciation.* This cryptic diversity provides considerable potential for local adaptation of populations and for rapid speciation and diversification when local conditions change or when members of populations disperse to new environments.
4. *Speciation can be rapid.*

## Discussion: adaptive accumulations, changing circumstances

Our review is intended to elucidate the evolution of jellyfish accumulations by identifying the taxa, and thereby also implicating relevant characteristics, of jellyfishes that occur en masse. We also hope to clarify that accumulations of medusae could be advantageous for scyphozoans and not just annoying to humans. To this end, our analysis took three conceptually linked but logically independent steps: compilation of traditional taxonomic lists of species, independent validation of those lists by molecular phylogenetic analyses, and consideration of evolutionary diversification within taxa. In our companion paper (Dawson & Hamner, this volume), we build on this foundation by mapping the evolution of mass occurrences and associated organismal and environmental characteristics on a phylogeny of Medusozoa.

Given the number of scyphozoan taxa that aggregate, bloom, and/or swarm, given that they often occur in clades of related species, and given that those clades are, in general, more diverse than clades of medusae that do not obviously accumulate, we conclude that occurrence en masse is the result of evolutionary adaptation. That mass occurrence has evolved multiple times in independent evolutionary lineages under similar circumstances is evidence of strong natural selection (e.g., Foster, 1999) for traits that enable occurrence en masse, which may in large part be an evolutionary response to conditions that predominate in shallow-water coastal environments (see Dawson and Hamner, this volume).

Some of the clearest patterns to emerge from our analyses relate to the evolution of the shallow-water Rhizostomeae, on which we focus during the closing remarks, in part to draw attention again to this much understudied but ecologically and economically important clade (Omori & Nakano, 2001), and in part because they illustrate well the utility of our approach. Our phylogenetic tree for Rhizostomeae (Fig. 4A) is derived exclusively from molecular information but it is, in general, very similar to the systematic trees described during the last century by classical systematists using only morphology (Figs. 1, 2). It is intellectually reassuring that these two completely independent datasets are easily integrated (Fig. 4B), because, as clarified by Schluter (2000), a reasonably accurate phylogeny of the taxonomic group of concern is critical for any evaluation of the traits associated with, or the environmental factors responsible for, subsequent adaptive radiation. Thus, traditional taxonomic and recent molecular information support recognition of Kolpophorae and Daktyliophorae as valid clades (Figs. 3, 4), and therefore the attributes of blooming and swarming primarily characterize those species within the sub-order Daktyliophorae. It will be important in the future to learn more about the muscular and circulatory characteristics used by Stiasny (1921; see Kikinger, 1992), as well as other correlated traits such as scapulae, which distinguish the clades within Rhizostomeae to determine if these differences reflect quite different patterns of ecology, including swimming behavior and feeding ecology. The scapulate medusae (families Rhizostomatidae and Stomolophidae) are particularly interesting because they are the jellyfishes that are most frequently fished and also include important bloomers such as *Nemopilema*.

The analyses also illustrate conditions under which medusae in typically non-blooming, non-swarming clades may become peculiarly abundant (Fig. 4B). *Mastigias*, for example, occurs at abundances of $10^5$–$10^6$ times greater in isolated brackish tropical lakes than it does in nearby semi-enclosed seawater coves or the adjacent coastal ocean. *Phyllorhiza* occurs in abundances troublesome for humans in the Gulf of Mexico when it invaded from the western Pacific (Graham et al., 2003). These are patterns seen in diverse other taxa including the scyphozoans *Aurelia*, *Periphylla*, and *Rhizostoma* and invasive and/or lake-locked hydromedusae.

Thus, clearly, the evolutionary differences among medusozoans, their shaping of modern ecological interactions, and the potential for better understanding of mass occurrences achievable through phylogenetic approaches are of considerable current consequence. Exploring further the various environment–organism interactions that are correlated with distinctive patterns of diversification and adaptation is the subject of our companion paper (Dawson and Hamner, this volume). It is important to emphasize that we do not advocate any particular set of ecological or phyletic relationships that appear herein (or in Dawson and Hamner, this volume), because most of the necessary datasets are currently barely sufficient for this kind of analysis. Rather, we advocate that this approach readily generates a variety of hypotheses that need further testing.

**Acknowledgments** M. Arai, P. Hamner, L. Martin, K. Pitt, J. Purcell, and two anonymous reviewers critiqued, improved, and suggested how to organize the content and presentation of this manuscript. WMH thanks the organizers of the 2nd International Jellyfish Blooms Symposium, K. Pitt and J. Seymour, for the invitation to present a plenary address at the meeting, and we thank the Editors for their patience and industry in preparing this volume. This work was supported in part by grant DEB-0717078 from the US National Science Foundation.

## References

Albert, D. J., 2005. Reproduction and longevity of *Aurelia labiata* in Roscoe Bay, a small bay on the Pacific coast of Canada. Journal of the Marine Biological Association of the United Kingdom 85: 575–581.

Albert, D. J., 2007. *Aurelia labiata* medusae (Scyphozoa) in Roscoe Bay avoid tidal dispersion by vertical migration. Journal of Sea Research 57: 281–287.

Arai, M. N., 1992. Active and passive factors affecting aggregations of hydromedusae: a review. Scientia Marina 56: 99–108.

Arai, M. N., 1997. A Functional Biology of Scyphozoa. Chapman and Hall, London.

Arai, M. N., 2001. Pelagic coelenterates and eutrophication: a review. Hydrobiologia 451 (Developments in Hydrobiology 155): 69–87.

Båmstedt, U., S. Kaartvedt & M. Youngbluth, 2003. An evaluation of acoustic and video methods to estimate the abundance and vertical distribution of jellyfish. Journal of Plankton Research 25: 1307–1318.

Barraclough, T. G., S. Nee & P. H. Harvey, 1998. Sister-group analysis in identifying correlates of diversification. Evolutionary Ecology 12: 751–754.

Barz, K., H.-H. Hinrichson & H.-J. Hirche, 2006. Scyphozoa in the Bornholm Basin (central Baltic Sea) the role of advection. Journal of Marine Systems 60: 167–176.

Bickford, D., D. J. Lohman, N. S. Sodhi, P. K. L. Ng, R. Meier, K. Winker, K. K. Ingram & I. Das, 2007. Cryptic species as a window on diversity and conservation. Trends in Ecology and Evolution 22: 148–156.

Bigelow, H. B., 1926. Plankton of the offshore waters of the Gulf of Maine. Bulletin of the Bureau of Fisheries,

Washington 40, 1924, part II, document number 968, pp. 1–509. figs. 1–134.
Billett, D. S. M., B. J. Bett, C. L. Jacobs, I. P. Rouse & B. D. Wigham, 2006. Mass deposition of jellyfish in the deep Arabian Sea. Limnology and Oceanography 51: 2077–2083.
Bolton, T. F. & W. M. Graham, 2004. Morphological variation among populations of an invasive jellyfish. Marine Ecology Progress Series 278: 125–139.
Bone, Q., 2005. Gelatinous animals and physiology. Journal of the Marine Biological Association of the United Kingdom 85: 641–653.
Bouillon, J. & F. Boero, 2000. Phylogeny and classification of Hydroidomedusae. The Hydrozoa: a new classification in the light of old knowledge. Thalassia Salentina 24: 1–46.
Brodeur, R. D., H. Sugisaki & G. L. Hunt Jr., 2002. Increases in jellyfish biomass in the Bering Sea: implications for the ecosystem. Marine Ecology Progress Series 233: 89–103.
Buecher, E., J. Goy, B. Planque, M. Etienne & S. Dallot, 1997. Long-term fluctuations of *Liriope tetraphylla* in Villefranche bay between 1966 and 1993 compared to *Pelagia noctiluca* populations. Oceanologica Acta 20: 145–157.
Calder, D. R., 1973. Laboratory observations on the life history of *Rhopilema verrilli* (Scyphozoa : Rhizostomeae). Marine Biology 21: 109–114.
Carroll, S. P., A. P. Hendry, D. N. Reznick & C. W. Fox, 2007. Evolution on ecological time-scales. Functional Ecology 21: 387–393.
Cartwright, P., S. L. Halgedahl, J. R. Hendricks, R. D. Jarrard, A. C. Marques, A. G. Collins & B. S. Lieberman, 2007. Exceptionally preserved jellyfishes from the Middle Cambrian. PLoS ONE 2(10): e1121.
Colin, S. P., J. H. Costello & H. Kordula, 2006. Upstream foraging by medusae. Marine Ecology Progress Series 327: 143–155.
Collins, A. G., 2002. Phylogeny of Medusozoa and the evolution of cnidarian life cycles. Journal of Evolutionary Biology 15: 418–432.
Collins, A. G., P. Schuchert, A. C. Marques, T. Jankowski, M. Medina & B. Schierwater, 2006. Medusozoan phylogeny and character evolution clarified by new large and small subunit rDNA data and an assessment of the utility of phylogenetic mixture models. Systematic Biology 55: 97–115.
Coyne, J. A. & H. A. Orr, 2004. *Speciation*. Sinauer Associates, Sunderland, MA.
Daly, M., M. R. Brugler, P. Cartwright, A. G. Collins, M. N. Dawson, D. G. Fautin, S. C. France, C. S. McFadden, D. M. Opresko, E. Rodriguez, S. Romano & J. Stake, 2007. The phylum Cnidaria: a review of phylogenetic patterns and diversity three hundred years after Linnaeus. Zootaxa 1668: 127–182.
DalyYahia, M. N., J. Goy & O. DalyYahia-Kéfi, 2003. Distribution et écologie des Méduses (Cnidaria) du golfe de Tunis (Méditerranée sud occidentale). Oceanologica Acta 26: 645–655.
Daryanabard, R. & M. N. Dawson, 2008. Jellyfish blooms: *Crambionella orsini* (Scyphozoa, Rhizostomeae) in the Gulf of Oman, Iran, 2002–2003. Journal of the Marine Biological Association of the United Kingdom 88: 477–483.
Dawson, M. N., 2003. Macro-morphological variation among cryptic species of the moon jellyfish, *Aurelia* (Cnidaria: Scyphozoa). Marine Biology 143: 369–379.
Dawson, M. N., 2004. Some implications of molecular phylogenetics for understanding biodiversity in jellyfishes, with emphasis on Scyphozoa. Hydrobiologia 530(531): 249–260.
Dawson, M. N., 2005a. Incipient speciation of *Catostylus mosaicus* (Scyphozoa, Rhizostomeae, Catostylidae), comparative phylogeography and biogeography in south-eastern Australia. Journal of Biogeography 32: 515–533.
Dawson, M. N., 2005b. Morphologic and molecular redescription of *Catostylus mosaicus conservativus* (Scyphozoa, Rhizostomeae, Catostylidae) from southeast Australia. Journal of the Marine Biological Association of the United Kingdom 85: 723–732.
Dawson, M. N., 2005c. *Cyanea capillata* is not a cosmopolitan jellyfish: morphological and molecular evidence for *C. annaskala* and *C. rosea* (Scyphozoa, Semaeostomeae, Cyaneidae) in southeast Australia. Invertebrate Systematics 19: 361–370.
Dawson, M. N., 2005d. Five new subspecies of *Mastigias* (Scyphozoa, Rhizostomeae: Mastigiidae) from marine lakes, Palau, Micronesia. Journal of the Marine Biological Association of the United Kingdom 85: 679–694.
Dawson, M. N., 2005e. Morphological variation and systematics in the Scyphozoa: Mastigias (Rhizostomeae, Mastigiidae)—a golden unstandard? Hydrobiologia 537: 185–206.
Dawson, M. N. & W. M. Hamner, 2003. Geographic variation and behavioral evolution in marine plankton: the case of *Mastigias* (Scyphozoa: Rhizostomeae). Marine Biology 143: 1161–1174.
Dawson, M. N. & W. M. Hamner, 2005. Rapid evolutionary radiation of marine zooplankton in peripheral environments. Proceedings of the National Academy of Sciences of the USA 102: 9235–9240.
Dawson, M. N & W. M. Hamner, this volume. A character-based analysis of the evolution of jellyfish blooms: adaptation and exaptation. Hydrobiologia (Developments in Hydrobiology). doi:10.1007/s10750-008-9591-x.
Dawson, M. N. & D. K. Jacobs, 2001. Molecular evidence for cryptic species of *Aurelia aurita* (Cnidaria, Scyphozoa). Biological Bulletin 200: 92–96.
Dawson, M. N. & L. E. Martin, 2001. Geographic variation and ecological adaptation in *Aurelia* (Scyphozoa: Semaeostomeae): some implications from molecular phylogenetics. Hydrobiologia/Dev. Hydrobiologia 451: 259–273.
Dawson, M. N., K. A. Raskoff & D. K. Jacobs, 1998. Preservation of marine invertebrate tissues for DNA analyses. Molecular Marine Biology and Biotechnology 7: 145–152.
Dawson, M. N., A. S. Gupta & M. H. England, 2005. Coupled biophysical global ocean model and molecular genetic analyses identify multiple introductions of cryptogenic species. Proceedings of the National Academy of Sciences of the USA 102: 11968–11973.
Decker, M. B., C. W. Brown, R. R. Hood, J. E. Purcell, T. F. Gross, J. C. Matanoski, R. O. Bannon & E. M. Setzler-Hamilton, 2007. Predicting the distribution of the scyphomedusa *Chrysaora quinquecirrha* in Chesapeake Bay. Marine Ecology Progress Series 329: 99–113.

Ellner, S. & N. G. Hairston, 1994. Role of overlapping generations in maintaining genetic variation in a fluctuating environment. American Naturalist 143: 403–417.

Folmer, O., M. Black, W. Hoeh, R. Lutz & R. Vrijenhoek, 1994. DNA primers for amplification of mitochondrial cytochrome *c* oxidase subunit I from diverse metazoan invertebrates. Molecular Marine Biology and Biotechnology 3: 294–299.

Foster, S., 1999. The geography of behaviour: an evolutionary perspective. Trends in Ecology and Evolution 14: 190–195.

Gershwin, L., 2001. Systematics and biogeography of the jellyfish *Aurelia labiata* (Cnidaria: Scyphozoa). Biological Bulletin 201: 104–119.

Gershwin, L. & A. G. Collins, 2002. A preliminary phylogeny of Pelagiidae (Cnidaria, Scyphozoa), with new observations of *Chrysaora colorata* comb. nov. Journal of Natural History 36: 127–148.

Gili, J.-M., J. Bouillon, F. Pages, A. Palanques & P. Puig, 1999. Submarine canyons as habitats of prolific plankton populations: three new deep-sea Hydroidomedusae in the western Mediterranean. Zoological Journals of the Linnean Society 125: 313–329.

Goy, J., P. Morand & M. Etienne, 1989. Long-term fluctuations of *Pelagia noctiluca* (Cnidaria, Scyphomedusa) in the western Mediterranean Sea. Prediction by climatic variables. Deep-Sea Research 36: 269–279.

Graham, W. M., 2001. Numerical increases and distributional shifts of *Chrysaora quinquecirrha* (Desor) and *Aurelia aurita* (Linné) (Cnidaria: Scyphozoa) in the northern Gulf of Mexico. Hydrobiologia 451 (Developments in Hydrobiology 155): 97–111.

Graham, W. M. & J. L. Largier, 1997. Upwelling shadows as nearshore retention sites: the example of northern Monterey Bay. Continental Shelf Research 17: 509–532.

Graham, W. M., Pagès, F. & W. M. Hamner, 2001. A physical context for gelatinous zooplankton aggregations: a review. Hydrobiologia 451 (Developments in Hydrobiology 155): 199–212.

Graham, W. M., D. L. Martin, D. L. Felder, V. L. Asper & H. M. Perry, 2003. Ecological and economic implications of a tropical jellyfish invader in the Gulf of Mexico. Biological Invasions 5: 53–69.

Greenberg, N., R. L. Garthwaite & D. C. Potts, 1996. Allozyme and morphological evidence for a newly introduced species of *Aurelia* in San Francisco Bay California. Marine Biology 125: 401–410.

Hamner, W. M., 1982. Strange world of Palau's salt lakes. National Geographic Magazine 161: 264–282.

Hamner, W. M., 1995. Predation, cover, and convergent evolution in epipelagic oceans. Marine and Freshwater Behaviour and Physiology 26: 71–89.

Hamner, W. M. & P. P. Hamner, 1998. Stratified marine lakes of Palau (Western Caroline Islands). Physical Geography 19: 175–220.

Hamner, W. M. & I. R. Hauri, 1981. Long-distance horizontal migrations of zooplankton (Scyphomedusae: *Mastigias*). Limnology and Oceanography 26: 414–423.

Hamner, W. M. & R. M. Jenssen, 1974. Growth, degrowth, and irreversible cell differentiation in *Aurelia aurita*. American Zoologist 14: 833–849.

Hamner, W. M., L. P. Madin, A. L. Alldredge, R. W. Gilmer & P. P. Hamner, 1975. Underwater observations of gelatinous zooplankton: Sampling problems, feeding biology, and behavior. Limnology and Oceanography 20: 907–917.

Hamner, W. M., R. W. Gilmer & P. P. Hamner, 1982. The physical, chemical, and biological characteristics of a stratified, saline, sulfide lake in Palau. Limnology and Oceanography 27: 896–909.

Hamner, W. M., P. P. Hamner & S. W. Strand, 1994. Sun compass migration by *Aurelia aurita* (Scyphozoa): population persistence versus dispersal in Saanich Inlet, British Columbia. Marine Biology 119: 347–356.

Hansson, L. J., 1997. Capture and digestion of the scyphozoan jellyfish *Aurelia aurita* by *Cyanea capillata* and prey response to predator contact. Journal of Plankton Research 19: 195–208.

Hansson, L. J., 2006. A method for in situ estimation of prey selectivity and predation rate in large plankton, exemplified with the jellyfish *Aurelia aurita* (L.). Journal of Experimental Marine Biology and Ecology 328: 113–126.

Hansson, L. J. & Kiørboe, 2006. Effects of large gut volume in gelatinous zooplankton: ingestion rate, bolus production and food patch utilization by the jellyfish *Sarsia tubulosa*. Journal of Plankton Research 28: 937–942.

Hay, S., 2006. Marine ecology: gelatinous bells may ring change in marine ecosystems. Current Biology 16: 679–682.

Hendry, A. P., P. Nosil & L. H. Rieseberg, 2007. The speed of ecological speciation. Functional Ecology 21: 455–464.

Herring, P., 2002. The Biology of the Deep Ocean. Oxford University Press, New York.

Hewitt, C. L., M. L. Campbell, R. E. Thresher, R. B. Martin, S. Boyd, B. F. Cohen, D. R. Currie, M. F. Gomon, M. J. Keough, A. J. Lewis, M. M. Lockett, N. Mays, M. A. McArthur, T. D. O'Hara, G. C. B. Poore, D. J. Ross, M. J. Storey, J. E. Watson & R. S. Wilson, 2004. Introduced and cryptogenic species in Port Phillip Bay, Victoria, Australia. Marine Biology 144: 183–202.

Holland, B. S., M. N. Dawson, G. L. Crow & D. K. Hofmann, 2004. Global phylogeography of *Cassiopea* (Scyphozoa: Rhizostomae): molecular evidence for cryptic species and multiple Hawaiian invasions. Marine Biology 145: 1119–1128.

Houghton, J. D. R., T. K. Doyle, M. W. Wilson, J. Davenport & G. C. Hays, 2006. Jellyfish aggregations and leatherback turtle foraging patterns in a temperate coastal environment. Ecology 87: 1967–1972.

Hsieh, Y.-H. P., F.-M. Leong & J. Rudloe, 2001. Jellyfish as food. Hydrobiologia 451 (Developments in Hydrobiology 155): 11–17.

Ishii, H. & U. Båmstedt, 1998. Food regulation of growth and maturation in a natural population of *Aurelia aurita* (L.). Journal Plankton Research 20: 805–816.

Jablonka, E. & M. J. Lamb, 2005. Evolution in four dimensions: genetic, epigenetic, behavioral, and symbolic variation in the history of life. MIT Press, Cambridge.

Jarms, G., U. Båmstedt, H. Tiemann, M. B. Martinussen & J. H. Fosså, 1999. The holopelagic life cycle of the deep-sea medusa *Periphylla periphylla* (Scyphozoa, Coronatae). Sarsia 84: 55–65.

Jarms, G., H. Tiemann & U. Båmstedt, 2002. Development and biology of *Periphylla periphylla* (Scyphozoa: Coronatae) in a Norwegian fjord. Marine Biology 141: 647–657.

Jeanmougin, F., J. D. Thompson, M. Gouy, D. G. Higgins & T. J. Gibson, 1998. Multiple sequence alignment with ClustalX. Trends in Biochemical Science 23: 403–405.

Johnson, D. R., H. M. Perry & W. M. Graham, 2005. Using nowcast model currents to explore transport of non-indigenous jellyfish into the Gulf of Mexico. Marine Ecology Progress Series 305: 139–146.

Kawahara, M., S. Uye, K. Ohtsu & H. Iizumi, 2006. Unusual population explosion of the giant jellyfish *Nemopilema nomurai* (Scyphozoa: Rhizostomeae) in East Asian waters. Marine Ecology Progress Series 307: 161–173.

Kikinger, R., 1992. *Cotylorhiza tuberculata* (Cnidaria: Scyphozoa)—Life history of a stationary population. Marine Ecology 13: 333–362.

Kiortsis, V., 1965. Planctonological survey of the North-Aegean Sea. Technical Report for U.S. Navy Contract N62558.

Kingsford, M. J., K. A. Pitt & B. M. Gillanders, 2000. Management of jellyfish fisheries, with special reference to the order Rhizostomeae. Oceanography and Marine Biology Annual Review 38: 85–156.

Kramp, P. L., 1961. Synopsis of the medusae of the world. Journal of the Marine Biological Association of the United Kingdom 40: 1–469.

Kramp, P. L., 1965. Some medusae (mainly scyphomedusae) from Australian coastal waters. Transcripts of the Royal Society of South Australia 89: 257–278.

Kramp, P. L., 1968. The scyphomedusae collected by the Galathea expedition 1950–52. Videnskabelige Meddelelser fra Dansk Naturhistorisk Forening I Kjoebenhavn 131: 67–98.

Kramp, P. L., 1970. Zoogeographical studies on Rhizostomeae (Scyphozoa). Videnskabelige Meddelelser fra Dansk Naturhistorisk Forening I Kjoebenhavn 133: 7–30.

Kremer, P., J. Costello, J. Kremer & M. Canino, 1990. Significance of photosynthetic endosymbionts to the carbon budget of the scyphomedusa *Linuche unguiculata*. Limnology and Oceanography 35: 609–624.

Larson, R. J., 1976. Cubomedusae: feeding—functional morphology, behavior and phylogenetic position. In Mackie, G. O. (ed.), Coelenterate Ecology and Behavior. Plenum, New York: 237–245.

Larson, R. J., 1986. Pelagic scyphomedusae (Scyphozoa: Coronatae and Semaeostomeae) of the Southern Ocean. Biology of the Antarctic Seas. Antarctic Research Series 41: 59–165.

Larson, R. J., 1987. First report of the little-known scyphomedusa *Drymonema dalmatinum* in the Caribbean Sea, with notes on its biology. Bulletin of Marine Science 40: 437–441.

Larson, R. J., 1992. Riding Langmuir circulations and swimming in circles: a novel form of clustering behavior by the scyphomedusa *Linuche unguiculata*. Marine Biology 112: 229–235.

Linneaus, C., 1758. Systema Naturae, 10th ed. Holmiae L. Salvii, Stockholm.

Lipton, J., 1991. An Exaltation of Larks: The Ultimate Edition. Viking Press, New York.

Lotan, A., R. Ben-Hillel & Y. Loya, 1992. Life cycle of *Rhopilema nomadica*: a new immigrant Scyphomedusan in the Mediterranean. Marine Biology 112: 237–242.

Lotan, A., M. Fine & R. Ben-Hillel, 1994. Synchronization of the life cycle and dispersal pattern of the tropical invader scyphomedusan *Rhopilema nomadica* is temperature dependent. Marine Ecology Progress Series 109: 59–65.

Lucas, C. H., 1996. Population dynamics of *Aurelia aurita* (Scyphozoa) from an isolated brackish lake, with particular reference to sexual reproduction. Journal of Plankton Research 18: 987–1007.

Lucas, C. H., 2001. Reproduction and life history strategies of the common jellyfish, *Aurelia aurita,* in relation to its ambient environment. Hydrobiologia 451(Developments in Hydrobiology 155): 229–246.

Lucas, C. H. & J. A. Williams, 1994. Population dynamics of the scyphomedusa *Aurelia aurita* in Southampton water. Journal of Plankton Research 16: 879–895.

Lukoschek, V. & J. S. Keogh, 2006. Molecular phylogeny of sea snakes reveals a rapidly diverged adaptive radiation. Biological Journal of the Linnaean Society 89: 523–539.

Lynam, C. P., S. J. Hay & A. S. Brierley, 2004. Interannual variability in abundance of North Sea jellyfish and links to the North Atlantic oscillation. Limnology and Oceanography 49: 637–643.

Lynam, C. P., S. J. Hay & A. S. Brierley, 2005. Jellyfish abundance and climatic variation: contrasting responses in oceanographically distinct regions of the North Sea, and possible implications for fisheries. Journal of the Marine Biological Association of the United Kingdom 85: 435–450.

Lynam, C. P., M. J. Gibbons, B. E. Axelsen, C. A. J. Sparks, J. Coetzee, B. G. Heywood & A. S. Brierley, 2006. Jellyfish overtake fish in a heavily fished ecosystem. Current Biology 16: 492–493.

Maddison, W. P. & D. R. Maddison, 1989. Interactive analysis of phylogeny and character evolution using the computer program MacClade. Folia Primatologica (Basel) 53: 190–202.

Malej, A., 1989. Behaviour and trophic ecology of the jellyfish *Pelagia noctiluca* (Forsskål, 1775). Journal of Experimental Marine Biology and Ecology 126: 259–270.

Marques, A. C. & A. G. Collins, 2004. Cladistic analysis of Medusozoa and cnidarian evolution. Invertebrate Biology 123: 23–42.

Martin, J. W., L. Gershwin, J. W. Burnett, D. G. Cargo & D. A. Bloom, 1997. *Chrysaora achlyos*, a remarkable new species of scyphozoan from the Eastern Pacific. Biological Bulletin 193: 8–13.

Martin, L. E., M. N. Dawson, L. J. Bell & P. L. Colin, 2005. Marine lake ecosystem dynamics illustrate ENSO variation in the tropical western Pacific. Biology Letters 2: 144–147.

Masilamoni, J. G., K. S. Jesudoss, K. Nandakumar, K. K. Satpathy, K. V. K. Nair & J. Azariah, 2000. Jellyfish ingress: a threat to the smooth operation of coastal power plants. Current Science 79: 567–569.

Matsumoto, G. I., K. A. Raskoff & D. J. Lindsay, 2003. *Tiburonia granrojo* n. sp., a mesopelagic scyphomedusa from the Pacific Ocean representing the type of a new subfamily (class Scyphozoa: order Semaeostomeae:

family Ulmaridae: subfamily Tiburoniinae subfam. nov.). Marine Biology 143: 73–77.

Mayer, A. G., 1910. Medusae of the World, III—the Scyphomedusae. Carnegie Institute, Washington.

Medlin, L., H. J. Elwood, S. Stickel & M. L. Sogin, 1988. The characterization of enzymatically amplified eukaryotic 16S-like ribosomal RNA-coding regions. Gene 71: 491–499.

Mianzan, H. W. & P. F. S. Cornelius, 1999. Cubomedusae and Scyphomedusae. In Boltovskoy, D. (ed.), South Atlantic Zooplankton. 1. Backhuys Press, Leiden: 513–559.

Mianzan, H., D. Sorarrain, J. W. Burnett & L. L. Lutz, 2000. Mucocutaneous junctional and flexural paresthesias caused by the holoplanktonic trachymedusa *Liriope tetraphylla*. Dermatology 201: 46–48.

Mills, C. E., 2001. Jellyfish blooms: are populations increasing globally in response to changing ocean conditions? Hydrobiologia 451 (Developments in Hydrobiology 155): 55–68.

Mills, C. E. & J. Goy, 1988. In situ observations of the behavior of mesopelagic *Solmissus* narcomedusae (Cnidaria, Hydrozoa). Bulletin of Marine Science 43: 739–751.

Mitchell, C. E. & A. G. Power, 2003. Release of invasive plants from fungal and viral pathogens. Nature 421: 625–627.

Miyake, H., K. Iwao & Y. Kakinuma, 1997. Life history and environment of *Aurelia aurita*. South Pacific Study 17: 273–285.

Möller, H., 1980. A summer survey of large zooplankton, particularly scyphomedusae, in North Sea and Baltic. Meeresforschung 28: 61–68.

Møller, L. F. & H. U. Riisgård, 2007. Population dynamics, growth and predation impact of the common jellyfish *Aurelia aurita* and two hydromedusae, *Sarsia tubulosa*, and *Aequorea vitrina* in Limfjorden (Denmark). Marine Ecology Progress Series 346: 153–165.

Moore, J. & P. Willmer, 1997. Convergent evolution in invertebrates. Biological Review 72: 1–60.

Morand, P., C. Carré & D. C. Biggs, 1987. Feeding and metabolism of the jellyfish *Pelagia noctiluca* (scyphomedusae, semaeostomae). Journal of Plankton Research 9: 651–665.

Morandini, A. C., F. da Silveira & G. Jarms, 2004. The life cycle of *Chrysaora lactea* Eschscholtz, 1829 (Cnidaria, Scyphozoa) with notes on the scyphistoma stage of three other species. Hydrobiologia 530(531): 347–354.

Morris, A. K., 2006. Zooplankton aggregations in California coastal zones, Ph.D. thesis, UCLA, 309 pp.

Norris, R. D., 2000. Pelagic species diversity, biogeography, and evolution. Paleobiology 26: 236–258.

Omori, M. & E. Nakano, 2001. Jellyfish fisheries in southeast Asia. Hydrobiologia 451 (Developments in Hydrobiology 155): 19–26.

Parrish, J. K. & W. M. Hamner (eds), 1997. Animal Groups in Three Dimensions. Cambridge University Press, Cambridge.

Pennak, R. W., 1956. The fresh-water jellyfish *Craspedacusta* in Colorado with some remarks on its ecology and morphological degeneration. Transactions of the American Microscopical Society 75: 324–331.

Pertsova, N. M., K. N. Kosobokova & A. A. Prudkovsky, 2006. Population size structure, spatial distribution, life cycle of the hydromedusa Aglantha digitale (O. F. Muller, 1766) in the White Sea. Oceanology 46: 228–237.

Pitt, K. A., 2000. Life history and settlement preferences of the edible jellyfish *Catostylus mosaicus* (Scyphozoa: Rhizostomeae). Marine Biology 136: 269–279.

Pitt, K. A. & M. J. Kingsford, 2000. Geographic separation of stocks of the edible jellyfish *Catostylus mosaicus* (Rhizostomeae) in New South Wales, Australia. Marine Ecology Progress Series 196: 143–155.

Pitt, K. A. & M. J. Kingsford, 2003. Temporal variation in the virgin biomass of the edible jellyfish, *Catostylus mosaicus* (Scyphozoa, Rhizostomeae). Fisheries Research 63: 303–313.

Posada, D. & K. A. Crandall, 1998. Modeltest: testing the model of DNA substitution. Bioinformatics 14: 817–818.

Purcell, J. E., 1989. Predation on fish larvae and eggs by the hydromedusa *Aequorea victoria* at a herring spawning ground in British Columbia. Canadian Journal Fisheries and Aquatic Sciences 46: 1415–1427.

Purcell, J. E., 2003. Predation on zooplankton by large jellyfish, *Aurelia labiata*, *Cyanea capillata* and *Aequorea aequorea*, in Prince William Sound, Alaska. Marine Ecology Progress Series 246: 137–152.

Purcell, J. E., 2005. Climate effects on formation of jellyfish and ctenophore blooms. Journal of the Marine Biological Association of the United Kingdom 85: 461–476.

Purcell, J. E., U. Båmstedt & A. Båmstedt, 1999. Prey, feeding rates, and asexual reproduction rates of the introduced oligohaline hydrozoan *Moerisia lyonsi*. Marine Biology 134: 317–325.

Purcell, J. E., D. L. Breitburg, M. B. Decker, W. M. Graham, M. J. Youngbluth & K. A. Raskoff, 2001. Pelagic cnidarians and ctenophores in low dissolved oxygen environments: a review. In Rabalais, N. N. & R. E. Turner (eds), Coastal Hypoxia: Consequences for Living Resources and Ecosystems. Coastal Estuarine Studies 58: 77–100.

Purcell, J. E., S. Uye & W. Lo, 2007. Anthropogenic causes of jellyfish blooms and their direct consequences for humans: a review. Marine Ecology Progress Series 350: 153–174.

Rakow, K. C. & W. M. Graham, 2006. Orientation and swimming mechanics by the scyphomedusa *Aurelia* sp. in shear flow. Limnology and Oceanography 51: 1097–1106.

Rottini Sandrini, L. & M. Avian, 1983. Biological cycle of *Pelagia noctiluca:* morphological aspects of the development from planula to ephyra. Marine Biology 74: 169–174.

Rottini Sandrini, L. & M. Avian, 1989. Feeding mechanism of *Pelagia noctiluca* (Scyphozoa: Semaeostomeae); laboratory and open sea observations. Marine Biology 102: 49–55.

Rottini Sandrini, L. & M. Avian, 1991. Reproduction of *Pelagia noctiluca* in the central and northern Adriatic Sea. Hydrobiologia 216(217): 197–202.

Russell, F. S., 1970. The Medusae of the British Isles. II Pelagic Scyphozoa with a Supplement to the First Volume on Hydromedusae. Cambridge University Press, Cambridge.

Santhakumari, V. & V. R. Nair, 1999. Distribution of hydromedusae from the exclusive economic zone of the west and east coasts of India. Indian Journal of Marine Sciences 28: 150–157.

Sasaki, A. & S. Ellner, 1997. Quantitative genetic variance maintained by fluctuating selection with overlapping generations: variance components and covariances. Evolution 51: 682–696.

Sasaki, A. & G. De Jong, 1999. Density dependence and unpredictable selection in a heterogeneous environment: compromise and polymorphism in the ESS reaction norm. Evolution 53: 1329–1342.

Schluter, D., 1996. Adaptive radiation along genetic lines of least resistance. Evolution 50: 1766–1774.

Schluter, D., 2000. The Ecology of Adaptive Radiation. Oxford University Press, New York.

Schneider, G., 1988. Larvae production of the common jellyfish *Aurelia aurita* in the Western Baltic 1982–1984. Kieler Meeresforschungen 6: 295–300.

Schneider, G., 1989. The common jelly-fish *Aurelia aurita:* standing stock, excretion, and nutrient regeneration in the Kiel Bight, Western Baltic. Marine Biology (Berlin) 100: 507–514.

Schroth, W., G. Jarms, B. Streit & B. Schierwater, 2002. Speciation and phylogeography in the cosmopolitan marine moon jelly, *Aurelia* sp. BioMed Central Evolutionary Biology 2: 1–10.

Shanks, A. L. & W. M. Graham, 1987. Orientated swimming in the jellyfish *Stomolophus meleagris* L. Agassiz (Scyphozoan: Rhizostomida). Journal of Experimental Marine Biology and Ecology 108: 159–169.

Sørnes, T. A., D. L. Aksnes, U. Båmstedt & M. J. Youngbluth, 2007. Causes for mass occurrences of the jellyfish *Periphylla periphylla*: an hypothesis that involves optically conditioned retention. Journal of Plankton Research Advance Access: 1–28. doi:10.1093/plankt/fbm003

Sparks, C. A. J., E. Buecher, A. S. Brierley, B. E. Axelsen, H. Boyer & M. J. Gibbons, 2001. Observations on the distribution, and relative abundance of *Chrysaora hysoscella* (Cnidaria, Scyphozoa) and *Aequorea aequorea* (Cnidaria, Hydrozoa) in the northern Benguela ecosystem. Hydrobiologia 451 (Developments in Hydrobiology 155): 275–286.

Stephens, L. D. & D. R. Calder, 2006. Seafaring scientist: Alfred Goldsborough Mayor, pioneer in Marine Biology. University of South Carolina Press, Columbia.

Stiasny, G., 1921. Studien über rhizostomeen. In van Oort, E. D. (ed.), Capita Zoologica. Martinus Njhoff, Gravenhage.

Strand, S. W. & W. M. Hamner, 1988. Predatory behavior of *Phacellophora camtschatica* and size-selective predation upon *Aurelia aurita* (Scyphozoa: Cnidaria) in Saanich Inlet, British Columbia. Marine Biology 99: 409–414.

Swofford, D. L., 2002. *PAUP*: Phylogenetic Analysis Using Parsimony (*and Other Methods). Version 4.* Sinauer Associates, Sunderland.

Takahashi, D. & T. Ikeda, 2006. Abundance, vertical distribution and life cycle patterns of the hydromedusa *Aglantha digitale* in the Oyashio region, western Subarctic Pacific. Plankton and Benthos Research 1: 91–96.

Titelman, J., L. Riemann, T. A. Sørnes, T. Nilsen, P. Griekspoor & U. Båmstedt, 2006. Turnover of dead jellyfish: stimulation and retardation of microbial activity. Marine Ecology Progress Series 325: 43–58.

Tomascik, T., A. J. Mah, A. Nontji & M. K. Moosa, 1997. The Ecology of the Indonesian Seas. I. Periplus Editions, Hong Kong.

Ueno, S. & A. Mitsutani, 1994. Small-scale swarm of a hydrozoan medusa *Liriope tetraphylla* in Hiroshima Bay, the Inland Sea of Japan. Bulletin of the Plankton Society of Japan 41: 165–166.

Vanhöffen, E. 1888. Untersuchungen uber semaostome und rhizostome medusen. Bibliotheca Zoologica.

von Lendenfeld, R., 1884. Local colour-varieties of scyphomedusae. Proceedings of the Linnean Society of New South Wales 9: 925–928.

von Lendenfeld, R., 1887. Descriptive catalogue of the medusae of the Australian seas. The Australian Museum, Sydney.

Williams, E. H. Jr., L. Bunkley-Williams, C.-G. Lilyestrom, R. J. Larson, N. A. Engstrom, E. A. R. Ortiz-Corps & J. H. Timber, 2001. A population explosion of the rare tropical/subtropical purple sea mane, *Drymonema dalmatinum*, around Puerto Rico in the summer and fall of 1999. Caribbean Journal of Science 37: 127–130.

Xian, W., B. Kang & R. Liu, 2005. Jellyfish blooms in the Yangtze Estuary. Science 307: 41.

Yasuda, T., 1968. Ecological studies on the jelly-fish, *Aurelia aurita,* in Urazoko Bay, Fukuii Prefecture—II. Occurrence pattern of ephyrae. Bulletin of the Japanese Society of Scientific Fisheries 34: 983–987.

Yasuda, T., 1969. Ecological studies on the jelly-fish, *Aurelia aurita,* in Urazoko Bay, Fukui Prefecture—III Occurrence pattern of the medusa. Bull. Japanese Society of Scientific Fisheries 35: 1–6.

Yasuda, T., 1971. Ecological studies on the jelly-fish, *Aurelia aurita,* in Urazoko Bay, Fukuii Prefecture—IV. Monthly change in bell-length composition and breeding season. Bulletin of the Japanese Society of Scientific Fisheries 37: 364–370.

Youngbluth, M. J. & U. Båmstedt, 2001. Distribution, abundance, behavior and metabolism of *Periphylla periphylla*, a mesopelagic coronate medusa in a Norwegian fjord. Hydrobiologia 451 (Developments in Hydrobiology 155): 321–333.

Zaitsev, Y. P. & V. Mamaev, 1997. Marine biological diversity in the Black Sea: a study of change and decline. United Nations Publications, New York.

Zavodnik, D., 1987. Spatial aggregations of the swarming jellyfish *Pelagia noctiluca* (Scyphozoa). Marine Biology 94: 265–269.

Hydrobiologia (2009) 616:193–215
DOI 10.1007/s10750-008-9591-x

JELLYFISH BLOOMS

# A character-based analysis of the evolution of jellyfish blooms: adaptation and exaptation

**Michael N Dawson · William M. Hamner**

Published online: 11 November 2008


**Abstract** Mass occurrence—aggregation, blooming, or swarming—is a remarkable feature of a subset of usually diverse scyphozoan clades, suggesting it is evolutionarily beneficial. If so, it should be associated with one or more phenotypic characteristics that are advantageous and which facilitate occurrence *en masse*. Here, we examine the evolution of morphological, ecological, and life history characteristics of medusozoans, focusing on the taxa that occur *en masse*. By tracing the evolution of aggregating, blooming, and swarming phenotypes, organismal traits, and environmental settings on an up-to-date synoptic phylogeny of classes and orders of Medusozoa, we are able to hypothesize circumstances that enable taxa to occur *en masse*. These include character states and character complexes related to podocyst formation, strobilation, oral arms, large size, and shallow-water habitat. These evolutionarily advantageous traits may be adaptations that evolved in response to selection for individual traits such as survival during periods of few resources, feeding on pulsed resources, and fecundity. These adaptations were apparently subsequently coopted by selection for reproductive success which favored mass occurrence. By considering the distribution of traits describing other phylogenetic lineages—when appropriately detailed ecological and systematic descriptions become available—it may be possible to predict which species are evolutionarily predisposed to form problematic blooms if environmental conditions permit.

**Keywords** Ecology · Evolution · Environment · Morphology · Phylogeny · Phenotype · Scyphozoa

Guest editors: K. A. Pitt & J. E. Purcell
Jellyfish Blooms: Causes, Consequences, and Recent Advances

**Electronic supplementary material** The online version of this article (doi:10.1007/s10750-008-9591-x) contains supplementary material, which is available to authorized users.

M. N Dawson (✉)
School of Natural Sciences, University of California, Merced, CA 95344, USA
e-mail: mdawson@ucmerced.edu

W. M. Hamner
Department of Ecology and Evolutionary Biology, University of California, Los Angeles, CA 90095-1606, USA

## Introduction

Large accumulations of jellyfish—aggregations, blooms, and swarms—are natural phenomena that apparently occur unpredictably spatially and temporally. Yet, taxonomically and phylogenetically, there are systematic patterns to their occurrence suggesting a certain suite of characteristics may predispose some medusozoans to occur *en masse* (Hamner & Dawson, 2008). If true, this would suggest that, assuming relatively consistent phenotype-environment interactions, mass occurrences of jellyfish should be predictable.

The physical mechanisms that generate temporary aggregations of medusae have been reviewed (Arai, 1992, 1997; Graham et al., 2001; Purcell, 2005; see also Decker et al., 2007) and environmental correlations with blooming proposed for eutrophication (Arai, 2001), overfishing (Mills, 2001; Brodeur et al., 2002), and species introductions (Graham et al., 2003; Holland et al., 2004; see also Purcell et al., 2007). Specific organismal traits that relate to mass occurrence, though, have received less attention. Our goal here is to begin to assess whether certain organismal traits may be associated with accumulation.

To explore this proposition, we first examine morphological, ecological, and life-history characters in an explicitly phylogenetic framework. By plotting attributes of extant taxa on a phylogeny it is possible to trace back in time, into ancestral chronospecies (see glossary of terms in Electronic Supplementary Material, Appendix A), the likely (i.e., most parsimonious) origin and routes of inheritance of phenotypic characters. Phenotypic character states that apparently evolved at similar times may be causally related. To further examine the postulate that scyphozoan accumulations are evolutionary adaptations to planktonic life (Hamner & Dawson, 2008), we take two additional steps in this study. The second step is to identify possible adaptive benefits with respect to mass occurrence. The third step is to assess the relative chronology of their evolution and thus distinguish adaptations from exaptations. This permits us, finally, to consider whether the observation that "species that have always been present suddenly experience severe increases of 'blooms', often with little evidence of what caused the population increase" (Mills, 2001) may in fact be evidence that jellyfish blooms are exaptations to the currently rapidly changing world.

## Materials and methods

To infer the evolution of aggregating, blooming, and swarming within Scyphozoa, we used MacClade v.4.08 (Maddison & Maddison, 1989) to trace the states of these characters (Hamner & Dawson, 2008) on a phylogeny of medusozoans, emphasizing scyphozoans, that is consistent with the weight of evidence from molecular phylogenetic, morphologic phylogenetic, and morphologic taxonomic hypotheses (see Hamner & Dawson, 2008; also, e.g., Stiasny, 1921; Kramp, 1961; Collins, 2002; Dawson, 2004; Collins et al., 2006; Daryanabard & Dawson, 2008). The phylogeny in Fig. 1 was created by writing Newick format trees to describe well-supported subtrees in phylogenies published in the peer-reviewed literature, redrawing graphical representations of these Newick trees as cladograms in FigTree v.1.1.1 (A. Rambaut, 2007: http://www.tree.bio.ed.ac.uk/software/figtree/), and then using an Adams consensus or supertree type approach to splice together the graphical subtrees in Adobe Photoshop CS3 v.10.0.1; this composite consensus tree was rewritten in Newick format and imported into MacClade v.4.08. The seven hierarchically integrated Newick format subtrees drawn from published results were as follows: ((Staurozoa1,Staurozoa2),((Hydrozoa1,Hydrozoa2),(Cubozoa1,Cubozoa2),((Coronatae1,Coronatae2),(Semaeostomeae1,(Semaeostomeae2,Rhizostomeae)))) from analyses of 18S rDNA by Hamner & Dawson (2008; see also Collins, 2002, Fig. 3). ((Coronatae1, Coronatae2),((Rhizostomeae1,Rhizostomeae2,Ulmaridae1,Ulmaridae2),Cyaneidae,Pelagiidae)) from analyses of 28S rDNA by Hamner & Dawson (2008; see also Collins et al., 2006, Fig. 3). ((((Mastigiidae, Thysanostomatidae),Cassiopeidae),Cepheidae),((Rhizostomatidae,Stomolophidae),Catostylidae)) from analyses of concatenated 18S, 28S, and COI by Hamner & Dawson (2008). (Atollidae,Atorellidae, Linuchidae,Nausithoidae,Paraphyllinidae,Periphyllidae) and (Cepheidae,(Mastigiidae,Versurigidae,Thysanostomatidae),Cassiopeidae) and ((Rhizostomatidae, Stomolphidae),(Lychnorhizidae,Lobonematidae,Catostylidae)) from Kramp (1961; see also Stiasny, 1921). (Other_Hydrozoa,(Other_leptothecates,Aequoreidae)) from Collins (2002; see also Collins et al., 2006). We also traced the evolution of canal structure, strobilation type, and photosynthesis across this consensus supertree. The coding of character states in extant taxa is shown by the boxes between branches and taxon names in the reconstructed phylogenies in Figs. 1 and 3.

To infer the evolution of morphological, ecological, and life-history characters possibly correlated with aggregating, blooming, and swarming in scyphozoans, we used MacClade v.4.08 (Maddison & Maddison, 1989) to trace the evolution of character

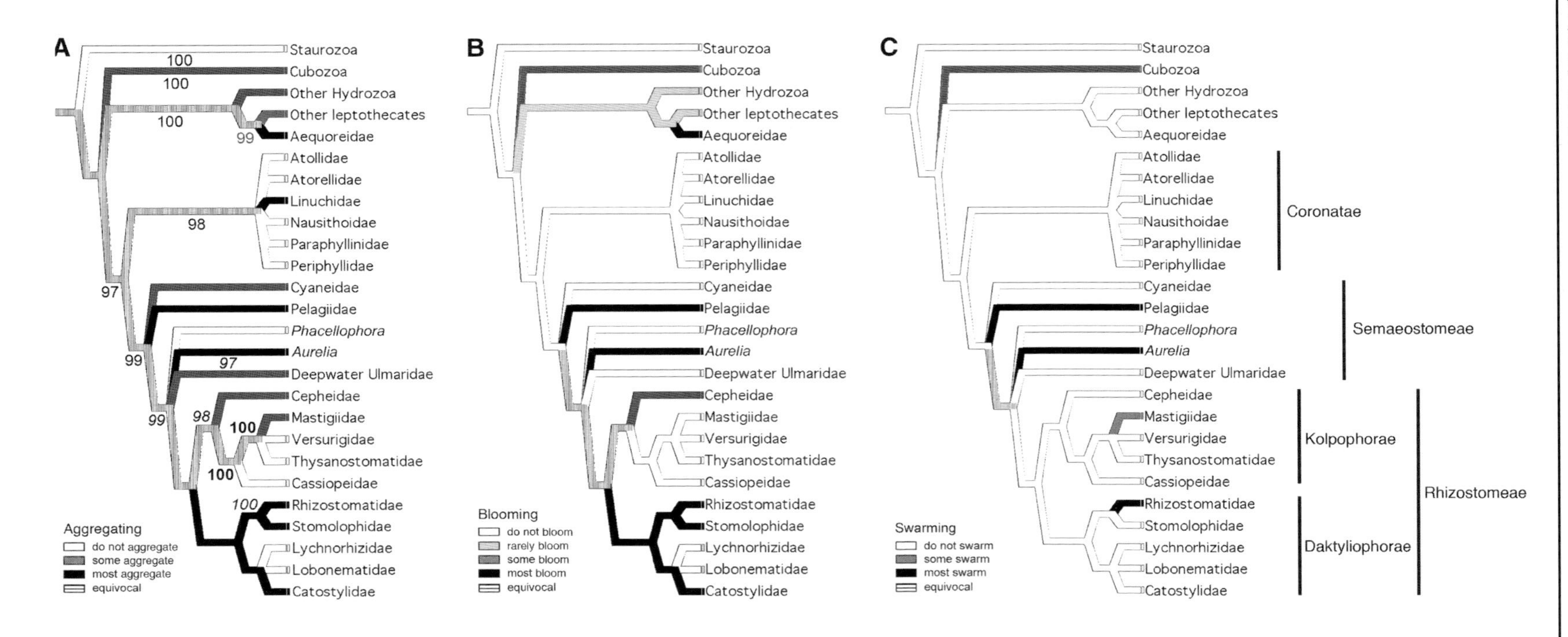


**Fig. 1** The phylogenetic distribution of the (**A**) aggregating, (**B**) blooming, and (**C**) swarming phenotypes in Medusozoa. Hypothetically, these phenotypes result from interactions among multiple component traits, ranging from largely environmental [aggregating], through largely life history [blooming], to largely organismal [swarming]. The 'supertree' is ultrametric, a consensus of the phylogenetically nested molecular analyses of Hamner & Dawson (2008), morphologic taxonomy of Kramp (1961), and molecular analyses of Collins (2002) and Collins et al. (2006). Bootstrap values, shown only in (**A**), are taken from the source phylogenies (gray, Collins (2002); black roman, 18S rDNA in Hamner & Dawson (2008); black italic, 28S rDNA in Hamner & Dawson (2008); black bold, concatenated 18S, 28S, and COI analyses in Hamner & Dawson (2008)) and thus indicate support for branches in the component subtrees, as annotated. Aggregations are defined as an accumulation of individuals that consequently likely interact. Blooms are restricted to 'true' blooms which are defined as normally and abnormally abundant seasonal appearance of jellyfish directly attributable to population growth. Swarms are defined as dense aggregations due to behavior. These three levels of accumulation are classified here as distinct character states for simplicity in our initial analyses, but may be artificially discrete classifications of a character complex whose states are continuously distributed. Small blocks between terminal branches and taxon names indicate how the character state was coded for that taxon; no block indicates coded as unknown and, therefore, any state indicated by the branch color is inferred rather than observed. Ancestral character states about which phylogenetic inference is ambiguous are shown as 'equivocal'

states coded in the datamatrix of van Iten et al. (2006). Although Marques & Collins (2004) and van Iten et al. (2006) have previously used parsimony analyses to infer phylogeny and patterns of evolution of anatomical, morphological, and life-history traits across classes and orders of medusae, such analyses can be compromised by homoplasy and tautology because phylogeny is reconstructed using the same characters whose pattern of evolution is then inferred from the phylogeny (Armbruster, 1992). These problems have been much debated (e.g., de Queiroz, 1996; Luckow & Bruneau, 1997), and may not substantially affect results if the true phylogenetic signal is strong or if particular methods are used (e.g., Ronquist, 2004), but cladistic approaches are compromised when the shape of the phylogeny is sensitive to inclusion or exclusion of the evolutionary hypothesis or characters (Zrzavy, 1997; Brooks & McClennan, 2002, p. 58). The results of Marques & Collins (2004) and van Iten et al. (2006), which differ greatly in topology due to re-weighting of a subset of characters, suffer from at least the latter. Thus, we reanalyzed the phenotypic data using a constraint cladogram derived from the molecular analyses of Collins et al. (2006; see also analyses of 18S rDNA and 28S rDNA by Hamner & Dawson, 2008), as follows: (Staurozoa,(Hydrozoa,(Cubozoa,(Coronatae, (Semaeostomeae,(Rhizostomeae)))))).

**Fig. 2** Schematic of the phylogeny of Tesserazoa (i.e., non-anthozoan cnidarians) on which are plotted the evolution of mass occurrences and characters potentially correlated or causally related to aggregating, blooming, and swarming. The phylogenetic relationships are constrained to conform to those indicated by recent molecular analyses (e.g., Collins et al., 2006; Hamner & Dawson, 2008). The evolution of 87 characters documented by Van Iten et al. (2006; modified from Marques & Collins, 2004) were traced on the tree, and branch lengths are proportional to the numbers of unambiguous character changes (minimum tree length = 79 steps, for reference the branch to Staurozoa is four steps). Only characters synapomorphic for (i.e., shared by members of a clade, but not their most recent common ancestor) the clades Cubozoa + Scyphozoa, Scyphozoa, Discomedusae, or Rhizostomeae are plotted in the figure. Directionality of character evolution within Medusozoa is polarized by rooting the phylogeny using outgroups Anthozoa (not shown) and Staurozoa. The relative timing and direction of evolutionary transitions are then indicated by branches annotated as follows "Character: ancestral state ⇒ derived state." The non-homoplastic synapomorphies (□) represent the key evolutionary transitions associated with evolution of occurrence *en masse*. Also shown are homoplasious character state changes (◇, ∇, Δ) which are unlikely to have a large impact on the ability of the species to occur *en masse* because they also occur in clades that do not bloom. The approximate proportions of families with aggregating [A], blooming [B], or swarming [S] species were estimated from Hamner & Dawson (2008) and are indicated by the pie charts embedded within the tree. The state of mesoglea (non-cellular ⇒ cellular) is homoplasious with the character state in Anthozoa; heterotrichous anisorhizas are coded polymorphic for the clade Anthothecata + Siphonophorae (van Iten et al., 2006); strobilation is coded as fixed for monodisc strobilation in all Rhizostomeae following van Iten et al. (2006) although this is the case only for Kolpophorae (see Fig. 3A)

## Results

Mass occurrences of medusae are not distributed evenly phylogenetically. Although species that aggregate can be found in all medusozoan classes, species that bloom or swarm are restricted almost entirely to Class Cubozoa and to Subclass Discomedusae within Class Scyphozoa (Fig. 1). Evolution of aggregations appears to precede evolution of blooms and/or swarms.

Considering the evolution of morphology and life history, we found up to 13 'synapomorphic' characteristics of Scyphozoa potentially related to their ability to form blooms and swarms (Fig. 2). Of these, only seven were unique synapomorphies for Scyphozoa or a subclade thereof in our ordinal-level analysis (Fig. 2): podocysts (can be formed), strobilation (absent, present), strobilation type (monodisc, polydisc), ephyral life-history stage (present), circular canal (present either partially or complete), radial canals (present, complex), and oral arms (with suctorial mouths). Other life-history, ecological, and distributional attributes of extant medusae are also clustered in more basal or more derived taxa (Fig. 3). Circulatory canals, for example, originate in Cyaneidae and tend to be more complexly anastomosing in more highly derived taxa (Ulmaridae and particularly Rhizostomeae; Fig. 4).

## Discussion

Aggregation, which is widespread in Medusozoa, is apparently an evolutionarily basal trait and a precursor to the blooming and/or swarming phenotypes which most often characterize cubozoans and scyphozoans (Fig. 1; see also Dawson & Hamner, 2008). The implied evolutionary trajectory is intuitively appealing. It is easy to conceive that selection for

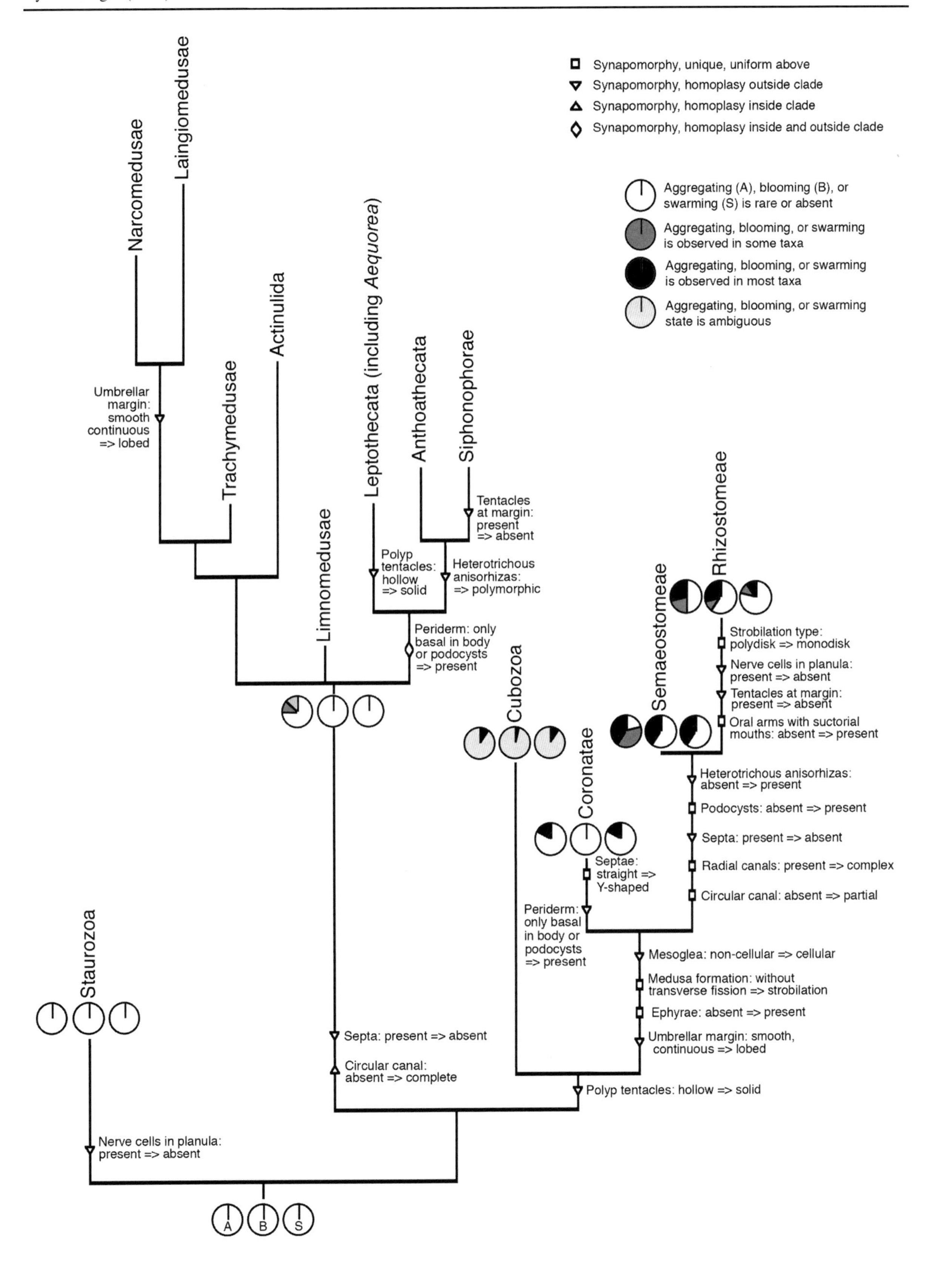

Synapomorphy, unique, uniform above
Synapomorphy, homoplasy outside clade
Synapomorphy, homoplasy inside clade
Synapomorphy, homoplasy inside and outside clade
Aggregating (A), blooming (B), or swarming (S) is rare or absent
Aggregating, blooming, or swarming is observed in some taxa
Aggregating, blooming, or swarming is observed in most taxa
Aggregating, blooming, or swarming state is ambiguous
Narcomedusae
Laingiomedusae
Umbrellar margin: smooth continuous => lobed
Trachymedusae
Actinulida
Limnomedusae
Leptothecata (including Aequorea)
Anthoathecata
Siphonophorae
Tentacles at margin: present => absent
Polyp tentacles: hollow => solid
Heterotrichous anisorhizas: => polymorphic
Periderm: only basal in body or podocysts => present
Cubozoa
Coronatae
Semaeostomeae
Rhizostomeae
Strobilation type: polydisk => monodisk
Nerve cells in planula: present => absent
Tentacles at margin: present => absent
Oral arms with suctorial mouths: absent => present
Heterotrichous anisorhizas: absent => present
Podocysts: absent => present
Septa: present => absent
Radial canals: present => complex
Circular canal: absent => partial
Septae: straight => Y-shaped
Periderm: only basal in body or podocysts => present
Mesoglea: non-cellular => cellular
Medusa formation: without transverse fission => strobilation
Ephyrae: absent => present
Umbrellar margin: smooth, continuous => lobed
Staurozoa
Septa: present => absent
Circular canal: absent => complete
Polyp tentacles: hollow => solid
Nerve cells in planula: present => absent
A
B
S

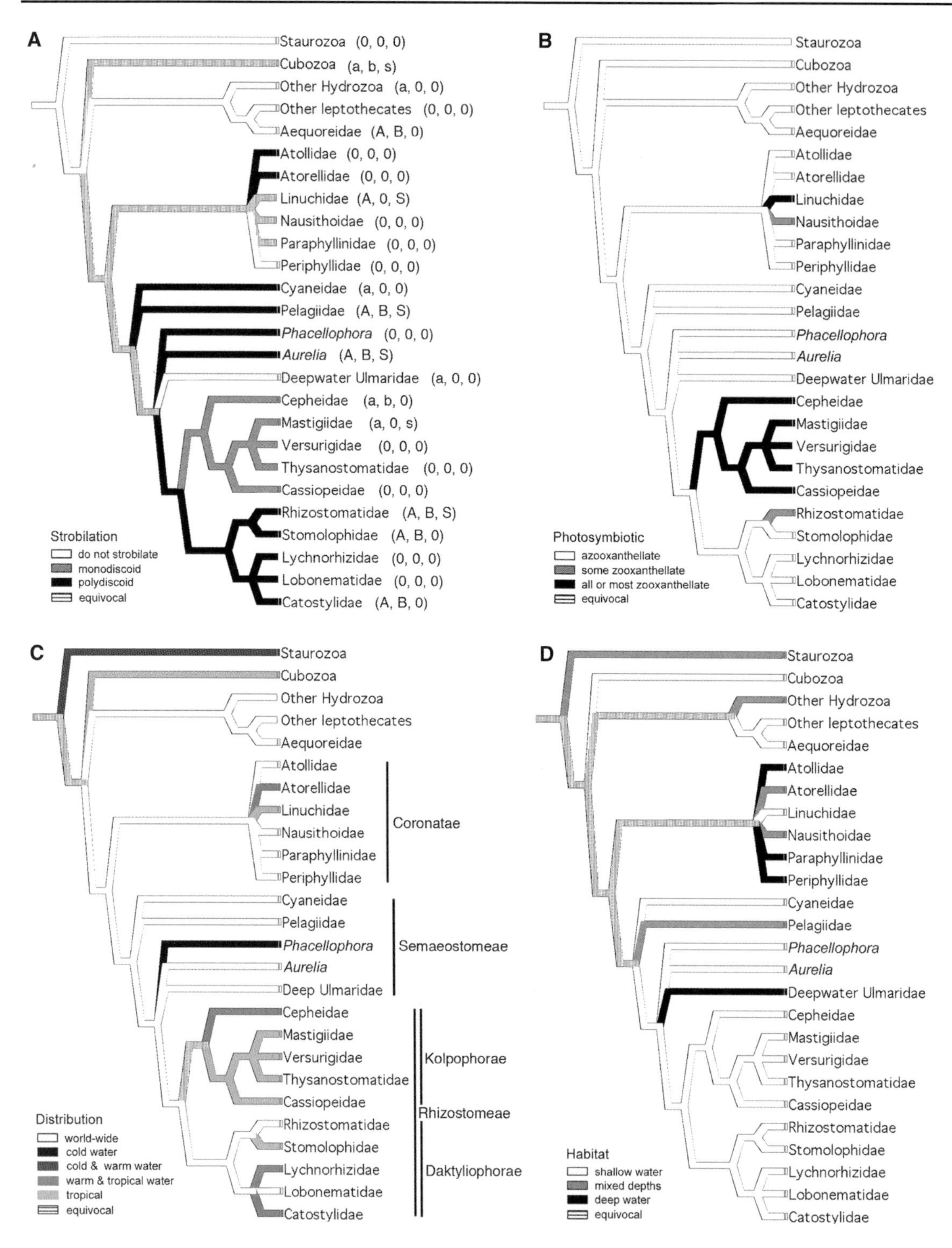


traits favoring aggregation, a phenomenon caused primarily by environmental characteristics exaggerated by directional swimming elicited in response to the immediate environment, is possible in many diverse taxa including ancestral medusozoans that likely were as anatomically complex as modern-day medusozoans (Cartwright et al., 2007). Such changes anticipate benefits of accumulation that could increase as a consequence of either further selection on newly evolved adaptations or new selection on

◀**Figure 3** Characteristics of organisms and environment that may affect the occurrence of aggregations, blooms, or swarms. (**A**) Strobilation; note that this rendition codes asexual reproduction of cubomedusae from polyps as a form of strobilation, following Straehler-Pohl & Jarms (2005), although following Werner (1973) the branch leading to Cubozoa would be colorless, indicating 'do not strobilate.' (**B**) Whether medusae are photosymbiotic or not. (**C**) The predominant geographic, i.e., latitudinal and longitudinal extent, distribution of medusae. (**D**) The predominant depth zone, or zones, of occurrence of medusae. Small blocks between terminal branches and taxon names indicate how the character state was coded for that taxon; no block indicates coded as unknown and, therefore, any state indicated by the branch color is inferred rather than observed. States about which phylogenetic inference was ambiguous are shown as 'equivocal.' Letters after taxa in (A) indicate whether some/most aggregate 'a/A,' bloom 'b/B,' or swarm 's/S,' a '0' indicates none show that kind of accumulation. Higher taxonomy is indicated in (C) for convenient reference

exaptations (i.e., other traits, evolved previously, that acquired new evolutionary value in the novel context of aggregations) leading to evolution of blooming and/or swarming. Concomitantly, the most complex aggregative behavior, swarming, is observed in the fewest taxa. This indicates that accumulation likely results from interactions among multiple characters, rather than being a consequence of single 'Mendelian' traits. Accumulation therefore may be modified by natural selection either on component characters or on the complex behavior as a whole.

Our molecular maximum likelihood phylogeny (Hamner & Dawson, 2008), like the morphological parsimony analysis of van Iten et al. (2006) but in contrast to Marques & Collins (2004), places Staurozoa basal to Medusozoa rather than as a derived clade within Medusozoa. This apparently major rearrangement of phylogeny does not affect inference of the evolution of the states of, or phenotypes that may favor, accumulation of jellyfish for three reasons: (1) few characters unambiguously distinguish Staurozoa from Medusozoa, (2) the relationships of Scyphozoa to putative sister taxa, Cubozoa or perhaps Hydrozoa, are not altered, and (3) the relationships within Scyphozoa are not altered. Thus, in addition to the giant fiber nerve net and statocysts characteristic of pelagic motile medusae (Marques & Collins, 2004) which can aggregate, 13 'synapomorphic' characteristics of Scyphozoa are potentially related to their ability to form blooms and swarms (Fig. 2; see also Marques & Collins, 2004; Collins et al., 2006; van Iten et al., 2006) but only seven unique synapomorphies for Scyphozoa or a subclade thereof are indicated to be the characters [and states] that promote blooms and/or swarms. Below, we discuss these characters and states (listed in *Results*), in some cases grouping them into likely character complexes (e.g., circular and radial canals are both components of the gastrovascular canal system) and in others expanding the list to include related characters (e.g., tentacles, the canal system necessitated by large size, and large size which enables improved swimming ability) reconstructed ambiguously by, or not in the matrices of, Marques & Collins (2004) or van Iten et al. (2006).

## Podocyst formation

Podocyst formation is documented in all semaeostome families and in the rhizostome families Rhizostomatidae and Stomolophidae (Arai, 1997, p. 164). Podocyst formation is therefore also probably a life-history strategy conserved in other rhizostomes not yet studied in this regard.

Podocysts are small aggregations of epidermal cells and amoebocytes surrounded by a chitinous cuticle. They are formed in several ways, and one or many may be produced by a single scyphistoma. They have low metabolic rates and may have exceptional longevity; those of *Chrysaora quinquecirrha* (Desor) can be viable for 2 years or more (Arai, 1997, p. 164). On excystment, each forms a new scyphistoma. Podocysts may, therefore, contribute to blooms in two ways. One, by allowing populations to resist long harsh conditions that would decimate polyp or medusa populations and, when conditions improve, metamorphosing into scyphistoma that may produce many medusae. Two, by leading, as the products of asexual reproduction, to increased scyphistoma population size and therefore, via strobilation, potentially also to increased medusa population size.

## Strobilation, strobilation type, and ephyral life-history stage

Although metamorphosis of cubozoan polyps into cubomedusae may be a modified form of strobilation (Straehler-Pohl & Jarms, 2005), strobilation and the concomitant production of ephyrae is unequivocally present only in Scyphozoa. Within Scyphozoa, the

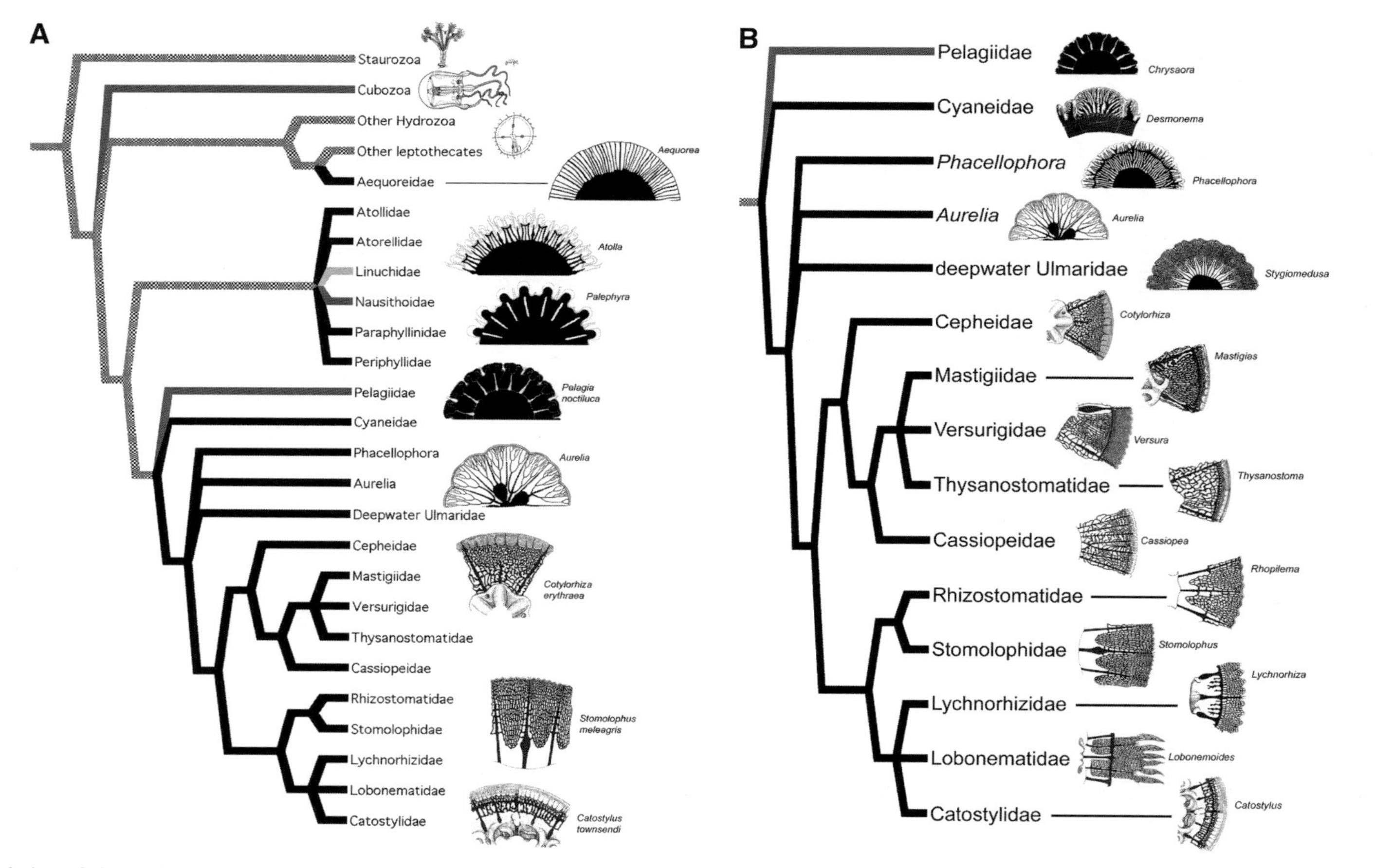


**Fig. 4** Evolution of size and canal structure in medusae: (**A**) general patterns across the Medusozoa and (**B**) more detail showing family-level variation within Discomedusae. Shading of branches in the trees indicates patterns of size evolution of medusa: light gray, predominantly small medusae; dark gray, mixed sizes of medusae; black, predominantly large medusae; hatched, phylogenetic inference of size is ambiguous. Illustrations of canal complexity are redrawn from Stiasny (1921), Kramp (1961), Larson (1986), Bouillon (1999), BIODIDAC (http://www.biodidac.bio.uottawa.ca/; image by Ivy Livingstone: Animalia—Cnidara—Cubozoa), and http://www.hermes.mbl.edu/BiologicalBulletin/KEYS/INVERTS/3/img/H.auricula1.JPG

biphasic life history and therefore strobilation and ephyrae predominate, with exceptions restricted to a single neritic species *Pelagia noctiluca* (Forskål) and a polyphyletic group of deepwater medusae (Fig. 3A), which have direct development (Larson, 1986; Jarms et al., 2002).

Almost all species that occur *en masse* do strobilate, whereas most species that do not occur *en masse* do not strobilate, suggesting that strobilation (or a correlated factor) does predispose jellyfish to occur *en masse*. However, *Pelagia noctiluca*, which is renowned for mass occurrence, develops directly from planula to ephyra (Rottini Sandrini & Avian, 1983) demonstrating that the polyp stage and strobilation are not necessary for bloom formation (a metagenetic life cycle is also not present in ctenophores or planktonic chordates which do form blooms).

Which species bloom may furthermore be influenced by the type of strobilation. Families in the rhizostome suborder Kolpophorae have monodisc strobilation (Holst et al., 2007) and rarely bloom or swarm (Fig. 1). In contrast, *Aurelia* spp., *Chrysaora* spp., and Rhizostomatidae all have polydisc strobilation and are the most notable blooming species. We hypothesize that the apparent association between polydisc strobilation and blooms is because polydisc strobilation probably produces more ephyrae per polyp per unit time, or more ephyrae per polyp at a particular time, than monodisc strobilation. However, Watanabe & Ishii (2001) may describe the only attempt to measure polyp productivity in relationship to strobilation, which means quantitative comparisons are not yet possible.

Whether strobilation, be it monodisc or polydisc, is itself a major factor contributing to blooms also depends on the number of medusae produced being impacted less by other processes, such as asexual reproduction via podocysts, planuloids, the preceding year's medusa population size and individual fecundity, or the survival of ephyrae. Yet, it is intuitive that simultaneous strobilation, particularly polydisc strobilation, could generate large cohorts of ephyrae leading to blooms of adult medusae.

## Canal system, medusa size, and dependent attributes

In stauromedusae and scyphistomae, the gastric system is divided in the suboral reaches by four septa along which run ciliary currents that distribute food and reverse to remove waste. Similar simple circulatory systems can be found in Cubozoa (e.g., Hamner et al., 1995) and Coronatae (Fig. 4). The gastric system is relatively simple in the phylogenetically basal scyphomedusae such as Pelagiidae and Cyaneidae in which septa completely subdivide the gastric cavity of the medusa into numerous pouches through which cilia move nutriment and wastes. In Cyaneidae, the pouches may additionally have distal canal-like protrusions and, in *capillata*-group medusae, pit-like intrusions into the deeply pleated subumbrellar circular and radial musculature (Stiasny & van der Maaden, 1943).

It is only in Ulmaridae and Rhizostomeae that the gastric system consists of a complex network of canals. In the basal *Phacellophora*, the system consists largely of multitudinous septae which create a complex pattern of radiating canals (see Figs. 2, 4). The remaining ulmarids have complex anastomosing networks of canals permeating the subumbrella. In *Aurelia* (Russell, 1970; reviewed by Arai, 1997, p. 100), the flow of food particles and waste is unidirectional except through the largest canals or through the mouth where ciliary currents are driven in opposite direction at different positions within the same tube. Little is known of the flow through the gastrovascular systems of other ulmarids or rhizostomes, but the complexly anastomosing canals of rhizostomes, particularly the dense meshes in Kolpophorae, appear at first glance to be likely to cause inefficient flow. We suggest, though, that directional flow through canals will be discovered when circulation is more carefully examined in all large medusae because efficient internal flows are required to regularly and completely flush the canals and to provide food and oxygen to densely packed and often folded tissues. All large organisms have mass-transport or circulatory system that move resources to, or waste products from, sites of origin to distant parts of the organism (Ruppert & Carle, 1983). The most relevant comparison is *Aequorea*, the largest hydromedusan, which has radial canals and ciliarly-driven flow (Hyman, 1940; Arai & Chan, 1989) evolutionarily analogous to the canals and flows in Ulmaridae and Rhizostomeae.

In terms of blooms and swarms, the importance of gastrovascular canals is perhaps not so immediately obvious as that of the rate of podocyst formation or type of strobilation by scyphistomae because gastrovascular structure is an attribute only of individuals that cannot contribute directly to the numerical magnitude of a bloom. Yet differences in canal structure are correlated with the ability to aggregate, bloom, and swarm (Figs. 1, 2, 4). Primarily taxa with extensive canal systems occur *en masse* whereas those without usually do not. We hypothesize the circulatory system is one of the most important anatomical features enabling bloom formation because it allows individual medusae to grow to large size. Large size enables attributes that promote occurrence *en masse*: increased feeding, increased fecundity, and natatory movement against the predominant flow.

### *Size*

Jellyfish blooms are characterized by rapid, seemingly sudden, increase in abundance of medusae. Blooms are therefore necessarily initiated by production of large numbers of ephyrae, and blooms are typically thought of in terms of rapid increases in numerical abundance. However, numerical abundance alone is demonstrably insufficient to constitute a bloom. First, blooms that draw attention are never of ephyrae although, assuming remotely normal life-tables, most ephyrae must die before developing into medusae, so the problematic true blooms of scyphomedusae that are reported must comprise a reduced number of individuals. Second, blooms of hydromedusae, which are almost exclusively small, are rare (see Hamner & Dawson, 2008). The principal exceptions are *Aequorea* (e.g., Purcell, 1989), which is the largest hydrozoan, and introduced species such as *Craspedacusta* landlocked in freshwater lakes (Pennak, 1956) and *Moerisia lyonsi* in Chesapeake Bay (e.g., Purcell et al., 1999). Thus, the aspect of true blooms that makes them remarkable is the rapid growth of the individual medusae that survive, which produces a remarkable increase in population biomass. The mass of mature scyphomedusae is thousands to hundreds of thousands of times greater than the mass of ephyrae, an increase that can be achieved through growth in just a few weeks to few months but which would take many orders of magnitude greater duration to achieve through strobilation. Thus, using food to promote somatic growth may increase population biomass much faster than increasing the number of juveniles of small size (Alldredge, 1984; Arai, 1997).

### *Growth rate, feeding rate*

The rapid growth of semaeostome and rhizostome scyphomedusae is enabled in two ways: (1) by construction of mesoglea to achieve large increase in size at relatively low metabolic cost and (2) by use of their large external surface as their primary feeding structure. Because scyphomedusae are diploblastic (but see Seipel & Schmid, 2006), ingested materials are allocated to increasing the 2-dimensional surface areas of the endoderm and ectoderm rather than, as in many 'higher' metazoans, fueling a disproportionately metabolically demanding 3-dimensional mesoderm. Increasing the third dimension—depth and thus body volume—in jellyfish is effected primarily by construction of mesoglea composed almost entirely of highly dispersed collagen molecules with exceptionally low metabolic requirements (Chapman, 1966; for other functions of mesoglea see e.g., deBeer & Huxley, 1924; Denton & Shaw, 1962; Chapman, 1966; Hamner & Jenssen, 1974; Thuesen et al., 2005). Exponential growth in biomass is thus possible, as documented in linear rates of bell diameter increase (e.g., Arai, 1997: Figs. 7.2, 7.3) and therefore cubic increases in volume in *Aurelia*.

The rate of growth of medusae via deposition of mesoglea is impressive not only in terms of its efficient allocation of resources (although estimates of conversion efficiency in medusae vary greatly from 2% to 37%; Arai, 1997, p. 185) but also in its effect on acquisition of new resources. Many semaeostome and all rhizostome scyphomedusae use a significant portion of their large external surface as their primary structure for prey capture; shallow-water cyaneids and ulmarids have extensive tentaculate surfaces while rhizostomes possess oral arms covered in suctorial mouths. When these medusae invest resources in exponential growth of their surface area, therefore, they also allocate those resources for exponential increase of their feeding potential. Individual feeding success of *Aurelia* therefore increases proportional to the area of the bell (Martin, 1999, p. 54) and exponentially with diameter (e.g., Bailey & Batty, 1983; de Lafontaine & Leggett, 1987). This in turn underwrites a rapid, non-linear, increase in prey capture (Arai, 1997,

Table 3.5) and, further, exponentially increased allocation to growth. We note this may not be the case for medusae that are photosymbiotic (e.g., García & Durbin, 1993) or others such as box jellyfishes, coronates, and pelagiids or deepwater cyaneids, which 'fish' for prey with few long tentacles, but expect that many other scyphomedusae may grow exponentially, given sufficient food, until they reallocate resources from growth to reproduction.

*Fecundity*

At maturation, the energy supporting exponential increase in body mass and feeding capacity is redirected into production of eggs and/or sperm, generating a rapid volumetric increase in reproductive capacity. A doubling of bell diameter in *Rhopilema esculenta* (Kishinouye) results in a 20-fold increase in daily egg production (Huang et al., 1985), and the number of eggs and larvae brooded by *Aurelia* increases linearly with the wet-weight of the medusa (Schneider, 1988). Large size thus greatly increases reproductive output of individuals and large individuals should, therefore, have a fitness advantage in terms of reproductive success. This selective benefit is one instance of a more general phenomenon known as "fecundity advantage" (Shine, 1988, 1989) first proposed by Charles Darwin (1874, p. 332) as a mechanism for maintaining larger female than male size and probably best known as an explanation for protandric hermaphroditism. Fecundity advantage has garnered considerable, but not unanimous, support (Shine, 1988) from studies of diverse organisms including copepods, cladocera, and rotifers (Gilbert & Williamson, 1983) but to our knowledge has not previously been tested in scyphomedusae. (N.B. Reanalysis of data describing bell diameters of *Aurelia* caught in net hauls on 13 occasions between September 1995 and December 1998 in a marine lake, Palau, reveals the largest medusae were usually female [$\chi^2 = 4.76$, d.f. $= 1$, $P < 0.0145$ one-tailed]. Female *Aurelia aurita* (Linné) are also larger than contemporaneous male *A. aurita* in Horsea Lake, England (see Lucas & Lawes, 1998, Figs. 2, 5), and female *Catostylus mosaicus* (Quoy & Gaimard) are predominantly larger than contemporaneous males in Botany Bay, New South Wales, Australia (Pitt & Kingsford, 2000, Fig. 3B). Thus, it is clear that the pulsed life-history style of scyphozoans has potential to create large populations of planulae and tiny scyphistomae, which can reproduce asexually *en masse* to provide the raw material—numbers of individuals—of which the growth could turn into the biomass of the next year's bloom.

*Natation, Reynolds number*

In addition to the life-history factors discussed above, the fact that small medusae may occur in large numbers but, unlike large medusae, apparently rarely form exceptional blooms ('population explosions' or 'booms') and do not swarm (but see Hamner & Dawson (2008) for some apparent exceptions) is in part a function of their ability to actively aggregate despite environmental factors that serve to diffuse them. Whether medusae have the potential to aggregate despite their immediate fluid environment depends upon their tendency to move forward, i.e., inertia, generally exceeding viscous forces retarding them, a relationship summarized by the Reynolds number.

Reynolds number describes how fluid density ($\rho$) and dynamic viscosity ($\mu$) interact with an organism's morphology (size and shape, but particularly length in the direction of flow, $l$) and speed ($U$), as follows: $Re = \rho * l * U * 1/\mu$. If $Re$ approaches unity or less, drag forces dominate and bring medusae to an abrupt halt at the end of each power stroke, as evident in videos of the saltational movement of ephyrae just liberated from strobilae and the movement of small prolate hydromedusae such as *Proboscidactyla flavicirrata* Brandt or small oblate hydromedusae such as *Mitrocoma cellularia* (A. Agassiz) (e.g., Colin & Costello, 2002). If $Re >> 1$, the organism's momentum dominates and it continues to glide with no discernible deceleration between power-strokes, as is the case for the large scyphomedusa *Stomolophus meleagris* L. Agassiz (Arai, 1997, p. 51). It is apparent from the equation that the characteristic length of medusae, which may vary three orders of magnitude, from a few millimeters to over 1 m, is the principal factor influencing a medusa's movement; speed varies by only one order of magnitude, and fluid density and dynamic viscosity of seawater are to all intents and purposes constant. Thus, an ephyra swimming at less than 1 cm $s^{-1}$ has $Re \approx 25$, a small *Pelagia* swimming at 2 cm $s^{-1}$ has $Re \approx 600$, a medium-sized rhizostome such as *Mastigias* swimming at 8 cm $s^{-1}$

has $Re \approx 2,000$, and a large rhizostome such as *Rhizostoma pulmo* (Macri) moving at 15 cm $s^{-1}$ has $Re \approx 100,000$ (estimates based on information from Aleyev, 1977; Arai, 1997; Dawson & Hamner, 2003, 2008).

Aleyev (1977), in addition to calculating organisms' *Re*, also classified swimming organisms in terms of their ratio of frontal diameter (*d*)/characteristic length (*l*). Aleyev's *d/l* is the inverse of 'fineness,' which is more commonly used in studies of the mechanics of jellyfish swimming (e.g., Costello & Colin, 1995; Colin & Costello, 2002; Dabiri et al., 2007). However, *d/l* is somewhat more holistic because it is affected by the form of oral arms and other appendages that trail the bell, whereas 'fineness' is typically calculated only for the bell of medusae that have no, or few, pendant oral structures (e.g., Colin & Costello, 2002). According to Aleyev's (1977) scheme, organisms with $Re < \sim 10^4$, irrespective of their *d/l*, and organisms with $Re < \sim 10^5$ and $d/l > 0.4$ are euplankton. Organisms with $10^4 < Re < 10^5$ and $d/l < 0.4$ are planktonekton. Organisms with $10^5 < Re < 10^7$ and $d/l < 0.4$ are nektoplankton. It is clear that, according to these generic descriptions, most scyphomedusae are euplankton, but that some semaeostomes and a considerable proportion of large rhizostomes, the principal clades of blooming and swarming species, are planktonekton or nektoplankton and thus have reasonable ability to determine their local position. For example, *Chrysaora* in Monterey Bay can swim directionally with astonishing accuracy even at night (Morris, 2006), presumably facilitating population maintenance (Sinclair, 1988). *Stomolophus meleagris* can navigate along continental shelves (Shanks & Graham, 1987), possibly using shear forces to orient, as is the case in *Aurelia* (Rakow & Graham, 2006).

It is important to note that these broad classifications only suggest tendencies because local hydrographic conditions may interact with medusa morphology in important ways. For example, in Saanich Inlet (Vancouver Island, British Columbia), during Spring, *Aurelia labiata* Chamisso & Eysenhardt ephyrae and small medusae are dispersed by tidal currents throughout the fjord because they are too small and too slow to avoid dispersal by ambient flow (Hamner et al., 1994). Yet when the jellyfish grow to bell diameters of about 4–6 cm they can make headway against reduced tidal currents within the fjord, and they swim directionally and horizontally toward the south-east during daylight hours using a time-compensated sun-compass navigation system, often swimming hundreds of meters to the south-east each day; bigger medusae migrate faster, and all accumulate eventually into small coves on the south-east side of Saanich Inlet in late summer (Hamner et al., 1994). Once in the embayments the jellyfish maintain their geographic position by changing their behavior, swimming vertically instead of horizontally during the day as well as at night. Tidal currents do not advect the jellyfish away because tidal currents tend to shear across the mouths of the coves. Thus, medusae accumulate, begin to spawn in late summer, and these specific behaviors are thus apparently adaptations that maintain the population within the specific physical context of Saanich Inlet. To the north, in Roscoe Bay, British Columbia, *Aurelia labiata* behave quite differently, swimming predominantly westward during sunny afternoons (Albert, 2005, 2007, 2008).

It is possible that passive dispersal of tiny ephyrae and small juveniles followed by active reaggregation by sexually maturing medium-to-large medusae is a common theme in the larger scyphomedusae. If true, this process likely has many facets, being aided by rhopalia, circadian rhythms, neural processing for a time-adjusted sun compass, and ontogenetic change in swimming ability perhaps attributable to changes in musculature and oral arm form.

## Form of oral arms and tentacles

The oral arms and tentacles have evolved dramatically within Scyphozoa, particularly Discomedusae. These modifications likely have an important influence on flow regimes, feeding potential (or 'clearance rate'), and prey type ('selectivity').

Coronates all have oral appendages that are limited to simple lips on a short manubrium and, with few exceptions, an array of robust aborally arcing marginal tentacles. In the case of *Atolla*, the marginal tentacles are complimented by a remarkable trailing hypertrophied tentacle (Larson, 1979; Hunt & Lindsay, 1998). Few in situ observations are available, but it is likely that the mesopelagic coronates use the aborally arcing marginal tentacles to capture zooplankton including small crustaceans and fishes (Larson, 1979) supplemented in *Atolla* by use of the hypertrophied tentacle to capture gelatinous zooplankton such as the physonect hydrozoan *Nanomia* (Hunt & Lindsay, 1998). While shallow-water tropical *Nausithoe* also have aborally arcing marginal tentacles and a predatory diet (Colin et al., 2006), these tentacles are greatly reduced in the epipelagic photosymbiotic *Linuche* (Mayer, 1910, plate 59).

The basal pelagiid semaeostomes, like coronates, possess 8 to 40 long ribbon- or thread-like nematocyst-bearing contractile feeding tentacles, but unlike coronates, these are all trailing and supplemented by four highly convoluted oral arms (Gershwin & Collins, 2002). Other semaeostomes, depending on their family and habitat, have few thick ribbon-like tentacles and flouncy oral arms (deepwater cyaneids *Desmonema* and *Drymonema*), curtains of fine thread-like tentacles supplemented by extravagant draping oral arms (shallow-water *Cyanea*, *Phacellophora*), many fine tentacles and digitate oral arms (Aureliinae), or large oral arms and no tentacles (deepwater ulmarids). This morphological variation is matched by, and likely causes, differences in diet among these medusae. Pelagiids capture relatively few but relatively large active zooplankton prey such as euphausiids and also juvenile fish with its ribbon like tentacles; *Chrysaora fuscescens* Brandt occasionally captures and eats juvenile *Loligo opalescens* Berry in Monterey Bay (Hamner pers. observations). They may also catch microzooplankton on their oral arms. The cyaneids and *Phacellophora* may capture large numbers of gelatinous zooplanktonic prey, engulfing them in their curtain-like oral arms. In contrast, the neritic ulmarid *Aurelia* relies almost solely on its curtain of fine tentacles to capture micro- and mesozooplankton, its oral arms being used to transport boluses of food from the oral margin to the central mouth.

These observations suggest an evolutionary trend within Scyphozoa and Semaeostomeae from capturing large prey items through contact-fishing to feeding on smaller prey captured using a fine filtration-type apparatus. Interestingly, though, food handling in all species involves extra-gastric digestion of captured prey, thus speeding passage through the gastrovascular cavity. Gelatinous prey, fish, or squid too large to ingest are digested and/or reduced in volume outside the stomach within the foliose oral arms of pelagiids and cyaneids (Strand & Hamner, 1988; Arai, 1997, p. 96; Hansson, 1997).

Rhizostome medusae, with no marginal tentacles, continue the trend of increasing specialization on more numerous smaller prey but evolved a quite different morphological solution for exploiting small prey. Rhizostomes have eight oral arms, pendant behind the bell, over which water is forced during swimming (e.g., D'Ambra et al., 2001). These large highly reticulated external feeding surfaces have myriads of tiny suctorial mouths with which they filter and capture very many meso- and microzooplankton prey (e.g., D'Ambra et al., 2001; Peach & Pitt, 2005). The form of the oral arms, sometimes with appendages, creates a prolate morphology, with longer characteristic length ($l$) relative to diameter ($d$) compared to the generally oblate morphology of, for example, *Aurelia*, with a potentially large effect on their swimming ability and feeding. Large prolate medusae must have much larger $Re$, faster swimming speeds, and more consistent swimming motion, than oblate medusae of similar bell diameter, indicating they should be able to capture more, smaller, faster prey (D'Ambra et al., 2001). This simple hypothesis, though, belies greater complexity because the dimensions and presentations of oral arms show considerable variation, including the flat spray of *Cassiopea*, the relatively squat oral arms of *Cephea* and *Mastigias* which in the latter case are appended by long mouthless terminal clubs, and the long coursing oral arms of *Thysanostoma*. These different morphologies suggest multiple functions of the oral arms, not restricted to prey capture alone, such as increasing surface area for photosymbiosis in many Kolpophorae or increasing swimming efficiency via streamlining or hydrodynamic stabilization in natatory medusae such as *Mastigias* (including the terminal clubs; see, e.g., Dawson & Hamner, 2003), *Thysanostoma*, and possibly other fast-swimming species (e.g., Catostylidae).

Photosymbiosis, lack of photosymbiosis

Photosymbiosis between Scyphozoa and dinoflagellate zooxanthellae has evolved at least three times (Fig. 3B). Notably, none of the photosymbiotic medusae—*Cassiopea*, *Mastigias*, *Rhizostoma*, *Linuche*, *Cephea*, *Cotylorhiza*, *Catostylus*, and *Phyllorhiza*—form blooms, with one remarkable exception: invasive, aposymbiotic *Phyllorhiza* in the Gulf of Mexico (Bolton & Graham, 2004). In contrast, the azooxanthellate *Aurelia*, *Pelagia*, *Catostylus*, *Cephea*, *Crambionella*, *Nemopilema*, and *Rhopilema* (and the large azooxanthellate hydrozoan *Aequorea*) all aggregate and bloom.

This suggests that the exploitation of two quite different sources of energy, or the potentially more constant supply of energy available to photosymbionts (Arai, 1997, p. 114), may diminish the tendency of jellyfish populations to bloom. Trophic generalism achieved through exploitation of a broad resource base or resource switching may insure a predator population against large fluctuations in one or another target prey (Hanski et al., 1991; Ricklefs & Miller, 1999, pp. 450, 461; Behmer & Joern, 2008) allowing the predator to achieve a temporally more even, or 'balanced,' energetic income. However, the dynamics of even seemingly facile one-consumer-two-resource systems can be complex and may not lead to relatively stable, moderate density, predator population sizes (e.g., Bonsall & Hassell, 1997; Kooi et al., 2004; Ives et al., 2008). Despite some evidence that cnidarian's symbiont genotypes may be 'shuffled' to suit different circumstances (Baker, 2003), others have argued that this is a rare ability (Goulet, 2006) and there is evidence that host dependencies on symbiont and prey cannot be switched (Tanner, 2002). Furthermore, predator–prey dynamics may proceed to extreme abundances due to temporal environmental variation, spatial heterogeneity, and dispersal (e.g., Gotelli, 2001) and while symbiotic cnidarian populations are well known to 'bust' (or 'crash') during bleaching events under extreme environmental conditions (e.g., Dawson et al., 2001), we have not found evidence that photosymbiotic scyphomedusae 'boom.'

An alternative explanation for the lack of blooms of photosymbiotic scyphomedusae, perhaps more robust than relying on a seemingly precarious balance between symbionts, might instead draw on evidence that symbioses involve trade-offs for both host and symbiont. This would be more in keeping with the implications of the adaptive bleaching hypothesis (Fautin & Buddemeier, 2004) and with theory that suggests symbioses are, in general, slightly maladaptive (e.g., Thompson et al., 2002). Thus, although high availability of specific food types could be inferred as an important potential environmental cause of blooms, we instead hypothesize that the more important facet influencing whether a species can bloom is its inherent ability to exploit available food. Photosymbiosis might limit the ability of host medusae to assimilate abundant prey through perhaps physiological mechanisms (for which we propose invasive *Phyllorhiza* as a model), or via trade-offs against particulate food capture (Arai, 1997, p. 116). Other hypotheses for the dearth of blooms of photosymbiotic scyphomedusae might include other limiting factors of the symbiosis, such as the availability of zooxanthellae to inoculate new polyps that require the symbiosis for strobilation (e.g., Sugiura, 1964).

## Habitats potentially related to accumulation

Only shallow water medusae occur *en masse* (Fig. 3D), often in habitats broadly distributed geographically, and with no phylogenetic bias beyond that which exists due to other ecological, life-history, or anatomical reasons (e.g., Figs. 2, 3A, B, 4). Shallow-water *Aequorea* bloom, the coronate *Linuche* resists dispersion between times when Langmuir cells force it to aggregate (Larson, 1992), and many semaeostomes and non-photosymbiotic rhizostomes include species that form blooms. In contrast, deepwater coronates and deepwater semaeostomes, also often distributed widely, do not bloom. As described by Omori & Nakano (2001), for commercially fished species, the "places where great numbers of…jellyfish occur are characterized by having a large tidal range, shallow depth, semi-enclosed water mass, freshwater inflow through river systems and development of mangrove swamps. Such factors apparently create favorable conditions for settling of polyps and recruitment." These factors or their analogues (e.g., non-mangrove shallow-water habitats in Europe and North America), or their correlates, may still be favorable for jellyfish even if all do not occur together. Yet, for the most part, the effects of these environmental variables and the mechanisms by which they might increase jellyfish abundance remain to be quantified (but see, e.g., Hofmann & Crow, 2002).

Strong seasonality, whether temperate or monsoonal, seems likely to be the ultimate source of favorable, and unfavorable, conditions which provide resources for successful growth and reproduction during only a fraction of each year. This has the potential to compress life cycles, and thus life-history stages, into relatively discrete periods which enforce coordinated cohorts. Such environmental forcing is expected to affect the scyphozoan populations and also their prey. Unusually successful cohorts, i.e., population explosions or 'booms,' may therefore be explained by Hjort's "critical stage" (Ottersen & Loeng, 2000; Hjort, 1914; cited in Stenseth et al., 2002) or Cushing's (1975, 1983, 1984) "match/mismatch" hypotheses, wherein large cohorts of predators result when peaks in feeding larval stages coincide with abundant resources; annual blooms would result from good matches, and 'crashes' or 'busts' from mismatches.

Seasonal abundance may also be exaggerated over interannual timescales by the predominant 'color' of variation in coastal environments. Environmental variance in marine environments commonly occurs as 'red noise,' i.e., consecutive measurements made on short time scales, commonly perceived as 'ecological' timescales (although these are also microevolutionary timescales), are autocorrelated. Autocorrelation can lead to multiannual periods of 'good years' or 'bad years,' a pattern that is expected to lead to population 'booms' (or unusual 'blooms') and population 'busts' (Lawton, 1988; Cuddington & Yodzis, 1999; Brodeur et al., 2002) depending on the organisms' abilities to respond to such fluctuations. This is not the case for deepwater habitats, in which the magnitude of variation is less than that in shallow-water coastal environments.

## Summary of patterns

Mass occurrences of jellyfish involve a phylogenetic subset of Medusozoa. The current analyses of higher taxa are limited by small sample size and incomplete information for almost all taxa, therefore we do not attempt detailed statistical analyses which would have no power to establish phenotype-environment correlations. Rather, our six preliminary conclusions (below) represent hypotheses requiring further research at the species and genus levels that may allow identification of causal versus coincidental correlations. Some of these may not be independent of each other, but rather may positively feedback increasing the potential of certain species to aggregate, bloom, or swarm.

### 1. Character complexes

Multiple organismal traits contribute to the formation of aggregations, blooms, and swarms. Many of these traits appear to co-occur when their evolution is traced on the medusozoan tree of life, occasionally originating together independently in Hydrozoa and Scyphozoa. Subsets of these traits likely form 'complexes' of characters that cannot be separated functionally and are not evolutionarily independent. These include many anatomical attributes associated with large size, behavioral attributes of large size, and morphological traits affecting prey capture.

### 2. Phenotype-environment correlates

Multiple components of phenotype and environment must interact to cause mass occurrences of scyphomedusae. No single factor alone is sufficient because no species, nowhere, blooms all the time (to our knowledge). The deepwater ulmarids, for example, are large enough to be probably nektoplankton or planktonekton and have highly reticulated canal systems to distribute nutriment, yet they do not form blooms—possibly because of attributes associated with life in the deep sea including low fecundity, direct development, slow growth, a relatively stable environment, and perhaps few cues to promote aggregation. The kolpoform rhizostomes, which have many anatomical and distributional commonalities with the daktylioform rhisozostomes, do not bloom possibly because they are photosymbiotic. *Pelagia* may be the best bloomer in Pelagiidae because the canal pouch may be relatively efficient for medusae of its relatively small size, but not for many of the larger *Chrysaora* species. Probably the best-known examples of the multifaceted nature of mass occurrences are the many documented cases of aggregations resulting from the interaction of intrinsic behavior and the local hydrographic environment (e.g., Hamner et al., 1994; Graham et al., 2001; Rakow & Graham, 2006).

### 3. Asexual reproduction by polyps increases the potential for blooms of medusae

Podocyst formation, strobilation, and potentially other modes of asexual reproduction by polyps can greatly increase the potential size of any subsequent medusa population. Such potential is generated in the form of small life-history stages that generally go unnoticed and/or have little ecological impact thus emphasizing, by stark contrast, the apparent effect when medusae occur *en masse*. In particular, polydisc strobilation is associated with swarming. For example, *Aurelia* and Daktyliophorae engage in polydisc strobilation and are often blooming species; Kolpophorae do not engage in polydisc strobilation and are not blooming species (confoundingly, they are also photosymbiotic).

4. Large size is the principal organismal attribute leading to blooms and swarms

Small jellyfish (e.g., hydromedusae except *Aequorea*), or small life-history stages, cannot form blooms or swarms, except under exceptional circumstances, irrespective of whether they have other attributes that could promote occurrence *en masse*. Large size is important for medusae to

a. *overcome fluid dynamic constraints* and swim strongly enough to reaggregate behaviorally,
b. *have capacity to consume prey items with such rapidity that population biomass can increase exponentially* such that prey populations can be depleted and the medusa population will
c. *become hyper-fecund*, potentially increasing population size of small polyps dramatically.

5. Feeding strategy can enable or inhibit bloom formation

Jellyfish blooms are favored if medusae have feeding morphologies that enable them to take advantage of rapid increases and high abundances of prey and that, conversely, limit population size when target prey abundance is low. Thus, blooms are most likely to occur in species that

a. *are not photosymbiotic*, because zooxanthellae may physiologically constrain the host's ability to take advantage of rapid increases in prey availability, and
b. *specialize on lower trophic levels*, such as microzooplankton, in which energy is more abundant, more easily renewed, and more likely to have large variations in availability.

6. Habitat can enable or inhibit aggregation, bloom, or swarm formation

Local environmental conditions may ultimately govern whether a population of a particular species can bloom or not. For example, coastal environments provide pulsed seasonal resources and large amounts of habitat for benthic polyp stages; many embayments provide more placid hydrodynamic regimes in which even medium-sized prolate medusae may be able to aggregate behaviorally. There are insufficient resources in the deep sea to support the huge biomass associated with surface swarms, but long-lived *Periphylla periphylla* Péron & Lesueur can accumulate in Scandinavian fjords because medusae are not advected away and because food is more plentiful than in the deep sea.

## Synthesis and caveat

Jellyfish that aggregate, bloom, or swarm are clustered taxonomically and phylogenetically; the taxa that include blooming and swarming species are usually more diverse than their non-accumulating sister taxa (Hamner & Dawson, 2008). We have shown here that possession of a suite of life-history and phenotypic traits—podocysts, strobilation (particularly polydisc strobilation), ephyrae, circular and radial canals, large size, high fecundity, oral arms modified for suctorial feeding, fine filtering tentacles, high feeding capacity, lack of photosymbiosis, rapid growth—appears phylogenetically correlated with the species-rich clades that bloom or swarm. From a functional perspective, there is reason to expect that these traits confer the ability to occur *en masse* and some related evolutionary benefit (i.e., diversification). We have also outlined environmental situations—protected shallow-water coastal habitats—that most often facilitate or promote mass occurrences of jellyfishes (see also Omori & Nakano, 2001). Consequently, because mass occurrences of jellyfish that have occurred are largely explainable, we hypothesize that those yet to occur should be mostly predictable, assuming relatively consistent phenotype-environment interactions.

Unfortunately, our list of phenotype-environment correlations currently must be incomplete and somewhat poorly bounded due to the spatially and temporally haphazard historical record of accumulations (and their absences), phylogenetically incomplete sampling, and descriptions of species of medusae still lacking morphological, ecological, and physiological details. Our analyses therefore serve to constrain, rather than define, the suites of characters and conditions that favor jellyfish blooms. With that admonition in hand, we believe that the suites of permissive characters remain informative and intriguing and that the tools of evolutionary biology can provide novel insight into the occurrence of jellyfish blooms and swarms.

## Adaptive accumulations, changing circumstances, and exaptive excesses?

Important transitions in the evolution of jellyfish accumulations—from aggregations to blooms and swarms—coincide with the evolution of select life-history and morphological characters. These characters are, by way of correlation, inferred to have enabled blooming and swarming and are, because of the evolutionary success of the clades in which they occur, inferred to be adaptive, either individually or in concert. Are blooms also adaptive? That is, are blooms and swarms themselves adaptions, as opposed to simply the sum of their component parts? And are recent unusual or unnatural blooms adaptive, or perhaps exaptive?

Many medusozoans differ from most other marine animals (e.g., species of bivalves, sea urchins, reef fishes) because the large sexually reproducing stage, rather than the tiny larval stage, has long pelagic duration. The larval and subsequent early developmental stages are mostly benthic in cubozoans and scyphozoans but mostly pelagic in other non-brooding invertebrates and vertebrates. Thus, although the cubozoans, scyphozoans, marine invertebrate broadcast spawners, and many reef fishes liberate massive numbers of very tiny individuals (cubomedusae, ephyrae, or larvae) into the water column, their life-history strategies subsequently diverge. The cubozoans grow and ephyrae metamorphose into planktonic juvenile medusae which remain in the water column through reproduction and senescence; in contrast, the larvae of other taxa metamorphose into late-stage larvae or juveniles and settle quickly into benthic communities. These two groups of tiny plankton thus have different selective forces to navigate if they are to meet demographic and evolutionary metrics of success, i.e., contributing offspring with their alleles to the next generation. The problem for medusae is to stay with their conspecifics, oftentimes near the location in which they were spawned, whereas the problem for other larvae is to find conspecifics in distant places despite the diluting effect of oceanic long-distance dispersal. While both may benefit from some kind of gregarious behavior, which has been demonstrated abundantly in planktonic dispersers such as barnacles, oysters, and tubeworms (Toonen & Pawlik, 2001; Dreanno et al., 2006; Tamburri et al., 2007) but for which there is still no convincing evidence of full or half sibs settling together distant from the parental location (Veliz et al., 2006), it is clear that medusae would not require gregariousness *per se* if they shared an appropriate suite of favorable life-history and behavioral traits. If all medusae in a demographic population started their planktonic life as ephyrae in the same approximate location at the same approximate time, a suite of shared responses to local environmental stimuli could lead, without any intervening interaction between individuals, a reasonable proportion of the population to still be in relatively close proximity at maturation. That said, a plausible mechanism is still required through which natural selection could maintain coordinated life histories and behaviors. While traits that enable medusae to originate in a short period of time, avoid advection away from the population (excepting those fortunate enough to be advected to another population elsewhere), avoid predation, and find and capture and consume food, all have proximate causes and immediate benefits to the individual (e.g., temporal 'matching' with food supply, retention in areas proven favorable for parents [not necessarily medusae in the preceding year]), they can never sum to more than growth and survival of individuals if the medusae do not subsequently encounter mates and reproduce sexually. In contrast, the medusae that do find mates can reproduce, pass on alleles that enhance mass occurrence, and have offspring that are 'fitter' than average and which in turn are also more likely to reproduce. Thus, we hypothesize that blooming and swarming are key adaptations that exert strong selective pressure and secure together the many component traits into one complex adaptation.

The evolution of aggregation, blooming, and swarming is thus apparently an emergent property subsequently favored by natural selection, a 'key innovation' (Donoghue, 2005). The many component traits were already adaptations in their own right for, say, increased feeding potential, increased fecundity, avoidance of advection, etc., before they also became exaptations for mass occurrence resulting from selection for breeding accumulations. An unusual example may illustrate this point. In Palau, *Mastigias* medusae in ancestral lagoonal populations and derived lake populations exhibit diel horizontal migrations that result in accumulations in all locations during the early morning and late afternoon. These accumulations could be reproductive swarms, but in fact result from selection for patterns of migration that maximize exposure of their photosymbionts to sunlight or minimize deaths caused by predatory anemones (Hamner & Hauri, 1981; Dawson & Hamner, 2003). The evolution of these accumulations is therefore not selection for reproduction, but a consequence of accumulation may be increased reproductive success.

Therefore, it is possible that for semaeostomes the evolutionarily basal and predominantly open coastal and oceanic pelagiids may occur *en masse* through a different balance of mechanisms under different selective pressures than those favoring mass occurrence of medusae such as *Aurelia* in semi-enclosed coastal environments. For example, *Chrysaora* can occur at high densities (16 $m^{-3}$) in narrow salinity and temperature regimes in the Chesapeake Bay (Decker et al., 2007) but are typically more disperse in open coastal environments (e.g., 1 to 100 per 10,000 $m^{-2}$, Doyle et al., 2007; also K. Bayha, pers. comm.), so *Chrysaora* may aggregate in suitable habitat in highly structured coastal waters (a pattern exaggerated by death if advected into neighboring unsuitable habitat), or bloom in coastal waters due to high rates of strobilation and dense or expansive polyp fields, but in neither case truly swarm. Similarly, *Pelagia* in the open ocean among West Papua's islands were observed over a spatial extent of at least 40 km on several different occasions spanning several weeks, but never in densities of more than a few individuals per cubic meter except where they were concentrated by local topography (M. Ammer, M. Dawson, B. Hoeksema, L. Martin, pers. obs.).

Thus, although we identified three classes of mass occurrence—aggregation, blooming, and swarming—at the beginning or our analyses (Hamner & Dawson, 2008), we recognize that these may be artificially discrete classifications of a character complex whose states are continuously distributed. Coarsely similar accumulations could result from convergence of different evolutionary pathways (see discussion in Donoghue, 2005), which should be apparent from detailed analyses of factors contributing to particular mass occurrences. Thus phylogenetic analyses of adaptations, exaptations, and key innovations require nuanced analyses of rich datasets that consider changes that accumulate across a nested series of nodes in the evolutionary tree (Donoghue, 2005). The first step toward this goal is a more explicit empirical approach to describing factors influencing, or potentially influencing, jellyfish accumulations, including places of origin, places of occurrence, physical environment, and life history. Also needed are more explicit descriptions of blooms themselves in terms of number and biomass, perhaps based on rates of increase relative to initial conditions and long-term population dynamics. This will not only enable more robust phylogenetic macroevolutionary analyses, but also improve microevolutionary studies of mass accumulations through detailed comparisons of potentially closely related populations (e.g., Hamner et al., 1994; Lucas, 2001; Albert, 2005, 2007, 2008).

Our inferences suggest an evolutionary explanation for why "species that have always been present suddenly experience severe increases or 'blooms', often with little evidence of what caused the population increase" (Mills, 2001). Simply, they have sufficient requisite exaptations for blooming such that, when environmental circumstances change to promote blooms, they are able to take advantage of the new conditions. Thus, over-fishing and eutrophication, which may lead to increases of microzooplankton populations, can cascade into jellyfish blooms; introduced species that can compete well for or exploit poorly used but abundant resources in the invaded location have increased potential to bloom; changed environmental conditions due to global warming, such as altered phenology or hydrography, may make new resources available or enable aggregations in new places and times.

The applied question that interests us is whether such changes will predictably lead to blooms. The predictions that attract most attention generally project oceans dominated by jellyfish (e.g., Dybas, 2002; http://www.shiftingbaselines.org/videos/index.html). However, as described by Mills (2001) and Purcell (2005), and as evident in our analyses, the answer to this question depends on the species, environment, and changes concerned. For example, if the local environment changes so that food availability is increased and relatively strongly pulsed in spring—for example, through increased rainfall, run-off, and nutrient enrichment—then we expect this to provide the opportunity for blooms to form. But, blooms will only form if there is a jellyfish species present that has the morphological and behavioral characteristics that enable it to take advantage of those resources such that many small ephyrae can grow rapidly to large size. In a reasonable proportion of cases, we can expect phenological changes to disfavor populations that already exploit pulsed resources. Hjort's (1914) first feeding and Cushing's (1975, 1983, 1984) match/mismatch hypotheses, wherein large cohorts of predators result when peaks in predator and prey populations coincide, cut both ways in times of environmental change.

## Closing remark

An important point worth emphasis is that our discussion so far has focused on the ecological responses of jellyfish to environmental change due to previously evolved characteristics. Yet many jellyfish species likely have considerable diversity within the very large populations of which they are comprised and therefore should have considerable potential for evolution when stabilizing selection is disrupted or redirected. Consequently, it is also important to consider what future evolutionary consequences may be caused by environmental change. We currently have no knowledge of the strength of selection on any traits, or of how the strength or direction of selection may be altered by changed conditions. Consequently, we believe that descriptive research must be complimented by experimentation, including the standard tools of evolutionary ecology and ecological genetics such as reciprocal transplants (when ethically appropriate), 'common garden' experiments, and field-based or laboratory-based manipulations. Such experiments may be guided by the many models of climate change now available that describe possible ranges, and likely trajectories, of environmental perturbations, at least on regional scales although local changes may vary.

Some data are available in the literature, and abundant data are collectable from nature, that could be used to corroborate laboratory (or field) experiments. Unfortunately, these data are often sporadic, as illustrated by our review (Hamner & Dawson, 2008), and for most taxa most data simply are not yet available. A global perspective on jellyfish ecology and evolution, of the kind only achievable through concerted coordinated effort, is required to rapidly reduce deficits in our knowledge of and ability to explain jellyfish blooms (and aggregations and swarms). Initiatives for regional and global programs focusing on the ecology of diverse scyphozoan taxa, identified accurately and placed phylogenetically, seem an essential next step to meet current needs for increased understanding of the past and future evolution of jellyfish blooms.

**Acknowledgments** Detailed critiques by M. Arai, P. Hamner, L. Martin, K. Pitt, J. Purcell, H. Swift, and two anonymous reviewers, and discussion with K. Bayha, L. Gomez Daglio, J. Lehman, and J. Vo helped improve and organize the content and presentation of this manuscript. WMH thanks the organizers of the 2nd International Jellyfish Blooms Symposium, K. Pitt and J. Seymour, for the invitation to present a plenary address at the meeting, and we thank the Editors for their patience and industry in preparing this volume. Allen G. Collins kindly provided the datamatrix for analyses in Fig. 2. This work was supported in part by grant DEB-0717078 from the US National Science Foundation to MND and AGC.

## References

Albert, D. J., 2005. Reproduction and longevity of *Aurelia labiata* in Roscoe Bay, a small bay on the Pacific coast of Canada. Journal of the Marine Biological Association of the United Kingdom 85: 575–581.

Albert, D. J., 2007. *Aurelia labiata* medusae (Scyphozoa) in Roscoe Bay avoid tidal dispersion by vertical migration. Journal of Sea Research 57: 281–287.

Albert, D. J., 2008. Adaptive behaviours of the jellyfish *Aurelia labiata* in Roscoe Bay on the west coast of Canada. Journal of Sea Research 59: 198–201.

Aleyev, Yu. G., 1977. Nekton. Dr. W. Junk, The Hague, vi + 435p

Alldredge, A. L., 1984. The quantitative significance of gelatinous zooplankton as pelagic consumers. In Fasham, M. J. R. (ed.), Flows of Energy and Materials in Marine Ecosystems: Theory and Practice. Plenum, New York: 407–433.

Arai, M. N., 1992. Active and passive factors affecting aggregations of hydromedusae: a review. Scientia Marina 56: 99–108.

Arai, M. N., 1997. A Functional Biology of Scyphozoa. Chapman and Hall, London.

Arai, M. N., 2001. Pelagic coelenterates and eutrophication: a review. Hydrobiologia 451 (Developments in Hydrobiology) 155: 69–87.

Arai, M. N. & I. M. Chan, 1989. Two types of excretory pores in the hydrozoan medusa *Aequorea victoria* (Murbach and Shearer, 1902). Journal of Plankton Research 11: 609–614.

Armbruster, W. S., 1992. Phylogeny and the evolution of plant-animal interactions. BioScience 42: 12–20.

Bailey, K. M. & R. S. Batty, 1983. Laboratory study of predation by *Aurelia aurita* on larvae of cod, flounder, plaice and herring: development and vulnerability to capture. Marine Biology 83: 287–291.

Baker, A. C., 2003. Flexibility and specificity in coral-algal symbiosis: diversity, ecology, and biogeography of *Symbiodinium*. Annual Reviews in Ecology Evolution and Systematics 34: 661–689.

Behmer, S. P. & A. Joern, 2008. Coexisting generalist herbivores occupy unique nutritional feeding niches. Proceedings of the National Academy of Sciences of the USA 105: 1977–1982.

Bolton, T. F. & W. M. Graham, 2004. Morphological variation among populations of an invasive jellyfish. Marine Ecology Progress Series 278: 125–139.

Bonsall, M. B. & M. P. Hassell, 1997. Apparent competition structures ecological assemblages. Nature 388: 371–373.

Bouillon, J., 1999. Hydromedusae. In Boltovskoy, D. (ed.), South Atlantic Zooplankton. Backhuys, Leiden: 385–465.

Brodeur, R. D., H. Sugisaki & G. L. Hunt Jr, 2002. Increases in jellyfish biomass in the Bering Sea: implications for the ecosystem. Marine Ecology Progress Series 233: 89–103.

Brooks, D. R. & D. A. McClennan, 2002. The nature of diversity–an evolutionary voyage of discovery. University of Chicago Press, Chicago.

Cartwright, P., S. L. Halgedahl, J. R. Hendricks, R. D. Jarrard, A. C. Marques, A. G. Collins & B. S. Lieberman, 2007. Exceptionally preserved jellyfishes from the Middle Cambrian. PLoS ONE 2(10): e1121.

Chapman, G., 1966. The structure and function of the mesoglea. In Rees, W. J. (ed.), The Cnidaria and their Evolution. Academic Press, London: 147–168.

Colin, S. P. & J. H. Costello, 2002. Morphology, swimming performance and propulsive mode of six co-occurring hydromedusae. The Journal of Experimental Biology 205: 427–437.

Colin, S. P., J. H. Costello & H. Kordula, 2006. Upstream foraging by medusae. Marine Ecology Progress Series 327: 143–155.

Collins, A. G., 2002. Phylogeny of Medusozoa and the evolution of cnidarian life cycles. Journal of Evolutionary Biology 15: 418–432.

Collins, A. G. P., Marques A. C. Schuchert, T. Jankowski, M. Medina & B. Schierwater, 2006. Medusozoan phylogeny and character evolution clarified by new large and small subunit rDNA data and an assessment of the utility of phylogenetic mixture models. Systematic Biology 55: 97–115.

Costello, J. H. & S. P. Colin, 1995. Flow and feeding by swimming scyphomedusae. Marine Biology 124: 399–406.

Cuddington, K. M. & P. Yodzis, 1999. Black noise and population persistence. Proceedings of the Royal Society of London B 266: 969–973.

Cushing, D. H., 1975. The natural mortality of the plaice. Journal du Conseil International pour l'Exploration de la Mer 36: 150–157.

Cushing, D. H., 1983. Are fish larvae too dilute to affect the density of their food organisms? Journal of Plankton Research 5: 847–854.

Cushing, D. H., 1984. The gadoid outburst in the North Sea. Journal du Conseil International pour l'Exploration de la Mer 41: 159–166.

Dabiri, J. O., S. P. Colin & J. H. Costello, 2007. Morphological diversity of medusan lineages constrained by animal-fluid interactions. The Journal of Experimental Biology 210: 1868–1873.

D'Ambra, I., J. H. Costello & F. Bentivegna, 2001. Flow and prey capture by the scyphomedusa *Phyllorhiza punctata* von Lendenfeld, 1884. Hydrobiologia 451: 223–227.

Darwin, C. R., 1874. The descent of man, and selection in relation to sex, 2nd ed. Appleton, New York.

Daryanabard, R. & M. N. Dawson, 2008. Jellyfish blooms: *Crambionella orsini* (Scyphozoa, Rhizostomeae) in the Gulf of Oman, Iran, 2002–2003. Journal of the Marine Biological Association of the UK 88: 477–483.

Dawson, M. N., 2004. Some implications of molecular phylogenetics for understanding biodiversity in jellyfishes, with emphasis on Scyphozoa. Hydrobiologia 530(531): 249–260.

Dawson, M. N. & W. M. Hamner, 2003. Geographic variation and behavioral evolution in marine plankton: the case of *Mastigias* (Scyphozoa: Rhizostomeae). Marine Biology 143: 1161–1174.

Dawson, M. N. & W. M. Hamner, 2008. A biophysical perspective on dispersal and the geography of evolution in marine and terrestrial systems. Journal of the Royal Society Interface 5: 135–150.

Dawson, M. N., L. E. Martin & L. K. Penland, 2001. Jellyfish swarms, tourists, and the Christ-child. Hydrobiologia 451 (Developments in Hydrobiology) 155: 131–144.

deBeer, G. R. & J. S. Huxley, 1924. Studies in dedifferentiation. V. Dedifferentiation and reduction in Aurelia. Quarterly Journal of Microscopical Science 68: 471–479.

Decker, M. B., C. W. Brown, R. R. Hood, J. E. Purcell, T. F. Gross, J. C. Matanoski, R. O. Bannon & E. M. Setzler-Hamilton, 2007. Predicting the distribution of the scyphomedusa *Chrysaora quinquecirrha* in Chesapeake Bay. Marine Ecology Progress Series 329: 99–113.

de Lafontaine, Y. & W. C. Leggett, 1987. Effect of container size on estimates of mortality and predation rates in experiments with macrozooplankton and larval fish. Canadian Journal of Fisheries and Aquatic Sciences 44: 1534–1543.

Denton, E. J. & T. I. Shaw, 1962. The buoyancy of gelatinous marine animals. Journal of Physiology 161: 14P–15P.

de Queiroz, K., 1996. Including the characters of interest during tree reconstruction and the problem of circularity and bias in studies of character evolution. American Naturalist 148: 700–708.

Donoghue, M. J., 2005. Key innovations, convergence, and success: macroevolutionary lessons from plant phylogeny. Paleobiology 31: 77–93.

Doyle, T. K., J. D. R. Houghton, S. M. Buckley, G. C. Hays & J. Davenport, 2007. The broad-scale distribution of five jellyfish species across a temperate coastal environment. Hydrobiologia 579: 29–39.

Dreanno, C., K. Matsumura, N. Dohmae, K. Takio, H. Hirota, R. R. Kirby & A. S. Clare, 2006. An alpha(2)-macroglobulin-like protein is the cue to gregarious settlement of the barnacle *Balanus amphitrite*. Proceedings of the National Academy of Sciences of the USA 103: 14396–14401.

Dybas, C. L., 2002. Jellyfish 'blooms' could be sign of ailing seas. Washington Post 06 May 2002: A09.

Fautin, D. G. & R. W. Buddemeier, 2004. Adaptive bleaching: a general phenomenon. Hydrobiologia 530(531): 459–467.

García, J. R. & E. Durbin, 1993. Zooplanktivorous predation by large scyphomedusae *Phyllorhiza punctata* (Cnidaria: Scyphozoa) in Laguna Joyuda. Journal of Experimental Marine Biology and Ecology 173: 71–93.

Gershwin, L. & A. G. Collins, 2002. A preliminary phylogeny of Pelagiidae (Cnidaria, Scyphozoa), with new observations of *Chrysaora colorata* comb. nov. Journal of Natural History 36: 127–148.

Gilbert, J. J. & C. E. Williamson, 1983. Sexual dimorphism in zooplankton (Copepoda, Cladocera, and Rotifera). Annual Review of Ecology and Systematics 14: 1–33.

Gotelli, N. J., 2001. A primer of ecology, 3rd ed. Sinauer, Sunderland.

Goulet, T. L., 2006. Most corals may not change their symbionts. Marine Ecology Progress Series 321: 1–7.

Graham, W. M., D. L. Martin, D. L. Felder, V. L. Asper & H. M. Perry, 2003. Ecological and economic implications of a tropical jellyfish invader in the Gulf of Mexico. Biological Invasions 5: 53–69.

Graham, W. M., F. Pagès & W. M. Hamner, 2001. A physical context for gelatinous zooplankton aggregations: a review. Hydrobiologia 451 (Developments in Hydrobiology) 155: 199–212.

Hamner, W. M & M. N Dawson, 2008. A systematic review of the evolution of jellyfish blooms: advantageous aggregations and adaptive assemblages. Hydrobiologia. doi: 10.1007/s10750-008-9620-9.

Hamner, W. M., P. P. Hamner & S. W. Strand, 1994. Sun compass migration by *Aurelia aurita* (Scyphozoa): population persistence versus dispersal in Saanich Inlet, British Columbia. Marine Biology 119: 347–356.

Hamner, W. M. & I. R. Hauri, 1981. Long-distance horizontal migrations of zooplankton (Scyphomedusae: *Mastigias*). Limnology and Oceanography 26: 414–423.

Hamner, W. M. & R. M. Jenssen, 1974. Growth, degrowth, and irreversible cell differentiation in *Aurelia aurita*. American Zoologist 14: 833–849.

Hamner, W. M., M. S. Jones & P. P. Hamner, 1995. Swimming, feeding, circulation, and vision in the Australian box jellyfish, *Chironex fleckeri* (Cnidaria; Cubozoa). Marine and Freshwater Research 46: 985–990.

Hanski, I., L. Hansson & H. Henttonen, 1991. Specialist predators, generalist predators, and the microtine rodent cycle. Journal of Animal Ecology 60: 353–367.

Hansson, L. J., 1997. Capture and digestion of the scyphozoan jellyfish *Aurelia aurita* by *Cyanea capillata* and prey response to predator contact. Journal of Plankton Research 19: 195–208.

Hjort, J., 1914. Fluctuations in the great fisheries of Northern Europe viewed in light of biological research. Rapports et Procès-Verbeaux des Réunions du Conseil International pour l'Exploration de la Mer 20: 1–228.

Hofmann, D. K. & G. Crow, 2002. Induction of larval metamorphosis in the tropical scyphozoan *Mastigias papua*: striking similarity with upside down-jellyfish *Cassiopea* spp. (with notes on related species). Vie et Milieu 52: 141–147.

Holland, B. S., M. N. Dawson, G. L. Crow & D. K. Hofmann, 2004. Global phylogeography of *Cassiopea* (Scyphozoa: Rhizostomae): Molecular evidence for cryptic species and multiple Hawaiian invasions. Marine Biology 145: 1119–1128.

Holst, S., I. Sötje, H. Tiemann & G. Jarms, 2007. Life cycle of the rhizostome jellyfish *Rhizostoma octopus* (L.) (Scyphozoa, Rhizostomeae), with studies on cnidocysts and statoliths. Marine Biology 151: 1695–1710.

Huang, M., J. Hu & Y. Wang, 1985. Preliminary study on the breeding habits of edible jellyfish in Hangzhou Wan Bay. Journal of Fisheries of China 9: 239–246. In Chinese; English abstract.

Hunt, J. C. & D. J. Lindsay, 1998. Observations on the behavior of *Atolla* (Scyphozoa: Coronatae) and *Nanomia* (Hydrozoa: Physonectae): use of the hypertrophied tentacle in prey capture. Plankton Biology and Ecology 45: 239–242.

Hyman, L. H., 1940. Observations and experiments on the physiology of medusae. Biological Bulletin (Woods Hole) 79: 282–296.

Ives, A. R., Á. Einarsson, V. A. A. Jansen & A. Gardarsson, 2008. High-amplitude fluctuations and alternative dynamical states of midges in Lake Myvatn. Nature 452: 84–87.

Jarms, G., H. Tiemann & U. Båmstedt, 2002. Development and biology of *Periphylla periphylla* (Scyphozoa: Coronatae) in a Norwegian fjord. Marine Biology 141: 647–657.

Kooi, B. W., L. D. J. Kuiper & S. A. L. M. Kooijman, 2004. Consequences of symbiosis for food web dynamics. Journal of Mathematical Biology 49: 227–271.

Kramp, P. L., 1961. Synopsis of the medusae of the world. Journal of the Marine Biological Association of the United Kingdom 40: 1–469.

Larson, R. J., 1979. Feeding in coronate medusa (Class Scyphozoa, Order Coronatae). Marine Behaviour and Physiology 6: 123–129.

Larson, R. J., 1986. Pelagic scyphomedusae (Scyphozoa: Coronatae and Semaeostomeae) of the Southern Ocean. Biology of the Antarctic Seas. Antarctic Research Series 41: 59–165.

Larson, R. J., 1992. Riding Langmuir circulations and swimming in circles: a novel form of clustering behavior by the scyphomedusa *Linuche unguiculata*. Marine Biology 112: 229–235.

Lawton, J., 1988. More time means more variation. Nature 334: 563.
Lucas, C. H., 2001. Reproduction and life history strategies of the common jellyfish, *Aurelia aurita*, in relation to its ambient environment. Hydrobiologia 451: 229–246.
Lucas, C. H. & S. Lawes, 1998. Sexual reproduction of the scyphomedusa *Aurelia aurita* in relation to temperature and variable food supply. Marine Biology 131: 629–638.
Luckow, M. & A. Bruneau, 1997. Circularity and independence in phylogenetic tests of ecological hypotheses. Cladistics 13: 145–151.
Maddison, W. P. & D. R. Maddison, 1989. Interactive analysis of phylogeny and character evolution using the computer program MacClade. Folia Primatologica (Basel) 53: 190–202.
Marques, A. C. & A. G. Collins, 2004. Cladistic analysis of Medusozoa and cnidarian evolution. Invertebrate Biology 123: 23–42.
Martin, L. E., 1999. The Population Biology and Ecology of *Aurelia* sp. (Scyphozoa: Semaeostomeae) in a Tropical Meromictic Marine lake in Palau, Micronesia. Ph.D. thesis, University of California, Los Angeles: 250 pp.
Mayer, A. G., 1910. Medusae of the World, III: the Scyphomedusae. Carnegie Institute, Washington.
Mills, C. E., 2001. Jellyfish blooms: are populations increasing globally in response to changing ocean conditions? Hydrobiologia 451 (Developments in Hydrobiology) 155: 55–68.
Morris, A. K., 2006, Zooplankton Aggregations in California Coastal Zones, Ph.D. thesis, University of California, Los Angeles: 309 pp.
Omori, M. & E. Nakano, 2001. Jellyfish fisheries in southeast Asia. Hydrobiologia 451 (Developments in Hydrobiology) 155: 19–26.
Ottersen, G. & H. Loeng, 2000. Covariability in early growth and year-class strength of Barents Sea cod, haddock, and herring: the environmental link. ICES Journal of Marine Science 57: 339–348.
Peach, M. B. & K. A. Pitt, 2005. Morphology of the nematocysts of the medusae of two scyphozoans, *Catostylus mosaicus* and *Phyllorhiza punctata* (Rhizostomae): implications for capture of prey. Invertebrate Biology 124: 98–108.
Pennak, R. W., 1956. The fresh-water jellyfish *Craspedacusta* in Colorado with some remarks on its ecology and morphological degenerataion. Transactions of the American Microscopical Society 75: 324–331.
Pitt, K. A. & M. J. Kingsford, 2000. Reproductive biology of the edible jellyfish *Catostylus mosaicus* (Rhizostomeae). Marine Biology 137: 791–799.
Purcell, J. E., 1989. Predation on fish larvae and eggs by the hydromedusa Aequorea victorea at a herring spawning ground in British Columbia. Canadian Journal of Fisheries and Aquatic Sciences 46: 1415–1427.
Purcell, J. E., 2005. Climate effects on formation of jellyfish and ctenophore blooms. Journal of the Marine Biological Association of the United Kingdom 85: 461–476.
Purcell, J. E., U. Bamstedt & A. Bamstedt, 1999. Prey, feeding rates, and asexual reproduction rates of the introduced oligohaline hydrozoan *Moerisia lyonsi*. Marine Biology 134: 317–325.
Purcell, J. E., S. Uye & W.-T. Lo, 2007. Anthropogenic causes of jellyfish blooms and their direct consequences for humans: a review. Marine Ecology Progress Series 350: 153–174.
Rakow, K. C. & W. M. Graham, 2006. Orientation and swimming mechanics by the scyphomedusa *Aurelia* sp. in shear flow. Limnology and Oceanography 51: 1097–1106.
Ricklefs, R. E. & G. L. Miller, 1999. Ecology, 4th ed. W. H. Freeman, New York.
Ronquist, F., 2004. Bayesian inference of character evolution. Trends in Ecology and Evolution 19: 475–481.
Rottini Sandrini, L. & M. Avian, 1983. Biological cycle of *Pelagia noctiluca:* morphological aspects of the development from planula to ephyra. Marine Biology 74: 169–174.
Ruppert, E. E. & K. J. Carle, 1983. Morphology of metazoan circulatory systems. Zoomorphology 103: 193–208.
Russell, F. S., 1970. The Medusae of the British Isles. II Pelagic Scyphozoa with a Supplement to the First Volume on Hydromedusae, Cambridge University Press, Cambridge.
Schneider, G., 1988. Larvae production of the common jellyfish *Aurelia aurita* in the Western Baltic 1982–1984. Kieler Meeresforschungen 6: 295–300.
Seipel, K. & V. Schmid, 2006. Mesodermal anatomies in cnidarian polyps and medusae. International Journal of Developmental Biology 50: 589–599.
Shanks, A. L. & W. M. Graham, 1987. Orientated swimming in the jellyfish *Stomolophus meleagris* L. Agassiz (Scyphozoan: Rhizostomida). Journal of Experimental Marine Biology and Ecology 108: 159–169.
Shine, R., 1988. The evolution of large body size in females: a critique of Darwin's "fecundity advantage" model. The American Naturalist 131: 124–131.
Shine, R., 1989. Ecological causes for the evolution of sexual dimorphism: a review of the evidence. The Quarterly Review of Biology 64: 419–461.
Sinclair, M., 1988. Marine Populations. Washington Sea Grant Program, Seattle.
Stenseth, N. C., A. Mysterud, G. Ottersen, J. W. Hurrell, K.-S. Chan & M. Lima, 2002. Ecological effects of climate fluctuations. Science 297: 1292–1296.
Stiasny, G., 1921. Studien über rhizostomeen. In van Oort, E. D. (ed.), Capita Zoologica. Martinus Njhoff, Gravenhage.
Stiasny, G. & H. van der Maaden, 1943. Über scyphomedusen aus dem Ochotskishen und Kamtschatka Meer nebst einer kritik der Genera Cyanea und Desmonema. Zoologische Jahrbücher Abteilung für Systematik 76: 227–266.
Straehler-Pohl, I. & G. Jarms, 2005. Life cycle of *Carybdea marsupialis* Linnaeus, 1758 (Cubozoa, Carybdeidae) reveals metamorphosis to be a modified strobilation. Marine Biology 147: 1271–1277.
Strand, S. W. & W. M. Hamner, 1988. Predatory behavior of *Phacellophora camtschatica* and size-selective predation upon *Aurelia aurita* (Scyphozoa: Cnidaria) in Saanich Inlet, British Columbia. Marine Biology 99: 409–414.
Sugiura, Y., 1964. On the life-history of rhizostome medusae. II. Indispensability of zooxanthellae for strobilation in *Mastigias papua*. Embryologia 8: 223–233.
Tamburri, M. N., R. K. Zimmer & C. A. Zimmer, 2007. Mechanisms reconciling gregarious larval settlement with adult cannibalism. Ecological Monographs 77: 255–268.

Tanner, J. E., 2002. Consequences of density-dependent heterotrophic feeding for a partial autotroph. Marine Ecology Progress Series 227: 293–304.

Thompson, J. N., S. L. Nuismer & R. Gomulkiewicz, 2002. Coevolution and maladaptation. Integrative and Comparative Biology 42: 381–387.

Thuesen, E. V., L. D. Rutherford Jr., P. L. Brommer, K. Garrison, M. A. Gutowska & T. Towanda, 2005. Intragel oxygen promotes hypoxia tolerance of scyphomedusae. Journal of Experimental Biology 208: 2475–2482.

Toonen, R. J. & J. R. Pawlik, 2001. Foundations of gregariousness: a dispersal polymorphism among the planktonic larvae of a marine invertebrate. Evolution 55: 2439–2454.

van Iten, H., J. Moraes Leme, M. G. Simões, A. C. Marques & A. G. Collins, 2006. Reassessment of the phylogenetic position of conulariids (?Ediacaran–Triassic) within the subphylum Medusozoa (phylum Cnidaria). Journal of Systematic Palaeontology 4: 109–118.

Veliz, D., P. Duchesne, E. Bourget & L. Bernatchez, 2006. Genetic evidence for kin aggregation in the intertidal acorn barnacle (*Semibalanus balanoides*). Molecular Ecology 15: 4193–4202.

Watanabe, T. & H. Ishii, 2001. *In situ* estimation of ephyrae liberated from polyps of *Aurelia aurita* using settling plates in Tokyo Bay, Japan. Hydrobiologia 451 (Developments in Hydrobiology) 155: 247–258.

Werner, B., 1973. New investigations on systematics and evolution of the class Scyphozoa and the phylum Cnidaria. Publications of the Seto Marine Biological Laboratory 20: 35–61.

Zrzavy, J., 1997. Phylogenetics and ecology: all characters should be included in the cladistic analysis. Oikos 80: 186–192.

Hydrobiologia (2009) 616:217–228
DOI 10.1007/s10750-008-9590-y

JELLYFISH BLOOMS

# A new Taqman© PCR-based method for the detection and identification of scyphozoan jellyfish polyps

Keith M. Bayha · William M. Graham

Published online: 23 September 2008


**Abstract** While blooms of large scyphomedusae and cubomedusae receive most public attention owing to effects on tourism (e.g., stinging swimmers), commerce, and fisheries, relatively little attention is given to the inconspicuous benthic polypoid stage. This is particularly troubling when considering the widespread translocation of some invasive marine jellyfish. The transport of benthic polyps (via ships, barges, and offshore drilling platforms) is theorized to be the most likely way in which invasive jellies are globally transported. Yet given the extremely small size and cryptic nature of most benthic polyps, identifying and tracking them in the field amongst the larger communities of fouling organisms is extremely difficult. To this end, we have developed a rapid molecular assay for detecting benthic jellyfish polyps from three scyphozoan genera in the Gulf of Mexico. One of these (*Phyllorhiza* spp.) is an invasive scyphozoan established in the Gulf of Mexico and is theorized to have been spread worldwide as a fouling organism on the hulls of cargo ships, while the other two (U.S. *Chrysaora* sp. and Gulf of Mexico *Aurelia* spp.) are local blooming animals that have shown recent numerical increases in the Gulf of Mexico. This method involves a multiplex Real-Time Polymerase Chain Reaction (PCR) assay using Taqman© probes that can be run on DNA extracted from whole-community scrapings of benthic surfaces, such as boat hulls, dock pilings, oilrigs, and settling plates. Specificity tests indicated that all Taqman© probes were successful against all individuals of target taxa, but not against 17 non-target local and worldwide scyphozoan and hydrozoan species. Tests showed all probes to be extremely sensitive, reacting to as few as 50 copies of template DNA, with one (*Chrysaora* sp.) reacting to as few as 10 copies. The assay correctly identified individual polyps of *Aurelia* sp. and *Chrysaora* sp. The use of this Taqman© assay on tissue collected from whole benthic scrapings should allow screening of incoming ships to the Gulf of Mexico for the invasive *P. punctata,* and locating and studying the cryptic benthic stages of northern Gulf of Mexico jellyfish, which will lead to a better understanding of the overall population distribution and bloom dynamics of medusae.

**Keywords** *Aurelia* · *Phyllorhiza punctata* · *Chrysaora* · Scyphistoma · Genetics

Guest editors: K. A. Pitt & J. E. Purcell
Jellyfish Blooms: Causes, Consequences, and Recent Advances

K. M. Bayha (✉)
University of California, Merced, 5200 North Lake Rd., Merced, CA 95348, USA
e-mail: kbayha@ucmerced.edu

W. M. Graham
Dauphin Island Sea Lab, 101 Bienville Blvd., Dauphin Island, AL 36528, USA

## Introduction

Large, conspicuous medusae (particularly scyphomedusae and cubomedusae) garner significant public attention owing to the impacts that these animals have on tourism, fisheries, and commerce (reviewed by Mills, 2001; Brodeur et al., 2002; Kawahara et al., 2006). Beginning in the 1980s with the introduction of the ctenophore *Mnemiopsis leidyi* (Agassiz, 1865) into the Black Sea (Vinogradov et al., 1989; reviewed by Kideys, 2002), gelatinous zooplankton gained further notoriety as potentially devastating marine invaders. While a modest number of invasive jellyfish species have been documented (reviewed by Graham & Bayha, 2007), the actual rate of invasion is likely much higher due to the cryptic nature of invasions and to the taxonomic vagaries that may only be resolved by molecular genetics (Holland, 2000; Dawson & Jacobs, 2001; Holland et al., 2004; Bayha, 2005; Graham & Bayha, 2007).

The majority of scyphozoan jellyfish exhibit a complex bipartite life history with a small benthic, asexually reproducing polypoid (=scyphistoma) stage alternating with a pelagic, sexually reproducing medusa stage (Arai, 1997; Schiariti et al., this volume). Most bioinvasion studies, or general jellyfish ecological investigations for that matter, have focused on the medusa stage of scyphozoans. In contrast, comparatively little work has addressed the benthic polyp despite the obvious role polyp populations have in medusae blooms. The presence of polyp populations in either native or novel environments is extremely difficult to assess owing to their small and cryptic nature. Even if these minute life stages can be detected, visual identification of polyps is often nearly impossible. This is further complicated in dispersive open shelf ecosystems as opposed to retentive enclosed or semi-enclosed systems, such as bays and estuaries, where the majority of field polyp studies have been done (e.g., Cargo & Schultz, 1966; Colin & Kremer, 2002; Miyake et al., 2002).

Among marine ecologists, the medusa stage of scyphozoans and cubozoans is equated with dispersal and the polyp stage is equated with retention. While this may be the case for local dispersal influenced only by physical transport (Bayha, 2005; Dawson, 2005; Dawson et al., 2005), it is likely that the polyp stage, in association with fouling communities on ships, mobile oil platforms, or the aquarium trade, is the primary vector for invasions. One species that is thought to have been transported in this way is *Phyllorhiza punctata* (von Lendenfeld 1884), a native to the Indo-Pacific that has invaded all major ocean bodies except the Southern Ocean (reviewed by Graham et al., 2003; Bolton & Graham, 2004; Abed-Navandi & Kikinger, 2007). In 2000, *P. punctata* appeared in large numbers off the coast of the northern Gulf of Mexico and has established a permanent population off the coast of Louisiana, with costs attributed to this invasion >$10 million (Graham et al., 2003 and references therein). While Graham et al. (2003) and Bolton & Graham (2004), suggested that a population brought into the region via hull fouling likely contributed to the 2000 (and subsequent) blooms, an alternate hypothesis by Johnson et al. (2005) suggested that the Loop Current transported these medusae from the Caribbean Sea. To help resolve this, and perhaps to understand other invasions, it is critical that a tool must be developed to assay incoming ships and structures for the presence of scyphozoan polyps.

Given the small and cryptic nature of benthic scyphozoan polyps, we developed a molecular method for identifying scyphozoan polyps in samples taken from benthic surfaces. This assay uses the Real-Time PCR method, mediated by taxon-specific Taqman© probes. This is an established method for species identification in the marine realm (e.g., Taylor et al., 2002; Popels et al., 2003) and incorporates taxon-specific Taqman© probes within the framework of a typical Polymerase Chain Reaction (PCR). If the Taqman© probe anneals (matches) during the PCR process, detectable fluorescence is emitted at a brightness level proportional to the original template copy number. If the probes are taxon-specific, fluorescence will only be given off if DNA from the target taxon is present. By labeling multiple probes with fluorescent molecules that can be differentiated, multiple species/taxa can be simultaneously screened for in the same tube (i.e., in multiplex). In this study, we have designed a multiplex Taqman© Real-Time PCR assay for the detection of jellyfish of three genera in the Gulf of Mexico (*Phyllorhiza* sp., *Chrysaora* sp. [formerly *C. quinquecirrha* Desor, 1848] and *Aurelia* spp.). This technique should allow future screening of boat hull surfaces for polyps of the invasive jellyfish *P. punctata* and field benthic surfaces for the polyps of resident scyphozoan species (*Aurelia* sp. and *Chrysaora* sp.). This technique will ultimately provide improved insight into invasion pathways, polyp settlement, and jellyfish bloom dynamics.

## Methods

### DNA sequence collection

Before any successful species-specific genetic probe can be designed, a large amount of DNA sequencing must be performed to ensure probes are truly species-specific. Probes must be based on a genetic sequence/region that does not vary within the target species or the probe will not identify all individuals of the species. To meet this requirement, and ensure that our species-specific Taqman© probes amplified all individuals of our target taxa, we strove to include sequence data from as many individuals and geographic regions as possible during the design process. We extracted DNA from the gonad tissue of 39 *Aurelia* sp. individuals originating from the northern Gulf of Mexico (Dauphin Island, AL [U.S.A.]) and the Florida Keys (U.S.A.). DNA was extracted using a modified CTAB protocol that allowed for two rounds of concentrated CTAB (Ausubel et al., 1989). The mitochondrial 16S region was PCR amplified using cnidarian 16S primers from Bridge et al. (1992) (Table 1). PCR conditions consisted of 8:00 at 95°C, then 38 cycles of 45 s at 95°C, 1 min at 51°C, and 1:30 at 72°C, followed by 10 min at 72°C. PCR products were then cleaned and cycle sequenced unidirectionally using the primer KMT-47 (Bayha, 2005). Sequencing products were ethanol precipitated and read on an Applied Biosystems 3730XL DNA Sequencer (University of Washington High Throughput Genomics Unit; Seattle, WA). DNA sequence data from a total of 25 individuals of *Chrysaora* sp. from seven geographic regions, including six U.S. Atlantic estuaries and the Gulf of Mexico, were taken from Bayha (2005). This *Chrysaora* sp. (formerly known as *C. quinquecirrha*) occurs in U.S. Atlantic estuaries and the Gulf of Mexico and is different from the U.S. coastal Atlantic *C. quinquecirrha* (Bayha, 2005). Likewise, 16S sequence data from 41 *Phyllorhiza* sp. individuals representing three native and five invasive populations worldwide were taken from an ongoing project (Bayha et al., unpublished data).

### Taqman© probe design

To identify within-species polymorphism, all 16S sequences from each of the targeted species were

**Table 1** Sequences of primers and probes used in this study

| Primer/Probe | Primer sequence | Species |
|---|---|---|
| Primer 1[a] | TCGACTGTTTACCAAAAACATAGC | Universal cnidarian |
| Primer 2[a] | ACGGAATGAACTCAAATCATGTAAG | Universal cnidarian |
| DMT-95 | AATTGGTGACTGGAATGAAT | *Phyllorhiza* sp. |
| DMT-96 | CCCCAACCAAACTAATAGG | *Phyllorhiza* sp. |
| DMT-102 | TGTCACCTAATTAGTGAATGGT | *Chrysaora* sp. |
| DMT-103 | CCCAACCAAACTGTCTTACT | *Chrysaora* sp. |
| DMT-112 | AAATGACAGTGAAGATGCTGT | *Aurelia* sp.[b] |
| DMT-113 | CAGTAAGGCATTTTTCATTTG | *Aurelia* sp.[b] |
| DMT-120 | TTAGAGTTCCCCTGATACTTTC | *Aurelia* sp.[c] |
| DMT-121 | TCTGTCTCCAAGAAAATTCC | *Aurelia* sp.[c] |
| DMT-78 | FAM—ACGAATCCCCAACTGTCT-BHQ1 | *Phyllorhiza* sp. |
| DMT-79 | TR—ACGAATCCCCAACTGTCT-BHQ2 | *Chrysaora* sp. |
| DMT-109 | HEX—TAAAAGGGTGTTAACCTGCA-BHQ1 | *Aurelia* sp.[b] |
| DMT-122 | CY5—CTTACAGTTGCTTGACAGCAT-BHQ2 | *Aurelia* sp.[c] |

FAM (6-carboxyfluorescin), HEX (6-carboxy-2′,4,4′,5′,7,7′-hexachlorofluorescein), TR(Texas Red), and CY5 (CY dye) are 'reporter dyes', while BHQ1 (Black Hole Quencher 1) and BHQ2 (Black Hole Quencher 2) are 'quencher' molecules. [a] Primers 1 and 2 are taken from Bridge et al. (1992). [b] Primers and probes designed to cover *Aurelia* sp. Clade 1. [c] Primers and probes designed to cover *Aurelia* sp. Clade 2

aligned using MegAlign 5.53 (DNAStar, Inc.). Consensus 16S sequences (with polymorphic bases labeled) were then aligned against 16S sequences of 46 scyphozoan and 46 hydrozoan species in order to identify potential species-specific regions i.e., ones that consistently vary significantly among species, but not within the target species (sequence data from NCBI GenBank; Bayha, 2005; Bayha, unpublished; Collins, unpublished). Genetic variation within *Chrysaora* sp. was very small (0–0.9%), indicating a single species in the Gulf of Mexico and U.S. Atlantic estuaries. However, in the case of both *Phyllorhiza* and *Aurelia*, 16S data showed large genetic differences indicative of multiple cryptic species/sub-species, a fact that was taken into account during the probe design process.

All taxon-specific Taqman© probes were designed using Beacon Designer 7.01 (Premier Biosoft, Inc.). Traditional Taqman© probes incorporate relatively long probes (24–30 base pairs) with high melting temperatures ($T_m$). However, the relative long length and high melting temperatures often make these probes insensitive to minor sequence mismatches (i.e., probes will anneal even when 1–2 base pair mismatches occur). Since earlier studies on jellyfish probes indicated that traditional Taqman© probes were typically too long to be sufficiently sequence-specific, we employed the techniques of Ugozzoli et al. (2002), in which shorter, more specific primers and probes with lower $T_m$ values were designed. We designed short probes (18–21 base pairs; $T_m \cong 55.5$–57°C) coupled with short primers (19–22 base pairs; $T_m \cong 54$–55°C) in an effort to increase sequence-specificity. Taqman© probes were designed to span species-specific regions, ensuring that areas of sequence mismatches occurred in the middle 1/3 of the probe (to increase sequence discrimination). Also, primers were designed so as to be as species-specific as possible by placing the 3′ end of the primer on a species-specific base wherever possible. Probes were compared to all cnidarian sequences on hand and in GenBank to ensure specificity using NCBI BLAST-N (Altschul et al., 1997) or the program Amplify 3.1.4 (Bill Engels, University of Wisconsin).

In the end, single *Chrysaora* sp. (DMT-79) and *Phyllorhiza* sp. (DMT-78) probes were designed, while two *Aurelia* sp. probes (DMT 109 and DMT-122) were designed (Table 1). The *Chrysaora* sp. probe (DMT-79) targets a single *Chrysaora* species that is present in U.S. Atlantic estuaries and the Gulf of Mexico (Bayha, 2005), while the *Phyllorhiza* probe (DMT-78) was designed to target two separate *Phyllorhiza* species/sub-species that have invaded worldwide (Bayha et al., unpublished data). In the case of Gulf of Mexico *Aurelia*, the two probes (DMT-109 and DMT-122) each target one of two distinct northern Gulf of Mexico *Aurelia* species/subspecies that were detected by DNA sequencing. As the *Phyllorhiza* sp. probe (DMT-78) was designed to detect two distinct *Phyllorhiza* species/subspecies with a single probe, probes will be referred to as taxon-specific from here on. Probes were labeled with fluorophore molecules that could be differentiated from each other by the Bio-Rad iQ5 (FAM. HEX, Texas Red and CY5) and ordered from MWG Biotech, Inc.

### Real-Time PCR with Taqman© probes

All Real-Time PCR runs were completed on a Bio-Rad iQ5 Real-Time PCR detection system using Bio-Rad iQ Multiplex Powermix polymerase mixture with optimal reaction conditions determined by trial and error. Optimal assay conditions consisted of 1 cycle of 8:00 at 95°C, followed by 40 cycles of 15 s at 95°C and 1 min at 62°C with probes DMT78, DMT79, and DMT-122 at 0.2 μM, probe DMT-109 at 0.9 μM, primers DMT-112 and DMT-113 at 0.9 μM, and the rest at 0.4 μM (Table 1). At the end of each run, threshold cycle ($C_t$) and maximum relative fluorescence units (RFU) were recorded and reactions were deemed positive or negative based on the Endpoint program in the Bio-Rad iQ Optical System Software (Version 1.1.1442.0), with corrections by eye.

### Taqman© probe specificity and sensitivity

The multiplex assay was run against 25 *Chrysaora* sp. individuals from U.S. Atlantic estuaries and the Gulf of Mexico coast, 25 *Phyllorhiza* sp. individuals from native and invasive regions worldwide, and 20 *Aurelia* sp. individuals from the northern Gulf of Mexico to verify successful amplification over a wide range of individuals (Table 2). The taxon-specificity of the assay was tested by running the assay against DNA extracts of 17 local and worldwide species of scyphozoa and hydrozoa (Table 2).

**Table 2** Species used to test species-specific Taqman© probes

| | Nearest location | Number (*n*) screened and assay identification | | | | |
|---|---|---|---|---|---|---|
| | | *Phyllorhiza* sp. probe (FAM) | *Chrysaora* sp. probe (TR) | *Aurelia* 1 probe (HEX) | *Aurelia* 2 probe (CY5) | Null reaction |
| Target species | | | | | | |
| *Phyllorhiza punctata* | Sydney, Australia | 5 | | | | |
| *Phyllorhiza punctata* | San Diego, USA | 5 | | | | |
| *Phyllorhiza punctata* | Louisiana, USA | 5 | | | | |
| *Phyllorhiza punctata* | Puerto Rico | 5 | | | | |
| *Phyllorhiza* sp. | Sao Sebastiao, Brazil | 5 | | | | |
| *Chrysaora* sp. | Rhode Island, USA | | 5 | | | |
| *Chrysaora* sp. | New Jersey, USA | | 5 | | | |
| *Chrysaora* sp. | Virginia, USA | | 5 | | | |
| *Chrysaora* sp. | Georgia, USA | | 5 | | | |
| *Chrysaora* sp. | Alabama, USA | | 5 | | | |
| *Aurelia* sp. | Alabama, USA | | | 7 | 13 | |
| Non-target species | | | | | | |
| *Stomolophus meleagris* (Agassiz, 1862) | Alabama, USA | | | | | 1 |
| *Stomolophus meleagris* | Atlantic Florida, USA | | | | | 1 |
| *Rhopilema verilli* (Fewkes, 1887) | Alabama, USA | | | | | 1 |
| *Drymonema* sp. | Alabama, USA | | | | | 1 |
| *Drymonema dalmatina* (Haeckel, 1880) | Foca, Turkey | | | | | 1 |
| *Linuche unguiculata* (Schwartz, 1788) | Bermuda | | | | | 1 |
| *Cyanea* sp. | Alabama, USA | | | | | 1 |
| *Cyanea* sp. | Niantic Bay, CT | | | | | 1 |
| *Cyanea* sp. | Niantic River, CT | | | | | 1 |
| *Cyanea capillata* | Norway | | | | | 1 |
| *Chrysaora quinquecirrha* | Cape Henlopen, DE | | | | | 1 |
| *Chrysaora lactea* (Eschscholtz, 1829) | Rio de Janiero, Brazil | | | | | 1 |
| *Chrysaora hysoscella* (Linnaeus, 1766) | Namibia | | | | | 1 |
| *Chrysaora colorata* (Russell, 1964) | Aquarium[a] | | | | | 1 |
| *Chrysaora fuscescens* (Brandt, 1835) | Aquarium[a] | | | | | 1 |
| *Chrysaora melanaster* (Brandt, 1835) | Aquarium[b] | | | | | 1 |
| *Catostylus mosaicus* (Quoy and Gaimard, 1824) | Australia | | | | | 1 |
| *Aequorea* sp. | Delaware, USA | | | | | 1 |

Nearest location listed for all species. [a]denotes specimens donated by the Aquarium of the Americas, while [b]denotes specimens donated by the Monterey Bay Aquarium

We evaluated probe sensitivity (low detection limit) for each primer/probe combination using plasmids of known copy number containing the target mitochondrial 16S region. For each probe template, 16S PCR amplifications were cloned using the TOPO-TA cloning kit for sequencing (Invitrogen, Inc.) and positive clones were cultured and DNA extracted. Copy numbers were calculated from DNA spectrophotometer readings. Tenfold serial dilutions were made in TE (10 mM Tris–HCl pH 8.0; 100 mM EDTA) to end with a range of 0 to 100,000,000 copies/μl. These samples were then used in duplicate Real-Time PCR reactions and lowest concentration detected and average $C_t$ values recorded.

## Tests of primer and probe competition

A problem of running multiplex Real-Time PCR reactions with more than one template at a time is the possibility of template competition for reactants (dNTP's, *Taq*, etc.). If one or more template is present at a significantly higher concentration than another, there is the possibility that the lesser template might be outcompeted and not detected. To combat this, we used a Real-Time PCR polymerase mix designed for multiplex reactions (Bio-Rad iQ Multiplex Power mix). To test for possible reactant competition, we ran the multiplex assay against a range of known plasmid template copy numbers (Table 3). Template copy numbers were chosen so as to test whether or not each template could be detected when other templates were at 10 or 100 times concentration. In addition, since the two *Aurelia* subspecies are most likely to coexist, we performed *Aurelia*-only tests in which each template (*Aurelia* 1 or *Aurelia* 2) was at 10 or 100 times the other.

## Scyphozoan polyp tests

To test the multiplex Taqman© assay against real scyphozoan polyps, we extracted DNA from a known

**Table 3** Tests of Primer/Probe competition

| Species and plasmid copies added | | | | Assay results | | | |
|---|---|---|---|---|---|---|---|
| *P. punctata* | *Chrysaora* sp. | *Aurelia* 1 | *Aurelia* 2 | *P. punctata* (FAM) | *Chrysaora* sp. (Texas Red) | *Aurelia* 1 (HEX) | *Aurelia* 2 (CY5) |
| $1 \times 10^5$ | $1 \times 10^6$ | $1 \times 10^6$ | $1 \times 10^6$ | + | + | + | + |
| $1 \times 10^6$ | $1 \times 10^5$ | $1 \times 10^6$ | $1 \times 10^6$ | + | + | + | + |
| $1 \times 10^6$ | $1 \times 10^6$ | $1 \times 10^5$ | $1 \times 10^6$ | + | + | + | + |
| $1 \times 10^6$ | $1 \times 10^6$ | $1 \times 10^6$ | $1 \times 10^5$ | + | + | + | + |
| $1 \times 10^5$ | $1 \times 10^5$ | $1 \times 10^5$ | $1 \times 10^6$ | + | + | + | + |
| $1 \times 10^5$ | $1 \times 10^5$ | $1 \times 10^6$ | $1 \times 10^5$ | + | + | + | + |
| $1 \times 10^5$ | $1 \times 10^6$ | $1 \times 10^5$ | $1 \times 10^5$ | + | + | − | + |
| $1 \times 10^6$ | $1 \times 10^5$ | $1 \times 10^5$ | $1 \times 10^5$ | + | + | + | + |
| $1 \times 10^5$ | $1 \times 10^7$ | $1 \times 10^7$ | $1 \times 10^7$ | + | + | + | + |
| $1 \times 10^7$ | $1 \times 10^5$ | $1 \times 10^7$ | $1 \times 10^7$ | + | + | + | + |
| $1 \times 10^7$ | $1 \times 10^7$ | $1 \times 10^5$ | $1 \times 10^7$ | + | + | − | + |
| $1 \times 10^7$ | $1 \times 10^7$ | $1 \times 10^7$ | $1 \times 10^5$ | + | + | + | + |
| $1 \times 10^5$ | $1 \times 10^5$ | $1 \times 10^5$ | $1 \times 10^7$ | + | + | − | + |
| $1 \times 10^5$ | $1 \times 10^5$ | $1 \times 10^7$ | $1 \times 10^5$ | + | + | + | + |
| $1 \times 10^5$ | $1 \times 10^7$ | $1 \times 10^5$ | $1 \times 10^5$ | + | + | + | + |
| $1 \times 10^7$ | $1 \times 10^5$ | $1 \times 10^5$ | $1 \times 10^5$ | + | + | + | + |
| − | − | $1 \times 10^5$ | $1 \times 10^6$ | N/A | N/A | + | + |
| − | − | $1 \times 10^5$ | $1 \times 10^7$ | N/A | N/A | − | + |
| − | − | $1 \times 10^6$ | $1 \times 10^5$ | N/A | N/A | + | + |
| − | − | $1 \times 10^7$ | $1 \times 10^5$ | N/A | N/A | + | + |

Various mixtures of known template copy numbers were tested with the multiplex Taqman© assay. Values at left indicate the composition of each tested sample and values on the right (+/−) indicate whether or not that jellyfish template was detected

number of cultured polyps of *Aurelia* (separate extractions of 1 polyp and 3 polyps) and *Chrysaora* sp. (1 polyp) using the DNA extraction methods from above and subsequently ran extracts against the multiplex Taqman© assay as described earlier.

## Results

### Probe specificity and sensitivity

Our multiplex Taqman© assay correctly identified DNA from all individuals the probes were designed for, regardless of geographical region (Table 2). Probe specificity was confirmed by the fact that all 17 local and worldwide non-target scyphozoan and hydrozoan species showed null reactions (no increase in fluorescence).

We determined the sensitivity of each primer/probe combination by using them in Real-Time PCR runs against known copy number plasmids containing the target 16S region. The low detection limit for the *Phyllorhiza* sp. probe was 50 copies ($R^2 = 0.998$; Fig. 1A), while the *Chrysaora* sp. probe's low detection limit was 10 copies ($R^2 = 0.993$; Fig. 1B). Both *Aurelia* sp. probes (*Aurelia* 1 and *Aurelia* 2) showed low detection limits of 50 copies ($R^2$ values 0.996 and 0.998, respectively, Fig. 1C, D). In general, final RFU (Relative Fluorescent Units) values were very low for the *Aurelia* 1 probe (DMT-109), though they were far enough above the background to detect valid threshold cycle ($C_t$) values, which indicate successful reactions.

### Tests of primer and probe competition

Tests showed no presence of reactant competition for three of the four taxon-specific Taqman© probes (*Phyllorhiza* sp., *Chrysaora* sp., and *Aurelia* 2 probes) when other species' template DNA was present at

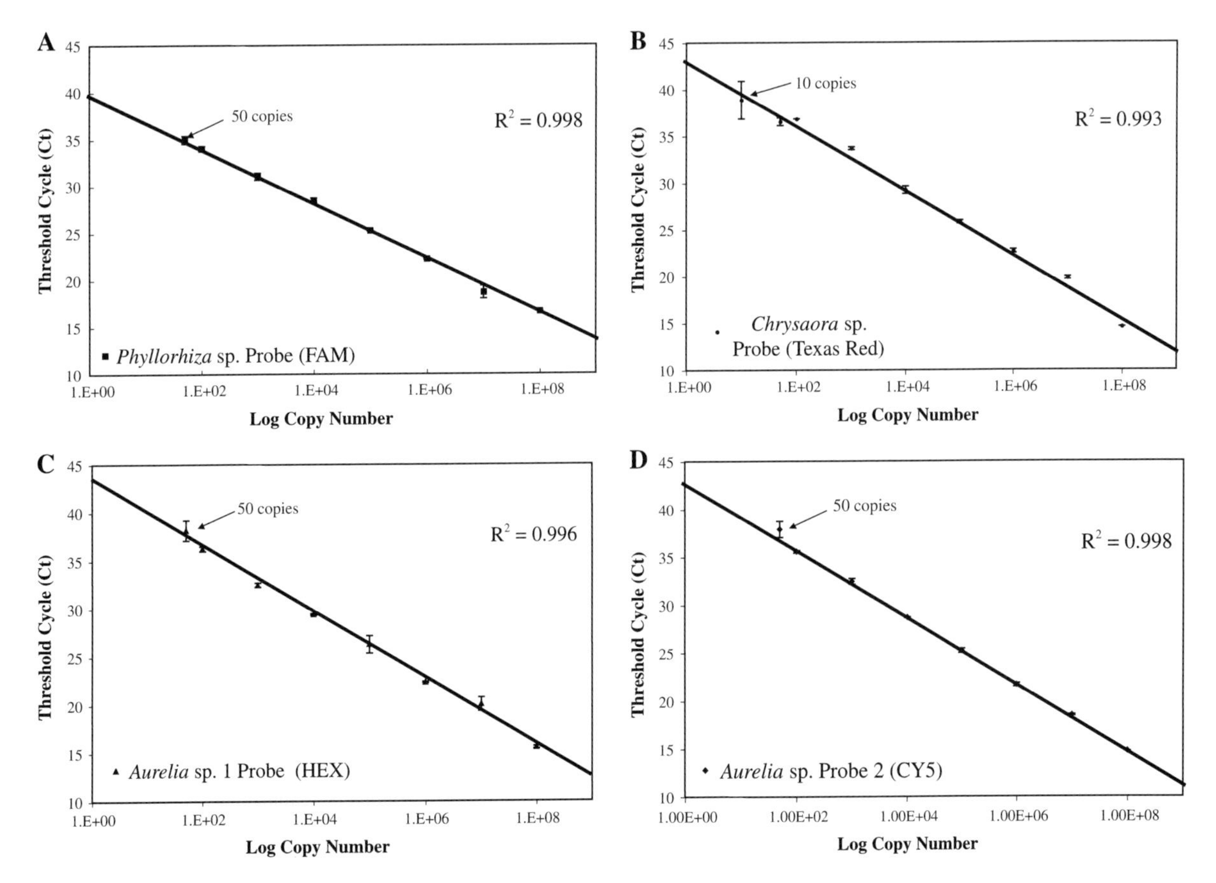


**Fig. 1** Sensitivity of multiplex Taqman© Real-Time PCR assay for *Phyllorhiza* sp. (**A**), *Chrysaora* sp. (**B**), *Aurelia* sp. clade 1 (**C**), and *Aurelia* sp. clade 2 (**D**). All samples were run in duplicate and error bars represent standard deviations

10 or 100 times their template concentration (Table 3). However, in some extreme circumstances, the *Aurelia* 1 probe did not detect its template DNA. The *Aurelia* 1 probe did not detect its template DNA in some cases when all four templates were present, some at 10–100 times concentration. Likewise, when only *Aurelia* 1 and *Aurelia* 2 templates were added, the *Aurelia* 1 probe did not detect its template when *Aurelia* 2 was present at 100 times concentration.

### Cultured polyp tests and assessment of field method

The multiplex Taqman© assay correctly identified cultured *Aurelia* sp. polyps, with a positive reaction by the *Aurelia* 2 probe (Fig. 2C) and *Chrysaora* sp., with a positive reaction by the *Chrysaora* sp. probe (Fig. 2B). As expected, DNA extracted from 3 *Aurelia* sp. polyps showed a lower $C_t$ value than for DNA samples taken from individual polyps (Fig. 2C).

## Discussion

### Primer and probe design and testing

The multiplex Real-Time PCR method using Taqman© probes described here allowed for the detection of jellyfish DNA from three genera found in the northern Gulf of Mexico of which one is a worldwide invader (*Phyllorhiza* sp.) and two are common nuisance jellyfish (*Chrysaora* sp. and *Aurelia* sp.). Our Taqman© probes are specific to their target taxa, successfully identifying all individuals of the taxa targeted, with no cross-reactivity from non-target species. This is attributed to two factors: (1) care in determining the total amount of within-species polymorphism during probe design by sequencing a large number of individuals from geographically wide regions and (2) comparison of all consensus sequences to a large number of published and unpublished scyphozoan (46) and hydrozoan (46) species during the design process.

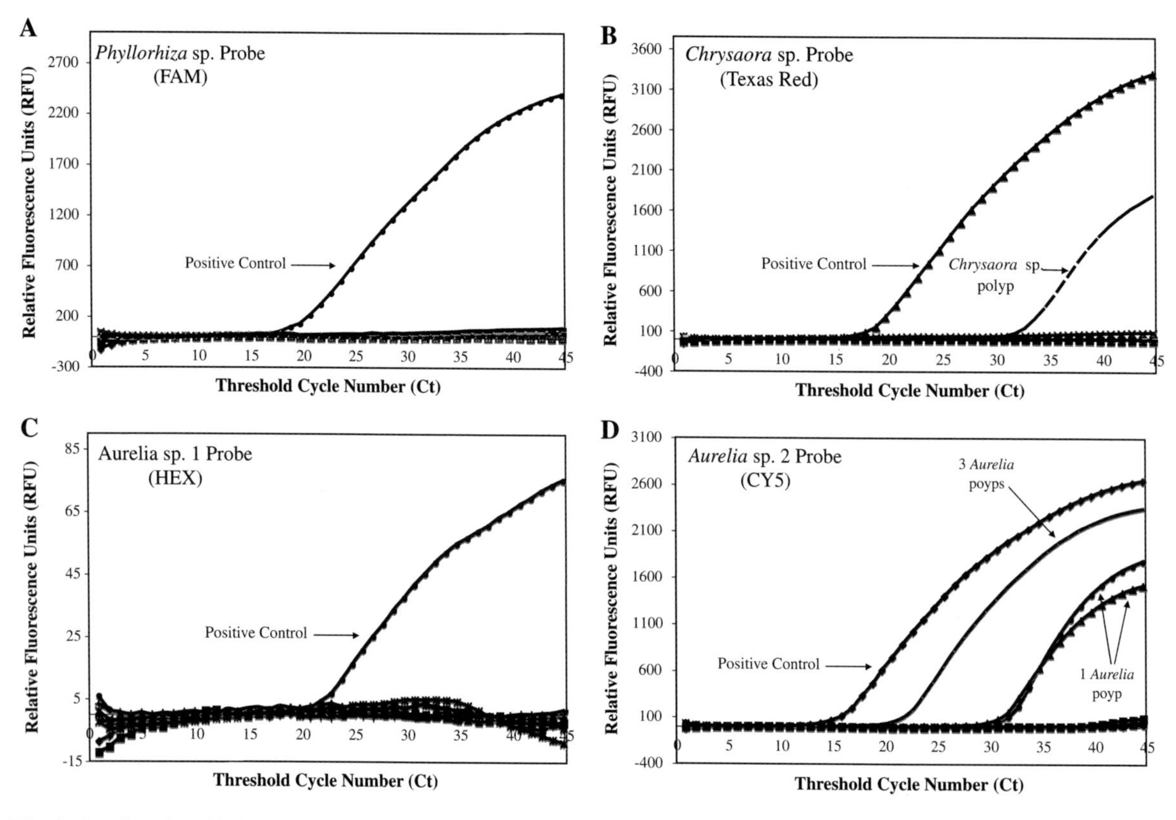


Fig. 2 Results of multiplex Real-Time PCR analysis of *Aurelia* sp. polyps (1 polyp, 3 polyps) and a single *Chrysaora* sp. polyp. Traces representing positive control samples and polyp samples are labeled appropriately

For this study, we decided to place our taxon-specific probes within the mitochondrial ribosomal 16S region. Mitochondrial 16S appears to be an ideal region for species markers in the scyphozoa, due to its general lack of variation within species, but high degree of genetic differentiation between species of scyphozoans (Schroth et al., 2002; Bayha, 2005; Bayha et al., unpublished data). This is apparently not the case for one of the more commonly used genetic regions for species and taxon-specific probes (nuclear ribosomal 18S), which appears to be relatively invariant between some scyphozoan species (NCBI GenBank sequences). More commonly, the cytochrome *c* oxidase subunit I (COI) region is being used as a species marker (Hebert et al. 2003). However, an earlier study (Bayha, 2005) found between-species 16S sequence variation in two scyphozoan genera (*Chrysaora* and *Cyanea*) to be relatively high, but within-species 16S variation to be much lower than in COI. Likewise, intraspecific polymorphic regions tend to cluster more in 16S, leaving for large invariant regions within species ideal for probe placement, while polymorphic regions are more uniformly distributed in COI, leaving for fewer opportunities for probe placement. This is probably due to the fact that COI variation occurs mostly at silent DNA base substitutions (i.e., ones that do not change the amino acid), which are extremely frequent and homogeneously distributed along the region. For these reasons, we would argue that the ribosomal 16S region is better suited for these purposes, at least within the Scyphozoa.

Although we attempted to correct for effects of reactant competition in our multiplex assay, testing showed evidence of its effects in certain circumstances, all involving the *Aurelia* 1 probe (HEX). This probe showed generally low final RFU values in all cases, probably a result of the degree of self-complementarity in the probe (i.e., self-annealing). However, the degree of polymorphism in the consensus region it was built on, and the degree of interspecific variation, meant that the region the probe spanned was really the only species-specific region available for placement of this type of probe. In some cases when all 4 templates were present, some at higher concentrations that the *Aurelia* 1 probe template, or when the *Aurelia* 2 probe template was at 100 times concentration, the *Aurelia* 1 probe did not detect its template. However, it is unlikely that all 4 templates would be present in any field sample. Likewise, while the two *Aurelia* species/subspecies do co-occur, it may be equally unlikely that one would be present at 100 times the quantity of the other. In these extreme cases, the *Aurelia* 1 probe would incorrectly show a null reaction. During the course of this study, a new type of Taqman© probes called Locked Nucleic Acid (LNA) Taqman© probes were marketed for general use (Ugozzoli et al., 2004). These probes incorporate nucleic acid bases with modified ribose moieties, allowing for probes with generally high $T_m$ values and stability typical of traditional Taqman©, but ones that are short and sequence-specific like the ones we developed for this project. We will most likely develop LNA probes for future work, probes that may be more successful than our current *Aurelia* 1 probe (i.e., capable of being placed in other species-specific areas of 16S).

The Taqman© Real-Time PCR method of species detection detailed here is preferable to other molecular detection methods due to its high specificity and sensitivity, and comparatively low cost. Compared to gel-based PCR methods of species detection (e.g., Graves et al., 1990; Frischer et al., 2000), this multiplex Real-Time PCR assay shows a higher degree of specificity and sensitivity. The addition of multiple species-specific primers and Taqman© probes adds another level of sequence discrimination over simple PCR with species-specific primers, and also allows for easier detection of multiple species at a time. In general, Real-Time PCR is sensitive to a much smaller amount of DNA template than traditional PCR and our assay has been shown to react to as few as 10–50 copies of template DNA. This is advantageous in that the method could identify scyphozoan polyps even when they are present in extremely low numbers. In addition, the above method can be run in $<2$ h with no need for gel electrophoresis or other further steps. Also, this method is generally lower in cost than methods, such as DNA sequencing and DNA chip microarray analysis.

Our Real-Time PCR method requires a single PCR step with fluorescence detection for species identification of the three target genera. High-throughput DNA sequencing, coupled with bioinformatic programs have been used for fish egg and larva identification (Richardson et al., 2007) but these methods entail high instrumentation and reagent costs, since both PCR and sequencing reactions are necessary. An extremely promising future technique involves microarray analysis using DNA chips, in which a glass slide is marked with a large number of species-specific oligonucleotides that bind and identify PCR products from target species (reviewed in Naimudden & Nishigaki, 2003). However, at present, costs are extremely high for this technology ($100–200+ per chip), running a large number of field samples is impractical and the method detailed here is more viable when screening for a small number of species. However, prices associated with this promising technology will undoubtedly come down in the future, though a large degree of effort still must be undertaken during the design phase (mass sequencing and testing with large numbers of individuals and species) to ensure that the species-specific oligonucleotides anneal to all individuals of the target species.

Field applications

Our method of screening hull surfaces of cargo ships entering busy seaports, such as those in the northern Gulf of Mexico, for polyps of the invasive jellyfish *Phyllorhiza* sp. will be a worthy tool. In pilot studies, we have employed a simple sampling device based on nylon scrubbing pads attached to the ends of flexible metal poles. These can be scraped along a surface (trapping benthic tissue in the pad) and then cut into pieces, with DNA extracted from each piece and tested against the multiplex assay. Testing of the device is ongoing, but DNA extracted from pads scraped along culturing plates with low densities of *Aurelia* sp. polyps showed positive identification by the *Aurelia* 2 probe (CY5) in initial tests (Bayha & Graham, unpublished data). It is known that moving from lab assays to field application is sometimes problematic and further testing against field samples is ongoing. These tests will include positive controls added to extracts in the form of an exogenous artificial DNA standard (pGEM-z vector DNA) that can be tested post-reaction via PCR with specific primers to ensure that no compounds that might inhibit PCR are present.

The multiplex Taqman© method should also be useful for identifying scyphozoan polyp populations in the field, whether by testing natural surfaces or settling plates. This will be extremely useful in waters where benthic polyps are not easily located due to heavy fouling by other benthic macrofauna, as is the case for the northern Gulf of Mexico. Rapid screening over large areas using our molecular assay virtually eliminates the need for time-consuming morphological species identification that may be of dubious accuracy due to the small size and cryptic nature of the polyps. Furthermore, understanding critical polyp habitat will ultimately allow researchers to monitor the relationship between polyp population dynamics and bloom dynamics of nuisance or invasive medusae.

Two of our Taqman© probes (for *Chrysaora* sp. and *Phyllorhiza* sp.) should be useful over a wide geographic range, while two (*Aurelia* 1 and *Aurelia* 2) may be useful only in the Gulf of Mexico. The *Chrysaora* sp. probe (DMT-79) correctly identified the target species against DNA of animals from the Gulf of Mexico and U.S. Atlantic estuaries north to Rhode Island (however, not the coastal *C. quinquecirrha*). The *Phyllorhiza* sp. probe correctly identified DNA of animals from the native range in Australia, and from animals in the Pacific, Atlantic, Gulf of Mexico, and Caribbean. Therefore, both probes should be useful for assays on samples from a wide geographic range. However, both *Aurelia* sp. primers (DMT-109 and DMT-122) were only positive against Gulf of Mexico *Aurelia*. As the probes were not tested against U.S. Atlantic *Aurelia* and no 16S data from animals in this range have been deposited in NCBI GenBank, more data will be necessary to determine whether or not they will be useful elsewhere.

In conclusion, we have developed a rapid and specific multiplex Real-Time PCR assay using taxon-specific Taqman© probes for identifying the polyps of three jellyfish genera found in the Gulf of Mexico (*Phyllorhiza* sp., *Chrysaora* sp., and *Aurelia* sp.). Given the small and cryptic nature of jellyfish polyps, and the general difficulty of finding them in the field, this method will be useful in two regards: (1) screening cargo ship hulls for the presence of the invasive jellyfish *Phyllorhiza* sp., which has invaded worldwide and (2) for screening field benthic samples and settling plates for polyps of *Phyllorhiza* sp., *Chrysaora* sp., and *Aurelia* sp. species mentioned in this article. The information generated may provide better understanding of worldwide transport mechanisms, regions of polyp settlement, and aggregation and general blooming dynamics of nuisance jellyfish, and may result in enhanced detection and, ultimately, prevention of marine jellyfish bioinvasions.

**Acknowledgments** This work was partially funded by Mississippi-Alabama Sea Grant (R/CEH-8; R/CEH-16) to William M. Graham. We would especially like to thank L. Chiaverano for supplying all *Aurelia* sp. medusae tissue, M. Miller for *Aurelia* sp. polyps, and G. Jarms for *Chrysaora* sp. polyps. The authors would also like to thank the Aquarium of the Americas (New Orleans, U.S.A.), the Barataria-Terrebonne National Estuary Program, T. Bolton, E. Buecher, A. Collins, R. Condon, N. Crochet, M. Davis, M. Dawson, E. Demir, J. Garcia-Sais, M. Gibbons, P. Kremer, V. Levenesque, D. Martin, M. Martinussen, M. Massimi, G. Matsumoto, J. Menadue, H. Mianzan, the Montery Bay Aquarium, A. Morandini, T. Moss, K. Pitt, J. Purcell, K. Rippingale, J. Seymour, S. Spina, and K. Stierhoff for their help with sample collection. Keith M. Bayha would like to thank K. Coyne and M. Vickery for their invaluable help with probe design and implementation.

## References

Abed-Navandi, A. & R. Kikinger, 2007. First record of the tropical scyphomedusa *Phyllorhiza punctata* von Lendenfeld, 1884 (Cnidaria: Rhizostomeae) in the central Mediterranean Sea. Aquatic Invasions 2: 391–394.

Altschul, S. F., T. L. Madden, A. A. Schäffer, J. Zhang, Z. Zhang, W. Miller & D. J. Lipman, 1997. Gapped BLAST and PSI-BLAST: a new generation of protein database search programs. Nucleic Acids Research 25: 3389–3402.

Arai, M. N., 1997. A Functional Biology of Scyphozoa. Chapman and Hall, London.

Ausubel, F. M., R. Brent, R. F. Kingston, D. D. Moore, J. G. Seidman, J. A. Smith & K. Struhl (eds), 1989. Current Protocols in Molecular Biology. Wiley and Sons, New York.

Bayha, K. M., 2005. The molecular systematics and population genetics of four coastal ctenophores and scyphozoan jellyfish of the U.S. Atlantic and Gulf of Mexico. PhD Dissertation, The University of Delaware, Newark.

Bolton, T. F. & W. M. Graham, 2004. Morphological variation among populations of an invasive jellyfish. Marine Ecology Progress Series 278: 125–139.

Bridge, D., C. W. Cunnigham, B. Schierwater, R. DeSalle & L. W. Buss, 1992. Class-level relationships in the phylum Cnidaria: evidence from mitochondrial genome structure. Proceedings of the National Academy of Sciences USA 89: 8750–8753.

Brodeur, R. D., H. Sugisaki & G. L. Hunt, 2002. Increases in jellyfish biomass in the Bering Sea: implications for the ecosystem. Marine Ecology Progress Series 233: 89–103.

Cargo, D. G. & L. P. Schultz, 1966. Notes on the biology of the sea nettle, *Chrysaora quinquecirrha*, in Chesapeake Bay. Chesapeake Science 7: 95–100.

Colin, S. P. & P. Kremer, 2002. Population maintenance of the scyphozoan *Cyanea* sp. settled planulae and the distribution of medusae in the Niantic River, Connecticut, USA. Estuaries 25: 70–75.

Dawson, M. N., 2005. Incipient speciation of *Catostylus mosaicus* (Scyphozoa, Rhizostomeae, Catostylidae), comparative phylogeography and biogeography in southesat Australia. Journal of Biogeography 32: 515–533.

Dawson, M. N. & D. K. Jacobs, 2001. Molecular evidence for cryptic species of *Aurelia aurita* (Cnidara, Scyphozoa). Biological Bulletin 200: 92–96.

Dawson, M. N., A. Sen Gupta & M. H. England, 2005. Coupled biophysical global ocean model and molecular genetic analyses identify multiple introductions of cryptogenic species. Proceedings of the National Academy of Sciences USA 102: 11968–11973.

Frischer, M. E., J. M. Danforth, L. C. Tyner, J. R. Leverone, D. C. Marelli, W. S. Arnold & N. J. Blake, 2000. Development of an *Argopecten*-specific 18S rRNA targeted genetic probe. Marine Biotechnology 2: 11–20.

Graham, W. M. & K. M. Bayha, 2007. Biological invasions by marine jellyfish. In Nentwig, W. (ed.), Ecological Studies, Volume 193: Biological Invasions. Springer-Verlag, Berlin Heidelberg: 239–255.

Graham, W. M., D. L. Martin, D. L. Felder, V. L. Asper & H. M. Perry, 2003. Ecological and economic implications of a tropical jellyfish invader in the Gulf of Mexico. Biological Invasions 5: 53–69.

Graves, J. E., M. J. Curtis, P. A. Oeth & R. S. Waples, 1990. Biochemical genetics of southern California basses of the genus *Paralabrax*: specific identification of fresh and ethanol-preserved individual eggs and early larvae. Fisheries Bulletin 88: 59–66.

Hebert, P. D. N., S. Ratnasingham & J. R. DeWaard, 2003. Barcoding animal life: cytochrome *c* oxidase subunit 1 divergences among closely elated species. Proceedings of the Royal Society B, Biological Sciences 270: S96–S99.

Holland, B. S., 2000. Genetics of marine bioinvasions. Hydrobiologia 420: 63–71.

Holland, B. S., M. N. Dawson, G. L. Crow & D. K. Hofmann, 2004. Global phylogeny of *Cassiopea* (Scyphozoa: Rhizostomeae): molecular evidence for cryptic species and multiple invasions of the Hawaiian Islands. Marine Biology 145: 1119–1128.

Johnson, D. R., H. M. Perry & W. M. Graham, 2005. Using nowcast model currents to explore transport of non-indigenous jellyfish into the Gulf of Mexico. Marine Ecology Progress Series 305: 139–146.

Kawahara, M., S. Uye, K. Ohtsu & H. Iizumi, 2006. Unusual population explosion of the giant jellyfish *Nemopilema nomurai* (Scyphozoa: Rhizostomae) in East Asian waters. Marine Ecology Progress Series 307: 161–173.

Kideys, A. E., 2002. Fall and rise of the Black Sea ecosystem. Science 297: 1482–1484.

Mills, C. E., 2001. Jellyfish blooms: are populations increasing globally in response to changing ocean conditions? Hydrobiologia 451: 55–68.

Miyake, H., M. Terazaki & Y. Kakinuma, 2002. On the polyps of the common jellyfish *Aurelia aurita* in Kagoshima Bay. Journal of Oceanography 58: 451–459.

Naimudden, M. & K. Nishigaki, 2003. Genome analysis technologies: towards species identification by genotype. Briefings in Functional Genomics and Proteomics 1: 356–371.

Popels, L. C., S. C. Cary, D. A. Hutchins, R. Forbes, F. Pustizzi, C. J. Gobler & K. J. Coyne, 2003. The use of quantitative polymerase chain reaction for the detection and enumeration of the harmful alga *Aureococcus anophagefferens* in environmental samples along the United States East Coast. Limnology and Oceanography: Methods 1: 92–102.

Richardson, D. E., J. D. Vanwye, A. M. Exum, R. K. Cowen & D. L. Crawford, 2007. High-throughput species identification: from DNA isolation to bioinformatics. Marine Ecology Notes 7: 199–207.

Schiariti, A., M. Kawahara & H. W. Mianzan, Life cycle of the jellyfish *Lychnorhiza lucernea* Haeckel 1880 (Scyphozoa, Rhizostomeae). This volume.

Schroth, W., G. Jarms, B. Streit & B. Schierwater, 2002. Speciation and phylogeography in the cosmopolitan marine moon jelly, *Aurelia* sp. BMC Evolutionary Biology 2: 1.

Taylor, M. I., C. Fox, I. Rico & C. Rico, 2002. Species-specific Taqman© probes for simultaneous identification of cod (*Gadus morhua* L.), haddock (*Melanogrammus aeglefinus* L.) and whiting (*Merlangius merlangus* L.). Molecular Ecology Notes 2: 599–601.

Ugozzoli, L. A., D. Chinn & K. Hamby, 2002. Fluorescent multicolor multiplex homogeneous assay for the simultaneous analysis of the two most common hemochromatosis mutations. Analytical Biochemistry 307: 47–53.

Ugozzoli, L. A., D. Latorra, R. Pucket, K. Arar & K. Hamby, 2004. Real-time genotyping with oligonucleotide probes containing locked nucleic acids. Analytical Biochemistry 324: 143–152.

Vinogradov, M. E., E. A. Shushkina, E. I. Musayeva & P. Y. Sorokin, 1989. A newly acclimated species in the Black Sea: the ctenophore *Mnemiopsis leidyi* (Ctenophora: Lobata). Oceanology 29(2): 220–224.

Hydrobiologia (2009) 616:229–239
DOI 10.1007/s10750-008-9596-5

JELLYFISH BLOOMS

# Comparative analysis of nuclear ribosomal DNA from the moon jelly *Aurelia* sp.1 (Cnidaria: Scyphozoa) with characterizations of the 18S, 28S genes, and the intergenic spacer (IGS)

**Jang-Seu Ki · Il-Chan Kim · Jae-Seong Lee**

Published online: 21 October 2008


**Abstract** Nuclear ribosomal DNAs (rDNA) constitute a multi-gene family with tandemly arranged units linked by an intergenic spacer (IGS). Here we present the complete DNA sequence (7,731 bp) of a single repeat unit of an rDNA sequence from the moon jelly *Aurelia* sp.1 (Cnidaria: Scypozoa). The tandemly repeated rDNA units consisted of coding and non-coding regions, whose arrangement was 18S rDNA (1,814 bp, 46.2% of GC content)-internal transcribed spacer 1 (ITS1: 272 bp, 39.7%)-5.8S rDNA (158 bp; 50.7%)-ITS2 (278 bp, 51.4%)-28S rDNA (3,606 bp, 49.7%)-IGS (1,603 bp, 45.6%). GC composition in the single unit of rDNA was 47.8%. None of the 5S rDNA was found in the repeat units. Putative structures of a termination transcription signal (poly(T) tract) and promoter-like bi-repeats within the non-coding region were also identified. A block of minisatellites with five repeats was detected within the IGS. Comparative analyses of parsimony and dot plots showed that the IGS was highly informative. The sequence revealed here was the first completion of rDNA from the phylum Cnidaria, using as a model of rDNA for making molecular comparisons of jellyfish members.

**Keywords** Jellyfish · Nuclear rDNA · Transcription repeat unit · Minisatellite

Guest editors: K. A. Pitt & J. E. Purcell
Jellyfish Blooms: Causes, Consequences, and Recent Advances

J.-S. Ki
Department of Molecular and Environmental Bioscience, Graduate School, Hanyang University, Seoul 133-791, South Korea

I.-C. Kim
Polar BioCenter, Korea Polar Research Institute, Korea Ocean Research and Development Institute, Incheon 406-840, South Korea

J.-S. Lee (✉)
Department of Chemistry, and the National Research Lab of Marine Molecular and Environmental Bioscience, College of Natural Sciences, Hanyang University, Seoul, 133-791, South Korea
e-mail: jslee2@hanyang.ac.kr

## Introduction

The moon jelly *Aurelia* Péron & Lesueur, 1810 (Cnidaria; Scyphozoa), is one of the most common and widely distributed species of jellyfish (Arai, 1997). At least 12 species of *Aurelia* have been described based on morphological differences of the medusae (Mayer, 1910; Kramp, 1961). At present, however, only three species, *A. aurita* Linnaeus, 1758, *A. labiata* Chamisso & Eysenhardt, 1821, and *A. limbata* Brandt, 1835, are accepted by systematists. Recent molecular tools, particularly DNA sequencing, can discriminate the species more clearly, allowing taxonomic revisions and molecular phylogenetic inference studies of the jellyfish (Schroth et al., 2002; Dawson et al., 2005; Collins et al., 2006).

For the moon jelly *Aurelia*, Dawson and colleagues (Dawson & Jacobs, 2001; Dawson & Martin, 2001; Dawson, 2003; Dawson et al., 2005) have constructed a more acceptable taxonomic system by combining their morphological characteristics and DNA sequences. Genetic information is generally accepted as a useful molecular tool to discriminate between jellyfish species.

The ribosomal DNA (rDNA) is the region of the genome coding for the RNA component of ribosomes. Eukaryotic nuclear rDNA is tandemly organized, with copy numbers up to the order of 10,000 (Schlötterer, 1998). Each repeat unit consists of the genes coding for the 18S, 28S, and the 5.8S rDNA. These coding regions are separated from each other in the primary transcript by the internal transcribed spacer (ITS) as well as by intergenic spacer (IGS). The rDNA coding regions have remained relatively constant within the same taxa, making these DNA regions the important information sources for the study of phylogenetic relationships. The ITS and IGS rDNA as non-coding regions can be useful to reconstruct relatively recent evolutionary events (Hillis & Dixon, 1991). Thus, complete rDNA sequences provide us with various options depending on the variability of the molecules. By searching all databases, rDNAs from jellyfish have been partially sequenced in a region spanning the 18S and partial 28S (Dawson et al., 2005). None of the complete structure on the transcription repeating unit of rDNA from Cnidarian members, including jellyfish, has been reported.

In Korean coastal environments, the moon jelly *Aurelia* was recently found and its dense blooms caused economic losses for fisheries and power plants (data from the National Fisheries Research and Development Institute, http://nfrdi.re.kr/). Recently, Ki et al. (2008) confirmed that all the moon jellies blooming in different areas of Korea had an identical genotype (*Aurelia* sp.1) using mitochondrial COI gene and nuclear ITS-5.8S rDNA sequences. Here, we present the entire nucleotide sequence of a single unit of rDNA from Korean *Aurelia* sp.1 and a characterization of the rDNA. Comparative analyses of parsimony and dot plot were performed with some known complete rDNA sequences in order to better understand the rDNA relationships among other eukaryotes.

## Materials and methods

### Sample collection

Five specimens of *Aurelia* sp.1 were collected from different localities in western Korean coastal waters (Incheon; 37°26′23″N, 126°22′40″E). After sampling, the individuals were immediately transferred into absolute ethanol to dehydrate and stored at room temperature until use. Genomic DNA was isolated using a previously described procedure (Lee, 2000). The DNA was purified with the DNeasy tissue kit (Qiagen, Valencia, CA) according to the manufacturer's instructions.

### Long and typical polymerase chain reaction

Polymerase chain reaction (PCR) primers (Table 1) were designed for amplification of the entire rDNA from *Aurelia*, based on a comparison of several eukaryotic rDNA sequences. Specifically, we developed primers targeting 28S rDNAs, by comparing sequences of *Atolla vanhoeffeni* Russell, 1957 (GenBank accession no. AY026368), *Catostylus* sp. Agassiz, 1862 (AY920777), *Chrysaora melanaster* Brandt, 1835 (AY920780), *Craterolophus convolvulus* Johnson, 1835 (AY920781), *Haliclystus sanjuanensis* Hyman, 1940 (AY920782), *Nausithoe rubra* Vanhöffen, 1902 (AY920776), and *Phacellophora camtschatica* Brandt, 1838 (AY920778).

Long and accurate (LA)-PCR amplification was carried out from purified genomic DNA and three primer pairs (e.g., JF-18F24, JF-28R1; JF-28F0, JF-28R3.3K; JF-18R70, JF-28F3.2K). PCR reactions were carried out in 1× PCR buffer (10 mM Tris–HCl, 50 mM KCl, 1.5 mM $MgCl_2$, 0.001% gelatin; pH 8.3) with <0.1 μg genomic DNA template, 200 μM each of the four dNTPs, 0.5 μM of each primer and 0.2 units of LA *Taq* polymerase (TaKaRa, Japan). Using a Thermoblock (iCycler, Bio-Rad, CA), PCR thermocycling parameters were as follows: 95°C for 3 min; 35 cycles of denaturation at 95°C for 20 s, annealing at 55°C for 30 s and extension at 68°C for 5 min, and a final extension at 72°C for 10 min. The PCR products (2 μl) were analyzed by 1.0% agarose gel electrophoresis according to a standard method.

**Table 1** Primers used for PCR and sequencing (Seq.) of complete rDNA unit from *Aurelia* sp.1

| Primer | Nucleotide sequence (5′–3′) | Location[b] | Application(s) |
|---|---|---|---|
| JF-18F24[a] | TGGTTGATCCTGCCAGTAG | 18S (5–23) | PCR/Seq. |
| JF-18R70 | CGCAGTTTCACAGTACAAGTGC | 18S (90–69) | PCR/Seq. |
| JF-28F3.2K | AGGGAACGTGAGCTGGGTTTAG | 28S (3155–3176) | PCR/Seq. |
| JF-28R1 | ACGCTTCTCGAGACTACAATTCGC | 28S (164–141) | PCR/Seq. |
| JF-28R3.3K | ATCTGCGGTTCCTCTCGTAC | 28S (3261–3242) | PCR/Seq. |
| JF-28F0 | AAGGATTCCCTCAGTAACGG | 28S (72–91) | PCR/Seq. |
| MJ-18F900 | TTCTTGGATTTACGAAAGAC | 18S (921–940) | Seq. |
| MJ-18F1778 | CCGAGAAGTCGCTCTAGTTC | 18S (1730–1749) | Seq. |
| JF-IGSR1 | GACTACTGGCAGGATCAACC | 18S (25–6) | Seq. |
| JF-28F1 | AGTCGGGTTGCTTGGGAATGCAGC | 28S (269–292) | Seq. |
| JF-28F2 | CGATAGCGAACAAGTACCGTGAG | 28S (341–363) | Seq. |
| JF-28F1390 | GAACCGAACGCTGAGTTAAG | 28S (968–986) | Seq. |
| MJ-IGS-F1 | GTTGCATTGCGGCTGAGTG | IGS (128–146) | Seq. |
| MJ-IGS-F900 | TTAGGATACCGAAATCAC | IGS (513–530) | Seq. |
| MJ-IGS-F1390 | TGGCTCGAGGCTGACATTG | IGS (1356–1375) | Seq. |
| MJ-IGS-R1k | AAAGTGAACCTGGCAGAC | IGS (680–663) | Seq. |

[a] F, forward primer, R, reverse primer, S, sequencing primer

[b] Locations refer to the *Aurelia* rDNA numbering (GenBank No. EU276014) revealed here

For direct DNA sequencing, PCR amplicons were purified with QIAquick PCR purification Kit (Qiagen GmbH, Germany). DNA sequencing reactions were performed in a ABI PRISM® BigDye™ Terminator Cycle Sequencing Ready Reaction Kit (P/N 4303149, PE Biosystems, CA) using the PCR products (2 μl) as the template and 10 picomoles of sequencing primers (Table 1). Labeled DNA fragments were analyzed on an automated DNA sequencer (Model 3700, Applied Biosystems, CA).

## Gene identification and secondary structures

Editing and contig assembly of the rDNA sequence fragments were carried out with Sequencher 4.7 (Gene Codes, MI, USA). The coding rDNA sequences were identified by considering those of other eukaryotes in the NCBI database and typical nucleotide sequences such as "TAT CTG G" for the start of 18S rDNA, "TTT GT" for the end of 28S rDNA (see Ki & Han, 2007). The sequence determined here has been deposited to GenBank with the accession number EU276014.

Putative secondary structures of assumed termination and promoter signals within the rDNA IGS were estimated using the program Mfold, version 3.2 (http://www.bioinfo.rpi.edu/applications/mfold/old/rna/) according to Zuker (2003). With the default option (e.g., temperature setting, $T = 37°C$), Mfold predicted six secondary structures from the entire IGS sequence. Different parameter settings of, for example, temperature ($T = 10°C$, $T = 20°C$, $T = 37°C$) did not affect the general architecture, but did result in different energy levels for the secondary structures. Among them, we finally selected a general model of secondary structures of termination and promoter-like signals. Secondary structure models inferred here were redrawn with RNA structure ver. 4.5 (Mathews et al., 1999).

## Data analysis

General molecular features of the *Aurelia* rDNA were calculated by Genetyx ver.7.0 (Hitachi Engineering Co. Ltd., Japan) and MEGA 4.0 (Kumar et al., 2001). In addition, nucleic acid distribution, sequence complexity, and entropy across the entire rDNA nucleotides of *Aurelia* sp.1 were calculated with the BioAnnotator in Vector NTI Advance 10.3.0 (Invitrogen). Repeat sequence patterns in the rDNA IGS sequences were analyzed using the Genetyx 7.0 and Tandem Repeats Finder (http://tandem.bu.edu/trf/trf.basic.submit.html).

For comparative molecular features, the rDNA of *Aurelia* sp.1 was compared with four complete rDNAs from *Meloidogyne artiellia* Franklin, 1961 (Nematoda, AF248477), *Anopheles albimanus* Wiedemann, 1820 (Arthropoda, L78065), *Chironomus tentans* Fabricius, 1805 (Arthropoda, X99212), and *Herdmania momus* Savigny, 1816 (Ascidiacea, X53538), which were retrieved from DDBJ/EMBL/GenBank. For parsimony analysis, sequence alignment was performed with the five complete rDNAs, using Clustal W ver. 1.4 with the default settings for gap inclusion and extension. Various regions were further aligned manually using the BioEdit 5.09 (North Carolina State University, NC). Genetic distance values were calculated by using the aligned DNA sequences according to the Kimura 2-parameter model (Kimura, 1980), and molecular similarity was measured in BioEdit 5.0.6. Further, comparative analyses such as pairwise (*p*) distance and parsimony-informative (P-I) sites, transition/transversion ratio were implicated with MEGA 4.0 with the above data matrix. Dot-plot analysis was carried out using the MegAlign 5.01 software (DNAstar Inc., WI).

## Results

### General features of complete rDNA unit of Aurelia sp.1

New PCR primers ($N = 6$) were designed for the isolation of the entire rDNA of *Aurelia* sp.1 (Table 1). LA-PCRs, by employing three primer pairs selected for amplification of less than 4 kb, successfully amplified the expected sizes of PCR fragments from genomic DNA of *Aurelia* sp.1 (data not shown). Using the DNA fragments obtained by PCR amplifications, we could carry out complete sequencing of the entire rDNA sequence of *Aurelia* (Fig. 1). By BLAST search, partial DNA sequences obtained from 18S- and 28S-containing PCR amplicons were matched to some sequences revealed from Cnidaria, including *Aurelia*. We observed that the 18S sequence matched well with those of *A. aurita* (AY039208), *Aurelia* sp. (AY920770), *Chrysaora melanaster* (AF358099), and *Stomolophus meleagris* Agassiz, 1862 (AF358101), and the 28S sequence matched those of *Phacellophora camtschatica* Brandt, 1835 (AY920778), and *Chrysaora melanaster* (AY920780). This confirmed that the new PCR primers successfully amplified the expected rDNA from *Aurelia*.

A total length of single rDNA repeat unit from *Aurelia* sp.1 was determined to be 7,731 bp. It was organized in the typical eukaryotic fashion of rDNA, i.e. 18S-ITS1-5.8S-ITS2-28S-IGS (Fig. 1). Specifically, DNA sequence of *Aurelia* rDNA was recorded at 1,814 bp (18S rDNA), 272 bp (ITS1), 158 bp (5.8S rDNA), 278 bp (ITS2), 3,606 bp (28S rDNA), and 1,603 bp (IGS). Each rDNA transcription unit was separated by non-coding regions (i.e. IGS). The IGS was assumed to comprise 3′ external transcribed spacer (ETS), 5′ ETS, and non-transcribed spacer (NTS). Intron-like sequences were not detected in 18S and 28S rDNA coding regions.

Nucleotide frequencies of complete rDNA were recorded at A, 26.1%; T, 26.1%; G, 26.8%; C, 21.0% (Table 2). Among them, cytosine (C) content was the rarest and GC content was 47.8% (AT, 52.3%). Specifically, base frequencies of each rDNA locus were identical. In addition, each nucleotide composition was nearly identical among the rDNA coding and non-coding genes excluding ITS1. Base composition of ITS was recorded at A, 27.6%; T, 32.7%; G, 22.4%; and C, 17.3%.

### Characteristics of Aurelia *rDNA IGS*

The IGS region of *Aurelia* contained 1,603 nucleotides and its GC content was 46.2%. In some invertebrates (e.g., *Calanus finmarchicus* Gunner, 1765, X06056), 5S rDNA as the smallest subunit of the rRNA coding genes is located within the 18S–28S rDNA repeats (Drouin et al., 1987). Searching the 5S rDNA database (http://rose.man.poznan.pl/5SData/), no pattern of 5S rDNA sequences in IGS of *Aurelia* was found. The IGS consisted of putative transcription termination and bi-repeated sequences. We detected a poly(T) tract (TTT ATT TTT ATT) in 5′ ETS region adjacent to the end of 28S rDNAs, and AGG CCG T(A)G T(G)G T(G)G in 3′ ETS region located upstream of the 18S rDNA (Fig. 1B). As predicted by structural models, the Mfold analysis showed that nucleotides between 28S and the termination signal sequences formed a stem and loop structure (Fig. 2A). On the other hand, some nucleotides of the bi-repeat sequences formed a hair-pin structure (Fig. 2B). Interestingly, a block of

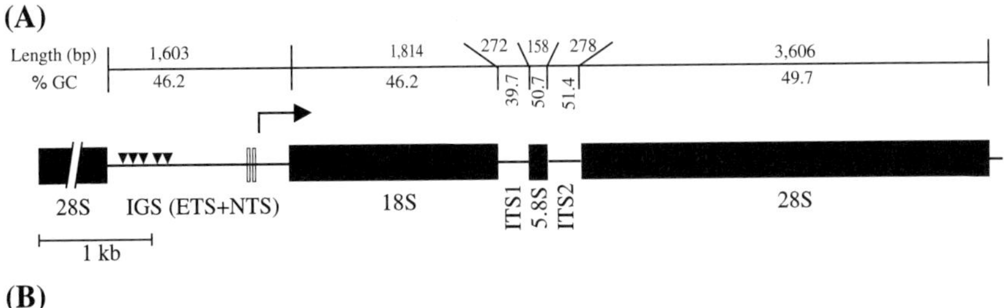


(B)

```
-70  TATTGTAAAAAGTAGAGTAGCCTTGTTGCTACGATCTTCTGAGATTAAGCCCTTCGTTCTACAGATTTGT
     |→ IGS
1    TAGCACTACGCGTGTTAACATTTATTTTTATTACAAAGTTTATAATATGTCCCATCAACAAAATCTAACT
71   TCTTGGTGCACGACATGGATGCATCTTGGTCGATGAGTTATGTTCCGAATGCCTCCGGTTGCATTGCGGC
141  TGAGTGGAGCCTTCCCGTTAGGGTTCCAGTTATGAGTTATTATAAAGTACGGTGTGCAGCTAACTAACCC
211  TAGCCCTAACCCTAACCCTAACCCTAACCCTAACCCTAACCCTAACCCTGCTGATGCTAACCCTAACCCT
281  AACCCTAGcTCTGCTAATGTTAACCCTAACCCTAGCTCTGCTGAGAAATATTATATATATATTTTCCCGT
351  TTTCATCGATGATGCGAATGTTGAAATTTTTCGCGGGAAAAACGGTCAAATATTTTGCATTCAATTAATC
421  AGCGAGGATTTATCAGAGAATAATAGTTGCTTATAATTTTTTTCTAAGTTGGGATAAAAGTTCAATAAAT
491  AAAGAATGAAAGAAAATATAGATTAGGATACCGAAATCACAAACATTATTGTCGCAGTGATAGATGTTCT
561  TGCGTGGTGATCGAGTTTCTATGAGTACTCGACATGGCATACATTTACCCTTCACATTTTTAAACTATAA
631  TTTTTATATATACGCCCCGTCAACAGATGTtTGTCTGCCAGGTTCACTTTGTATCATATTACCAGTGTAG
701  CCCCCCTGGGGGAGTTTTTTGAGGATGTTTCCACACCGTCGCATAGACAATGTTGTAAGAAAAAATTTAC
771  ATGCCAGAATGGTTGCGAGTTATTGCAGGCTAATGTGGTTAACAGCGAACTTTGTTCCGTCGCCTATCCG
841  TAAGCCAGTTGTCGTGACTAAGTAATGTGAGAGGCGAGCTGGGTAGTGATGAGGAAGCCTGGGAGAGGGG
911  CGCAGTCGGCATTCTGGCTGCGTGACCTGTTACCTGTTGGAATCGTCACGGTGATCTTGGCTCGAGGCTG
981  ACATTGGGCAGTCGACGGTTGGATGATGGATGGGTTAGAGCTGGATGTGTGCGTTGCATCTTTTTTAATC
1051 ATGCGGAGTACTCGGTCGGCTATCTGGTGATAGGCGAGGGCCGTACACGAGAGTATCGTGCGTGCCGAGG
1121 TTCGTTCATGCTGGAATTGTGTCGGTCCTCCGCCAACTCATGTTGGCTGACGGTGCCGTGAGTAAGTGCT
1191 ATAGTGCTAATTAGCAGCAGTATAGACATACGGATGGATGTTGTGTTAATGGTCGTTGACCAGCAGCCGT
1261 GGCTTTTGTAGTGCCGTGAAGATACATGCTTCTCCATCGGTCGCTCTCGCCGCCCTATTTGCTGCTCGCT
1331 TGCAAGATAGCTAGCTAGCTAGCCGATGAGGCCGTGTGTGTGAGTGCATCAGGCCGAGGGGGATGGAAAG
1401 ACGTGGGAAAGAAAAGATTAGACCTGACTGTAGTTATCGATCGGTGTACTTGTACTCCGCGTGGGTAACT
1471 CGGTGAGCTAGTCTTGGAAGTTGTCGATTAACAGACACCAACTACTAACTAACGACACTGTTGTTGTTCG
1541 TTACCGAGCAAATCGGGCGGCGGCAACAACAGTTACTCGGCGCCTAGTGCTGTGTGTGAATGTTATCTGG
                                                               IGS ←|
+8   TTGATCCTGCCAGTAGTCATATGCTTGTCTCAAAGATTAAGCCATGCATGT
```

**Fig. 1** Schematic presentation of the single unit of rDNA complex (**A**) and nucleotide sequence (**B**) of *Aurelia* sp. Solid boxes indicate the ribosomal RNA genes, and thin lines represent ITS or IGS. Nucleotide sequences in length and GC composition of each locus are represented on/under line by calculation from a single unit of rDNA. The putative transcription start site is represented by an arrow; solid inverted-triangles represent sub-repeats in IGS. In IGS sequence (**B**), bi-repeats are indicated in asterisks, and minisatellite-like nucleotides are marked in lines. A putative termination signal (poly(T) tract) is represented under a dot line. A box in IGS sequence represents a AT-rich region

**Table 2** Nucleotide composition and length of each rDNA locus from *Aurelia* sp.1

| Locus | Nucleotide composition (%) | | | | Length (bp) |
|---|---|---|---|---|---|
| | A | T | G | C | |
| 18S | 26.5 | 27.3 | 26.2 | 20 | 1,814 |
| ITS1 | 27.6 | 32.7 | 22.4 | 17.3 | 272 |
| 58S | 24.1 | 25.3 | 26.6 | 24.1 | 158 |
| ITS2 | 25.2 | 23.4 | 25.5 | 25.9 | 278 |
| 28S | 26.4 | 23.9 | 28.2 | 21.5 | 3,606 |
| IGS | 24.9 | 29.4 | 25.1 | 20.5 | 1,603 |
| Total | 26.1 | 26.2 | 26.8 | 21 | 7,731 |

minisatellite, "CTA ACC CTA GCC CTA ACC"-like nucleotide sequences with five sub-repeats, was detected between the transcription termination signal and the bi-repeat sequences (Fig. 2).

## Nucleic acid distribution and sequence complexity

Taking into account the new genetic information presented here, we constructed a map of rDNA organization and calculated their molecular characteristics, including GC content, thiamine (T) distribution, sequence complexity, and entropy (Fig. 3). Maps of characteristics along the entire rDNA were obtained using sliding windows of 100 nucleotides. The distribution of GC content varied around 50% across the complete rDNA. However, some zones of both ITS and IGS showed considerably low GC content. In contrast, the T distribution fluctuated. We found that the low GC content was caused by a high content of nucleotide T within the IGS. In addition to this, sequence variability was analyzed with sequence complexity and entropy

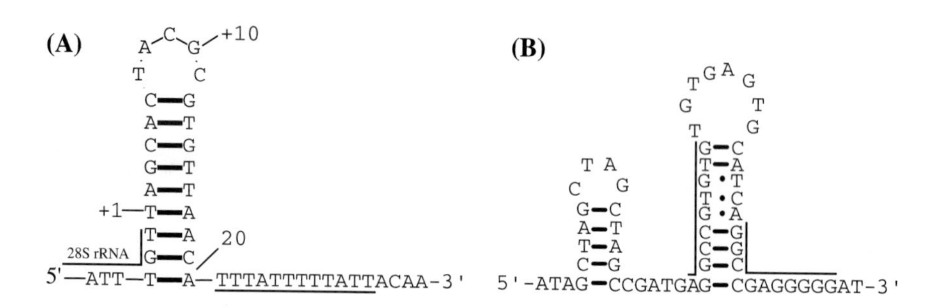


**Fig. 2** Putative secondary structure models for termination-associated sequence (**A**) and bi-repeats (**B**), which are marked in lines, within the IGS of *Aurelia* sp.1

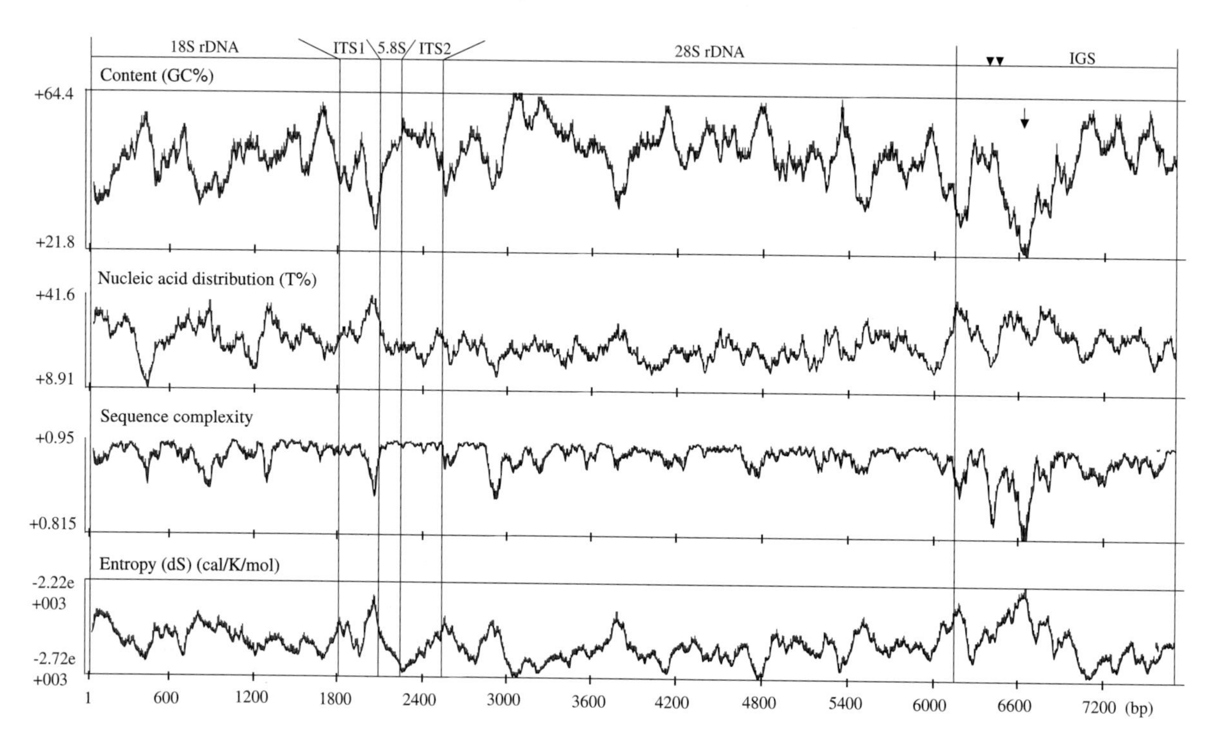


**Fig. 3** GC content (%), nucleic acid distribution (% thiamine), sequence complexity and entropy (dS) in 100-bp windows across the entire rDNA nucleotides of *Aurelia* sp.1. Reverse triangles represent the locations of minisatellites within the IGS rDNA. A downward arrow represents a termination transcription signal of poly(T) tract

plotting. Overall, the two variables fluctuated against one another along the rDNA. These observations show a clear difference in profiles of the coding and other non-coding regions. The sequence complexity was considerably higher in non-coding regions such as ITS and IGS, than the coding region. Notably, a zone recording the lowest complexity corresponded to the internal subrepeats and poly(T) regions within the IGS.

## Comparative analyses of entire rDNA

The rDNA sequence of *Aurelia* was compared with those of the other eukaryotic members, including *Meloidogyne artiellia* (Nematoda), *Anopheles albimanus* (Arthropoda), *Chironomus tentans* (Arthropoda), and *Herdmania momus* (Ascidiacea; Table 3). Comparative analysis showed that the transition:transversion ratio (Ts/Tv) was higher in coding genes than in non-coding genes; 1.039 in 18S rDNA, 0.951 in 28S, 0.84 in ITS and 0.809 in IGS. High *p*-distance, in contrast, was recoded at ITS (1.710) and IGS (1.509) when compared with those in 18S and 28S. In addition, parsimony analyses showed that the non-coding genes of rDNA contained most P–I sites; 15.1%, 18S; 27.9%, ITS and 19.5%, 28S; 29.9%, IGS. Within coding genes, the P–I of 28S rDNA was higher than that of 18S, probably due to hypervariable regions within the 28S (Hassouna et al., 1984). Variation of non-coding rDNAs was ~1.5 times greater than that of the coding genes as judged by the comparison of P–I sites.

**Table 3** Sequence characteristics of rDNA locus among five complete rDNAs from *Aurelia* sp.1 (Cnidaria, accession No. EU276014), *Meloidogyne artiellia* (Nematoda, AF248477), *Anopheles albimanus* (Arthropoda, L78065) *Chironomus tentans* (Arthropoda, X99212), and *Herdmania momus* (Ascidiacea, X53538), respectively. *p*-distances were calculated with the Kimura 2-parameter model

| Locus | Nn | Nc | Nv | Ts | Tv | Ts/Tv | P-I | %P-I | *p*-distance |
|---|---|---|---|---|---|---|---|---|---|
| 18S rDNA | 2044 | 1038 | 919 | 242 | 233 | 1.039 | 308 | 15.1 | 0.324 |
| ITS, 5.8S | 832 | 130 | 661 | 166 | 187 | 0.888 | 228 | 27.4 | 1.097 |
| ITS1, 2 | 664 | 63 | 567 | 137 | 163 | 0.84 | 185 | 27.9 | 1.710 |
| 28S rDNA | 4228 | 1838 | 2179 | 557 | 586 | 0.951 | 826 | 19.5 | 0.403 |
| IGS | 2245 | 203 | 1735 | 445 | 550 | 0.809 | 671 | 29.9 | 1.509 |

*Note*: Nn, total number of sites; Nc, total number of conserved sites; Nv, total number of variable sites; Ts, transition; Tv, transversion; P-I, parsimony-informative site

In order to determine whether there was any sequence homology between parallel sequence alignments, the rDNA of *Aurelia* was compared with those of the above eukaryotes by dot-matrix analysis (Fig. 4). The dot plots were obtained using sliding windows of 60 nucleotides along the entire rDNAs. The dots and lines in Fig. 4 represent regions of homology between sequence pairs. Overall, high similarities of the complete rDNAs between *Aurelia* and Nematoda, Insecta and Chordata were recorded at coding regions (e.g., 18S, 5.8S, 28S) rather than the other non-coding regions. No homologous sequence zones within non-coding rDNAs were detected.

## Discussion

In this study, the complete rDNA mapping of *Aurelia* sp.1 was achieved for the first time. The gene arrangement of *Aurelia* sp.1 was identical to those of the typical eukaryote rDNA in order (e.g., 18S-5.8S-28S rDNA, no 5S rDNA). Also, none of the insertion or deletion sites were found within the coding regions, when compared with rDNAs from other eukaryotes such as *Meloidogyne artiellia* (Nematoda, AF248477), *Anopheles albimanus* (Arthropoda, L78065) *Chironomus tentans* (Arthropoda, X99212), and *Herdmania momus* (Ascidiacea, X53538).

The 18S rDNA sequence of *Aurelia* contained 1,814 nucleotides, which was generally similar in size to that of other Cnidaria 18S rDNAs (e.g., *Antipathes galapagensis,* Anthozoa, AF100943, 1,815 bp and *A. aurita,* Scyphozoa, AY039208, 1,808 bp). However, it was shorter than that of *Atolla vanhoeffeni* (Anthozoa, AF100942, 1,825 bp) and longer than those of most Cnidaria such as *Cassiopea* sp. (Scyphozoa, AF099675, 1,802 bp), *Tripedalia cystophora* Conant, 1897 (Cubozoa, L10829, 1,802 bp), *Corynactis* sp. Allman, 1846 (Anthozoa, AJ133559; 1,798 bp), *Coryne pusilla* Gaertner, 1774 (Hydrozoa, AJ133558, 1,797 bp), *Anemonia sulcata* Pennant, 1777 (Anthozoa, X53498, 1,799 bp), *Tubastraea coccinea* Lesson, 1829 (Anthozoa, AJ133556, 1,797 bp), and *Parazoanthus axinellae* Schmidt, 1862 (Anthozoa, U42453, 1,795 bp). On the other hand, GC contents of the 18S rDNA sequences were nearly identical around 46% among the sequences of Cnidarian members.

So far, many sequences of *Aurelia* rRNA have been deposited in the public databases (e.g., EMBL/DDBJ/GenBank). However, most of them are either partial or complete sequences of individual rDNA sequences such as 18S, ITS-5.8S, and 28S (Dawson et al., 2005). None of the complete regions from the 18S to 28S rDNA loci has been sequenced as yet. Therefore, we could not compare sequences at different loci, due to intraspecific variations. The present nuclear rDNA sequence, composed of the 18S, ITS, and 28S rDNA sequences, can be used as a reference sequence to compare the different sequences of loci. By BLAST search, we could compare *Aurelia* sp.1 to other *Aurelia* genotypes from various geographical regions. A comparison of 18S rDNA 1,765 sites showed that DNA similarities between the current *Aurelia* sp.1 were 99.2% with *A. aurita* (AY039208), 99.6% with *A. aurita* (AY428815), and at 98.9% with *Aurelia* sp.2 (AY920770), suggesting that the 18S rDNA may be highly conserved within the genus *Aurelia.* In contrast, ITS comparisons showed that DNA similarities were quite different. For example, highest similarity

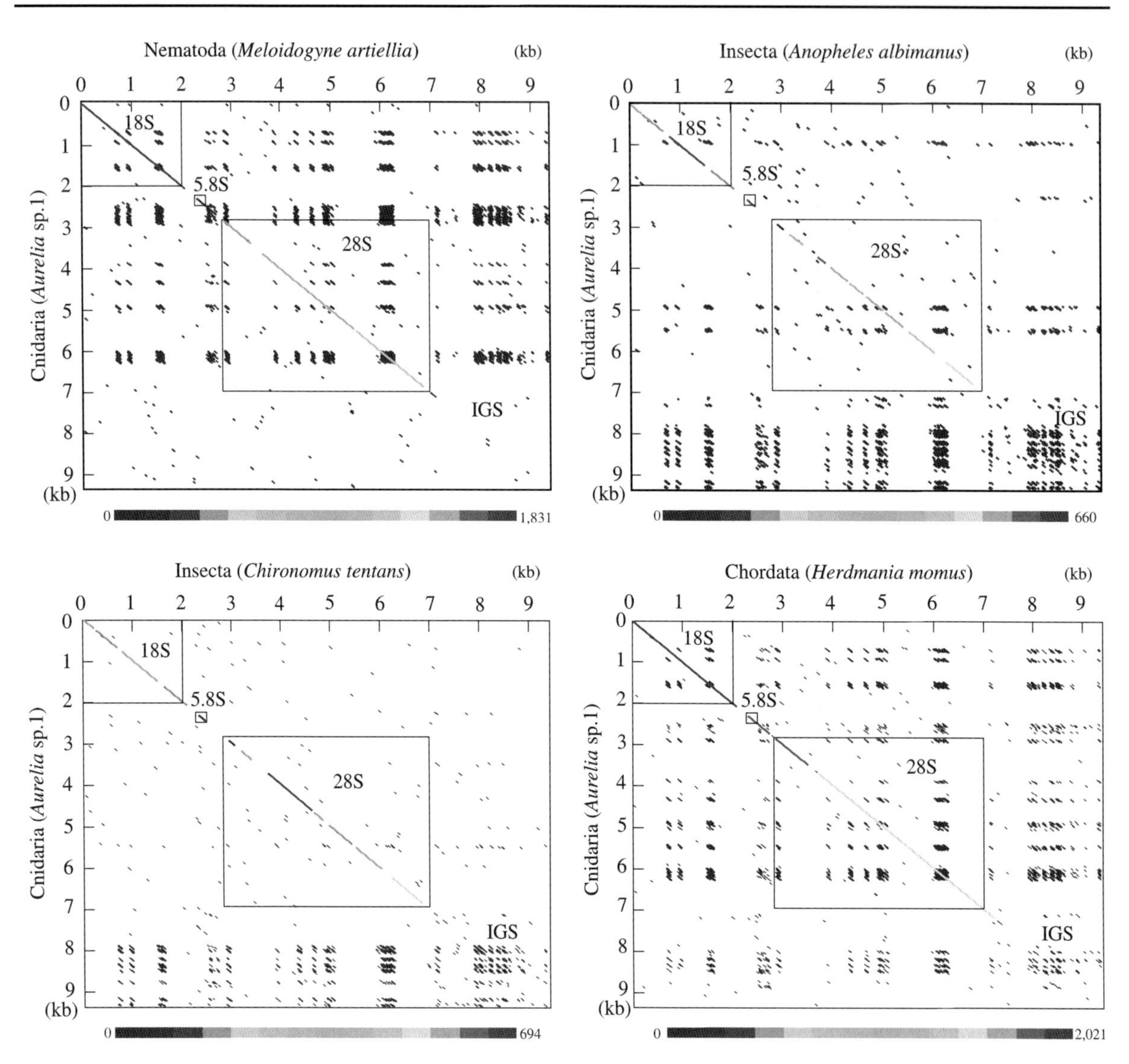


**Fig. 4** Dot matrix comparisons of rDNA sequences. Dot plots between *Aurelia* sp.1 with *Meloidogyne artiellia* (Nematoda, AF248477), *Anopheles albimanus* (Arthropoda, L78065), *Chironomus tentans* (Arthropoda, X99212), and *Herdmania momus* (Ascidiacea, X53538) sequences. Color scale bars represent consecutive sequence length of some regions detected similarly between the two sequence pairs. The open boxes in matrixes indicate rDNA coding regions such as 18S, 5.8S, and 28S

(99.7%) was recorded between current *Aurelia* sp. and *Aurelia* sp.1 (AY935214; isolation locality from Miyazu Bay, Japan), followed by 99.4% with *Aurelia* sp.1 (AY935203, California). Interestingly, DNA sequence similarities between our study and those reported by others for *Aurelia* spp. were less than 80% (mostly <70%). This suggests that *Aurelia* sp. studied by us could be assigned to *Aurelia* sp.1. In this context, the current data on *Aurelia* sp.1 were significantly different from those of *A. aurita* (AY935205, 69.2%; AY935206, 71.8%), *A. labiata* (AY935202, 70.8%), and *A. limbata* (AY935215, 79.4%). ITS data from the *Aurelia* sp.1 studied by us were congruent with previous findings (Dawson et al., 2005), where the genus *Aurelia* was distinguished into 13 genotypes that probably represent species.

The tandemly repeated rDNA units of *Aurelia* consisted of coding and non-coding regions. Of them, the IGS of *Aurelia* was more divergent than the other genes (Table 3, Figs. 3, 4) and apparently consisted of three components, i.e. 3′ ETS, NTS, and 5′ ETS.

As a putative RNA polymerase I transcription initiation site, many promoters of the RNA polymerase I are bipartite, consisting of a proximal promoter domain and an upstream control element (UCE) (Chen et al., 2000). In the present study, an internal repeat sequence of "AGG CCG TGT GTG." located at 1,359–1,370 and "AGG CCG AGG GGG." at 1,381–1,392 within the IGS of *Aurelia* was detected (see Figs. 1, 2). These kinds of motifs were consistent with those being proximal promoters and UCE motifs, suggesting a putative proximal domain for RNA polymerase I (Marilley & Pasero, 1996). Taking into consideration both the motif as promoter and transcription probably begins from 20 to 30 bp downstream of the 2nd motif, the 5′ ETS region can be ~190 bp long in *Aurelia* sp.1 Regarding the 5′ ETS, we found a poly(T) tract (TTT ATT TTT ATT) located at positions 21–32 from end side nucleotide of 5′ IGS, which is considered as a signal for termination of the rDNA transcript (Lang et al., 1994; Jeong et al., 1995; Mason et al., 1997). Indeed, a similar pattern of poly(T), which was a tract of "TTT TTT TTT T" and was located at positions 21–32 from end side nucleotide of 5′ IGS, was identified in the IGS of the cnidarian member (*Junceella fragilis*; AF154670). Therefore, the 5′ ETS is likely to be ~30 bp long. The remaining nucleotide sequence belonged to NTS, which was at least 1,350 bp long. Frequently, the NTS region contains many minisatellite sequences, such as "AG", "AC", "AT". However, *Aurelia* NTS has a block of minisatellite, "CTA ACC CTA GCC CTA ACC"-like nucleotide sequence, with five repeats within the IGS. This is the first finding in Cnidaria IGS, representing the number of repeats and nucleotide sequences that may be used as a genetic marker for population history studies.

With regard to IGS length variations, the rDNA IGS of *Aurelia* was short when compared with other eukaryotic IGSs (Table 3). For example, the IGS varies in length from about 2.0 kb (e.g., *Meloidogyne artiellia*, Nematoda, AF248477; *Herdmania momus*, Ascidiacea, X53538), to about 21 kb in mammals and, with very few exceptions, contains significant regions of internal repeating. A pair-wise comparison between *Aurelia* and other eukaryotes showed that low similarities were recorded in all coding and non-coding regions (Table 4). Of them, ITS and IGS showed remarkably low similarities (<47.2%) to the data set. For the comparison of Cnidaria rDNA, we found only one nucleotide sequence of complete rDNA IGS from Anthozoa (*Junceella fragilis* Walker & Bull 1983, AF154670). It contained partial 28S, complete IGS, and partial 18S rDNA. Upon comparison, DNA similarities between them were measured at 80.7% in 285 sites of 28S, at 40.3% in 1,063 sites of IGS, and at 90.3% at 339 sits of 18S. In addition, dot-matrix plots between *Aurelia* sp.1 and four other eukaryotes graphically showed that the sequences of IGS were highly variable with respect to those of other eukaryotic rDNA coding regions (Fig. 4). Furthermore, sequence characteristics such as P–I sites, *p*-distance suggested that the IGS was highly informative of the other rDNA regions (Table 3), although the data included here were sampled from relatively distant relatives such as Nematoda, Arthropoda, and Chordata.

**Table 4** Similarity scores (percentage) between *Aurelia* sp.1 and four other eukaryotes

| | *Aurelia* sp. | | | |
|---|---|---|---|---|
| | 18S | ITS+5.8S | 28S | IGS |
| *Anopheles albimanus* #L78065 | 66.5 | 33.9 | 56.1 | 25.4 |
| *Chironomus tentans* #X99212 | 65.2 | 33.9 | 56.1 | 29.0 |
| *Meloidogyne artiellia* #AF248477 | 69.5 | 32.9 | 67.4 | 29.2 |
| *Herdmania momus* #X53538 | 78.6 | 47.2 | 70.0 | 39.8 |

Analysis of sequence complexity, entropy, and nucleotide distribution provides an efficient way to detect simple sequence repeats in the rDNA. The most simple but frequent case of a low complexity zone is a region of simple sequence repeats. In the current analyses, some signals were detected in the IGS of *Aurelia*. Here, we present plots of complexity distribution along the rDNA using a window size of 100 bases (Fig. 4). By varying the window size, we could obtain a more complete picture of the repeat region. A small-size sliding window reveals relatively short repeats, while a long window may generate long dispersed repeats. In the case of *Aurelia*, a region of low complexity and high entropy corresponded to those of several poly(T) tracts and high AT regions. Overall, analysis of the rDNA revealed that IGS was more informative than the other rDNA regions. These results were highly congruent with the parsimony (Table 3) and dot matrix analyses (Fig. 4).

## Conclusion

In this study, we determined for the first time the single unit sequence of the rDNA of the widely distributed moon jelly *Aurelia* rDNA and presented a structural model of nuclear rDNA for molecular comparisons, particularly in jellyfishes. Using molecular analyses, we detected some useful characteristics such as a block of minisatellites, poly(T) tract, and bi-repeat patterns from the *Aurelia* rDNA IGS. Based on these facts, we conclude that rDNA IGS of *Aurelia* sp.1 is highly informative and is potentially useful as a marker for populations to study the global dispersal and expansion of *Aurelia*.

**Acknowledgments** We would like to thank Drs. Sheikh Raisuddin and Hans-Uwe Dahms for English editing on the manuscript. We are very grateful to Dr. Timothy J. Page and two anonymous reviewers for reading and critical comments. This work was supported by grants of KOSEF NRL (2006) and ETEP (2006) funded to Jae-Seong Lee, and by a grant of KOSEF (2007) funded to Heum Gi Park. Thus work was also supported by a grant from Korea Polar Research Institute (PE08050) funded to Il-Chan Kim.

## References

Arai, M. N., 1997. Functional biology of Scyphozoa. Chapman & Hall, New York, NY: 300.

Chen, C. A., D. J. Miller, N. V. Wei, C.-F. Dai & H.-P. Yang, 2000. The ETS/IGS region in a lower animal, the seawhip, *Junceella fragilis* (Cnidaria: Anthozoa: Octocorallia): compactness, low variation and apparent conservation of a pre-rRNA processing signal with fungi. Zoological Studies 39: 138–143.

Collins, A. G., P. Schuchert, A. C. Marques, T. Jankowski, M. Medina & B. Schierwater, 2006. Medusozoan phylogeny and character evolution clarified by new large and small subunit rDNA data and an assessment of the utility of phylogenetic mixture models. Systematic Biology 55: 97–115.

Dawson, M. N., 2003. Macro-morphological variation among cryptic species of the moon jellyfish, *Aurelia* (Cnidaria: Scyphozoa). Marine Biology 143: 369–379.

Dawson, M. N. & D. K. Jacobs, 2001. Molecular evidence for cryptic species of *Aurelia aurita* (Cnidaria, Scyphozoa). Biological Bulletin 200: 92–96.

Dawson, M. N. & L. E. Martin, 2001. Geographic variation and ecological adaptation in *Aurelia* (Scyphozoa: Semaeostomeae): some implications from molecular phylogenetics. Hydrobiologia 451: 259–273.

Dawson, M. N., A. Sen Gupta & M. H. England, 2005. Coupled biophysical global ocean model and molecular genetic analyses identify multiple introductions of cryptogenic species. Proceedings of the National Academy of Sciences of the United States of America 102: 11968–11973.

Drouin, G., J. D. Hofman & W. F. Doolittle, 1987. Unusual ribosomal RNA gene organization in copepods of the genus *Calanus*. Journal of Molecular Biology 196: 943–946.

Hassouna, N., B. Michot & J.-P. Bachellerie, 1984. The complete nucleotide sequence of mouse 28S rRNA gene. Implications for the process of size increase of the large subunit rRNA in higher eukaryotes. Nucleic Acids Research 12: 3563–3583.

Hillis, D. M. & M. T. Dixon, 1991. Ribosomal DNA: molecular evolution and phylogenetic inference. Quarterly Review of Biology 66: 411–453.

Jeong, S. W., W. H. Lang & R. H. Reeder, 1995. The release element of the yeast polymerase I transcription terminator can function independently of Reb1p. Molecular and Cellular Biology 15: 5929–5936.

Ki, J.-S. & M.-S. Han, 2007. Informative characteristics of 12 divergent domains in complete large subunit rDNA sequences from the harmful dinoflagellate genus, *Alexandrium* (Dinophyceae). Journal of Eukaryotic Microbiology 54: 210–219.

Ki, J.-S., D.-S. Hwang, K. Shin, W. D. Yoon, D. Lim, Y. S. Kang, Y. Lee & J. -S. Lee, 2008. Recent moon jelly (*Aurelia* sp.1) blooms in Korean coastal waters suggest global expansion: examples inferred from mitochondrial COI and nuclear ITS–5.8S rDNA sequences. ICES Journal of Marine Science 65: 443–452.

Kimura, M., 1980. A simple method for estimating evolutionary rates of base substitutions through comparative studies of nucleotide sequences. Journal of Molecular Evolution 16: 111–120.

Kramp, P. L., 1961. Synopsis of the medusae of the world. Journal of the Marine Biological Association of the United Kingdom 40: 337–342.

Kumar, S., K. Tamura, I. B. Jakobsen & M. Nei, 2001. MEGA3: Molecular Evolutionary Genetics Analysis software. Bioinformatics 17: 1244–1245.

Lang, W. H., B. E. Morrow, Q. Ju, J. R. Warner & R. H. Reeder, 1994. A model for transcription termination by RNA polymerase I. Cell 79: 527–534.

Lee, J.-S., 2000. The internally self-fertilizing hermaphroditic teleost *Rivulus marmoratus* (Cyprinodontiformes, Rivulidae) $\beta$-actin gene: amplification and sequence analysis with conserved primers. Marine Biotechnology 2: 161–166.

Marilley, M. & P. Pasero, 1996. Common DNA structural features exhibited by eukaryotic ribosomal gene promoters. Nucleic Acids Research 24: 2204–2211.

Mason, S., M. Wallisch & I. Grummt, 1997. RNA polymerase I transcription termination: similar mechanisms are employed by yeast and mammals. Journal of Molecular Biology 268: 229–234.

Mathews, D. H., J. Sabina, M. Zuker & D. H. Turner, 1999. Expanded sequence dependence of thermodynamic parameters improves prediction of RNA secondary structure. Journal of Molecular Biology 288: 911–940.

Mayer, A. G., 1910. Medusae of the World, Vol. 3. The Scyphomedusae. Carnegie Institution of Washington, Washington, DC: 603–630.

Schlötterer, C., 1998. Ribosomal DNA probes and primers. In Karp, A., P. G. Isaac & D. S. Ingram (eds), Molecular

Tools for Screening Biodiversity. Chapman & Hall, London: 267–276.

Schroth, W., G. Jarms, B. Streit & B. Schierwater, 2002. Speciation and phylogeography in the cosmopolitan marine moon jelly, *Aurelia* sp. BMC Evolutionary Biology 2: 1–10.

Zuker, M., 2003. Mfold web server for nucleic acid folding and hybridization prediction. Nucleic Acids Research 31: 3406–3415.

Hydrobiologia (2009) 616:241–246
DOI 10.1007/s10750-008-9588-5

JELLYFISH BLOOMS

# The potential importance of podocysts to the formation of scyphozoan blooms: a review

**Mary Needler Arai**

Published online: 4 November 2008


**Abstract** Podocysts are cysts with stored reserves of organic compounds produced beneath the pedal discs of polyps of scyphozoans in the orders Rhizostomae (suborder Dactyliophorae) and Semaeostomae. They excyst small polyps that develop into fully active polyps (scyphistomae) capable of further podocyst production and of medusa production by strobilation. They contribute to increasing the number of polyps and also to survival through seasonal periods of reduced food availability or predation. These attributes may help support scyphozoan blooms, but as yet there are few quantitative data.

**Keywords** Reproduction · Polyp · Medusa · Production · Excystment · Predation

## Introduction

Research on scyphozoan blooms has concentrated on the ephyrae and adult medusae (Purcell et al., 2007). More recently, there has been recognition that it is necessary to examine the benthic stages. This has centered on the sequence from planulae settling through polyps and scyphistomae to strobilation releasing ephyrae (Purcell, 2007). Much less work has been done on other modes of asexual reproduction of the polyps such as direct budding of new polyps, production of stolons and hence new polyps, production of planuloid swimming buds, longitudinal fission, and formation and excystment of cysts (Arai, 1997).

Cysts may be formed by encystment of whole polyps, by planulae as they settle (planulocysts), by pedal discs of polyps (podocysts), or by substrate contacting portions of stolons. Whereas the encystment of whole polyps or planulae and subsequent excystment does not lead to increased numbers of polyps, a single polyp may form several podocysts, and, thereby may increase the population of polyps. This note summarizes the present knowledge of such asexual reproduction by formation of podocysts and their subsequent excystment. I postulate that this type of reproduction contributes significantly to increasing the numbers of some species and hence potentially to the production of some medusan blooms.

## Podocyst anatomy, formation, and excystment

Podocysts are chitin-covered cysts that are formed against the substrate by the pedal discs of scyphistomae. After forming a podocyst, the polyp may form a stolon and then move to a new position and leave

Guest editors: K. A. Pitt & J. E. Purcell
Jellyfish Blooms: Causes, Consequences, and Recent Advances

M. N. Arai (✉)
Pacific Biological Station, Nanaimo, BC,
Canada V9T 6N7
e-mail: araim@island.net

the podocyst behind. It may then form another pedal disc, and ultimately another podocyst in the new position. In some species, such as *Catostylus mosaicus* (Quoy & Gaimard), cysts may also be formed by free tips of stolons applied directly to the substrate without polyp involvement. These stoloniferous cysts have also been designated as podocysts by some authors (e.g., Pitt, 2000), but it is not clear if they are equivalent structures.

Podocysts are formed by many Rhizostomae (Table 1) and Semaeostomae (Table 2). As noted by Holst et al. (2007), within the Rhizostomae podocyst production is limited to the suborder Dactyliophorae. The Kolpophorae primarily carry out asexual reproduction of the polyps by production of planuloid swimming buds.

The podocysts are typically irregular discs, most with a central depression on top (Holst et al., 2007). The cell migration involved in their formation has been investigated in some Semaeostomae, but is still incompletely understood (Chapman, 1970; Magnusen, 1980). The covering cuticle is formed of a chitin–protein complex tanned by phenolic substances (Blanquet, 1972b). The podocysts are initially opaque, ranging in color from brown to almost white (Tables 1 and 2). In *Cyanea* spp., in which both planulocysts and podocysts are produced, planulocysts may be clearly distinguished from the podocysts by the former's plano-convex shape, shiny surface, and larger diameter (Dong et al., 2006).

During excystment, a very small four-tentacled polyp is usually formed from the internal tissues that leaves the podocyst via its degraded top. Such polyps have been observed in *Nemopilema nomurai* (see Kawahara et al., 2006), *Rhopilema esculenta* (see Ding & Chen, 1981), *Rhizostoma octopus* (see Kühl, 1972), *R. pulmo* (see Paspaleff, 1938), *Aurelia aurita* (see Chapman, 1968), and *Chrysaora quinquecirrha* (see Cargo & Rabenold, 1980), but *C. fuscescens* polyps excysted with two tentacles (Widmer, 2008). These polyps grow into fully developed scyphistomae capable of forming podocysts or ephyrae (Cargo & Rabenold, 1980; Lotan et al., 1992). The result may be a clone of scyphistomae and podocysts produced from a single scyphistoma over several months (e.g., *Nemopilema nomurai*, see Kawahara et al., 2006).

## Timing and factors controlling production, survival, and excystment of podocysts

The literature on podocysts gives an overall impression of abundant reproduction by this process. There are a number of descriptions and illustrations of trails of several podocysts produced by a single scyphistoma. For example, a single polyp of *Chrysaora quinquecirrha* in culture formed 52 podocysts and six polyps in less than 3 months (Cargo & Schultz, 1966) and a single polyp of *C. fuscescens* formed 53 podocysts and 51 polyps in 8 months (Widmer, 2008). *Rhopilema nomadica* polyps in culture produced an average of 14 podocysts during 2 months (Lotan et al., 1992). There are, however, few quantitative data as yet on the factors influencing the rates and periods of podocyst production and survival.

The earliest literature speculated that podocysts were produced during poor conditions, and that they provided protection against predation or a limited food supply. More recent papers show that podocysts may indeed protect against predation (see below), but the rate of their production is usually positively

**Table 1** Characteristics of podocysts of scyphozoans in the Order Rhizostomeae (suborder Dactyliophorae)

| Species | Color | Diameter | References |
|---|---|---|---|
| *Catostylus mosaicus* (Quoy & Gaimard) | | | Pitt (2000) |
| *Nemopilema nomurai* Kishinouye | Whitish | ca. 300 μm | Kawahara et al. (2006) |
| *Rhizostoma octopus* (Linnaeus) | | 200–500 μm | Kühl (1972), Holst et al. (2007) |
| *Rhizostoma pulmo* (Macri) | | | Paspaleff (1938) |
| *Rhopilema esculenta* Kishinouye | | 200–500 μm | Ding & Chen (1981), Guo (1990) |
| *Rhopilema nomadica* (Galil, Spanier & Ferguson) | Greenish | 100–230 μm | Lotan et al. (1992) |
| *Rhopilema verrilli* (Fewkes) | Whitish | 296–511 μm | Cargo (1971), Calder (1973) |
| *Stomolophus meleagris* (L. Agassiz) | | | Calder (1982) |

**Table 2** Characteristics of podocysts of scyphozoans in the Order Semaeostomeae

| Species | Color | Diameter | References |
|---|---|---|---|
| *Aurelia aurita* (Linnaeus) | Yellowish-brown | 200–700 μm | Chapman (1966, 1968, 1970), Widersten (1969) |
| *Aurelia labiata* Chamisso & Eysenhardt | Greenish-yellow | | Widmer (2006, 2008) |
| *Chrysaora achlyos* Martin et al. | | | Schaadt et al. (2001) |
| *Chrysaora colorata* (Russell) | Greenish-yellow | 200–500 μm | Gershwin & Collins (2002) |
| *Chrysaora fuscescens* (Brandt) | Yellowish-brown | 674 μm | Widmer (2008) |
| *Chrysaora hysocella* (Linnaeus) | | 250–440 μm | Chuin (1930), Morandini et al. (2004) |
| *Chrysaora lactea* Eschscholtz | Yellowish-brown | 200–300 μm | Morandini et al. (2004) |
| *Chrysaora melanaster* (Brandt) | | 300–400 μm | Kakinuma (1967), Morandini et al. (2004) |
| *Chrysaora quinquecirrha* (Desor) | Reddish-brown | 350–450 μm | Truitt (1939), Littleford (1939), Cargo & Rabenold (1980), Magnusen (1980), Morandini et al. (2004) |
| *Cyanea capillata* (Linnaeus) | Greenish-brown | | Grondahl & Hernroth (1987) |
| *Cyanea lamarcki* (Péron & Lesueur) | | | Widersten (1969) |
| *Cyanea nozakii* Kishinouye | | | Dong et al. (2006) |

correlated with the availability of food in otherwise good conditions. Morandini et al. (2004) reported that *Chrysaora lactea* polyps produced podocysts without feeding, but the resulting podocysts did not excyst. Scyphistomae of *Chrysaora quinquecirrha* retained in the laboratory for long periods without food or water exchange decreased in size without encysting, whereas scyphistomae maintained with natural food may form podocysts (Littleford, 1939). When *Rhopilema esculenta* polyps were fed *Artemia* nauplii daily, they produced an average of 7.4 podocysts each in 35 days, and 28.6% of the podocysts excysted (Guo, 1990). When polyps were fed at various intervals up to 28 days, podocyst production rates decreased greatly when feeding intervals exceeded 4 days and excystment became very rare.

After production, podocysts may not survive longer in the laboratory than the polyps of the same species that are able to feed. Littleford (1939) reared some individual polyps of *Chrysaora quinquecirrha* in culture at 20–24°C through 4 years and found that they strobilated each summer. They also formed podocysts in the autumn, which either disintegrated or released polyps in a few months. Black et al. (1976) found survival of a portion of podocysts of that species to at least 25 months; of 536 podocysts at 25°C, 21 disintegrated and 284 excysted within the 25-month period. There was some excystment throughout the 25 months, with monthly rates varying from 1 to 12% of the population.

The yolk of the cells of the central storage zone of podocysts of *Aurelia aurita* and *Chrysaora quinquecirrha* contains carbohydrate, lipid, and protein reserves (Chapman, 1968; Black, 1981). Although metabolism is low, as shown by low oxygen uptake, half of the DNA, one-third of the protein, and one-fifth of the lipid are lost from *C. quinquecirrha* within a year (Black, 1981). There are few data on the maximum survival times of podocysts or polyps of this or other species; however, since podocysts are formed when food is available and the stored reserves are soon depleted, they are probably not useful for survival periods much beyond 2 years, but are more important for survival during the short seasonal periods when food becomes scarce for the active polyps.

The effects of other factors influencing production and excystment such as temperature, and salinity differ between species. In culture at a constant temperature of 20°C, the maximum number of podocysts produced per year by scyphistomae of *Rhizostoma octopus* (L.) was 37 (Holst et al., 2007); fewer podocysts were formed at lower constant temperatures. Similarly, podocysts of *Rhopilema esculenta* were not formed when the temperature was below 10°C or the salinity was less than 6 (Lu et al., 1997). The rate of production increased as the temperature was increased from 15 to 30°C and the optimum salinity was 20–22. In another study on that species, excystment also increased as the temperature increased from 15 to 30°C and optimum salinity for excystment was 18–22 (Jiang et al., 1993). In contrast, podocyst formation of *Chrysaora quinquecirrha* increased when temperatures cooled toward 2–4°C (Cargo & Schultz, 1967) and excystment occurred at 15–18°C (Cargo, 1974). Low temperature acclimation of this species increased the activity of glucose-6-phosphate dehydrogenase, which is necessary for fatty acid synthesis and presumably is important for lipid storage in the podocyst (Blanquet, 1972a).

Podocyst production does not usually occur in the same period as strobilation (although that was observed in *Rhopilema nomadica*, see Lotan et al., 1992). In the field, podocysts of some species may be produced by scyphistomae prior to the period of strobilation (e.g., *Rhopilema esculenta*, see Ding & Chen, 1981), and *Nemopilema nomurai* (see Kawahara et al., 2006). For *Cyanea capillata* from Chesapeake Bay and *Cyanea* sp. from Connecticut, podocyst formation occurred during warming temperatures and excystment occurs during cooling temperatures (Cargo, 1974; Brewer & Feingold, 1991). *Chrysaora quinquecirrha* from Chesapeake Bay formed podocysts in August, but excysted and strobilated in the spring with rising temperature (Cargo & Schultz, 1967; Cargo & Rabenold, 1980). Gröndahl (1988), who used the term podocyst incorrectly for planulocysts of *C. lamarckii*, observed in Sweden that true podocysts were formed after the strobilation period, specifically in winter for *Aurelia aurita* and in late summer or autumn for *C. capillata*; *C. capillata* then excysted in winter and *A. aurita* in spring. These data indicate differences in timing of production between species and also between different populations of the same species.

Other factors may influence the rate of podocyst production. Production and excystment of *Rhopilema esculenta* podocysts were greater in dark than natural light (Jiang et al., 1993). Cargo & Schultz (1966) found encystment by *Chrysaora quinquecirrha* polyps when oxygen was depleted with hydrogen sulphide, but Condon et al. (2001) did not find encystment in oxygen-depleted water.

There is some ability of podocysts to repair physical damage. If the cuticle was removed from dormant podocysts of *Chrysaora quinquecirrha*, it may be replaced by a thinner cuticle (Blanquet, 1972b; Black et al., 1976; but not Black, 1981). Further removal of the cyst coat causes unusually fast excystment of the four-tentacled polyp. In *Aurelia aurita*, the cell mass released may be still at the planuloid stage or already developed as usual into a small polyp (Chapman, 1968).

The podocyst structure is an effective deterrent against predation. Predators on scyphistomae include various arthropods and nudibranchs (Oakes & Haven, 1971). Aeolid nudibranchs such as *Cratena* spp. and *Coryphella* spp. store undischarged nematocysts of the polyps in cnidosacs at the tips of cerata, finger-like projections on the dorsal surface, and release the nematocysts if disturbed (Cargo & Burnett, 1982; Östman, 1997). In Chesapeake Bay, *Cratena* sp. fed on polyps of *Chrysaora quinquecirrha*, but did not attack the podocysts (Cargo & Schultz, 1967). Similarly, *Coryphella verrucosa* (Sars) ate most scyphistomae of *Aurelia aurita* on settling plates in Gullmar Fjord, Sweden during October, but cysts remained through the winter (Hernroth & Gröndahl, 1985).

## Possible importance of podocysts to blooms

If scyphozoan species known to produce podocysts (Tables 1 and 2) are compared with those involved in blooms (e.g., Purcell et al., 2007; Hamner & Dawson, 2008), a strong similarity is found. Familiar generic names such as *Aurelia*, *Chrysaora*, *Cyanea*, *Nemopilema*, *Rhopilema*, and *Rhizostoma* appear in both. This is especially striking when the rather meager knowledge of podocysts described above is considered. Blooms also are produced by scyphozoans without podocysts, such as the holoplanktonic *Pelagia noctiluca* (Forskål). Nevertheless, it appears probable that podocysts contribute in some part to the ability of a species to reach bloom size populations.

In order to understand the importance of podocysts to ecological situations, it will be necessary to investigate their functions in the field. Much of the present data have been obtained from laboratory cultures. Aquaria have developed husbandry techniques for production of polyp cultures of species such as *Chrysaora achlyos* (see Schaadt et al., 2001) and *C. fuscescens* (see Widmer, 2008), and aquaculture techniques have been developed for *Rhopilema esculenta* (see You et al., 2007). The small size of the podocysts restricts ability to find them in the field;

however, settling plates can be deployed. If polyps develop on the settling plates, it will then be possible and necessary to track populations of polyps and quantify cyst production and excystment under various conditions. In the Gullmar fjord of Sweden, *Cyanea capillata* polyps at least 1-year-old produced chains of podocysts during the autumn, but podocysts were not produced by the newly settled planulae (Gröndahl & Hernroth, 1987). Thus, the plates will need to be studied in periods other than during strobilation, as has been largely studied to date.

The value of the podocysts lies in increasing production of polyps, short-term survival during low food supply, and also in the protection against predators such as nudibranchs. These all require further investigation. As described above, the initial production rate of new individuals by encystment is seasonally higher than that by strobilation in many species. What is not understood is the high variation in time of excystment and of the numbers surviving, even among the cysts produced by a single polyp. Few studies have been conducted for more than a few months, so it is unknown how long encysted podocysts can survive depletion of the food reserves. It is also unknown whether any of the arthropod predators of the polyps are able to penetrate the chitin of the podocysts.

**Acknowledgment** I thank D. Welch for help in translating portions of Chinese papers.

## References


Arai, M. N., 1997. A Functional Biology of Scyphozoa. Chapman & Hall, New York: 316 pp.

Black, R. E., 1981. Metabolism and ultrastructure of dormant podocysts of *Chrysaora quinquecirrha* (Scyphozoa). Journal of Experimental Zoology 218: 175–182.

Black, R. E., R. T. Enright & L.-P. Sung, 1976. Activation of the dormant podocyst of *Chrysaora quinquecirrha* (Scyphozoa) by removal of the cyst covering. Journal of Experimental Zoology 197: 403–413.

Blanquet, R. S., 1972a. Temperature acclimation in the medusa, *Chrysaora quinquecirrha*. Comparative Biochemistry and Physiology 43B: 717–723.

Blanquet, R. S., 1972b. Structural and chemical aspects of the podocyst cuticle of the scyphozoan medusa, *Chrysaora quinquecirrha*. Biological Bulletin 142: 1–10.

Brewer, R. H. & J. S. Feingold, 1991. The effect of temperature on the benthic stages of *Cyanea* (Cnidaria: Scyphozoa), and their seasonal distribution in the Niantic River estuary, Connecticut. Journal of Experimental Marine Biology and Ecology 152: 49–60.

Calder, D. R., 1973. Laboratory observations on the life history of *Rhopilema verrilli* (Scyphozoa: Rhizostomeae). Marine Biology 21: 109–114.

Calder, D. R., 1982. Life history of the cannonball jellyfish, *Stomolophus meleagris* L. Agassiz, 1860 (Scyphozoa, Rhizostomida). Biological Bulletin 162: 149–162.

Cargo, D. G., 1971. The sessile stages of a scyphozoan identified as *Rhopilema verrilli*. Tulane Studies in Zoology and Botany 17: 31–34.

Cargo, D. G., 1974. Comments on the laboratory culture of Scyphozoa. In Smith, W. L. & M. H. Chanley (eds), Culture of Marine Invertebrate Animals. Plenum Publishing Corporation, New York: 145–154.

Cargo, D. G. & J. W. Burnett, 1982. Observations on the ultrastructure and defensive behavior of the cnidosac of *Cratena pilata*. Veliger 24: 325–327.

Cargo, D. G. & G. E. Rabenold, 1980. Observations on the asexual reproductive activities of the sessile stages of the sea nettle *Chrysaora quinquecirrha* (scyphozoan). Estuaries 3(1): 20–27.

Cargo, D. G. & L. P. Schultz, 1966. Notes on the biology of the sea nettle, *Chrysaora quinquecirrha*, in Chesapeake Bay. Chesapeake Science 7: 95–100.

Cargo, D. G. & L. P. Schultz, 1967. Further observations on the biology of the sea nettle and jellyfishes in Chesapeake Bay. Chesapeake Science 8: 209–220.

Chapman, D. M., 1966. Evolution of the scyphistoma. Symposia of the Zoological Society of London 16: 51–75.

Chapman, D. M., 1968. Structure, histochemistry and formation of the podocyst and cuticle of *Aurelia aurita*. Journal of the Marine Biological Association of the United Kingdom 48: 187–208.

Chapman, D. M., 1970. Further observations on podocyst formation. Journal of the Marine Biological Association of the United Kingdom 50: 107–111.

Chuin, T.-T., 1930. Le cycle évolutif du scyphistome de *Chrysaora*. Travaux de la Station Biologique de Roscoff 8: 1–174.

Condon, R. H., M. B. Decker & J. E. Purcell, 2001. Effects of low dissolved oxygen on survival and asexual reproduction of scyphozoan polyps (*Chrysaora quinquecirrha*). Hydrobiologia 451(Developments in Hydrobiology 155): 89–95.

Ding, G. & J. Chen, 1981. The life history of *Rhopilema esculenta* Kishinouye. Journal of Fisheries of China 5: 93–102. Pl. 1–2 (Chinese with English abstract).

Dong, J., C.-Y. Liu, Y.-Q. Wang & B. Wang, 2006. Laboratory observations on the life cycle of *Cyanea nozakii* (Semaeostomida, Scyphozoa). Acta Zoologica Sinica 52: 389–395.

Gershwin, L.-A. & A. G. Collins, 2002. A preliminary phylogeny of Pelagiidae (Cnidaria, Scyphozoa), with new observations of *Chrysaora colorata* comb. nov. Journal of Natural History 36: 127–148.

Gröndahl, F., 1988. A comparative ecological study on the scyphozoans *Aurelia aurita, Cyanea capillata* and *C. lamarckii* in the Gullmar Fjord, western Sweden, 1982 to 1986. Marine Biology 97: 541–550.

Gröndahl, F. & L. Hernroth, 1987. Release and growth of *Cyanea capillata* (L.) ephyrae in the Gullmar Fjord, western Sweden. Journal of Experimental Marine Biology and Ecology 106: 91–101.

Guo, P., 1990. Effect of nutritional condition on the formation and germination of the podocyst of scyphistomae of *Rhopilema esculenta* Kishinouye. Journal of Fisheries of China 14: 206–211. (Chinese with English abstract).

Hamner, W. M. & M. N. Dawson, 2008. A review and synthesis on the systematics and evolution of jellyfish blooms: advantageous aggregations and adaptive assemblages. Hydrobiologia. doi:10.1007/s10750-008-9620-9.

Hernroth, L. & F. Gröndahl, 1985. On the biology of *Aurelia aurita* (L.): 2. Major factors regulating the occurrence of ephyrae and young medusae in the Gullmar Fjord, western Sweden. Bulletin of Marine Science 37: 567–576.

Holst, S., I. Sötje, H. Tiemann & G. Jarms, 2007. Life cycle of the rhizostome jellyfish *Rhizostoma octopus* (L.) (Scyphozoa, Rhizostomeae), with studies on cnidocysts and statoliths. Marine Biology 151: 1695–1710.

Jiang, S., N. Lu & J. Chen, 1993. Effect of temperature, salinity and light on the germination of the podocyst of *Rhopilema esculenta*. Fisheries Science (Tokyo) 12: 1–4. (Japanese with English abstract).

Kakinuma, Y., 1967. Development of a scyphozoan, *Dactylometra pacifica* Goette. Bulletin of the Marine Biological Station of Asamushi 13: 29–33. pl. 1–3.

Kawahara, M., S.-I. Uye, K. Ohtsu & H. Iizumi, 2006. Unusual population explosion of the giant jellyfish *Nemopilema nomurai* (Scyphozoa: Rhizostomeae) in East Asian waters. Marine Ecology Progress Series 307: 161–173.

Kühl, H., 1972. Hydrography and biology of the Elbe estuary. Oceanography and Marine Biology an Annual Review 10: 225–309.

Littleford, R. A., 1939. The life cycle of *Dactylometra quinquecirrha,* L. Agassiz in the Chesapeake Bay. Biological Bulletin 77: 368–381.

Lotan, A., R. Ben-Hillel & Y. Loya, 1992. Life cycle of *Rhopilema nomadica*: a new immigrant scyphomedusan in the Mediterranean. Marine Biology 112: 237–242.

Lu, N., S. Jiang & J. Chen, 1997. Effect of temperature, salinity and light on the podocyst generation of *Rhopilema esculenta* Kishnouye. Fisheries Science 16: 3–8. (Chinese with English abstract).

Magnusen, J. E., 1980. Epidermal cell movement during podocyst formation in *Chrysaora quinquecirrha*. In Tardent, P. & R. Tardent (eds), Developmental and Cellular Biology of Coelenterates. Elsevier/North Holland Biomedical Press, Amsterdam: 435–440.

Morandini, A. C., F. L. da Siveira & G. Jarms, 2004. The life cycle of *Chrysaora lactea* Eschscholtz, 1829 (Cnidaria, Scyphozoa) with notes on the scyphistoma stage of three other species. Hydrobiologia 530(531): 347–354.

Oakes, M. J. & D. S. Haven, 1971. Some predators of polyps of *Chrysaora quinquecirrha* (Scyphozoa, Semaeostmae) in the Chesapeake Bay. Virginia Journal of Science 22: 45–46.

Östman, C., 1997. Abundance, feeding behavior and nematocysts of scyphopolyps (Cnidaria) and nematocysts in their predator, the nudibranch *Coryphella verrucosa* (Mollusca). Hydrobiologia 355: 21–28.

Paspaleff, G. W., 1938. Über die Entwicklung von *Rhizostoma pulmo* Agass. Arbeiten aus der Biologischen Meeresstation am Schwarzen Meer in Varna 7: 1–25.

Pitt, K. A., 2000. Life history and settlement preferences of the edible jellyfish *Catostylus mosaicus* (Scyphozoa: Rhizostomeae). Marine Biology 136: 269–279.

Purcell, J. E., 2007. Environmental effects on asexual reproduction rates of the scyphozoan *Aurelia labiata*. Marine Ecology Progress Series 348: 183–196.

Purcell, J. E., S.-I. Uye & W.-T. Lo, 2007. Anthropogenic causes of jellyfish blooms and their direct consequences for humans: a review. Marine Ecology Progress Series 350: 153–174.

Schaadt, M., L. Yasukochi, L. Gershwin & D. Wrobel, 2001. Husbandry of the black jelly (*Chrysaora achlyos*), a newly discovered scyphozoan in the eastern North Pacific Ocean. Bulletin de l'Institut Océanographique Monaco 20: 289–296.

Truitt, R. V., 1939. Stoloniferous, pedal disc and somatic budding in the common sea nettle, *Dactylometra quinquecirrha,* L. Asassiz. Bulletin of the Natural History Society Maryland 9: 38–39.

Widersten, B., 1969. Development of the periderm and podocysts in *Cyanea palmstruchi* Swartz 1809. Zoologiska Bidrag fran Uppsala 38: 51–60.

Widmer, C. L., 2006. Life cycle of *Phacellophora camtschatica* (Cnidaria: Scyphozoa). Invertebrate Biology 125: 83–90.

Widmer, C. L., 2008. Life cycle of *Chrysaora fuscescens* (Cnidaria: Scyphozoa) and a key to sympatric ephyrae. Pacific Science 62: 71–82.

You, K., C. Ma, H. Gao, F. Li, M. Zhang, Y. Qiu & B. Wang, 2007. Research on the jellyfish (*Rhopilema esculentum* Kishinouye) and associated aquaculture techniques in China: current status. Aquaculture International 15: 479–488.

Hydrobiologia (2009) 616:247–258
DOI 10.1007/s10750-008-9597-4

JELLYFISH BLOOMS

# Effects of temperature and light intensity on asexual reproduction of the scyphozoan, *Aurelia aurita* (L.) in Taiwan

**Wen-Cheng Liu · Wen-Tseng Lo · Jennifer E. Purcell · Hao-Hsien Chang**

Published online: 23 September 2008


**Abstract** Jellyfish blooms cause problems worldwide, and they may increase with global warming, water pollution, and over fishing. Benthic polyps (scyphistomae) asexually produce buds and small jellyfish (ephyrae), and this process may determine the population size of the large, swimming scyphomedusae. Environmental factors that affect the asexual reproduction rates include food, temperature, salinity, and light. In this study, polyps of *Aurelia aurita* (L.), which inhabit Tapong Bay, southwest Taiwan, were tested in nine combinations of temperature (20, 25, 30°C) and light intensity (372, 56, and 0 lux) in a 12 h light–12 h dark photoperiod. Production of new buds decreased with warmer temperature and stronger light intensity. Warm temperature accelerated strobilation and increased the daily production of ephyrae. The proportion of ephyrae of total asexual reproduction (new buds + ephyrae) increased dramatically in warmer temperature and more light. Survival was reduced in the highest temperature. Strobilation did not occur in the lowest temperature in darkness. All measures of total asexual reproduction indicated that mid- to high temperatures would lead to faster production of more jellyfish. Continuous high temperatures might result in high polyp mortality. Light affected asexual reproduction less than did temperature, only significantly accelerating the strobilation rate. Because the interactive effects of light and temperature were significant for the time period polyps survived and the potential production of jellyfish polyp$^{-1}$, combined light and temperature effects probably are important for strobilation in situ.

**Keywords** Jellyfish · Climate · Strobilation · Bloom

Guest editors: K. A. Pitt & J. E. Purcell
Jellyfish Blooms: Causes, Consequences, and Recent Advances

W.-C. Liu (✉) · W.-T. Lo · J. E. Purcell · H.-H. Chang
Department of Marine Biotechnology and Resources, National Sun Yat-Sen University, Kaohsiung 804, Taiwan, ROC
e-mail: m955020001@student.nsysu.edu.tw

J. E. Purcell
Shannon Point Marine Center, Western Washington University, 1900 Shannon Point Road, Anacortes, WA 98221, USA

## Introduction

The ecological importance of large gelatinous zooplankton, such as jellyfish and ctenophores, has been demonstrated for years (reviewed by Arai, 1988, 2001; Purcell, 1997; Mills, 2001). When jellyfish aggregate in great abundance and form large swarms, or ‘blooms’, they often cause significant environmental and economic impacts. For instance, they compete for food resources with zooplanktivorous fish and also consume fish eggs and larvae (Möller, 1980), which might lead to the depletion of fish stocks and reduction of harvest (reviewed by Purcell, 1985; Arai, 1988; Bailey & Houde, 1989; Purcell & Arai, 2001). In addition, they clog and break trawl nets (Shimomura, 1959; Uye & Ueta, 2004), and block the cooling water intake of coastal power plants (Rajagopa et al., 1989).

Increasing populations of jellyfish may reflect the success of asexual reproduction of the benthic polyps (scyphistomae), because they produce the young jellyfish (ephyrae). Several different methods of asexual reproduction of the benthic polyps have been reported (Kakinuma, 1975; Arai, 1997, this volume; Lucas, 2001; Watanabe & Ishii, 2001). Therefore, production and mortality of the polyp stage might greatly affect the population size of the medusa stage (Gröndahl, 1988; Lucas, 2001). To better understand and possibly forecast jellyfish blooms, it is essential to study the benthic stage in detail; however, compared with medusae, few field or laboratory studies have been done on the benthic stages of common blooming species, such as *A. aurita* (Linnaeus), especially in tropical regions.

Several environmental factors, including nutrients, temperature, salinity, and photoperiod are important in both the seasonal timing of reproduction and the reproductive success in marine invertebrates (Olive, 1985). Results from several studies suggested that temperature has decisive effects on jellyfish populations (reviewed in Purcell, 2005). In temperate regions, *A. aurita* polyps usually produce jellyfish (strobilate) in late winter and early spring when temperature and daylight just begin to increase (reviewed in Lucas, 2001; Miyake et al., 2002). Similarly, *Nemopilema nomurai* (Kishinouye) strobilated in the laboratory when water temperature was increased from 13 to 23°C (Kawahara & Uye, 2006). Total asexual reproduction and the proportion of ephyrae of total asexual reproduction (buds + ephyrae) of other temperate species (*Chrysaora quinquecirrha* Desor, *Moerisia lyonsi* Boulenger, and *A. labiata* Chamisso & Eysenhardt) also increased in warm conditions (Purcell et al., 1999; Ma & Purcell, 2005a, 2005b; Purcell, 2007). In contrast, in subtropical south Florida, *Cassiopea xamachana* Bigelow strobilate in late summer and autumn near the annual high temperature (Fitt & Costley, 1998). In tropical Palau, the population of *Mastigias* sp. was reduced in unusually warm, salty conditions (Dawson et al., 2001). Therefore, jellyfish populations in temperate regions may increase in higher temperatures due to global warming, while jellyfish populations in tropical regions may not.

Light also is an important factor in jellyfish ecology, even for the species without endosymbiotic algae (Hamner et al., 1994; Molinero et al., 2005); Purcell & Decker, 2005, but little is known about the significance of light for scyphozoan abundance. Custance (1964) suggested that light inhibited strobilation; however, the maximum ephyra production occurred in the group with light. In contrast, longer light periods accelerated strobilation in two species (Loeb, 1973; Purcell, 2007).

The common moon jellyfish, *A. aurita*, which occurs between 40°S and 70°N (Russell, 1970; Mianzan & Cornelius, 1999), is a cosmopolitan species (Mayer, 1910; Arai, 1997) with several sibling species identified in recent years (Dawson et al., 2005). Moon jellyfish blooms have been recorded worldwide in temperate (Kang & Park, 2003; Uye et al., 2003; Uye & Ueta, 2004) to tropical regions (Martin, 1999). Human problems with jellyfish blooms have increased recently, especially in Asia and Europe (reviewed in Purcell et al., 2000). The blooms may be related to the combinations of several factors (Purcell et al., 2000), including climate change, especially warming (Uye & Ueta, 2004; Purcell, 2005; Attrill et al., 2007), eutrophication (Arai, 2001), over fishing (Zaitsev, 1992; Kideys & Gücü, 1995; Ishii and Båmstedt 1998), and addition of hard surfaces to coastal waters (Lo et al., 2008). Polyps are known to occur on the undersides of docks in harbors (e.g., Kakinuma, 1975; Miyake et al., 2002; Hoover & Purcell, this volume), but little is known about the responses of the polyp stage to changes in environmental conditions.

No previous studies have been conducted on asexual reproduction rates of tropical non-zooxanthellate scyphozoans. The purpose of this study was twofold. First, effects of temperature and light can be additive in other organisms (Balzar & Hardeland, 1996), but this had not been tested for jellyfish. Second, we wanted to compare the responses to temperature and light of tropical *A. aurita* with those of temperate *A. labiata* (in Purcell, 2007). Taiwanese *A. aurita* live in much warmer waters with a greater temperature range (19–32°C) and more consistent light (11–13 h photoperiod) than *A. labiata* in the northeast Pacific (7–15°C, and 8–18 h photoperiod) and we suspected that their responses to temperature and light might differ. Thus, we measured the survival, polyp budding rates, and rates of ephyra production of *A. aurita* polyps to test the effects of combinations of different temperatures and light intensities on asexual reproduction rates and the possible consequences for bloom events.

## Materials and methods

All *A. aurita* polyps were obtained from the National Museum of Marine Biology and Aquarium, Pingtung, Taiwan. The polyps were produced by jellyfish collected from Tapong Bay, southwest Taiwan, where the temperature ranged from 19 to 32°C, and maintained at the aquarium for more than 2 years. Polyps of similar size were placed individually in 50-$cm^3$ wells of 6-well acrylic boxes and cultured in 110-μm filtered sea water. Polyps were kept in a darkened incubator at 20°C without food for one month before the experiment to prevent any previous conditions from stimulating strobilation.

The experiment consisted of two orthogonal treatments. The first was temperature with three levels (20, 25, 30°C) spanning the annual temperature range in situ (Chen, 2002). The second treatment was three levels of "light intensity" (light: 372 lux, shade: 56 lux, and dark: 0 lux). Unfortunately, no data were collected on the natural range of light intensities polyps of *A. aurita* experience in Taiwan. Therefore, light intensities for the experiment were chosen according to those reported for the related species, *A. labiata* (in Purcell, 2007). Light intensities underneath a dock where *A. labiata* polyps occur abundantly was ≤64 lux in January and ≤140 lux in June (Purcell, 2007). Because tropical latitudes would receive stronger light than temperate latitudes, we set the highest light intensity at 372 lux and the middle group at 56 lux, which is similar to Purcell's (2007) maximum intensity. In our study, 18 polyps were arbitrarily allocated to each of the nine combinations of temperature and light. The experiment began on 17 February 2007.

Three temperatures were maintained in three incubators (HIPOINT). Three light intensity treatments were created within each incubator: the dark treatment in opaque black boxes, the shade treatment in clear boxes with one layer of black plastic screen, and the light treatment in clear boxes with no screen. Light was provided by one natural-spectrum bulb in the top center of each incubator. A black plastic screen was set below the light bulb to reduce the light intensity. Photoperiod was maintained at 12 h light–12 h dark in each incubator, which is typical of light conditions in spring and autumn. Temperatures in each treatment were measured and recorded once hourly throughout the experiment by Onset light and temperature loggers (HOBO) (Table 1). The light intensity in each treatment also was measured by an INS digital lux meter (DX-100).

Polyps were moved into their treatment temperatures but kept in the dark and unfed for 1 week. Thereafter, all polyps were fed excess *Artemia* sp. nauplii and cleaned every 2 or 3 days. After the polyps fed for 2–3 h, every well was cleaned with swabs, and seawater with uneaten food was discarded and replaced with filtered seawater of the same temperature and salinity. This feeding protocol was intended to provide saturating prey briefly, thus providing equal food to all treatments and minimizing enhanced feeding at warmer temperature (Ma & Purcell, 2005b). During data collection and cleaning, polyps were exposed to microscope light and indirect ceiling fluorescent room light. Each box of six polyps was exposed to the same extraneous light for about 10 min twice a week. The numbers of new buds (#B) and ephyrae (#E) for each polyp were counted at each cleaning. After enumeration, ephyrae and new buds that had separated from initial polyps were removed.

For analyses, several characteristics were defined as follows. The duration between attachment to death of the polyps was the 'survival period' (SP); the duration between polyp attachment and the beginning of strobilation was the 'pre-strobilation period'; the duration between the beginning of strobilation to release of all ephyrae was the 'strobilation period'. The numbers of buds and ephyrae were divided by the SP for each polyp (buds $d^{-1}$ and ephyrae $d^{-1}$, respectively). The proportion of strobilating polyps of the total polyps in each treatment is the 'strobilation ratio'. The proportion of ephyrae of the total asexual reproduction was calculated as #E/(#B + #E) for each polyp. To estimate the possible population size of jellyfish for each treatment, we multiplied #B + 1 (the original polyp) by #E to calculate the 'potential yield'.

**Table 1** Experimental conditions in nine combinations of temperature (20, 25, 30°C) and light intensity (Light = 372 - lux, Shade = 56 lux, Dark = 0 lux) at local seawater salinity (33.5) with 12-h photoperiod during the experiment as measured with Onset data loggers

| Temperature (°C) and light intensity (lux) | 20°C | 25°C | 30°C | *H*- and *P*-values (Light) |
|---|---|---|---|---|
| Light treatments | 20.9 ± 0.01°C | 25.5 ± 0.01°C | 30.0 ± 0.02°C | $H_2 = 0.081; P = 0.96$ |
| | 372 ± 0.3 lux | 372 ± 0.2 lux | 372 ± 0.2 lux | |
| Shade treatments | 20.9 ± 0.02°C | 25.5 ± 0.02°C | 30.0 ± 0.01°C | $H_2 = 0.024; P = 0.99$ |
| | 56 ± 0.1 lux | 56 ± 0.1 lux | 56 ± 0.1 lux | |
| Dark treatments | 20.9 ± 0.01°C | 25.5 ± 0.01°C | 29.9 ± 0.07°C | Not tested |
| | BD | BD | BD | |
| *H*- and *P*-values (Temperature) | $H = 0.0125, P = 0.911$ | $H = 1.363, P = 0.243$ | $H = 0.865, P = 0.352$ | |

Treatments were tested by Kruskal–Wallis ANOVA on Ranks (*H*-statistic). $P < 0.05$ is considered significantly different

### Statistical analysis

Because the three light treatments were repeated within each of the three incubators, the treatments were not truly independent; however, because temperature and light were monitored independently by HOBO data loggers in each treatment, and the treatments did not differ significantly (Table 1), and we show that the treatments were effectively independent. Thus, parametric statistical tests were used. The numbers of strobilating polyps were transformed into relative proportions (strobilation ratio) and compared among treatments by a Chi-square test. The proportion of polyps that died without strobilating were analyzed by non-parametric Kruskal–Wallis one-way ANOVA on ranks because numerous zeros made Chi-square analysis inappropriate. Pairwise comparisons were by Dunn's method if the ANOVA results were significant. Other data (including SP, pre-strobilation period, strobilation period, #B/SP, #E/SP, #E/(#E + #B), and potential yield) were stabilized by ln(X + 1) transformation (Underwood, 1997) and then analyzed by Two-Way Analysis of Variance (ANOVA) with Sigma Stat. If the ANOVA results were significant, Bonferroni pairwise comparisons were calculated.

## Results

### Survival

Survival during the first week was high in all treatments (>95%). There was a significant interaction between temperature and light intensity ($P < 0.05$; Table 2). There was no significant difference among light intensities for polyps raised at 20 and 30°C, but at 25°C, the SP in the light (45 d) was shorter and significantly different from survival in the dark (62 d) ($P = 0.027$; Table 2; Fig. 1). Comparison of SPs among temperatures within each light intensity group indicated that, in the light, survival was significantly longer at 20°C (74 d) than 25°C (45 d) and 30°C (43 d) ($P < 0.001$). The SPs varied among all temperatures in both the shade ($P < 0.05$; 20°C, 79 d; 25°C, 60 d; 30°C, 31 d) and dark treatments ($P < 0.01$; 20°C, 86 d; 25°C, 62 d; 30°C, 35 d) (Fig. 1).

In 30°C-light, 34% of the polyps died without strobilating; it seems that warm temperature with high light intensity caused high mortality before strobilation. In addition, only 17% of the polyps in 25°C-shade died without strobilating, although, almost one-third of the polyps in 30°C-shade died without strobilating (Fig. 2). Similar results occurred at other light intensities. The differences in the proportion of polyps that died without strobilating among temperatures were significant ($P < 0.01$), but the differences were not significant among light intensities ($P > 0.05$) (Table 3).

### Budding rates

The interaction effects of temperature and light intensity on the numbers of buds polyp$^{-1}$ d$^{-1}$ were not significant. The greatest numbers of new buds were produced in the low temperature (20°C) treatments (Fig. 3). When these data were standardized by

**Table 2** The effects of temperature and light intensity on polyps were tested in combinations of temperature (20, 25, 30°C) and light intensity (L; light = 372 lux, S; shade = 56 lux, D; dark = 0 lux) at local seawater salinity (33.5) and 12-h photoperiod in an 86-day experiment

| Variable tested | Survival period (d) | Pre-strobilation period (d) | Strobilation period (d) | #B/SP (no./d) | #E/SP (no./d) | $\frac{\#E}{\#E+\#B}$ | Potential yield |
|---|---|---|---|---|---|---|---|
| Temperature | | | | | | | |
| Test statistic | $F_{2,157} = 83.50$ | $F_{2,85} = 35.68$ | $F_{2,85} = 20.34$ | $F_{2,157} = 84.57$ | $F_{2,85} = 6.3$ | $F_{2,157} = 31.75$ | $F_{2,157} = 20.66$ |
| $P$-value | $P < 0.001$ | $P < 0.001$ | $P < 0.001$ | $P < 0.001$ | $P = 0.003$ | $P < 0.001$ | $P < 0.001$ |
| Pairwise | | 30 25 20 | 30 25 20 | 30 25 20 | 20 25 30 | 20 25 30 | |
| Light | | | | | | | |
| Test statistic | $F_{2,157} = 0.54$ | $F_{2,85} = 0.29$ | $F_{2,85} = 1.87$ | $F_{2,157} = 1.74$ | $F_{2,85} = 1.21$ | $F_{2,157} = 2.62$ | $F_{2,157} = 1.16$ |
| $P$-value | NS | NS | NS | NS | NS | NS | NS |
| T × L | | | | | | | |
| Test statistic | $F_{4,155} = 3.17$ | – | – | $F_{4,155} = 1.05$ | – | $F_{4,155} = 1.49$ | $F_{4,155} = 5.02$ |
| $P$-value | $P = 0.018$ | – | – | NS | – | NS | $P < 0.001$ |
| | Pairwise comparison of SP | | | | Pairwise comparison of potential yield | | |
| | By light intensity: | By temperature: | | | By light intensity: | By temperature: | |
| | 20L$^a$ 20S$^a$ 20D$^a$ | 20L$^a$ 25L$^b$ 30L$^b$ | | | 20L$^a$ 20S$^a$ 20D$^b$ | 20L$^a$ 25L$^a$ 30L$^a$ | |
| | 25L$^a$ 25S$^{a,b}$ 25D$^b$ | 20S$^a$ 25S$^b$ 30S$^c$ | | | 25L$^a$ 25S$^a$ 25D$^a$ | 20S$^{a,b}$ 25S$^a$ 30S$^b$ | |
| | 30L$^a$ 30S$^a$ 30D$^a$ | 20D$^a$ 25D$^b$ 30D$^c$ | | | 30L$^a$ 30S$^a$ 30D$^a$ | 20D$^a$ 25D$^b$ 30D$^a$ | |

Two-way ANOVAs and Chi-square test showed significant effects. Significantly different pairwise comparisons for each ANOVA were indicated by disconnected underlines. Pre-strobilation period is from attachment to strobilation. Strobilation period is from beginning if strobilation to first ephyra release. #B = number of buds; #E = number of ephyrae. Potential yield estimated possible jellyfish yield [(#B + 1) × #E]. NS = not significant. Significantly different pair-wise comparisons for each ANOVA are indicated by different letters (a, b, c)

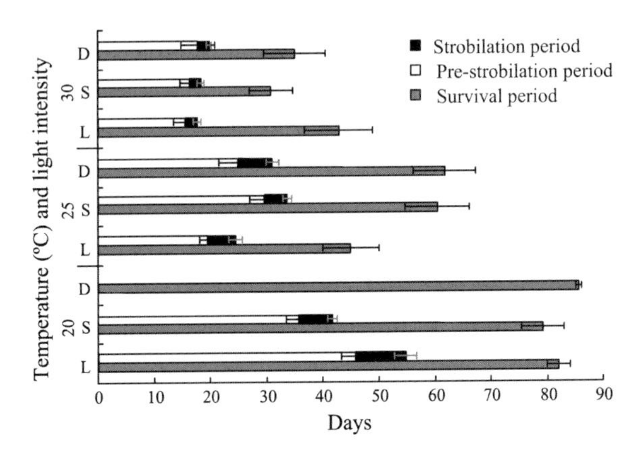


**Fig. 1** Average SPs (gray bars), pre-strobilation periods (white bars), and strobilation periods (black bars), in an 86-day experiment with combinations of temperature (20, 25, 30°C) and light intensity (L: light = 372 lux; S: shade = 56 lux; D: dark = 0 lux) at local seawater salinity (33.5) and 12-h photoperiod

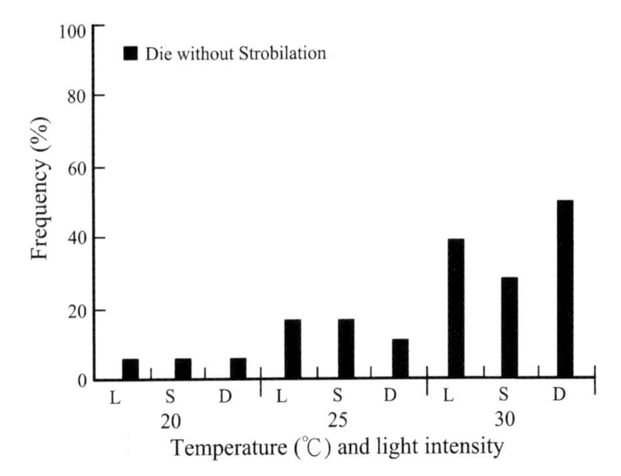


**Fig. 2** Percentages of polyps that died without strobilating in an 86-day experiment with combination of temperature (20, 25, 30°C) and light intensity (L: light = 372 lux; S: shade = 56 lux; D: dark = 0 lux) at local seawater salinity (33.5) and 12-h photoperiod

the days of survival (SP), the most new buds polyp$^{-1}$ d$^{-1}$ were in 20°C-shade (0.09), fewer in 25°C-shade (0.03), and fewest in 30°C-shade (0.01). Similar results were found in other treatments; the numbers of new buds polyp$^{-1}$ d$^{-1}$ were lower at warmer temperature, and the temperature effect was significant ($P < 0.001$, Table 2). Fewer new buds

**Table 3** The proportion of polyps that died without strobilating in nine combinations of temperature (20, 25, 30°C) and light intensity (Light = 372 lux, Shade = 56 lux, Dark = 0 lux) at local seawater salinity (33.5) with 12-h photoperiod

| Light intensity | Temperature | | |
|---|---|---|---|
| | 20°C | 25°C | 30°C |
| Propotion of polyps that diad without strobilating (%) | | | |
| Light | 0 | 17 | 50 |
| Light | 0 | 33 | 33 |
| Light | 17 | 0 | 20 |
| Shade | 17 | 17 | 33 |
| Shade | 0 | 33 | 17 |
| Shade | 0 | 0 | 33 |
| Dark | 0 | 33 | 50 |
| Dark | 0 | 0 | 20 |
| Dark | 17 | 0 | 67 |
| | Temperature | | Light |
| Statistical analysis | | | |
| Test statistic | $H_2 = 13.571$ | | $H_2 = 0.064$ |
| *P*-value | $P = 0.001$ | | NS |
| Dunn's method | 20 25 30 | | |

*H*: Kruskal–Wallis statistic. NS: not significant. Significant pairwise comparisons are indicated by disconnected underlines

polyp$^{-1}$ d$^{-1}$ were produced at higher light intensities, but the difference was not statistically significant.

## Strobilation ratios and rates

The proportions of polyps that strobilated (strobilation ratios) were higher at the two warmer temperatures than at 20°C (Table 4), with the highest ratios at 25°C. Strobilation in 25°C-shade (77.8%) was about 1.8 times that in 20°C-shade (44.4%). Although 30°C-shade did not show the highest strobilation ratio (66.7%), it was still about 1.5 times that in 20°C-shade. Similar results were found in other light treatments. The strobilation ratio also differed within light conditions. In the light and shade conditions, 56% and 63% of polyps strobilated,

**Table 4** The proportions of polyps strobilating (strobilation ratio) in nine combinations of temperature (20, 25, 30°C) and light intensity (Light = 372 lux, Shade = 56 lux, Dark = 0 lux) at local seawater salinity (33.5) with 12-h photoperiod

| Light intensity | Strobilation ratio (%) | | | | |
|---|---|---|---|---|---|
| | Temperature | | | Test statistic | *P*-value |
| | 20°C | 25°C | 30°C | | |
| Light | 50 | 83 | 33 | Temperature $\chi^2_{10} = 33.381$ | $P < 0.001$ |
| Light | 17 | 67 | 50 | | |
| Light | 50 | 100 | 80 | | |
| Shade | 50 | 83 | 67 | Light $\chi^2_8 = 21.546$ | $P = 0.006$ |
| Shade | 33 | 50 | 67 | | |
| Shade | 50 | 100 | 67 | | |
| *Dark* | *0* | *67* | *50* | *Temperature* $\chi^2_2 = 9.597$ | $P = 0.008$ |
| *Dark* | *0* | *100* | *60* | | |
| *Dark* | *0* | *100* | *33* | | |

Because there was no strobilation in 20°C-dark, we excluded that treatment, and compared the strobilation ratio among temperatures and among light and shade groups. Chi-square tests showed that there were significant differences among temperatures and light intensities. Within the dark group, 25°C- and 30°C-groups were significantly different (cells in italics)

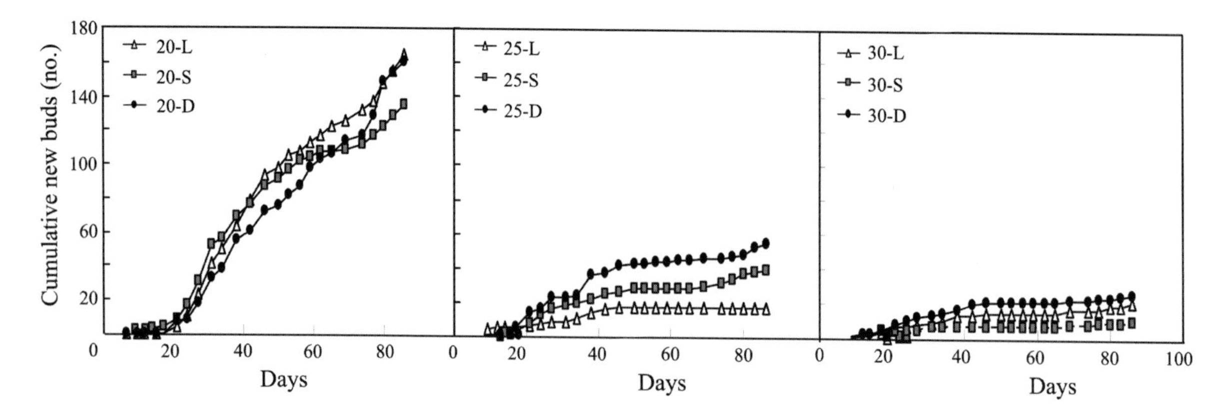


**Fig. 3** The cumulative numbers of new buds in treatments combining temperature (20, 25, 30°C) and light intensity (L: light = 372 lux; S: shade = 56 lux; D: dark = 0 lux) at local seawater salinity (33.5) and 12-h photoperiod in a 86-day experiment

respectively; however, only 45.3% of polyps strobilated in dark treatments, and no polyps strobilated in 20°C-dark (Table 4). Because there was no strobilation in 20°C-dark, we could not do a Chi-square test; therefore, we excluded 20°C-dark and analyzed the results in two parts. Tests for interaction were not possible. We first compared light and shade treatments in all temperatures (20, 25, 30°C); differences in strobilation ratio were significant among temperatures ($P < 0.001$) and light intensities ($P < 0.01$; Table 4). Second, differences among dark treatments at 25 and 30°C were significant for temperature ($P < 0.01$; Table 4).

Interaction effects of temperature and salinity on the pre-strobilation and strobilation periods could not be tested because of missing data (non-strobilating polyps). Both the pre-strobilation and strobilation periods were shorter in 25 and 30°C treatments than at 20°C (Figs. 1, 4). Polyps in all light intensities at the highest temperature (30°C) started strobilating 8 days after the experiment began, and both the pre-strobilation and strobilation periods were shortest at 30°C (Fig. 1). Polyps in 30°C-shade took one-third the number of days to strobilate and release ephyrae as those in 20°C-shade. Differences among treatments were significant for temperature ($P < 0.001$; Table 2). Light intensities did not have consistent effects, and the differences among treatments were not significant.

## Ephyra production

Because no strobilation occurred in the 20°C-dark group, interaction effects could not be tested. Strobilation was faster in warmer temperatures, which resulted in higher ephyra production polyp$^{-1}$ d$^{-1}$ (Table 5). When the numbers of ephyrae were standardized by the SPs, ephyra production

**Table 5** Ephyra production per day and the proportion of total asexual reproduction of each polyp that resulted in production of ephyrae in different combinations of temperature (20, 25, 30°C) and light intensity (Light = 372 lux, Shade = 56 lux, Dark = 0 lux) at local seawater salinity (33.5) and 12-h photoperiod in an 86-day experiment. B = buds; E = ephyrae; SP = survival period

| Temperature (°C) | Light intensity | #E/SP (#E polyp$^{-1}$ d$^{-1}$) | $\frac{\#E}{\#E + \#B}$ |
|---|---|---|---|
| 20 | Light | 0.02 ± 0.01 | 15 ± 5.1% |
| 20 | Light | 0.03 ± 0.01 | 19 ± 6.0% |
| 20 | Light | 0 ± 0.0 | 0 ± 0.0% |
| 25 | Shade | 0.09 ± 0.01 | 74 ± 8.6% |
| 25 | Shade | 0.08 ± 0.01 | 56 ± 8.3% |
| 25 | Shade | 0.08 ± 0.01 | 59 ± 6.7% |
| 30 | Dark | 0.06 ± 0.02 | 41 ± 11.3% |
| 30 | Dark | 0.05 ± 0.01 | 60 ± 10.9% |
| 30 | Dark | 0.04 ± 0.01 | 34 ± 10.1% |

Numbers are mean ± SE

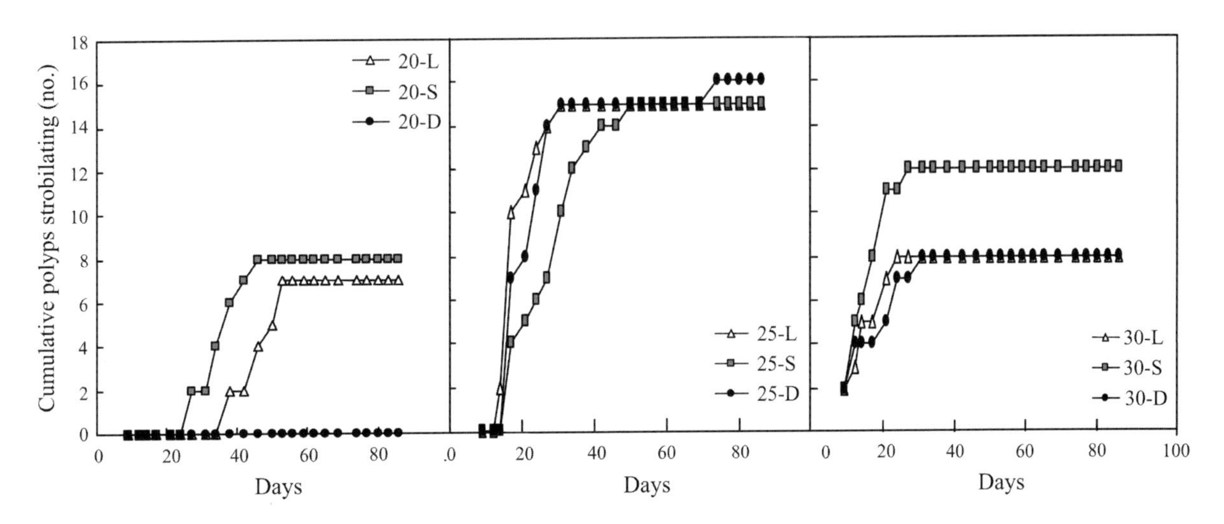


**Fig. 4** The cumulative numbers of strobilating polyps in treatments combining temperature (20, 25, 30°C) and light intensity (L: light = 372 lux; S: shade = 56 lux; D: dark = 0 lux) at local seawater salinity (33.5) and 12-h photoperiod in an 86-day experiment

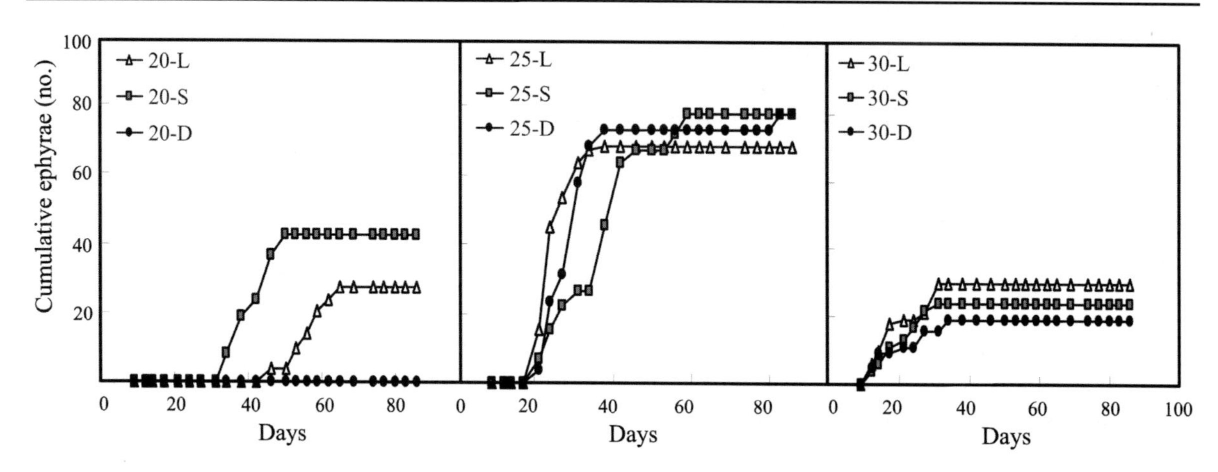


**Fig. 5** The cumulative numbers of ephyrae in treatments combining temperature (20, 25, 30°C) and light intensity (L: light = 372 lux; S: shade = 56 lux; D: dark = 0 lux) at local seawater salinity (33.5) and 12-h photoperiod in an 86-day experiment

polyp$^{-1}$ d$^{-1}$ in both 25°C-shade (0.08) and 30°C-shade (0.05) was about twice that in 20°C-shade (0.03) (Table 4). Similar results were found in other treatments, and the differences among groups were significant for temperature ($P < 0.01$, Table 2). More ephyrae were produced daily at the intermediate light intensity, and ephyrae were produced only in treatments with light at 20°C (Fig. 5); however, differences among groups were not statistically significant.

Interaction effects between temperature and light were not significant for the proportions of ephyrae of the total asexual reproduction (new buds + ephyrae) (Table 2). The proportions of ephyrae were higher in the two warmer temperatures than at 20°C (Table 5). The proportions of ephyrae in 25°C-shade (56%) and 30°C-shade (60%) were both much higher than in 20°C-shade (19%). Similar results were found in other treatments, and the differences among groups were significant for temperature ($P < 0.001$; Table 2). The Bonferroni test indicated that 20°C was different from the other temperatures. Higher proportions of ephyrae occurred in treatments with light than in dark (Table 5). The ratio of ephyrae of the total asexual reproduction in light and shade groups was greater than 40%, on average, but dark groups had <31% ephyrae; however, the differences among groups were not statistically significant (Table 2).

The potential yield of jellyfish showed significant interactive effects of temperature and light ($P < 0.001$; Table 2). There was no significant difference among light intensities for potential yield at 25°C and 30°C, but at 20°C, the potential yield in the dark (0) was none

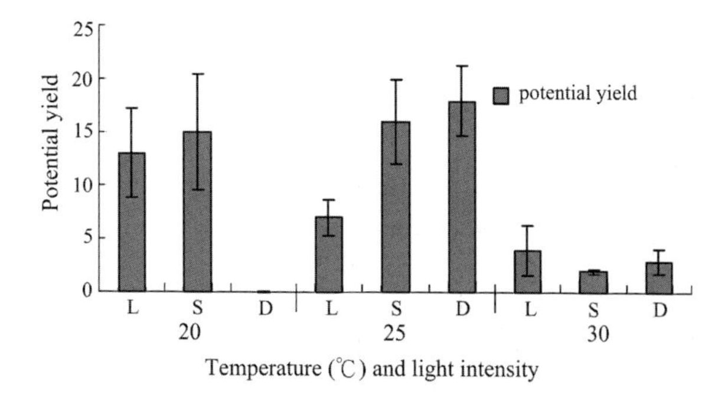


**Fig. 6** The potential yield of ephyrae [(# buds + 1) × # ephyrae] in combinations of temperature (20, 25, 30°C) and light intensity (L: light = 372 lux; S: shade = 56 lux; D: dark = 0 lux) at local seawater salinity (33.5) and 12-h photoperiod in a 86-day experiment

and significantly different from potential yield in the light (13) and shade (15) (Fig. 6; Table 2). Comparison of potential yield among temperatures within each light intensity group indicated that there was no significant difference among temperature for potential yield in light; both in the shade and dark, potential yield was significantly more at 25°C (Shade, 16; Dark, 18) than 20°C (Shade, 15; Dark, 0) and 30°C (Shade, 2; Dark, 3) ($P < 0.001$; Table 2).

## Discussion

### Interactive effects of temperature and light

The effects of warm temperature and light can work together to augment acceleration of seasonal cycles in organisms (Balzar & Hardeland, 1996). Both

Loeb (1973) and Purcell (2007) showed accelerated strobilation in treatments with more light, but temperature effects were not tested. In our experiment, interactive effects of light and temperature were significant for the SP and the potential yield of jellyfish (Table 2, Fig. 6). The low yield in 30°C treatments were caused by the high polyp mortality at 30°C. Although the results were not statistically significant, light and shade groups tended to strobilate earlier than dark groups. Strobilation occurred almost immediately in all treatments at the two highest temperatures (25 and 30°C), therefore, it was not possible to detect potential accelerating effects of light. There was greater separation of light treatment effects at 20°C where polyps required longer to strobilate; polyps in 20°C-dark did not strobilate (Fig. 4). That result provides the strongest suggestion that light and temperature might have additive effects on the timing of strobilation.

Effects of temperature

Temperature was a key factor in our research. The SP was shorter in higher temperatures, which probably reflects increased metabolism combined with limited food reserves (Fig. 1). For each 1°C warming, the SP became 4.7 days shorter. To compare fairly among temperatures, we divided numbers of ephyrae by SP for each polyp. Both the 25°C- and 30°C-groups produced twice the number of ephyrae $d^{-1}$ as in the 20°C-group (Table 5). Thus, at the warmer temperatures, more ephyrae were produced faster. Even though the total numbers were less, it was more like a 'bloom'.

At the warmest temperature (30°C), many polyps died without strobilating (Fig. 2). Heat stress at unusually high, prolonged temperatures has been observed in other scyphomedusae. Widmer (2005) showed that temperate *A. labiata* ephyrae grew best at 21°C; although they grew faster at 22.5–28°C, they showed irreversible heat stress after a week and ultimately perished. Thus, the jellyfish may have been unable to recover from heat lasting more than 7–10 days. Similarly, the thermal maximum of *A. labiata* polyps may be near 20°C, as suggested by less acceleration in strobilation and increase in ephyra production between 15°C and 20°C versus 10°C and 15°C (Purcell, 2007).

Warm temperature also strongly accelerated strobilation. Polyps immediately strobilated after they were moved to 30°C from 20°C, even though they had not yet been fed. Polyps were fed on the 8th day and most polyps in 30°C strobilated on the 10th day. It took twice as long at 25°C, and much longer at 20°C (Fig. 1). At warmer water temperatures, both pre-strobilation and strobilation periods decreased by about 3.6 days per 1°C warming in the light groups, and about 2.3 days per 1°C in shade and dark groups.

Polyps in 30°C did not produce the most ephyrae overall (Fig. 5). If we exclude the 20°C-dark treatment in which polyps did not strobilate, and compare 30°C with 20°C, ephyra production at 30°C was less than at 20°C. This result may be due to insufficient energy reserves, since the polyps were unfed for 1 month before the experiment. *A. aurita* medusae without food reduced somatic growth and allocated energy to reproduction (Ishii & Båmstedt, 1998). Polyps lived with no food for more than 30 days (Hiromi et al., 1995), and they still strobilated after 2 months of starvation, but released fewer ephyrae (Spangenberg, 1967). Our results suggest that high temperatures in poor food conditions would be detrimental to polyp populations in situ.

Even though polyps have the potential to produce more ephyrae in warm conditions, the population might not grow in continued warming. This study showed that polyps mostly bud in cool temperatures and strobilate in warm temperatures. In nature, polyps usually rest or bud before they receive the cues for strobilation. Thus, the "potential yield" of the jellyfish population should be calculated as the number of polyps multiplied by the number of ephyrae. In this study, the potential yield was lowest in 30°C-groups, and was even less than half of 20°C groups if we omit the 20°C-dark group, which never strobilated. Polyps at 25°C, which budded in intermediate numbers and strobilated at moderate speed, had the highest potential yield (Fig. 6). At high temperature, polyps may strobilate without budding and the polyp population would not be renewed. In addition, high temperature resulted in high polyp mortality, both before and after strobilation (Fig. 2). Constant warming may cause untimely strobilation, and polyps could have insufficient energy for revival, and thus, the polyp population would diminish. Strobilation by *Mastigias* sp. polyps was greatest at 31.5°C and was reduced or absent at higher temperatures (32.2–35.2°C; Dawson et al., 2001). These two cases suggest that high temperatures in the tropics may be near the upper tolerance limits for scyphozoans.

This study showed that warming stimulates jellyfish production, which is similar to the results for European *A. aurita* (in Van der Veer & Oorthuysen, 1985; Schneider, 1989; Lucas & Williams, 1994) and other temperate species (Purcell et al., 1999; Ma & Purcell, 2005a, b; Purcell, 2007); however, many jellyfish strobilate after temperatures decrease in situ (reviewed in Lucas, 2001). The abundance of *A. aurita* ephyrae in December, the coldest month in southwest Taiwan, suggested that field polyps began to strobilate when temperature decreased (Chen, 2002). The various responses to temperature around the world, even for species from the same area, suggest that changing temperature is a cue for strobilation.

We concluded that warmer temperature might induce faster strobilation and concentrated ephyra production, which might result in a more intense bloom; however, the jellyfish population could decrease in continued warming because of decreased budding and increased mortality.

Effects of light

Polyps that have been found in nature are attached where light levels are low (e.g., 67–140 lux or 2–6 $\mu$E m$^{-2}$ s$^{-1}$, where *A. labiata* polyps lived; Purcell, 2007), and planulae settle on undersurfaces (Brewer, 1978), so their habitat lacks bright conditions. In a preliminary experiment, with the same treatment combinations except with stronger light, only the 25°C- and 30°C-dark groups strobilated. Because only dark groups strobilated in that experiment, we reduced the light intensities by covering the light with a screen. Our results showed that only the proportion of polyps strobilating (strobilation ratio) was significantly affected by light (Fig. 4, Table 4). Generally, higher proportions of polyps in the intermediate light intensity (56 lux) strobilated than in the highest intensity (372 lux), suggesting that the higher light may inhibit strobilation. A laboratory study on *A. aurita* in Japan showed significant differences in the cumulative proportions of polyps strobilating in light (150 lux; 30%) and dark (94.8%) conditions, and polyps in the dark released eight times more ephyrae than polyps in light (Ishii & Shioi, 2003). On the contrary, in lower light intensities (0, 26, and 52 lux), the numbers of ephyrae produced were similar in light and dark treatments, but strobilation by *A. labiata* polyps was accelerated with increasing light; photoperiod, which provided greater differences in light exposure (6-fold) than did light intensity (2-fold), was more significant (Purcell, 2007). Thus, the greatest strobilation and ephyra production in *Aurelia* spp. may be at light intensities between 52 and 150 lux.

Comparisons of tropical and temperate *Aurelia* spp.

Taiwanese *A. aurita* experience a greater range of water temperatures seasonally (19–32°C) than do *A. labiata* in the northeast Pacific (7–15°C); nevertheless, warm temperature accelerated strobilation in both species. Taiwanese *A. aurita* appeared to be detrimentally affected by 30°C temperatures, however. Their polyp survival and strobilation were significantly less at the warmest temperature. Our results suggest that tropical *A. aurita* polyp populations could be reduced by ocean warming. In contrast, temperate *A. labiata* polyps had strong survival and strobilation even 3°C above temperatures encountered in situ (Purcell, 2007), which suggests that ocean warming could increase their populations in the northeast Pacific.

Photoperiod ranges from 8 to 18 h light d$^{-1}$ in Washington, USA, where *A. labiata* lives. In low light intensities (0, 26, 52 lux), the numbers of *A. labiata* ephyrae produced were not significantly different, but strobilation was accelerated with increasing light (Purcell, 2007). Light availability changes dramatically when *A. labiata* polyps strobilate in situ (February–March) and could provide a strong signal for the timing of their strobilation. In contrast, photoperiod differs less seasonally (10.7–13.2 h light d$^{-1}$) in tropical Taiwan for *A. aurita* than for temperate *A. labiata*. In our experiments, strobilation was not significantly faster in the two treatments with light; however, at the lowest temperature (20°C, the temperature in situ when polyps strobilate), strobilation may have been delayed beyond the duration of our experiment. We saw possible negative effects of light intensities (372 lux) on strobilation of *A. aurita*, which was not tested on *A. labiata*. We were unable to clearly determine if temperature and light effects were additive. Although the strength of seasonal light transitions differs in temperate and tropical waters, light affected strobilation for both tropical and temperate *Aurelia* spp. polyps.

**Acknowledgments** We would like to thank Mr. T. M. Hsiao of the National Museum of Marine Biology and Aquarium, who provided the polyps, and also all members of marine zooplankton laboratory of National Sun Yet-Sen University. This research was supported by grants from the National Science Council and the Ministry of Education of the Republic of China to Lo, W. T. [NSC95-2611-M 110-006, 95-C030220 (Kuroshio project)].

## References

Arai, M. N., 1988. Interactions of fish and pelagic coelenterates. Canadian Journal Zoology 66: 1913–1927.

Arai, M. N., 1997. A functional biology of Scyphozoa. Chapman & Hall, London: 316.

Arai, M. N., 2001. Pelagic coelenterates and eutrophication: a review. Hydrobiologia 451: 69–87.

Arai, M. N., This volume. Are podocysts important to the formation of scyphozoan blooms? Hydrobiologia.

Attrill, M. J., J. Wright & M. Edwards, 2007. Climate-related increases in jellyfish frequency suggest a more gelatinous future for the North Sea. Limnology and Oceanography 52: 480–485.

Bailey, K. M. & E. D. Houde, 1989. Predation on eggs and larvae of marine fishes and the recruitment problem. Advances in Marine Biology 25: 1–83.

Brewer, R. H., 1978. Larval settlement behaviour in the jellyfish *Aurelia aurita* (Linnaeus) (Scyphozoa: Semaeostomae). Estuaries 1: 121–122.

Chen, E. L., 2002. Population dynamics and feeding of the moon jellyfish (*Aurelia aurita*) in Tapong Bay, southwestern Taiwan. Masters thesis, National Sun Yat-sen University, Taiwan: 74 pp.

Custance, D. R. N., 1964. Light as an inhibitor of strobilation in *Aurelia aurita*. Nature 204: 1219–1220.

Dawson, M. N., L. E. Martin & L. K. Penland, 2001. Jellyfish swarms, tourists, and the Christ-child. Hydrobiologia 451: 131–144.

Dawson, M. N., A. Sen Gupta & M. H. England, 2005. Coupled biophysical global ocean model and molecular genetic analyses identify multiple introductions of cryptogenic species. Proceedings of the National Academy of Sciences 102:11968–11973. www.pnas.org_cgi.

Fitt, W. K. & K. Costley, 1998. The role of temperature in survival of the polyp stage of the tropical Rhizostome jellyfish *Cassiopea xamachana*. Journal of Experimental Marine Biology and Ecology 222: 79–91.

Gröndahl, F., 1988. A comparative ecological study on the scyphozoans *Aurelia aurita*, *Cyanea capillata*, *C. lamarckii* in the Gullmar Fjord, western Sweden, 1982 to 1986. Marine Biology 97: 541–550.

Hamner, W. M., P. P. Hamner & S. W. Strand, 1994. Sun-compass migration by *Aurelia aurita* (Scyphozoa): population retention and reproduction in Saanich Inlet, British Columbia. Marine Biology 119: 347–356.

Hiromi, J., T. Yamomoto, Y. Koyama & S. Kadota, 1995. Experimental study on predation of scyphopolyp *Aurelia aurita*. Bulletin of the College of Agriculture and Veterinary Medicine, Nihon University 52:126–130. (in Japanese; English abstract).

Hoover, R.A. & J. E. Purcell, This volume. Substrate preferences of scyphozoan *Aurelia labiata* polyps among common dock-building materials. Hydrobiologia.

Ishii, H. & U. Båmstedt, 1998. Food regulation of growth and maturation in a natural population of *Aurelia aurita* (L.). Journal of Plankton Research 20: 805–816.

Ishii, H. & H. Shioi, 2003. The effects of environmental light condition on strobilation in *Aurelia aurita* polyps. Sessile Organisms 20: 51–54.

Kakinuma, Y., 1975. An experimental study of the life cycle and organ differentiation of *Aurelia aurita* Lamarck. The Bulletin of the Marine Biological Station of Asamushi 15: 101–113.

Kang, Y. S. & M. S. Park, 2003. Occurrence and food ingestion of the moon jellyfish (Scyphozoa: Ulmariidae: *Aurelia aurita*) in the southern coast of Korea in summer. Journal of the Korean Society of Oceanography 8: 199–202. (in Korean with English abstract).

Kawahara, M., S. I. Uye, K. Ohtsu & H. Iizumi, 2006. Unusual population explosion of the giant jellyfish *Nemopilema nomurai* (Scyphozoa: Rhizostomeae) in East Asian waters. Marine Ecology Progress Series 307: 161–173.

Kideys, A. E. & A. C. Gücü, 1995. *Rhopilema nomadica*: a poisonous Indo-Pacific scyphomedusan new to the Mediterranean coast of Turkey. Israel Journal of Zoology 41: 615–617.

Lo, W.-T., J. E. Purcell, J.-J. Hung, H.-M. Su & P.-K. Hsu, 2008. Enhancement of jellyfish (*Aurelia aurita*) populations by extensive aquaculture rafts in a coastal lagoon in Taiwan. ICES Journal of Marine Science 65: 453–461.

Loeb, M. J., 1973. The effect of light on strobilation in the Chesapeake Bay sea nettle *Chrysaora quinquecirrha*. Marine Biology 20: 144–147.

Lucas, C. H., 2001. Reproduction and life history strategies of the common jellyfish, *Aurelia aurita*, in relation to its ambient environment. Hydrobiology 451: 229–246.

Lucas, C. H. & J. A. Williams, 1994. Population dynamics of the scyphomedusa *Aurelia aurita* in Southampton Water. Journal of Plankton Research 16: 879–895.

Ma, X. & J. E. Purcell, 2005a. Effects of temperature, salinity and predators on mortality of and colonization by the invasive hydrozoan, *Moerisia lyonsi*. Marine Biology 147: 215–224.

Ma, X. & J. E. Purcell, 2005b. Temperature, salinity and prey effects on polyp versus medusa bud production of the invasive hydrozoan, *Moerisia lyonsi*. Marine Biology 147: 225–234.

Martin, L. E., 1999. The population biology and ecology of *Aurelia* sp. (Scyphozoa: Semaeostomeae) in a tropical meromictic marine lake in Palau, Micronesia. Ph.D. thesis, University of California, Los Angeles: 250 pp.

Mayer, A. G., 1910. Medusae of the World. III. The Scyphomedusae. Carnegie Institute of Washington, Washington: 499–735.

Mianzan, H. W. & P. F. S. Cornelius, 1999. Cubomedusae and scyphomedusae. In Boltovskoy, D. (ed.), South Atlantic Zooplankton. I. Backhuys Publishers, Leiden: 513–559.

Mills, C. E., 2001. Jellyfish blooms: are populations increasing globally in response to changing ocean conditions? Hydrobiologia 451: 55–68.

Miyake, H., M. Terazaki & Y. Kakinuma, 2002. On the polyps of the common jellyfish *Aurelia aurita* in Kagoshima Bay. Journal of Oceanography 58: 451–459.

Molinero, J. C., F. Ibanez, P. Nival, E. Buecher & S. Souissi, 2005. North Atlantic climate and northwestern Mediterranean plankton variability. Limnology and Oceanography 50: 1213–1220.

Möller, H., 1980. Population dynamics of *Aurelia aurita* medusae in Kiel Bight, Germany (FRG). Marine Biology 60: 123–128.

Olive, P. J. W., 1985. Physiological adaptation and the concepts of optimal reproductive strategy and physiological constraint in marine invertebrates. In Laverack, M. S. (ed.), Physiological Adaptations of Marine Animals. Symposia of the Society for Experimental Biology 39: 267–300.

Purcell, J. E., 1985. Predation on fish eggs and larvae by pelagic cnidarians and ctenophores. Bulletin of Marine Science 37: 739–755.

Purcell, J. E., 1997. Pelagic cnidarians and ctenophores as predators; selective predation, feeding rates and effects on prey populations. Annales de l'Institut Oceanographique, Paris 73: 125–137.

Purcell, J. E., 2005. Climate effects on formation of jellyfish and ctenophore blooms. Journal of the Marine Biological Association of the United Kingdom 85: 461–476.

Purcell, J. E., 2007. Environmental effects on asexual reproduction rates of the scyphozoan, *Aurelia labiata*. Marine Ecology Progress Series 348: 183–196.

Purcell, J. E. & M. N. Arai, 2001. Interaction of pelagic cnidarians and ctenophores with fish: a review. Hydrobiologia 451: 27–44.

Purcell, J. E., E. D. Brown, K. D. E. Stokesbury, L. H. Haldorson & T. C. Shirley, 2000. Aggregations of the jellyfish *Aurelia labiata*: abundance, distribution, associations with age–0 walleye pollock, and behaviors promoting aggregation in Prince William Sound, Alaska. U.S.A. Marine Ecology Progress Series 195: 145–158.

Purcell, J. E. & M. B. Decker, 2005. Effects of climate on relative predation by scyphomedusae and ctenophores on copepods in Chesapeake Bay during 1987–2000. Limnology and Oceanography 50: 376–387.

Purcell, J. E., J. R. White, D. A. Nemazie & D. A. Wright, 1999. Temperature, salinity and food effects on asexual reproduction and abundance of the scyphozoan *Chrysaora quinquecirrha*. Marine Ecology Progress Series 180: 187–196.

Russell, F. S., 1970. The Medusae of the British Isles. II Pelagic Scyphozoa with a Supplement to the First Volume on Hydromedusae. Cambridge University Press, London: 284.

Schneider, G., 1989. The common jellyfish *Aurelia aurita*: standing stock, excretion and nutrient regeneration in the Kiel Bight, western Baltic. Marine Biology 100: 507–514.

Shimomura, T., 1959. On the unprecedented flourishing of 'Echizenkurage' *Stomolophus nomurai* (Kishinouye), in the Tsushima Warm Current regions in autumn, 1958. Bulletin of Japan Sea Regional Fisheries Research Laboratory 7: 85–107. (in Japanese with English abstract).

Spangenberg, D. B., 1967. Iodine induction of metamorphosis in *Aurelia*. Journal of Experimental Zoology 160: 1–10.

Underwood, A. J., 1997. Experiments in ecology: their logical design and interpretation using analysis of variance. Cambridge University Press, Cambridge.

Uye, S., N. Fujii & H. Takeoka, 2003. Unusual aggregations of the scyphomedusa *Aurelia aurita* in coastal waters along western Shikoku. Plankton Biology and Ecology 50: 17–21.

Uye, S. & Y. Ueta, 2004. Recent increase of jellyfish populations and their nuisance to fisheries in the Inland Sea of Japan. Bulletin of the Japanese Society of Fisheries Oceanography 68: 9–19. (in Japanese with English abstract).

Van Der Veer, H. W. & W. Oorthuysen, 1985. Abundance, growth and food demand of the scyphomedusan *Aurelia aurita* in the western Wadden Sea. Netherlands Journal of Sea Research 19: 38–44.

Watanabe, T. & H. Ishii, 2001. *In situ* estimation of the number of ephyrae liberated from polyps of *Aurelia aurita* on settling plates in Tokyo Bay, Japan. Hydrobiologia 451: 247–258.

Widmer, C. L., 2005. Effects of temperature on growth of north-east Pacific moon jellyfish ephyrae, *Aurelia labiata* (Cnidaria: Scyphozoa). Journal of the Marine Biological Association of the United Kingdom 85: 569–573.

Zaitsev, Y. P., 1992. Recent changes in the structure of the Black Sea. Fisheries Oceanography 1: 180–189.

Hydrobiologia (2009) 616:259–267
DOI 10.1007/s10750-008-9595-6

JELLYFISH BLOOMS

# Substrate preferences of scyphozoan *Aurelia labiata* polyps among common dock-building materials

**Richard A. Hoover · Jennifer E. Purcell**

Published online: 1 October 2008


**Abstract** New habitat on proliferating marine construction may increase jellyfish polyp populations, and thereby increase jellyfish populations worldwide. In this investigation, we examined planula settlement and polyp immigration rates of the scyphozoan *Aurelia labiata* Chamisso & Eysenhardt, 1821 on six common dock-building materials. The planulae and polyps preferred plastics (expanded polystyrenes, low and high density polyethylene) to rubber and treated wood when choosing habitat on man-made surfaces. Substrate surface texture and the presence/absence of anti-fouling chemicals are discussed as possible causes for these substrate preferences. This study illustrates the potential effects of different man-made structures on jellyfish populations, and provides useful information to coastal managers and port authorities for reduction of biofouling and jellyfish bloom effects.

Guest editors: K. A. Pitt & J. E. Purcell
Jellyfish Blooms: Causes, Consequences, and Recent Advances

R. A. Hoover (✉) · J. E. Purcell
Shannon Point Marine Center, Western Washington University, 1900 Shannon Point Rd, Anacortes, WA 98221, USA
e-mail: rhoover1@aol.com

J. E. Purcell
e-mail: purcelj3@wwu.edu

**Keywords** Jellyfish · Colonize · Recruitment · Biofouling · Polyethylene · Marine construction

## Introduction

Over the past several decades, increased numbers of jellyfish have been reported worldwide (Papathanassiou et al., 1987; Estes et al., 1998; Mills, 2001; Brodeur et al., 1999, 2002; Purcell, 2005; Xian et al., 2005; Purcell et al., 2007). Multiple causes may be responsible for the jellyfish blooms, including anthropogenic ecosystem degradation and climatic forcing. In turn, these large jellyfish populations can have important effects on ecosystem dynamics and human enterprise.

Jellyfish blooms cause a variety of problems for humans both directly and indirectly. One important consequence of jellyfish blooms is reduction in zooplankton and ichthyoplankton biomass (e.g., Möller, 1980; Lindahl & Hernroth, 1983; Olsson et al., 1992; Smayda, 1993; Schneider & Behrends, 1994, 1998; Lucas et al., 1997; Ishii & Tanaka, 2001; Uye & Ueta, 2004; Uye & Shimauchi, 2005) and the subsequent altering of trophic webs on which higher predators rely (Mills, 1995; Purcell & Arai, 2001; Purcell & Sturdevant, 2001; Purcell, 2003). In this way, jellyfish affect the health of coastal fisheries. Fisheries are also negatively affected by clogging of fishing nets and equipment. Large jellyfish populations also cause damages by clogging cooling-water intakes of coastal power plants (reviewed in Mills, 2001; Purcell et al., 2007), and by reducing tourism (UNEP, 1991).

The life cycles of most jellyfish are comprised of a benthic asexual stage (polyp) and a pelagic sexual stage (medusa). Polyps form when sexually created planulae attach to a benthic surface. Through a process termed strobilation, individual polyps then segment into stacks of distinct discs, which eventually detach and grow to become mature pelagic medusae. Meanwhile, the polyps also reproduce by budding more polyps, which can produce colonies of millions of individuals. Successful strobilation of great numbers of polyps creates large jellyfish blooms.

Due to a cosmopolitan distribution and widespread ecological and anthropological effects, jellyfish from the genus *Aurelia* were chosen for this study. Although *Aurelia* spp. are common jellyfish around the world, little is known about the factors that control their populations (Purcell, 2007). Factors that affect the inconspicuous polyp are probably very important in subsequent development of jellyfish blooms.

In most temperate waters, recruitment of *Aurelia* spp. planulae occurs following sexual reproduction in the fall (reviewed in Lucas, 2001; Wantanabe & Ishii, 2001). Settlement and metamorphosis of the polyps are controlled by a combination of many factors, including predation by benthic fauna, the physical characteristics of the substratum and location, contact with biofilms, and gregarious behavior (Brewer, 1978; Keen, 1987; Gröndahl, 1989; Lucas, 2001; Holst & Jarms, 2007). Polyps usually are found at shallow depths, seldom deeper than 20–25 m (Russell, 1970; Lucas, 2001; but see Miyake et al., 2004). With high abundances of suitable substrates and medusae in semi-enclosed bays, fjords, and inlets (Lucas et al., 1997; Ishii & Båmstedt, 1998; Miyake et al., 2002), the largest polyp colonies may occur in those areas. Polyps are found attached to a great variety of both living and nonliving natural substrata, including rocks, shells, polychaete tubes, ascidians, algae, and bryozoans (Russell, 1970; Grosberg, 1981; Miyake et al., 1997, 2002, 2004); however, they also readily attach to artificial substrates, such as plastic or ceramic settling plates and many dock-building materials (Hernroth & Gröndahl, 1985a, b; Wantanabe & Ishii, 2001; Ishii & Katsukoshi, this volume). In addition, the planulae attach almost exclusively to the underside of settling surfaces (Brewer, 1978; Holst & Jarms, 2007), suggesting that position is an important criterion for polyp survival.

Polyps employ a variety of strategies to ensure survival. An important first step is settlement in a suitable habitat. Keen (1987) cites positions in boundary layers and local shear stress as key factors in determining recruitment rates. Finding the most suitable habitats also involves a variable amount of inter- and intra-specific competition, likely resulting in high planula mortality (Lucas, 2001). After settlement, polyps are surprisingly mobile, using stolons to 'walk' across surfaces or detaching and floating to new locations. Both the sexual and asexual stages of the *Aurelia* life cycle are considered r-strategists, producing a large number of offspring and/or clones to offset mortality (Lucas, 2001). Polyps exhibit many types of asexual reproduction including budding from the base of the parent, longitudinal and transversal fission, pedal laceration, formation of cysts, and strobilation (reviewed in Lucas, 2001; Arai, this volume). Interestingly, the amount of food available directly affects the allocation of energy to different reproductive strategies. With adequate food present, both budding and stolonation increase. With less food, energy allocation shifts from budding to the production of free-swimming ephyrae (Gong, 2002).

Once established, the maintenance and growth of polyp colonies become a complex interplay between environmental conditions, mortality, further recruitment, and the size and number of clonal genets found within the colony. Colonial growth has been positively correlated with local temperatures, separate from feeding regimes (Robinson, 1970), and polyp mortality has been positively correlated with populations of nudibranch predators (Hernroth & Gröndahl, 1985a), yet little is known about how clonal growth affects further recruitment. The polyps of *Aurelia aurita* (Linnaeus) have been observed to prey on conspecifics during times of low food availability (Gröndahl, 1988; Miyake et al., 1997), and genetically distinct clones may work together to defend resources from other clones (Buss, 1990), but it is not known at what levels gregarious behavior turns aggressive.

Several species of scyphozoan were recently found to prefer artificial substrates to natural ones in the laboratory (Holst & Jarms, 2007). In situ this trend may be compounded by the fact that adequate natural substrates are often limited in size, by environmental

**Table 1** Dock-building materials used in testing the substrate preferences of settling planulae and immigrating polyps of the scyphozoan, *Aurelia labiata*

| Building material | Description and use of material |
|---|---|
| Styrofoam® brand expanded polystyrene foam (SEPS) | A closed cell (will not take on water) blue foam used as flotation material. ($R_m \approx 1$ mm) |
| Generic expanded polystyrene foam (GEPS) | White beaded closed cell foam used as flotation material. ($R_m \approx 1$ mm) |
| Ammoniacal copper zinc arsenate pressure treated lumber (wood) | Douglas fir is commonly used as structural material. Conforms to water quality standards for surface waters of the State of Washington, published in WAC 173-201A. ($R_m \approx 1$ mm) |
| Vulcanized rubber | Common tire rubber used as structural and flotation material. ($R_m < 1$ mm) |
| High density polyethylene (HDPE) | Hard plastic used to make ridged tub-like containers that contain polystyrene foam for structural and flotation material. ($R_m < 1$ mm) |
| Low density polyethylene (LDPE) | Flexible shrink-wrap sheeting used to contain polystyrene foam for flotation material. ($R_m < 1$ mm) |

$R_m$ = maximum roughness height, a measure of surface roughness defined as the height from the deepest trough to the highest peak

conditions and by competing organisms. A newly introduced structure, on the other hand, provides an abundance of much-sought-after sheltered surface area. Competition for new space is fierce, but through high rates of sexual reproduction, gregarious settlement of planulae, clonal expansion, and the ability to protect space by stinging competitors, scyphozoan polyps could be among the fittest of colonizing organisms. The building of new in-water marine structures may increase polyp colony size and therefore blooms of jellyfish, including *Aurelia* spp. (Miyake et al., 2002; Holst & Jarms, 2007; Lo et al., 2008).

At the Cornet Bay Marina on the north shore of Whidbey Island, WA (48°24′ N, 122°38′ W), the polyps of *Aurelia labiata* Chamisso & Eysenhardt, 1821 cover almost the entire underside of one of the marina's covered floating docks. There appear to be dozens of individual clone genets that vary in area from a few square centimeters to several square meters. The population is estimated to cover a total area of approximately 685 $m^2$, with over one hundred million individuals.

In Cornet Bay marina, polyp density appeared to differ with the type of dock-building material inhabited. Thus, the Cornet Bay populations provided an opportunity to study how different marine building materials affect the population dynamics of these cnidarians. Most floating docks in Washington State are made from different combinations of seven primary materials (Table 1; Brian Williams, Washington Dept. of Fish and Wildlife, pers. comm.). The Washington State Department of Ecology regulates acceptable chemical and physical features of those materials in an attempt to reduce colonization by organisms. In this study, we tested *A. labiata* planula settlement and polyp immigration rates onto those primary dock-building materials.

## Methods

### Settlement of *Aurelia labiata* planulae onto common dock materials in the laboratory

*Aurelia labiata* medusae were collected in August and October 2004 from East Sound, Orcas Island, WA (48°40′ N, 122°55′ W). Six specimens were gently scooped from the water with a colander and placed in a bucket filled with seawater. Upon return to Shannon Point Marine Center (SPMC), the medusae were transferred to a 200 L kreisel. The constant inflow of 50-μm-filtered seawater to the kreisel created a circular current, which protected the jellies from blunt surfaces and aided in food capture. The kreisel was covered by black plastic to reduce algal growth. In order to promote sexual production of planulae, the medusae were fed an excess of *Artemia* sp. napulii 3 times weekly.

To compare the settling rates of *Aurelia labiata* planulae on different dock materials (Table 1), eighteen 2 cm × 2 cm × 5 mm settling plates of each substrate were affixed with aquarium silicone to the bottom of a piece of laminated wood (38 cm ×

48 cm × 2 cm) that then was floated on the water surface in the kreisel. The number of newly settled polyps attached to each settling plate was counted with the aid of a magnifying glass each week for 1 month during October and November 2004. Although asexual budding was rare at this early stage of the life cycle, two polyps formed by budding were removed. Because the attachment of *Aurelia* spp. larvae to a settling surface has been shown to be position dependent (Brewer, 1978; Keen, 1987; Holst & Jarms, 2007), the substrate plates were arranged systematically to minimize potential error caused by any gradients within the kreisel. The plates were organized in a 12 by 9 array with the first row containing 2 sequences of the 6 materials, specifically, wood, GEPS, HDPE, LDPE, SEPS, and rubber; that pattern repeated in each subsequent row, but was offset by one substrate. Seawater averaged 10.1 ± 0.4°C temperature and 30.6 ± 0.3 salinity in the kreisel during the experiment. A one-way ANOVA was used to test for differences in settling rates, and a Tukey HSD pairwise comparison test was used to contrast settling rates among substrate types.

### Immigration of *Aurelia labiata* polyps onto common dock materials in the field

To compare the immigration rates (movement to and asexual budding) of *Aurelia labiata* polyps onto common dock materials, 5 cm × 7 cm × 3 mm plates of six different dock substrate types (Table 1; $n = 36$) were attached systematically to the underside of the Cornet Marina docks. To eliminate potential genetic and location biases, the plates were placed in three discrete groups of 12 (two plates of each substrate per group) within the boundaries of what appeared to be three separate clones. Three-inch stainless steel screws were used as attachment anchors and as framing reference points. During July and August 2004, the plates were checked every 2 weeks and photographed on the fourth, sixth, and eighth weeks.

A Nikonos 5 underwater SLR camera with a 35 mm lens, a 1:2 extension tube, and a Nikonos SB-105 strobe were used for all photographs. In order to photograph exactly the same areas biweekly, the lower left hand corner of a 5 cm × 7 cm metal framer was placed around each site-marking screw when the photographs were taken. All photographs were taken with an aperture of F 22 and minimum focal distance on Kodak EliteChrome 36 exposure slide film, ISO 200.

The photographs were analyzed to determine percent cover of *Aurelia labiata* polyps and other fouling organisms. The slides were first converted to a Bitmap digital format using an EPSON Perfection 2450 Photo scanner and processed using the OPTIMUS 6.0 image analysis software package. Due to low variance in color thresholds between polyps and substrates, a Wallis filter with a desired average of 100, a standard deviation of 75, and a variance threshold of 16 were applied to the images in order to better define edges and increase contrast between the polyps and the background. To further ensure reproducible measurement, color thresholds were standardized for each survey location, and each slide was separately scanned for the presence of other fouling organisms. The percent cover of polyps was determined by measuring the covered area falling within the range of the color thresholds. The percent coverage of other fouling organisms was measured by using an editing tool to trace and fill their respective area. A one-way ANOVA was used to test for differences in percent cover of polyps, and a Tukey HSD pairwise comparison test was used to contrast percent cover among substrate types.

## Results

### Settlement of *Aurelia labiata* planulae onto common dock materials in the laboratory

Recruitment rates of *Aurelia labiata* planulae onto 6 common dock-building materials were measured in the laboratory. Little recruitment was observed during the first 2 weeks of the experiment; however, during the third week, recruitment was observed on all of the substrates, with SEPS foam, GEPS foam, and LDPE plastic displaying their greatest recruitment during the month-long experiment. During the fourth week, HDPE, rubber, and wood all had their greatest recruitment, while recruitment to SEPS, GEPS, and LDPE slowed slightly (Fig. 1). After 1 month, the greatest total recruitment was observed on SEPS foam (41 recruits) followed by GEPS foam (33), HDPE plastic (33), LDPE plastic (27), wood (17), and vulcanized rubber (11). A one-way ANOVA showed significant differences among

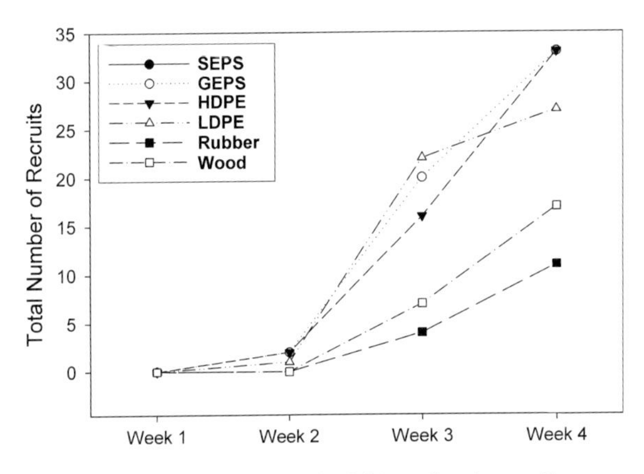


**Fig. 1** Total numbers of *Aurelia labiata* planulae settling onto different dock-building materials in the laboratory. Totals are sums for 18 replicates of each material. Abbreviations as in Table 1

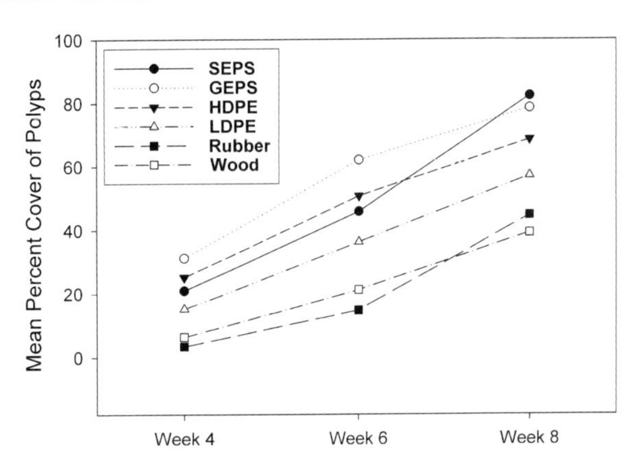


**Fig. 3** Mean immigration rates of *Aurelia labiata* polyps to different dock-building materials in Cornet Bay during the second month of the experiment. Error bars removed to reduce clutter. $n = 6$ replicates of each material. Abbreviations as in Table 1

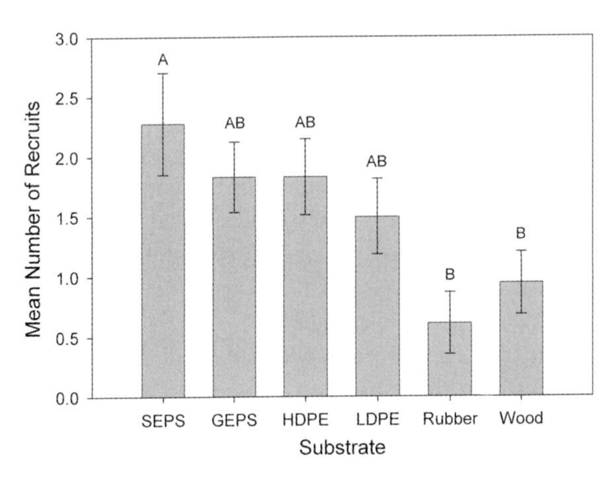


**Fig. 2** Mean numbers of *Aurelia labiata* planulae settling onto dock-building materials after 1 month in the laboratory. Significantly different groups are indicated by the letters A and B. Bars represent standard errors. $n = 18$ replicates of each material. Abbreviations as in Table 1

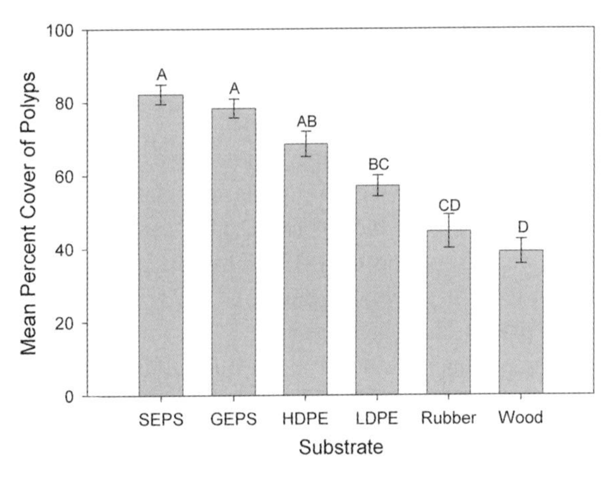


**Fig. 4** Mean percent cover of immigrating *Aurelia labiata* polyps on dock-building materials after 2 months (July and August, 2004) in Cornet Bay. Significantly different groups are indicated by the letters A, B, C, and D. $n = 6$ replicates of each material. Bars indicate standard error. Abbreviations as in Table 1

treatments ($F_{5,\ 107} = 3.84$, $P = 0.003$), and the null hypothesis that substrate type had no effect on recruitment was rejected. A Tukey HSD comparison of means separated the dock materials into the following two significantly different groups: (1) SEPS, GEPS, HDPE and LDPE; and (2) GEPS, HDPE, LDPE, rubber and wood (Fig. 2).

## Immigration of *Aurelia labiata* polyps onto common dock materials in the field

Immigration of *Aurelia labiata* polyps onto 6 common dock-building materials was measured in Cornet Bay. Due to camera error, no photographs exist for the second week; however, no appreciable immigration was observed at that time. By the fourth week, a hierarchy of polyp substrate preference was already apparent (Fig. 3). Between the fourth and eighth weeks, cover increased rapidly and at similar rates on all substrates (Fig. 3). After 2 months, the mean number of polyps that had colonized each substrate type differed significantly among substrates (Fig. 4). Polyp immigration after 2 months ranged from 82.2 ± 6.5% cover for the blue SEPS foam to 39.2 ± 8.3% on wood. A one-way ANOVA showed significant differences among treatments

**Table 2** Colonization by organisms (algae[a] and bryozoans[b]) other than *Aurelia labiata* polyps onto plates made from common dock-building materials (abbreviations in Table 1) after 2 months (July and August, 2004) in Cornet Bay

| Dock material | Plates (No.) | Cover (%) |
|---|---|---|
| SEPS | 1[a] | 2.0 ± 4.9 |
| GEPS | 0 | 0 |
| HDPE | 3[a] | 1.9 ± 3.2 |
| LDPE | 5[a] | 5.7 ± 5.7 |
| Rubber | 2[a] | 0.2 ± 0.4 |
| Wood | 1[a], 1[b] | 4.0 ± 9.4 |

$n = 6$ replicates of each material. Mean percent cover ± standard deviation (all plates)

($F_{5,35} = 23.85$, $P < 0.0001$); the null hypothesis that substrate type did not affect immigration was rejected. A Tukey HSD pairwise comparison test separated the dock materials into the following four significantly different groups: (1) SEPS, GEPS, and HDPE; (2) HDPE and LDPE; (3) LDPE and rubber; and (4) rubber and wood (Fig. 4).

During the 2-month experiment, algae and bryozoans also colonized the dock materials, but much less than did polyps (Table 2). These other fouling organisms were found on only 13 of the 36 immigration plates (36%), averaging only 6.3 ± 7.0% cover on those plates. The mean percent cover from these other fouling organisms was 2.3 ± 5.1% for all immigration plates. A one-way ANOVA showed no differences in the fouling organisms among treatments ($F_{5,35} = 1.11$, $P = 0.375$); therefore, competition for space by these other fouling organisms was considered negligible for this experiment.

## Discussion

Populations of scyphozoan polyps are notoriously difficult to find on natural substrates in situ, and nearly all field studies of *Aurelia* spp. have been conducted beneath floating, man-made structures (Hernroth & Gröndahl, 1985a; Miyake et al., 1997, 2002; Gong, 2002). *Aurelia* spp. polyps probably are found more often underneath docks because docks provide large sheltered areas within protected harbors. In order to determine if the use of different types of dock-building materials could influence colonization by *Aurelia* spp., the substrate preferences of *Aurelia labiata* were tested.

Clear differences were observed in planula settlement and polyp immigration onto 6 different dock materials that indicated preferences for synthetic plastics (EPS foams and polyethylene) over rubber and treated wood when choosing habitats on man-made structures. Nearly the same preference hierarchy was established for both settlement of planulae in the laboratory and immigration of polyps in the field (Figs. 2 and 4). The settlement and immigration were greatest on EPS foams, which have the softest and most varied texture of all the substrates ($R_m \approx$ 1 mm, see Table 1). Although Brewer (1978) found that the surface texture of microscope cover slips scratched in different patterns did not have an effect on planula settlement in the laboratory, the larger scale (mm versus μm) surface textures of materials tested in situ suggest that texture may influence colonization choices. One statistical model developed to explain how roughness affects settlement of planktonic larvae states that the rate of settlement should increase with the areal density of roughness features (Eckman, 1990). In addition, boundary layer thickness and local shear stress created by surface roughness are factors that affect recruitment rates of *Aurelia* sp. planulae (Keen, 1987). Our results are in accordance with those findings.

The presence of anti-corrosive and anti-fouling chemicals in the substrates also could have affected the choices of planulae and polyps for dock substrates. Although the colonization and settling plates made of wood were the most natural of the substrates presented, they also had the greatest amount of chemical treatment in the form of Ammoniacal Copper Zinc Arsenate (ACZA). By contrast, the plastics tested generally are considered to be benign (Fisheries and Oceans Canada, 2002). Therefore, it is not surprising that the ACZA-treated wood had significantly lower rates of recruitment and immigration than the synthetic plastics. ACZA treatment is a relatively new technology, developed in the 1920s and first incorporated into lumber production during the 1940s. Since its development, the use of ACZA-treated lumber in marine construction has increased greatly. It is considered to be a great improvement over previous chemical wood treatments because of reduced concentrations of arsenate and the treatment's ability to preserve the cellular structure of the wood. Studies show that while toxic trace metals do leach from newly installed ACZA lumber, most leaching occurs within 2 weeks and is in amounts that meet federal marine water quality criteria for all but the most stagnant waters (Brooks, 1996, 1999).

The majority (>90%) of the extensive population of *Aurelia labiata* polyps at Cornet Bay occurred on uncontained, expanded polystyrene (EPS) dock flotation blocks. The remainder of the colony was mostly on low-density polyethylene. EPS foam is inexpensive, lightweight, resistant to corrosion, and highly buoyant, making it the most cost-effective and commonly used flotation material today. In 1994, the Washington State Legislature made the use of uncontained-EPS foam in the building of over-water structures illegal; however, marinas and docks built prior to 1994 probably still have uncontained-EPS as a flotation material. To continue taking advantage of the cost effectiveness of EPS, many float manufactures now enclose EPS inside of high and low density polyethylene (HDPE and LDPE), which in our experiments were not significantly less preferred than EPS by *A. labiata* planulae and polyps. Thus, the use of EPS foams and polyethylenes for dock flotation provides a favorable habitat for jellyfish populations.

To our knowledge, only a few other studies have examined the settlement preferences of scyphozoans for man-made substrates. In experiments in the Mediterranean, Lotan et al. (1992) found that the larvae of *Rhopilema nomadica* Galil, 1990 settled on artificial substrates of ceramic, PVC, and glass, but did not settle on natural rocks covered in epibiota. Settlement trends for glass and against seagrass were reported for planulae of *Catostylus mosaicus* Quoy and Gaimard, 1824 in Australia (Pitt, 2000). Most recently, Holst & Jarms (2007) found that planulae of five scyphozoan species preferred to settle on artificial substrates (concrete, machined wood, polyethylene, and glass) over the natural substrate of shells. As in our experiments, plastic (polyethylene) was the preferred substrate in these experiments. Holst & Jarms (2007) speculated that settling planulae may be exhibiting preferences for different biofilms that develop on the various substrate types. The addition of hard substrates and marine litter to soft-bottomed coastal areas is a possible factor in jellyfish population expansion.

Successful competition for space with other fouling organisms is also important for population persistence and growth of scyphozoan polyps. Other fouling organisms occupied only 3% of the surface area on the 36 immigration plates; therefore, competition from these other fouling organisms was considered to be negligible during our study. The immigration plates were located within a large, established polyp colony, suggesting that once established, the polyps of *Aurelia labiata* can successfully occupy open space and dominate habitats. Further study of competition during and after settlement is needed to understand the ability of *A. labiata* to colonize new habitats.

The idea that marine structures could increase jellyfish populations is relatively new, but the potential effects are too great to be ignored. Through high rates of sexual reproduction, gregarious settlement of planulae, clonal expansion, and the plasticity to adapt rapidly to changing conditions, scyphozoan polyps are well suited to exploit these new habitats.

The increased exploitation of new marine structures is a global problem. Hundreds of new marine structures are built each year in the State of Washington and many thousands are installed worldwide. In Japan, large, problematic blooms of *Aurelia aurita* medusae are potentially linked to huge polyp populations found in nearby marinas and harbors (Miyake et al., 2002). The coastal waters of China and Southeast Asia have also experienced recent jellyfish population explosions (Xian et al., 2005). In Taiwan, aquaculture rafts enhanced *A. aurita* jellyfish populations, probably because of increased substrate availability, shading, and restricted water exchange in a coastal lagoon (Lo et al., 2008). Aquaculture is expanding rapidly in Asia, which may provide extensive favorable new habitat for scyphozoan polyps (Purcell et al., 2007). Floating marine structures such as small aquaculture pens and houseboats are likely to increase scyphozoan population growth.

Actions could be taken to minimize the effects of coastal development. More rigorous monitoring of marine construction is important. Permitting authorities, port managers, and contractors need to be informed about possible consequences of marine construction. Use of appropriate building materials, research and development of new environmentally friendly anti-fouling technologies, retrofitting of older marine structures, and more strict regulation of coastal development should be made priorities. Further research on the effects of marine structures on jellyfish populations also is needed.

**Acknowledgments** We would like to thank the director and staff of Shannon Point Marine Center for supplying the facilities, equipment, and support to make this study possible, The Charles and June Ross Foundation and the Biology department of Western Washington University for their financial support, Brian Williams of the Washington Department of Fish and Wildlife for his technical expertise, Mr. Dundee, owner of the Cornet Bay Marina for allowing us to conduct experiments at his facilities, Dr. Brian Bingham for advise concerning experimental design, and Nathan Schwarck for his support and friendship throughout. This research was funded in part by National Science Foundation ADVANCE Fellows Award OCE-0137419.

## References

Arai, M. N., this volume. Are podocysts important to the formation of scyphozoan blooms? Hydrobiologia.

Brewer, R. H., 1978. Larval settlement behaviour in the jellyfish *Aurelia aurita* (Linnaeus) (Scyphozoa: Semaeostomae). Estuaries 1: 121–122.

Brodeur, R. D., C. E. Mills, J. E. Overland, G. E. Walters & J. G. Schumacher, 1999. Evidence for a substantial increase in gelatinous zooplankton in the Bering Sea, with possible links to climate change. Fisheries Oceanography 8: 296–306.

Brodeur, R. D., H. Sugisaki & G. L. Hunt, Jr., 2002. Increases in jellyfish biomass in the Bering Sea: implications for the ecosystem. Marine Ecology Progress Series 233: 89–103.

Brooks, K. M., 1996. Evaluating the environmental risks associated with the use of chromated copper arsenate-treated wood products in aquatic environments. Estuaries 19: 296–305.

Brooks, K. M., 1999. Recommendations to the National Marine Fisheries Service for the use of CCA-C, ACZA and creosote treated wood products in aquatic environments where threatened or endangered species occur. Western Wood Preservers Institute, Vancouver, Washington.

Buss, L. W., 1990. Competition within and between encrusting clonal invertebrates. Trends in Ecology & Evolution 5: 352–356.

Eckman, J. E., 1990. A model for passive settlement by planktonic larvae onto bottoms of differing roughness. Limnology and Oceanography 35: 887–901.

Estes, J. A., M. T. Tinker, T. M. Williams & D. F. Doak, 1998. Killer whale predation on sea otters linking oceanic and nearshore ecosystems. Science 282: 473–476.

Fisheries and Oceans Canada, 2002. Saskatchewan fact sheet 3: what you should know about fish habitat and building materials. http://www.dfo-mpo.gc.ca/canwaters-eauxcan/infocentre/guidelines-conseils/factsheets-feuillets/manitoba/pdf/fact3e.pdf.

Gong, A. J., 2002. Allocations to clonal replication in a marine scyphozoan (*Aurelia*). Science & Engineering 62: 3516–3635.

Gröndahl, F., 1988. Interactions between polyps of *Aurelia aurita* and planktonic larvae of scyphozoans: an experimental study. Marine Ecology Progress Series 45: 87–93.

Gröndahl, F., 1989. Evidence of gregarious settlement of planula larvae of the scyphozoan *Aurelia aurita*: an experimental study. Marine Ecology Progress Series 56: 119–125.

Grosberg, R. K., 1981. Competitive ability influences habitat choice in marine invertebrates. Nature 290: 700–702.

Hernroth, L. & F. Gröndahl, 1985a. On the biology of *Aurelia aurita* (L.). 3. Predation by *Coryphella verrucosa* (Gastropoda, Opisthobranchia), a major factor regulating the development of *Aurelia* populations in the Gullmar Fjord, Western Sweden. Ophelia 24: 37–45.

Hernroth, L. & F. Gröndahl, 1985b. On the biology of *Aurelia aurita* (L.). 2. Major factors regulating the occurrence of ephyrae and young medusae in the Gullmar Fjord, Western Sweden. Bulletin of Marine Science 37: 567–576.

Holst, S. & G. Jarms, 2007. Substrate choice and settlement preferences of planula larvae of five Scyphozoa (Cnidaria) from German Bight, North Sea. Marine Biology 151: 863–871.

Ishii, H. & U. Båmstedt, 1998. Food regulation of growth and maturation in a natural population of *Aurelia aurita* (L.). Journal of Plankton Research 20: 805–816.

Ishii, H. & K. Katsukoshi, this volume. Distribution and changes in abundance of *Aurelia aurita* polyps on a pier in the innermost part of Tokyo Bay.

Ishii, H. & F. Tanaka, 2001. Food and feeding of *Aurelia aurita* in Tokyo Bay with an analysis of stomach contents and a measurement of digestion times. Hydrobiologia 451: 311–320.

Keen, S. L., 1987. Recruitment of *Aurelia aurita* (Cnidaria: Scyphozoa) larvae is position-dependent, and independent of conspecific density, within a settling surface. Marine Ecology Progress Series 38: 151–160.

Lindahl, O. & L. Hernroth, 1983. Phyto-zooplankton community in coastal waters of western Sweden – an ecosystem off balance. Marine Ecology Progress Series 10: 119–126.

Lo, W.-T., J. E. Purcell, J.-J. Hung, H.-M. Su & P.-K. Hsu, 2008. Enhancement of jellyfish (*Aurelia aurita*) populations by extensive aquaculture rafts in a coastal lagoon in Taiwan. ICES Journal of Marine Science 65.

Lotan, A., R. Ben-Hillel & Y. Loya, 1992. Life cycle of *Rhopilema nomadica*: a new immigrant scyphomedusan in the Mediterranean. Marine Biology 122: 237–242.

Lucas, C. H., 2001. Reproduction and life history strategies of the common jellyfish, *Aurelia aurita*, in relation to its ambient environment. Hydrobiologia 451: 229–246.

Lucas, C. H., A. G. Hirst & J. A. Williams, 1997. Plankton dynamics and *Aurelia aurita* production from two contrasting ecosystems: causes and consequences. Estuarine Coastal and Shelf Science 45: 209–219.

Mills, C. E., 1995. Medusae, siphonophores and ctenophores as planktivorous predators in changing global ecosystems. ICES Journal of Marine Science 52: 575–581.

Mills, C. E., 2001. Jellyfish blooms: are populations increasing globally in response to changing ocean conditions? Hydrobiologia 451: 55–68.

Miyake, H., J. Hashimoto, M. Chikuchishin & T. Miura, 2004. Scyphopolyps of *Sanderia malayensis* and *Aurelia aurita* attached to the tubes of vestimentiferan tube worm, *Lamellibrachia satsuma*, at submarine fumaroles in Kagoshima Bay. Marine Biotechnology 6: S174–S178.

Miyake, H., K. Iwao & Y. Kakinuma, 1997. Life history and environment of *Aurelia aurita*. South Pacific Studies 17: 273–285.

Miyake, H., M. Terazaki & Y. Kakinuma, 2002. On the polyps of the common jellyfish *Aurelia aurita* in Kagoshima Bay. Journal of Oceanography 58: 451–459.

Möller, H., 1980. Scyphomedusa as predators and food competitors of larval fish. Meeresforschung 28: 90–100.

Olsson, P., E. Granéli, P. Carlsson & P. Abreu, 1992. Structuring of a post spring phytoplankton community by manipulation of trophic interactions. Journal of Experimental Marine Biology and Ecology 158: 249–266.

Papathanassiou, E., P. Panayotidis & K. Anagnlstaki, 1987. Notes on the biology and ecology of the jellyfish *Aurelia aurita* Lam. In Elefsis Bay (Saronikos Gulf, Greece). Pubblicazioni della Stazione Zoologica di Napoli I. Marine Ecology 8: 49–58.

Pitt, K. A., 2000. Life history and settlement preferences of the edible jellyfish *Catostylus mosaicus* (Scyphozoa: Rhizostomeae). Marine Biology 136: 269–279.

Purcell, J. E., 2003. Predation on zooplankton by large jellyfish, *Aurelia labiata, Cyanea capillata* and *Aequorea aequorea* in Prince William Sound, Alaska. Marine Ecology Progress Series 246: 137–152.

Purcell, J. E., 2005. Climate effects on formation of jellyfish and ctenophore blooms. Journal of the Marine Biology Association of the U.K. 85: 1–16.

Purcell, J. E., 2007. Environmental effects on asexual reproduction rates of the scyphozoan, *Aurelia labiata*. Marine Ecology Progress Series 348: 183–196.

Purcell, J. E. & M. N. Arai, 2001. Interactions of pelagic cnidarians and ctenophores with fish: a review. Hydrobiologia 451: 27–44.

Purcell, J. E. & M. V. Sturdevant, 2001. Prey selection and dietary overlap among zooplanktivorous jellyfish and juvenile fishes in Prince William Sound, Alaska. Marine Ecology Progress Series 210: 67–83.

Purcell, J. E., S.-I. Uye & W.-T. Lo, 2007. Anthropogenic causes of jellyfish blooms and direct consequences for humans: a review. Marine Ecology Progress Series 350: 153–174.

Robinson, C. H., 1970. Density regulation in populations of scyphistomae. M.A. Thesis. University of California, Davis, USA.

Russell, F. S., 1970. The medusae of the British Isles. II Pelagic Scyphozoa with a supplement to the first volume on Hydromedusae. Cambridge University Press, London: 284 pp.

Schneider, G. & G. Behrends, 1994. Population dynamics and the trophic roles of *Aurelia aurita* medusae in the Kiel Bight and western Baltic. ICES Journal of Marine Science 51: 359–367.

Schneider, G. & G. Behrends, 1998. Top–down control in a neritic plankton system by Aurelia aurita medusae–a summary. Ophelia 48: 71–82.

Smayda, T., 1993. Experimental manipulations of phytoplankton + zooplankton + ctenophore communities, and foodweb roles of the ctenophore *Mnemiopsis leidyi*. ICES cm 1993/L: 68: 13 pp.

UNEP (United Nations Environmental Programme), 1991. Jellyfish blooms in the Mediterranean, Proceedings of II Workshop on Jellyfish in the Mediterranean Sea, Mediterranean Action Plan Technical Reports Series: 47.

Uye, S. & H. Shimauchi, 2005. Population biomass, feeding, respiration and growth rates, and carbon budget of the scyphomedusa *Aurelia aurita* in the Inland Sea of Japan. Journal of Plankton Research 27: 237–248.

Uye, S. & U. Ueta, 2004. Recent increase of jellyfish populations and their nuisance to fisheries in the inland Sea of Japan. Bulletin of the Japanese Society of Fisheries Oceanography 68: 9–19.

Wantanabe, T. & H. Ishii, 2001. *In situ* estimation of ephyrae liberated from polyps of *Aurelia aurita* using settling plates in Tokyo Bay, Japan. Hydrobiologia 451: 247–258.

Xian, W., B. Kang & R. Liu, 2005. Jellyfish blooms in the Yangtze estuary. Science 307: 41.

Hydrobiologia (2009) 616:269–277
DOI 10.1007/s10750-008-9598-3

JELLYFISH BLOOMS

# Ontogenetic changes in the ecological function of the association behavior between jack mackerel *Trachurus japonicus* and jellyfish

**Reiji Masuda**

Published online: 22 September 2008


**Abstract** Commensal behavior of jack mackerel *Trachurus japonicus* (Temminck & Schlegel) with jellyfishes has been widely observed but its ecological function is still unclear. The goal of the present research is to examine the function of association behavior with jellyfish in the laboratory and in field observations with an emphasis on ontogenetic changes. In the laboratory, jack mackerel juveniles (mean standard length (SL) = 11, 19, 38, and 55 mm) were placed in 500-l polycarbonate tanks with two live moon jellyfish, *Aurelia aurita* (Linné), and one artificial jellyfish made of silicon. Association behavior with either live or artificial jellyfish was visually observed under the following conditions: control, presence of a predator model, before and after feeding live *Artemia*, 1 h and 3 h after feeding, and at night. Jack mackerel at 11 mm SL associated with both the moon jellyfish and artificial jellyfish, unrelated to the presence of a predator model or feeding. Juveniles at 19 mm associated with moon jellyfish only in the presence of a predator model. Larger juveniles associated with moon jellyfish at 1 h and 3 h after feeding. Thus the ecological function of association was proposed to develop first from school formation, next as a hiding place from predators, and then as a food source. Underwater observations of jack mackerel associating with giant jellyfish *Nemopilema nomurai* (Kishinouye) in two different areas in the Sea of Japan supported this hypothesis. High predation pressure from benthic piscivorous fishes in the southern area (Tsushima) may encourage association with jellyfish, whereas pressure from pelagic predators in the northern area (Maizuru) may encourage settlement to rocky reef habitats in temperate waters. Thus the jellyfish may also function as a vehicle for the northward migration of this species.

**Keywords** Association behavior · Commensal behavior · *Trachurus japonicus* · *Aurelia aurita* · *Nemopilema nomurai* · Behavioral ontogeny

## Introduction

Both biotic and abiotic floating objects in the ocean attract fish juveniles to form flotsam fish assemblages (Kingsford, 1993). Man-made fish attraction devices (FADs) are commonly used for small-scale artisanal and industrial fisheries (Dempster & Taquet, 2004). Since some fish species are strongly attracted to both man-made and natural floating objects, this association behavior is likely to have impacts on the subsequent distribution and survival of the larvae

Guest editors: K. A. Pitt & J. E. Purcell
Jellyfish Blooms: Causes, Consequences, and Recent Advances

R. Masuda (✉)
Maizuru Fisheries Research Station, Kyoto University, Nagahama, Maizuru, Kyoto 625-0086, Japan
e-mail: reiji@kais.kyoto-u.ac.jp

and juveniles. Although association of fishes with flotsam are well-documented (Kingsford, 1993), the ecological function of the association remains to be clarified (Purcell & Arai, 2001). Castro et al. (2002) summarized its possible functions as protection from predators, safeguard for drifting, and as a meeting point for conspecifics. Jack mackerel *Trachurus japonicus* (Temminck & Schlegel) is one of the most abundant pelagic fish species in Japanese coastal waters. Juveniles of the family Carangidae are known to associate with jellyfish, and jack mackerel is no exception; Shojima (1962) reported jack mackerel juveniles associating with moon jellyfish *Aurelia aurita* (Linné) and *Chrysaora melanaster* Brandt, and Sassa et al. (2006) reported that the abundance of jack mackerel juveniles in their net sampling was positively correlated with that of *Pelagia noctiluca* (Forskål).

Jellyfish blooms are causing problems all over the world (reviewed by Purcell et al., 2007). In the seas off Japan, increases in moon jellyfish have been recognized since the 1960s in Tokyo Bay (Omori et al., 1995) and since the 1980s in the Seto Inland Sea (Uye et al., 2003). Both increases are likely to have been induced by the rapid development of the coastal areas resulting in eutrophication, and by overfishing competitors of, and predators upon, jellyfish (Purcell et al., 2007). Recently blooms of the giant jellyfish *Nemopilema nomurai* (Kishinouye) have caused extensive damage by clogging nets and reducing fish catches, especially along the west coast of Japan (Kawahara et al., 2006). Kawahara et al. (2006) reported that the suspected spawning area of giant jellyfish is along the coast of China and west Korean Peninsula, and Sassa et al. (2006) found that there is a major spawning ground in the southern area of the East China Sea (Fig. 1).

Here I hypothesize that jack mackerel make use of jellyfish blooms to enhance survival in their early life history. Indeed in Japanese coastal waters, jack mackerel has been relatively abundant in recent years compared to other pelagic or reef fishes (Masuda, 2008). In previous studies, we found that jack mackerel juveniles utilized moon jellyfish as a shelter when they were attacked by a predator, and that they also utilize moon jellyfish as a prey collector (Masuda et al., 2008). However, the importance of these functions at different developmental stages is unknown. The goal of this research was to investigate

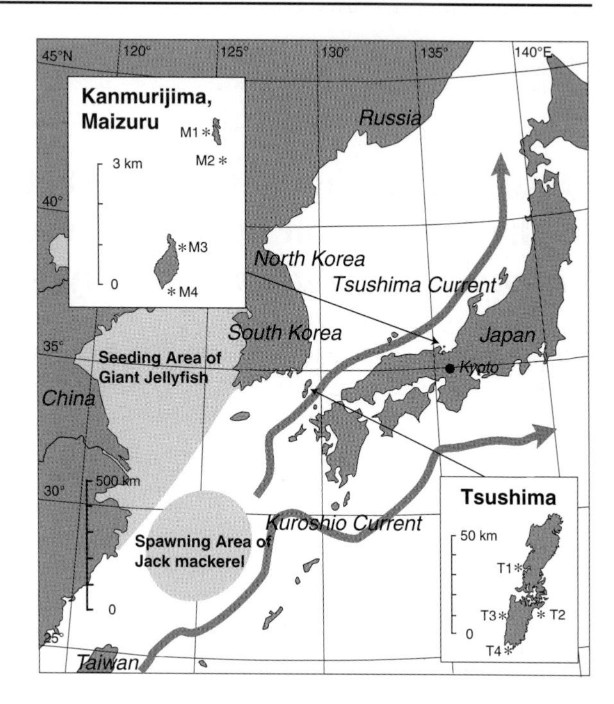


**Fig. 1** Field research area for the study of the association of juvenile jack mackerel with *Nemopilema nomurai*. Seeding area of giant jellyfish and spawning area of jack mackerel are based on Kawahara et al. (2006) and Sassa et al. (2006), respectively

ontogenetic changes of the ecological function of the association of jack mackerel with jellyfish. In the laboratory, jack mackerel at different developmental stages were kept with both live moon jellyfish and an artificial jellyfish model, and then association with live or artificial jellyfish was observed under different conditions, such as being threatened by a predator model and providing prey animals. Association of jack mackerel juveniles with giant jellyfish was also observed underwater in the Sea of Japan to verify the results of the laboratory experiment.

## Materials and methods

### Laboratory experiment using moon jellyfish and artificial jellyfish

Jack mackerel juveniles were collected off the Kanmurijima Islands, Maizuru, Kyoto, Japan (35.71°N, 135.44°E; Station M2 in Fig. 1) on October 20 and November 10, 2006. Juveniles associated with giant jellyfish were caught by a snorkeler using two handnets. About 100 individuals were collected on each sampling date. Fish were brought back to Maizuru Fisheries Research Station (MFRS) of Kyoto University and then fed defrosted and chopped krill until being used.

Moon jellyfish were collected in a bucket from a float at the MFRS facility. The bell diameter of the experimental jellyfish was 112.5 ± 21.5 mm (mean ± SD). Three artificial jellyfish (diameter 102.0 ± 1.0 mm) were made of silicon using a mold of plaster. Lead weights were added to make them neutrally buoyant in seawater.

A predator model was made of silicon from a frozen chub mackerel (*Scomber japonicus* Houttuyn; total length (TL) = 157 mm, standard length (SL) = 140 mm) that was molded in plaster. The surface was realistically painted, and stainless steel wire (60 cm length) was used to support the body from above.

Three identical 500-l transparent polycarbonate tanks (diameter 102 cm, water depth 66 cm) were used for the experiment. Two moon jellyfish and one artificial jellyfish were put in each tank. Experiments were repeated on four occasions and a different size class of fish was used each time (11, 19, 38, and 55 mm SL; Table 1). On each occasion, 30 juvenile jack mackerel of similar sizes were measured and 10 randomly selected individuals were gently introduced into each of the three experimental tanks at 16:00 h and then left overnight to acclimatize. Water temperature fluctuated between 16.9 and 21.2°C. Water was exchanged at 150 l $h^{-1}$. Light intensity was approximately 1000 lux at midday, and ranged from 0.03 to 0.15 lux at night.

Behavioral observations were conducted (1) immediately before threatening with the predator model, (2) immediately after the threat with the predator model, (3) immediately before feeding, (4) immediately after feeding, (5) at 1 h after feeding, (6) at 3 h after feeding, and (7) at night (21:00 h). During a threat, the chub mackerel model was moved in the tank at a speed of 30 cm $s^{-1}$ by hand for 1 min, and then the model was hung by the wire 15 cm below the surface and 20 cm distant from the tank wall. Predator threatening and feeding were conducted at 08:00 h and 13:00 h, respectively. Each suite of observations was repeated daily for five consecutive days. Preliminary observations revealed that both moving obstacles outside tanks and sound stimuli disturbed experimental fish, whereas a stationary observer did not. Therefore, I kept still during observations, except during threats with the predator model.

**Table 1** Fish standard lengths (SL in mm, mean ± SD), dates of capture and the start of trials in the experiment testing association of young jack mackerel with moon jellyfish

| SL of fish | Capture date | Start of exp. |
|---|---|---|
| 10.7 ± 0.8 | October 20, 2006 | October 22, 2006 |
| 18.7 ± 1.9 | October 20, 2006 | October 27, 2006 |
| 37.8 ± 6.1 | November 10, 2006 | November 13, 2006 |
| 54.7 ± 2.79 | November 10, 2006 | November 18, 2006 |

A fish was considered to be associated with a jellyfish when it stayed within a distance of one body length. The number of juvenile jack mackerels that associated with the jellyfish was determined using visual counts. Counts of 10 s duration were repeated 30 times for each real and artificial jellyfish. Since there were two live moon jellyfish and only one artificial jellyfish in each tank, the averages for the two moon jellyfish were calculated over 5 days of observations in each tank to determine the association index for moon jellyfish and artificial jellyfish. Averages of the association index among triplicate tanks were used for further analysis.

Statistical analysis of the association index was conducted after ln (x + 1) transformation to improve the normality of the distribution (Zar, 1996). The association indices of juvenile fish with live and artificial jellyfish, and indices before and after predator model threatening, were compared using paired *t*-tests. Indices immediately before and after feeding, and 1 h and 3 h after feeding were compared by repeated measures ANOVA followed by Tukey's HSD test. All statistical analyses were conducted using JMP computer software (ver. 5.0.1 J, SAS Institute Inc., Cary, NC, USA).

## Underwater observation of jack mackerel associated with giant jellyfish

Underwater observations of fish assemblages associated with giant jellyfish were conducted by SCUBA divers and snorkelers at four different sites off the Tsushima Islands, Nagasaki (34.11°N, 129.20°E–34.46°N, 129.28°E) and off the Kanmurijima Islands, Maizuru, Kyoto (35.68°N, 135.43°E–35.71°N, 135.44°E; Fig. 1). Observations were conducted on September 5 and 6, November 30, and December 1,

2006 in Tsushima with a total dive time of 11.1 h, and on August 24, October 27, November 10, and December 7, 2006 in Maizuru with 11.4 h of dive time. Visibility was at least 10 m in both areas. Bell diameter was visually estimated to the nearest 10 cm and recorded for all the encountered giant jellyfish. Species and body length (nearest cm) of fishes associated with giant jellyfish also were estimated visually and recorded. A scale on underwater plastic notebook helped with these visual estimations. Water temperature during the dives ranged from 20 to 28°C in Tsushima and from 17 to 29°C in Maizuru.

Visual censuses of fish assemblages were also conducted by SCUBA in both areas to find factors that might influence association between jack mackerel and jellyfish. The major focus was on predators of jack mackerel; therefore a transect method that sampled both benthic and pelagic habitats was applied. Transects (50 m by 2 m) were established at 54 locations in Stations T1, T3, and T4 of the Tsushima Islands on November 30 and December 1, 2006, and at 43 locations in Stations M1, M2, and M3 of the Kanmurijima Islands, Maizuru on October 20 and December 7, 2006, respectively. The length of each transect was estimated by the number of fin kicks (55 kicks) based on a preliminary measurement made by swimming along a known distance. All the fish encountered during SCUBA dives along each transect were recorded. The maximum depths were 26.5 m and 23.1 m, average water temperatures were 20.0°C and 19.2°C, and the censuses took 3.6 h and 3.35 h in Tsushima and Maizuru, respectively.

The total number of fish species, fishes, jack mackerel, pelagic predators, and benthic predators were compared between Tsushima and Maizuru using U-tests. Predators were defined as piscivorous fishes with the body larger than 15 cm SL that could potentially feed on recruited jack mackerel juveniles of 5 cm SL (personal observation).

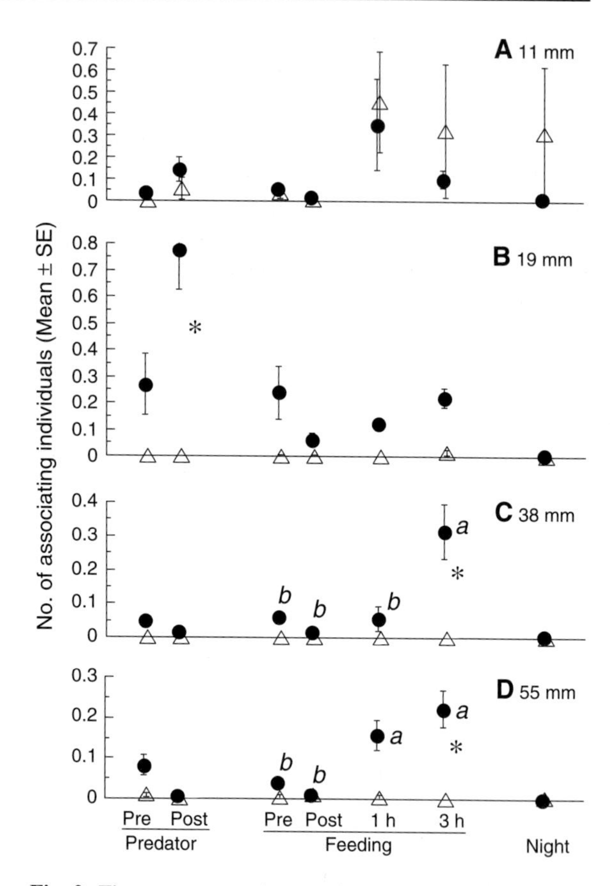


**Fig. 2** The average numbers (±SE) of 11 mm (**A**), 19 mm (**B**), 38 mm (**C**), and 55 mm (**D**) juvenile jack mackerels associated with live *Aurelia aurita,* (closed circles) and an artificial jellyfish (open triangles) before and after exposure to an artificial predator, before, immediately after and 1 h and 3 h after feeding, and at night. Asterisks represent differences between live moon jellyfish and artificial jellyfish ($P < 0.05$, *t*-test), and different letters represent significant differences among immediately before, immediately after, and 1 h and 3 h after feeding ($P < 0.05$, Tukey's HSD test)

## Results

### Laboratory experiment using moon jellyfish and artificial jellyfish

Jack mackerel at 11 mm SL was associated with both live and artificial jellyfish (Fig. 2A). Because of the high variation among observation times and tanks, no systematic trend was observed related to the presence of the predator model or feeding. Association with artificial jellyfish at night was also observed. Fish at 19 mm SL or larger were associated with live jellyfish, but not with an artificial jellyfish (Fig. 2B–D). Fish at 19 mm SL showed a strong association after the threat by the predator model ($P < 0.05$, paired *t*-test), but minimal association in relation to feeding (Fig. 2B). In contrast, fish at 38 mm and 55 mm did not associate with jellyfish when threatened by the predator model, but showed a strong association at 1 h and 3 h after feeding ($P < 0.05$, Tukey's HSD test).

### Underwater observation of jack mackerel associated with giant jellyfish

A total of 110 giant jellyfish were observed. Bell diameters of jellyfish in Tsushima were significantly smaller than those in Maizuru, and so were the number of jack mackerel associating with each jellyfish ($P < 0.05$, U-test; Table 2). The smallest size of associating jack mackerel was 6 mm SL in both areas (Figs. 3, 4A). Few small larvae were associated with jellyfish and there was a relatively large variation in sizes of fish associating with each jellyfish. Fish of 10–40 mm SL were commonly observed swimming between tentacles and then hiding within the bell cavity when disturbed. Fish larger than 40 mm often swam away from the jellyfish when they were disturbed by the observer. The largest specimen at 65 mm SL was found in Tsushima (Fig. 4D). Fish species other than jack mackerel that were associated with giant jellyfish included medusa fish, *Psenopsis anomala* (Temminck & Schlegel) (7–40 mm SL), and thread-sail filefish, *Stephanolepis cirrhifer* (Temminck & Schlegel) (6–11 mm SL), in Tsushima, and only medusa fish (12–90 mm SL) in Maizuru.

**Table 2** Total numbers of giant jellyfish (*Nemopilema nomurai*) observed, mean (±SD) bell diameter, frequency of jack mackerel association, and mean (±SD) number of jack mackerel associated with each jellyfish in Tsushima and Maizuru in 2006

| | Tsushima | Maizuru |
|---|---|---|
| No. of jellyfish | 39 | 71 |
| Bell diameter (cm) | 52 ± 35 | 113 ± 21 |
| Frequency of jack mackerel | 19 | 60 |
| No. of jack mackerel | 2.9 ± 5.7 | 11.0 ± 20.9 |

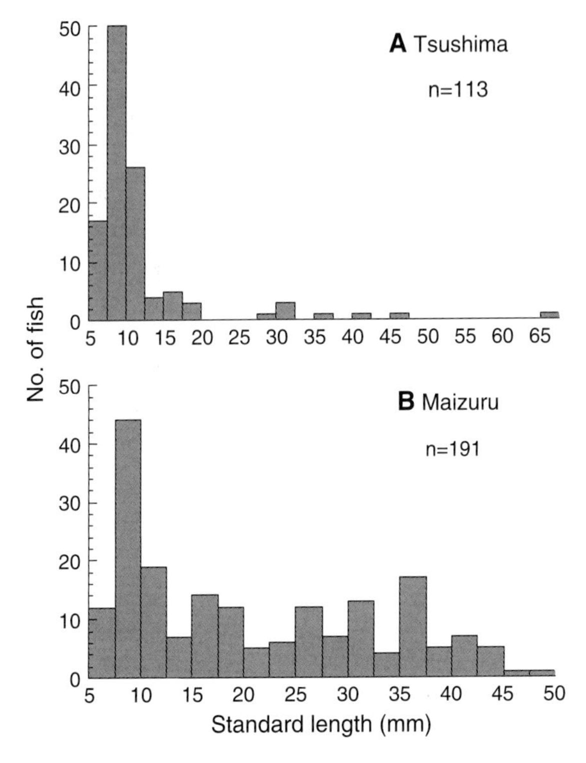


**Fig. 3** Size distributions of jack mackerel associated with *Nemopilema nomurai*, in the Tsushima (**A**) and Maizuru (**B**) study sites

A total of 60 fish species and 2565 individuals and 30 species and 4383 individuals were recorded in the visual censuses in Tsushima and Maizuru, respectively. Although the fish assemblage in Tsushima included more species because of the coral reef habitat, the number of species per transect was not significantly different between these two areas (Fig. 5A; $P = 0.19$, U-test). The number of individuals was higher in Maizuru, largely due to the presence of many jack mackerel (Fig. 5B, C). No jack mackerel was found in reef areas in Tsushima. Tsushima had more large (>15 cm) benthic predators, such as the groupers *Epinephelus chlorostigma* (Valenciennes) and *E. bruneus* Bloch, and lionfish, *Pterois lunulata* Temminck & Schlegel, whereas pelagic predators, such as yellowtail, *Seriola quinqueradiata* Temminck & Schlegel, predominated in Maizuru (Fig. 5D, E).

## Discussion

### Ontogeny of function in association behavior with jellyfish

Although there exists an extensive review on the association of fish with jellyfish (Purcell & Arai, 2001), most of the proposed ecological functions are based on circumstantial evidence in the field. Arai (2005) summarized 69 species of fish reported to feed on jellyfishes. Shojima (1962) revealed that although the carangid fishes, *Decapterus maruadsi* (Temminck & Schlegel) and *Trachurus japonicus,* associated with jellyfishes, such as *Aurelia aurita* and *Aequorea*

**Fig. 4** Underwater photographs of jack mackerel associated with giant jellyfish, *Nemopilema nomurai*. Estimated body lengths are 6–9 mm (**A**), 25 mm (**B**), 40 mm (**C**), and 65 mm SL (**D**). (**A**)–(**C**) were taken in Maizuru, and (**D**) in Tsushima, September–December, 2006

*coerulescens* (Brandt), none of them contained jellyfish in their gut contents, suggesting that these carangids use jellyfish for protection against predators rather than as food. Few researchers have tested the ecological function of association between fish and jellyfish through experimental approaches.

The smallest jack mackerel larvae that associated with giant jellyfish were 6 mm SL in both the Maizuru and Tsushima areas. Larvae of this size are estimated to be about 15 days post-hatching based on hatchery rearing data (Masuda, 2006). Sassa et al. (2006) reported that a major spawning area of jack mackerel occurs in the central part of the East China Sea in spring. The relatively slow speed of the Tsushima Current (i.e., 7 cm $s^{-1}$; Hase et al., 1999) would place the spawning location of the 6 mm SL larvae found in Maizuru only 90 km southwest of the sampling area. Castro et al. (2002) concluded that planktivorous fish associated with drifting objects to reach food-rich areas. I suggest that the early stage association of jack mackerel with jellyfish may also function to redistribute juveniles to new habitats.

In the experimental tanks, jack mackerel juveniles of 11 mm SL were associated with both live and artificial jellyfish but did not show a change in their strength of association in relation to the presence of a predator model or provision of prey items. This suggests that they were associated with jellyfish for reasons other than predator avoidance or prey collection. Small larvae or juveniles (6–20 mm SL) associating with giant jellyfish tended to have large variation in size, whereas juveniles at 20–40 mm SL tended to have less size variation. This may reflect that a solitary larva or juvenile might have been attracted to a jellyfish after some period of drifting in the ocean, where they met conspecifics and then formed schools. This is in contrast to *Trachurus novaezelandiae* Richardson, which may arrive at FADs in already-formed schools (Dempster & Kingsford, 2004).

Masuda & Tsukamoto (1999) proposed that association with floating objects by early juvenile carangid fishes may be partly to enable efficient school formation with conspecifics, since association with objects often precedes formation of proper schools. Jack mackerel may also make use of giant jellyfish as a meeting place, since finding jellyfish would be easier than finding a small conspecific in the open ocean. Fréon & Dagorn (2000) suggested that among 16 hypotheses proposed for the ecological

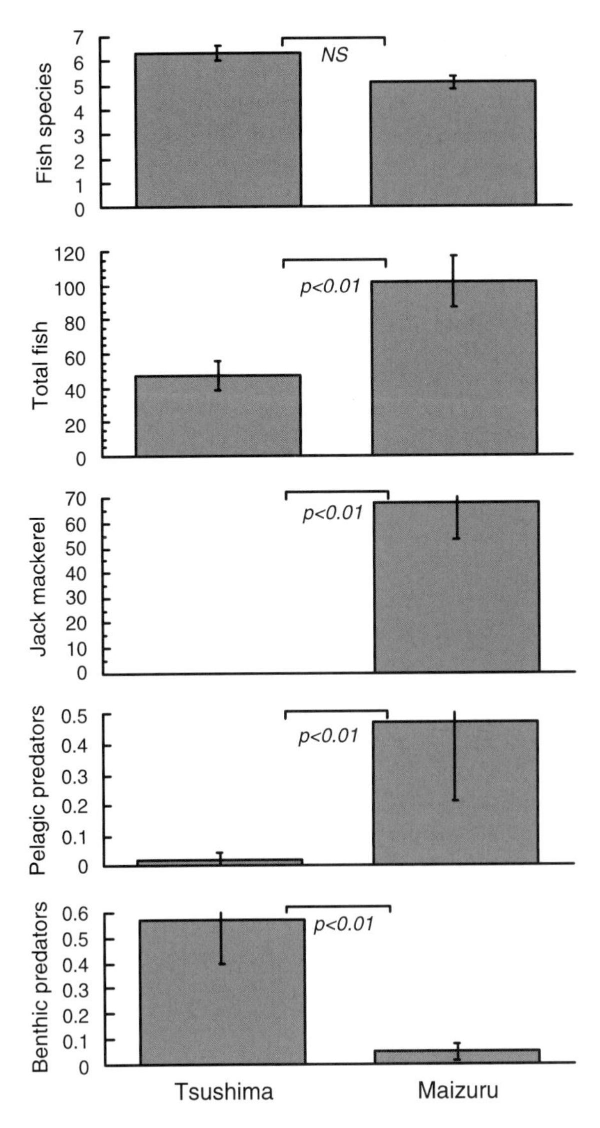


**Fig. 5** The numbers of fish species, fish specimens, jack mackerel, and pelagic and benthic predators larger than 150 mm SL found along 100 $m^2$ census transects. Means ± SE of 54 and 43 replicates in Tsushima and Maizuru study sites are presented, respectively

function of association behavior with FADs, the 'meeting point hypothesis' is most persuasive. The association of 6 mm jack mackerel larvae with giant jellyfish fits well with this hypothesis.

Fish at 19 mm showed association with moon jellyfish in the presence of a model predator, whereas this response was not observed in larger juveniles. Absence of an anti-predator response in larger juveniles may be partly due to the relatively small size of the predator model (140 mm SL) and the small size of the moon jellyfish (102 mm diameter) used in this experiment. In the underwater observations, jack mackerel at 40 mm or smaller often associated with giant jellyfish with the presence of a diver; however, larger juveniles tended to swim away from the jellyfish toward the sea bottom when they were disturbed.

Jack mackerel at 38 mm and 55 mm SL showed strong association with moon jellyfish at 1 h and 3 h after introducing *Artemia* sp. to the tank, perhaps as a mechanism to feed on the *Artemia* from the gut cavity of moon jellyfish (Masuda et al., 2008). Although *Trachurus trecae* Cadenat off West Africa and *T. murphyi* Nichols in the Southwestern Pacific were reported to contain jellyfish in their gut contents (reviewed by Arai, 1988, 2005), jellyfish itself is seldom a common prey item for *T. japonicus*. Mianzan et al. (2001) found that anchovies, *Engraulis anchoita* Hubbs & Marini, fed on the salps, *Iasis zonaria* (Pallas), when the densities of preferable prey, i.e., copepods, were scarce. However, in rearing conditions, juvenile *T. japonicus* ranging from 5.8 to 46 mm SL did not feed on moon jellyfish even when they were starved for 2 days (Masuda, 2006). Since the major prey of young *T. japonicus* is copepods (Hirota et al., 2004), they are more likely to use jellyfish as a prey collector.

Masuda & Tsukamoto (2000) found that association with floating objects by another carangid, the striped jack, *Pseudocaranx dentex* (Bloch & Schneider), was stronger at night than in daytime. Miyazaki et al. (2000) proposed that association at night may prevent dispersion of the school since their light intensity threshold for school formation is relatively high due to their small eye diameter. Here, association at night was observed only for 11 mm SL juveniles with artificial jellyfish. Light intensity in the laboratory ranged from 0.03 to 0.15 lux at night, which may be above the light intensity threshold for schooling in larger juveniles, which would not need jellyfish for association. Light intensity threshold for school formation was reported to be 0.05 in the striped jack at 20 and 65 mm TL (Miyazaki et al., 2000); jack mackerel is likely to have similar threshold value. Further experiments on association at low-light intensity are required to test the possibility of association with jellyfish in darkness.

Based on the laboratory experiments and field observations, the function of association with jellyfish in jack mackerel is suggested to change gradually from a meeting place with conspecifics, to a hiding place from predators, then to a prey collector of zooplankton while migrating in the oligotrophic offshore environment.

Ontogeny of mechanisms in associating with jellyfish

Masuda & Tsukamoto (2000) revealed that striped jack started to show association with both transparent and gray floating objects at 11.9 mm TL (ca. 11 mm SL), whereas this behavior was absent in smaller larvae (5.5, 8.3, or 10.3 mm TL). The discrepancy between their experiments and the presence of 6 mm SL jack mackerel larvae associating with giant jellyfish in the present study is attributable to the relatively early development of optokinetic responses. Masuda & Tsukamoto (1998) also demonstrated that an optokinetic response following a moving screen first appears at 6 mm TL in the striped jack, whereas the flotsam used in Masuda & Tsukamoto (2000) were fixed objects. Perhaps the optokinetic response to moving objects in the ocean is the primary mechanism for jack mackerel larvae to associate with jellyfish.

Jack mackerel larvae at 6 mm do not have scales and thus should be vulnerable to the nematocysts of jellyfish. Nevertheless, juvenile jack mackerel seem to be relatively immune to jellyfish nematocysts compared with other fish, such as red sea bream, *Pagrus major* (Temminck & Schlegel), chub mackerel, *Scomber japonicus* (in Masuda, 2006), and Japanese anchovy, *Engraulis japonicus* (Houttuyn) (Masuda, unpublished data). Indeed, the largest larvae to be eaten by a moon jellyfish in 5-min predation trials were 5.3 mm for jack mackerel and 23 mm for Japanese anchovy. This may guarantee a relative advantage for jack mackerel larvae in associating with jellyfish rather than associating with abiotic floating objects or drifting singly.

Dempster & Kingsford (2004) found that *Trachurus novaezelandiae* is the predominant fish species as juveniles associated with drifting objects off the east coast of Australia (34°S, 152°E). Based on the sudden appearance of large number of *T. novaezelandiae*, they suggested that this species arrived in schools, and that initiation of schooling behavior may be a necessary developmental precursor for association with drifting structures. This interpretation does not seem appropriate for the congener *T. japonicus* because larvae of only 6 mm SL associated with jellyfish, whereas school formation starts at 10 mm SL (personal observation). The presence of jellyfish may be a factor that facilitates associating with conspecifics and subsequent school formation.

Scales of juveniles are completely formed at 27 mm SL in jack mackerel (personal observation). This should increase their protection against nematocysts and thus it is not surprising that they would associate with jellyfish, especially when threatened with a predator model (Fig. 2B) or when disturbed by an observer in the vicinity of a jellyfish. Jack mackerels have tight scales compared with other pelagic fishes (personal observation). Although this may sacrifice cruse swimming capability (Masuda, 2006), it is likely to be adapted for protection against nematocysts. Associations with jellyfish become less strong as they grow beyond 40 mm SL perhaps due to a decreased dependence on jellyfish as an anti-predator strategy. A highly-organized, large school with matured capability of social communication (Nakayama et al., 2007) is likely to be their major anti-predator behavior at this stage.

The visual censuses suggested that the rocky reefs in Maizuru have more pelagic, but fewer benthic predators compared with the upstream locations in the Tsushima Current (Fig. 5D, E). The predation pressure from pelagic predators may cause jack mackerel to abandon giant jellyfish as a vehicle for migration and shift to a reef-associated habitat. It is notable that the smallest size of jack mackerel found in the rocky shore habitat along MFRS is 40 mm SL (Masuda et al., 2008).

**Acknowledgments** I thank Jenny Purcell, Kylie Pitt and two anonymous reviewers for providing constructive comments on the early version of the manuscript. This study was supported by the FRECS and STOPJELLY projects from the Fisheries Agency of Japan.

## References

Arai, M. N., 1988. Interactions of fish and pelagic coelenterates. Canadian Journal of Zoology 66: 1913–1927.

Arai, M. N., 2005. Predation on pelagic coelenterates: A review. Journal of the Marine Biological Association of the United Kingdom 85: 523–536.

Castro, J. J., J. A. Santiago & A. T. Santana-Ortega, 2002. A general theory of fish aggregation to floating objects: An

alternative to the meeting point hypothesis. Reviews in Fish Biology and Fisheries 11: 255–277.
Dempster, T. & M. J. Kingsford, 2004. Drifting objects as habitat for pelagic juvenile fish off New South Wales, Australia. Marine and Freshwater Research 55: 675–687.
Dempster, T. & M. Taquet, 2004. Fish aggregation device (FAD) research: gaps in current knowledge and future directions for ecological studies. Reviews in Fish Biology and Fisheries 14: 21–42.
Fréon, P. & L. Dagorn, 2000. Review of fish associative behaviour: toward a generalization of the meeting point hypothesis. Reviews in Fish Biology and Fisheries 10: 183–207.
Hase, H., J.-H. Yoon & W. Koterayama, 1999. The current structure of the Tsushima warm current along the Japanese Coast. Journal of Oceanography 55: 217–235.
Hirota, Y., S. Uehara & H. Honda, 2004. Ontogenetic changes of feeding selectivity in juvenile jack mackerel *Trachurus japonicus* collected off south-east Kyushu, Japan. Fisheries Science 70: 100–107.
Kawahara, M., S. Uye, K. Ohtsu & H. Iizumi, 2006. Unusual population explosion of the giant jellyfish *Nemopilema nomurai* (Scyphozoa: Rhizostomeae) in East Asian waters. Marine Ecology Progress Series 307: 161–173.
Kingsford, M. J., 1993. Biotic and abiotic structure in the pelagic environment: importance to small fishes. Bulletin of Marine Science 53: 393–415.
Masuda, R., 2006. Ontogeny of anti-predator behavior in hatchery-reared jack mackerel *Trachurus japonicus* larvae and juveniles: patchiness formation, swimming capability and interaction with jellyfish. Fisheries Science 72: 1225–1235.
Masuda, R., 2008. Seasonal and interannual variation of subtidal fish assemblages in Wakasa Bay with reference to the warming trend in the Sea of Japan. Environmental Biology of Fishes 82: 387–399. doi:10.1007/s10641-007-9300-z.
Masuda, R. & K. Tsukamoto, 1998. The ontogeny of schooling behaviour in the striped jack. Journal of Fish Biology 52: 483–493.
Masuda, R. & K. Tsukamoto, 1999. School formation and concurrent developmental changes in carangid fish with reference to dietary conditions. Environmental Biology of Fishes 56: 243–252.
Masuda, R. & K. Tsukamoto, 2000. Onset of association behavior in striped jack, *Pseudocaranx dentex*, in relation to floating objects. Fishery Bulletin, US 98: 864–869.
Masuda, R., Y. Yamashita & M. Matsuyama, 2008. Jack mackerel *Trachurus japonicus* juveniles utilize jellyfish for predator avoidance and as a prey collector. Fisheries Science 74: 282–290.
Mianzan, H., M. Pájaro, G. A. Colombo & A. Madirolas, 2001. Feeding on survival-food: gelatinous plankton as a source of food for anchovies. Hydrobiologia 451: 45–53.
Miyazaki, T., S. Shiozawa, T. Kogane, R. Masuda, K. Maruyama & K. Tsukamoto, 2000. Developmental changes of the light intensity threshold for school formation in the striped jack *Pseudocaranx dentex*. Marine Ecology Progress Series 192: 267–275.
Nakayama, S., R. Masuda & M. Tanaka, 2007. Onsets of schooling behavior and social transmission in chub mackerel *Scomber japonicus*. Behavioral Ecology and Sociobiology 61: 1383–1390.
Omori, M., H. Ishii & A. Fujinaga, 1995. Life history strategy of *Aurelia aurita* (Cnidaria, Scyphomedusae) and its impact on the zooplankton community of Tokyo Bay. ICES Journal of Marine Science 52: 597–603.
Purcell, J. E. & M. N. Arai, 2001. Interactions of pelagic cnidarians and ctenophores with fish: A review. Hydrobiologia 451: 27–44.
Purcell, E. J., S. Uye & W.-T. Lo, 2007. Anthropogenic causes of jellyfish blooms and their direct consequences for humans: A review. Marine Ecology Progress Series 350: 153–174.
Sassa, C., Y. Konishi & K. Mori, 2006. Distribution of jack mackerel (*Trachurus japonicus*) larvae and juveniles in the East China Sea, with special reference to the larval transport by the Kuroshio Current. Fisheries Oceanography 15: 508–518.
Shojima, Y., 1962. On the postlarvae and juveniles of carangid fishes collected together with the jellyfishes. Contributions from the Seikai Regional Fisheries Research Laboratory 147: 48–58.
Uye, S., N. Fujii & H. Takeoka, 2003. Unusual aggregations of the scyphomedusa *Aurelia aurita* in coastal waters along western Shikoku, Japan. Plankton Biology and Ecology 50: 1–21.
Zar, J. H., 1996. Biostatistical Analysis, 3rd ed. Prentice Hall, New Jersey.

Hydrobiologia (2009) 616:279–289
DOI 10.1007/s10750-008-9589-4

JELLYFISH BLOOMS

# Kinematic properties of the jellyfish *Aurelia* sp.

**Tom Bajcar · Vlado Malačič · Alenka Malej · Brane Širok**

Published online: 22 September 2008


**Abstract** A new, relatively simple method for determining the kinematic properties of jellyfish is presented. The bell movement of the scyphomedusa (*Aurelia* sp.) during its pulsation cycle was analysed using computer-aided visualization. Sequences of video images of individual *Aurelia* in a large aquarium were taken using a standard video camera. The images were then processed to obtain time series of the relative positions of selected points on the surface of the medusa's bell. The duration of the bell relaxation was longer than that of the bell contraction, thereby confirming published results. In addition, the area of the exumbrellar surface of *Aurelia* increased during bell relaxation by more than 1.3-times that of the exumbrellar surface area during the maximum contraction of the bell. The volume change during the bell pulsation cycle was also measured using the same visualization method. Significant changes, of up to 50%, in the subumbrellar cavity volume were revealed while, in contrast, the volume between the exumbrellar and subumbrellar surfaces generally remained unchanged during the entire pulsation cycle of the bell. Comparison of the time series of the exumbrellar surface area and of the subumbrellar cavity volume indicated that the change of volume takes place before the change of the surface area of the bell.

**Keywords** Scyphozoa · Computer-aided visualization · Bell pulsation

Guest editors: K. A. Pitt & J. E. Purcell
Jellyfish Blooms: Causes, Consequences, and Recent Advances

T. Bajcar (✉) · B. Širok
Faculty of Mechanical Engineering, University of Ljubljana, Aškerčeva 6, 1000 Ljubljana, Slovenia
e-mail: tom.bajcar@fs.uni-lj.si

B. Širok
e-mail: brane.sirok@fs.uni-lj.si

V. Malačič · A. Malej
National Institute of Biology, Marine Biology Station Piran, Fornače 41, 6330 Piran, Slovenia

V. Malačič
e-mail: malacic@mbss.org

A. Malej
e-mail: malej@mbss.org

## Introduction

Jellyfish is a key predator in many pelagic environments worldwide. Accumulating evidence indicates an increase in the frequency and intensity of gelatinous plankton outbreaks (Brodeur et al., 1999; Hays et al., 2005; Attril et al., 2007) as well as expansion of the range of their distribution (Graham, 2001; Graham et al., 2003; Malej & Malej, 2004). When very abundant, medusae play an important trophic role and seriously affect zooplankton populations such as ichthyoplankton and, consequently, fisheries (Purcell & Arai, 2001). The mechanics of prey capture vary between medusae, which influences prey selection

(Costello & Colin, 1995) and, consequently, impacts on the ecosystem (Malej et al., 2007). The importance of medusae jet propulsion for their displacement (swimming) and predation has been recognized since the early 1970s (Gladfelter, 1972, 1973), and our understanding of the swimming and feeding of medusae has increased substantially in last few decades (Mills, 1981; Costello & Colin, 1994; Ford et al., 1997; D'Ambra et al., 2001; Colin et al., 2006; Malej et al., 2007). Models have been developed to describe medusae jet propulsion (Daniel, 1983), swimming (Matanoski & Hood, 2006) and flow patterns generated by medusae (Dabiri et al., 2005, 2006). Costello et al. (2008) suggested two distinctive propulsion modes used by medusae: jet propulsion characteristic for small-sized organisms and rowing propulsive mode generally used by larger Scyphozoa. The differences in propulsion modes have important ecological consequences most notably on prey selection.

Among the Scyphomedusae, that recurrently appear in great densities, *Aurelia* is the most widespread and it swarms in both cold and temperate regions. Morphological characteristics affect swimming and foraging behaviour, and *Aurelia* has been classified as an oblate medusa having a rowing mode of propulsion and lower swimming performance, which create high fluid disturbance (Dabiri et al., 2005; Colin et al., 2006). The resulting fluid entrainment enhances prey encounter and affects the selection of prey organisms. Understanding the motion of *Aurelia* and the dynamics of a vortex ring (Dabiri et al., 2005) is thus very important for the species' ecology. The locomotive force depends very much on the shape and volume of the bell, the added-mass of the wake vortex tensor and the velocity of the wake.

While the Reynolds number is a parameter that relates only to the steady motion of fish, the Strouhal number relates the tail-flapping frequency to the forward speed, normalized by the amplitude of lateral motion of the tail (Lauder & Tytell, 2006). However, a similar 'generic' Strouhal number of a wake vortex was introduced for jellyfish, which is proportional to the wake vortex ratio (Dabiri et al., 2005). This is defined as the ratio of the wake vortex added-mass term to the vortex circulation term and shows whether the added-mass of wake vortices is to be considered in the swimming dynamics (the reaction or locomotive force experienced by the organism).

For a wide range of sizes of oblate medusae, the shape profile of the bell at its full contraction and relaxation is independent of size (Dabiri et al., 2005). This was demonstrated with plots of the bell profile in which dimensions were normalized by the cube root of the ejected bell fluid volume. Obviously, the geometry of the jellyfish bell plays a fundamental role. The sensitivity study of locomotion of medusae (Dabiri & Gharib, 2003) showed that, although the bell volume of medusae could be estimated from the bell profile with much greater precision than the bell aperture area, the errors of the former play a much greater role in the error of the dynamics, manifested in errors of acceleration, swimming velocity and position of medusae. A second geometric uncertainty, which was discussed in the same work, is related to the bell aperture radius. On contraction, this could be lower than the maximum radius of the bell, which is located somewhere between the bell margin and apex. This is related to the fact that the shape of the bell approaches a hemi-ellipsoid only when the medusa is in its relaxed state. Dabiri & Gharib (2003) also concluded that a single parameter was not sufficient to provide an accurate description of the animal's kinematics.

Kinematic analysis of swimming of two Cubomedusae species was done by Shorten et al. (2005) using laboratory-based digital video records. A computer analysis of the reference points identified on bell parts (nodes) enabled description of the swimming gaits used by two jellyfish species. However, the method allowed only a few nodes on the surface of the jellyfish to be identified, which were extracted manually from the image sequences.

The aim of our study was, therefore, to develop a new methodology for the research of kinematic properties of jellyfish based on computer-aided visualization. We have focussed on the velocities of the parts of the jellyfish bell that have not been addressed in previous research, and *Aurelia* was used as a model oblate medusa. The volume and surface area of the bell were also examined with this method.

## Methodology

### Methodology of determining bell contour from images

Computer-aided visualization was used to determine the bell contour from images of medusae swimming.

Standard video recordings of *Aurelia* individuals were made while they were swimming in a large aquarium (Cretaquarium, Thalassocosmos, Greece). Image acquisition frequency was 25 Hz, i.e. a time difference between two successive images of 0.04 s. The methodology used to determine cyclic alteration of the medusa bell shape was as follows: *transformation* of each video image in the sequence into a binary image, with the outer contour of the *Aurelia* sp. bell clearly visible; *selection* of points along the contour of the medusa bell, where the movement is to be observed; *tracking* the time series of the spatial position of each selected point on the outer contour of the bell.

The uncertainty of the method arises mainly from the position of the medusa on the images – the medusa can approach or move away from the camera or can turn around its lateral axis (parallel to the bell aperture) during the sequence, giving an isometric view of the bell instead of its profile. Therefore, selection of the appropriate sequence of images was essential. We selected sequences other than those in which visual inspection revealed that the medusa was swimming in a plane that is obviously tilted with respect to the vertical one. The latter was deduced by observing the contour of the apex (which should on average be close to a single line) and the presence of the top of the bell (if the top was not visible, the medusa was tilted). The number of images in the sequence was selected so that the entire cycle of the periodic bell cycle of the selected medusa was included. The duration of the cycles differed and varied from about 3.5 to 5.5 s. About 130 images covered one cycle of 5.2 s (see sample images in Fig. 1).

The changes in the bell shape can be observed in the $x$–$y$ plane (Fig. 1). Transformation of video images into binary ones was done using the Matlab program package. The Sobel method (Sobel, 1978) for edge detection was used to determine the outer contour of the medusa's bell. It performs a 2D spatial gradient measurement on an image. Typically, it is used to find the approximate absolute gradient magnitude at each point in an input greyscale image. The Sobel edge detector uses a pair of $3 \times 3$ convolution masks (much smaller than an actual image), one estimating the gradient in the $x$-direction (columns, Fig. 1) and the other in the $y$-direction (rows, Fig. 1). The mask is slid over the image, manipulating one square of pixels at a time. At each point in the image, the resulting approximated gradients ($G$) can be combined to give the gradient magnitude, using the Širok et al. (2002) expression:

$$G = \sqrt{G_x^2 + G_y^2}, \tag{1}$$

where $G_x$ and $G_y$ are gradients along the $x$- and $y$-directions. By adjusting the threshold value of the gradient $G$, it is possible to use this method to search for the edges between dark and bright regions (Fig. 2).

The gradient approach (Eq. 1) was applied to determine both the exumbrellar (outer) and subumbrellar (inner) contours of the medusa's bell (Fig. 3). The results were a little more uncertain in the case of the subumbrellar contour, since the grey level gradients between the subumbrellar contour and the subumbrellar cavity were smaller than those between the exumbrellar contour and the surrounding water.

## Methodology for quantifying bell contour movement

The outer contour was next divided into an appropriate number of points, whose movements were observed. The bell was assumed to be continuously axisymmetric around its central axis, since the medusa in the selected sequences of images generally did not move in ways other than straightforward.

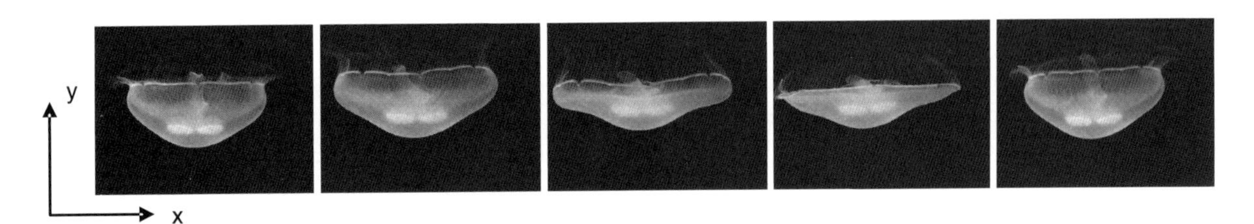


**Fig. 1** *Aurelia* sp. sample images of the sequence of changes in the form of the medusa's bell (time between two neighbouring sample images is 1.2 s)

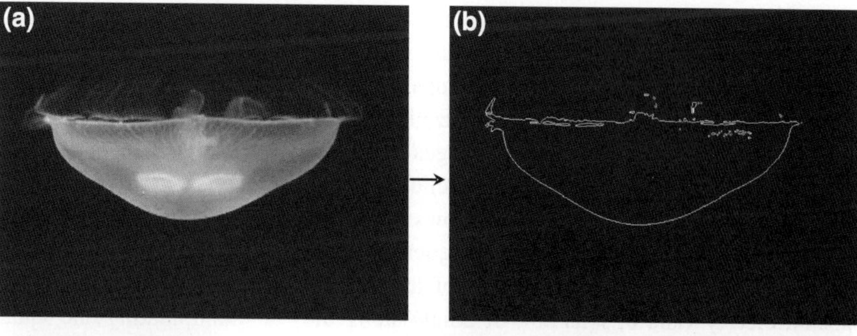


**Fig. 2** Transformation of a video image (**a**) into a binary image (**b**) defining the outer contour of the medusa's bell

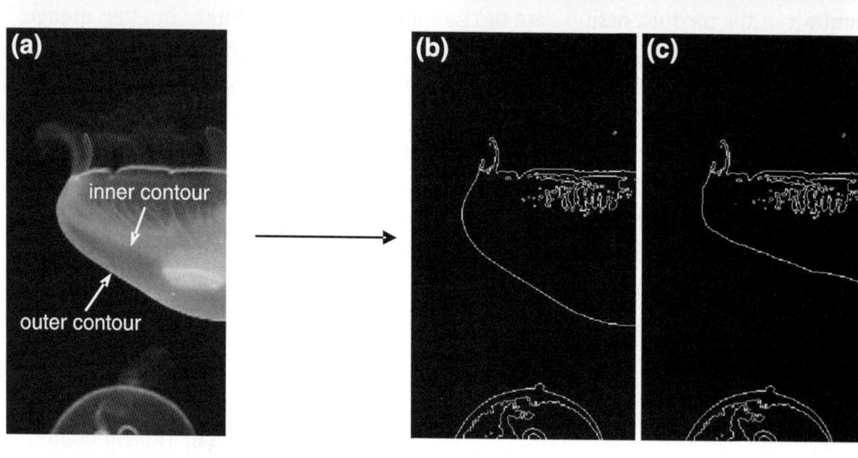


**Fig. 3** Processing of the medusa image (**a**) to determine the exumbrellar (outer) contour (**b**) and the subumbrellar (inner) contour (**c**)

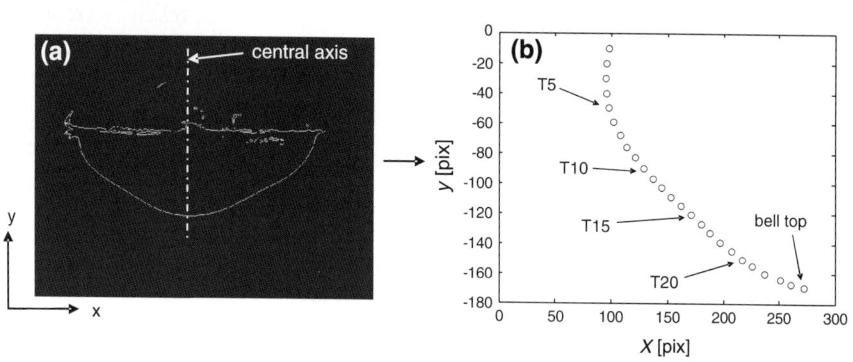


**Fig. 4** Transformation of the outer contour of the bell shape (**a**) into several points in the *x–y* plane (**b**)

Thus, only half the bell contour (Fig. 4a) can be observed and transformed into points. The part of the outer contour that lies on the left of the central axis was chosen, since this part was not affected by the presence of other medusae behind the one observed on the binary images (Figs. 2b and 4a).

The contour on the left side of the central axis (Fig. 4a) was transformed into several points, again by using the Matlab software (Fig. 4b). Velocities of every fifth point along the contour were studied, denoted as T5, T10, T15, T20 and as the 'bell top' (Fig. 4b) for the sake of easier orientation. Our analysis focussed on the bell contour; so the points along the tentacles at the bell margin were removed, since they were not present on all images of the sequence due to poor lighting. This procedure resulted in 25 points along the outer contour of the bell on the left side of the central axis (Fig. 4a). An effort was made to keep the points along the outer contour of the bell equidistant. However, due to the limited resolution of the images, this was not completely possible; therefore the average distance ($l_n$) between two neighbouring points on the whole image sequence was 11.4 ± 2.2 pixels. Pixel units, in which images are recorded, are appropriate for the measure of distances in this study, since many important kinematic properties are extracted with non-dimensional quantities, in which space dimensions cancel out. Moreover, as was pointed out in the Introduction, the shape profile of the bell at its full contraction and relaxation is size independent (Dabiri et al., 2005).

### Methodology for quantifying bell surface area

The intention of this study was to verify the stretching and contraction of the medusa's bell during the whole bell pulsation cycle. For this purpose, a computer-aided visualization method, similar to the one described above, was applied to determine the length of the outer bell contour, as depicted in Figs. 2b and 4a. Axisymmetry of the medusa's bell was again assumed, so only the left half of the bell was processed on each image (Fig. 4a). However, the contour was not transformed into a number of points. It was used rather to calculate the outer bell surface area $S_o$ applying again the Matlab program.

$$S_o = 2 \cdot \pi \cdot \sum_j \left(x_o - x_{\text{out},j}\right) \cdot h, \qquad (2)$$

where $j$ denotes the number of pixels in the vertical direction ($y$) from the top of the bell to its bottom, $x_o$ is the horizontal position of the vertical symmetry axis of the bell, $x_{\text{out},j}$ denotes the position of the outer contour at a specified vertical coordinate $j$, $h$ is the vertical height at each vertical position $j$ and has therefore a value $h = 1$ pix, which corresponds to the resolution of the image.

## Results

### Movement of the exumbrellar contour

During the power stroke of the medusa (contraction), the average distance $l_n$ (number of pixels) between two neighbouring points was generally up to 25% smaller than during the recovery stroke (relaxation). This indicated that there was either a certain level of stretching of the bell during relaxation or contraction during the power stroke; this is further investigated in the next section. The images were acquired at a fixed acquisition frequency; however, the power stroke was shorter than the recovery stroke (Fig. 5).

By processing the whole sequence of photographs, a time series of point coordinates was obtained (Fig. 6). All the selected points in Fig. 6 correspond to the points shown in Fig. 4b. The time series in Fig. 6a represent the horizontal movement of points on the bell contour, which was the result of both medusa displacement (i.e. the movement of the medusa's mass centre) and the relative movement

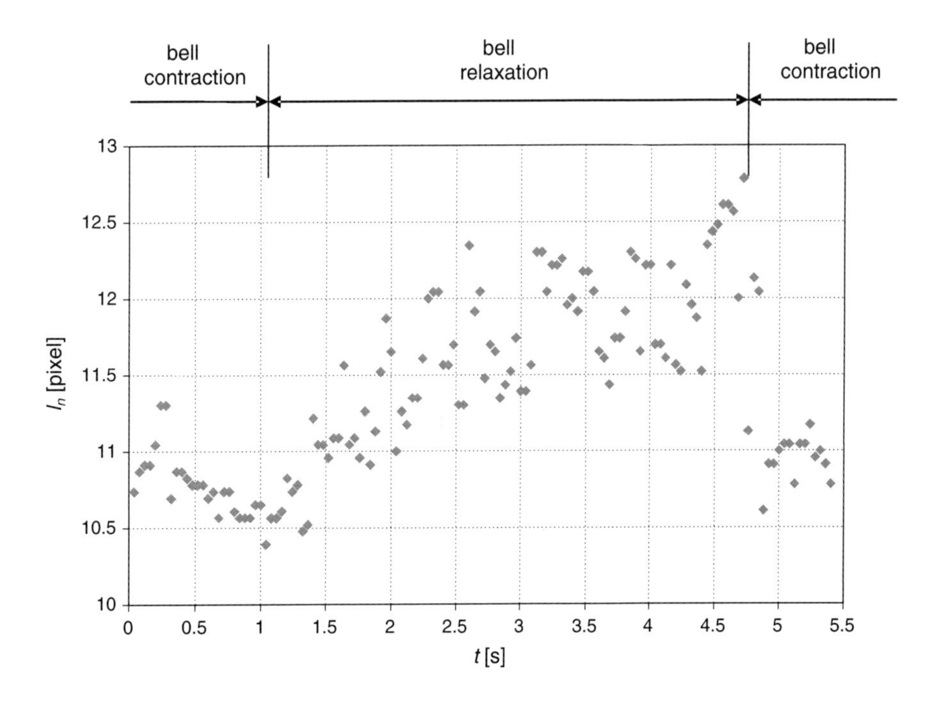


**Fig. 5** Average distance, in pixels, between two neighbouring points on the bell of the medusa for each image in the sequence

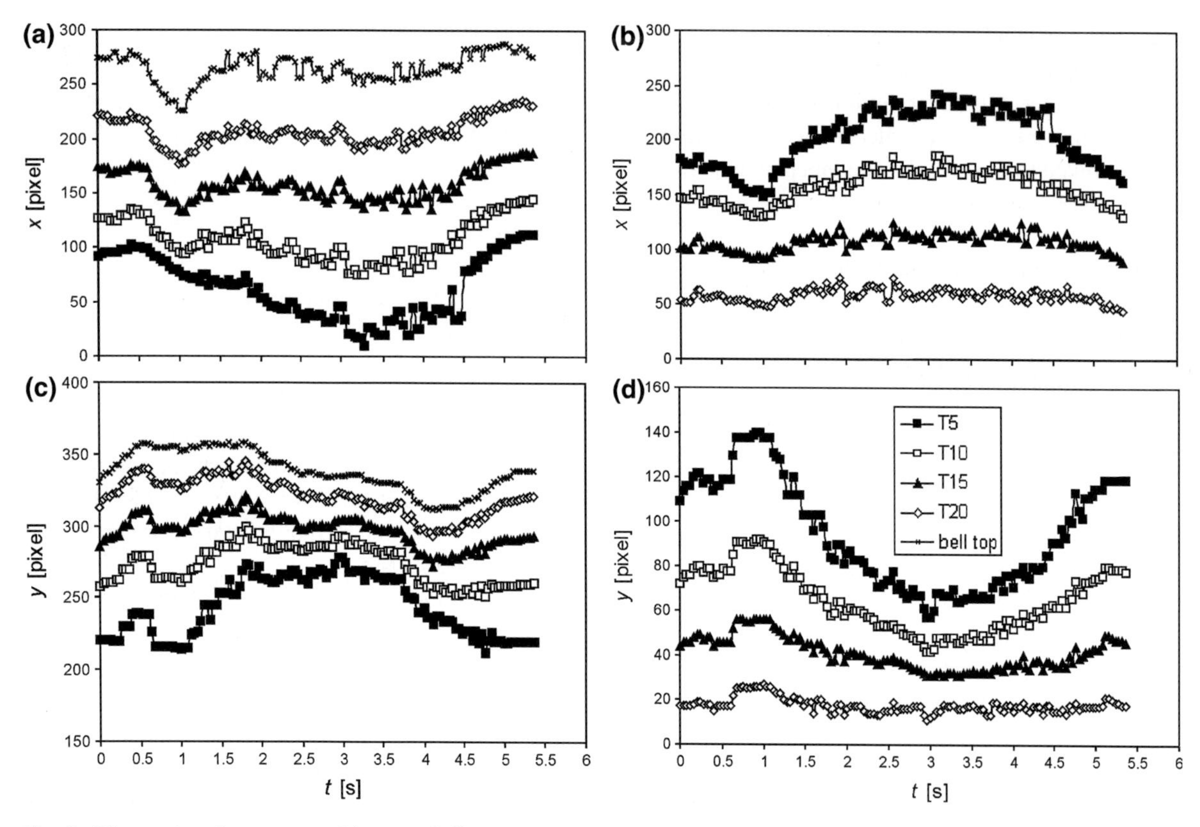


**Fig. 6** Time series of movement of the outer bell contour: (**a**) in $x$-direction (horizontal) for five different points on the outer contour of the medusa's bell; (**b**) in $x$-direction, relative to the bell top (fixed point); (**c**) in the $y$-direction (vertical) for five points on the outer contour of the medusa's bell; (**d**) in $y$-direction, relative to the bell top (fixed point)

of the points with respect to the bell top, which was assumed to move in the $x$-direction only when the medusa as a whole moved in that direction. Time series of the movement of the points in the $x$-direction on the bell outer contour relative to the bell top are shown in Fig. 6b.

The lines in Fig. 6a and b are not smooth. This was most probably due to the error in calculating the position of a particular point, which had to be repeated for each image in the sequence. Such a procedure does not guarantee that the points are exactly equidistant.

A similar procedure can be applied to extract the time series of the point movement in the $y$-direction, as shown in Fig. 6c and d. Movement in both the $x$- and $y$-directions is more pronounced near the bell margin (e.g. points T5 and T10, Fig. 6b, d) than in the vicinity of the bell top (Fig. 6b, d). In this way, it is possible to compare the relative movement between arbitrary points chosen on the outer contour of the medusa's bell.

## Changes of the bell surface area

Another sequence of images was used for determining alterations of the outer bell surface area during the bell pulsation cycle. This sequence was a little shorter than the first one and took less than 4 s to cover the complete pulsation cycle of the medusa. The area of the exumbrellar surface (i.e. the outer surface area of the bell) increased during relaxation and reached a maximum when the bell was in its most expanded state ($t$ = ca. 2.5 s, Fig. 7). The surface area of the outer bell at this point was ca. 1.34-fold greater than that at the state of maximum bell contraction ($t$ = ca. 3.7 s, Fig. 7).

## Changes of the bell volume

Alterations in volume of the medusa and, in particular, the volume of subumbrellar space during the bell pulsation were further investigated. Here, the visualization method was again applied in order to obtain the exumbrellar and the subumbrellar contours of the medusa's bell from each image in the sequence, as shown in Fig. 3. The two contours confined the body of the medusa in a 2D image. Assuming axisymmetry of the bell movement, one curve for each half-contour was obtained for every image in the sequence (Fig. 8).

The bell volume between the exumbrellar and subumbrellar surfaces, $V_{es}$ (depicted as two contours in a 2D diagram; Fig. 8), and the volume of the seawater in the subumbrellar cavity, $V_w$, were readily calculated for a particular image using Eqs. (3) and (4):

$$V_{es} = \pi \cdot \sum_j \left( (x_o - x_{out,j})^2 - (x_o - x_{in,j})^2 \right) \cdot h \quad (3)$$

$$V_w = \pi \cdot \sum_j (x_o - x_{in,j})^2 \cdot h, \quad (4)$$

where $x_{in,j}$ denotes the position of the inner contour at a specified vertical coordinate $j$. Other quantities are the same as those specified in Eq. 2.

There was no significant change in bell volume between the exumbrella and subumbrella during a bell pulsation cycle (Fig. 9a). The dispersion of the results is nevertheless large; the highest values were ca. 1.2-fold greater than the lowest values. However, this could have resulted from the inherent uncertainty of the method, which was mainly due to the lighting

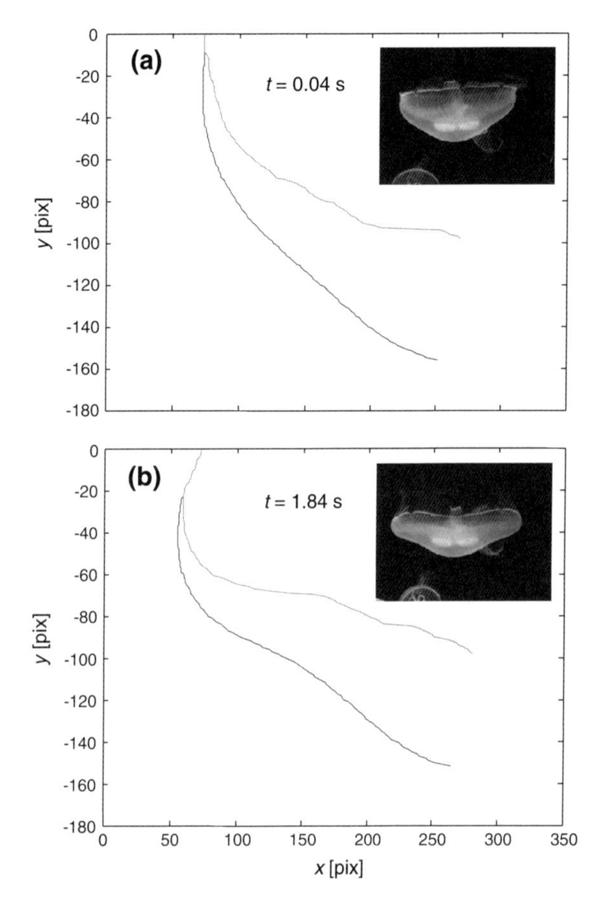


**Fig. 8** Outer and inner half-contours of the bell during phases of contraction (**a**) and relaxation (**b**) obtained by computer-aided visualization

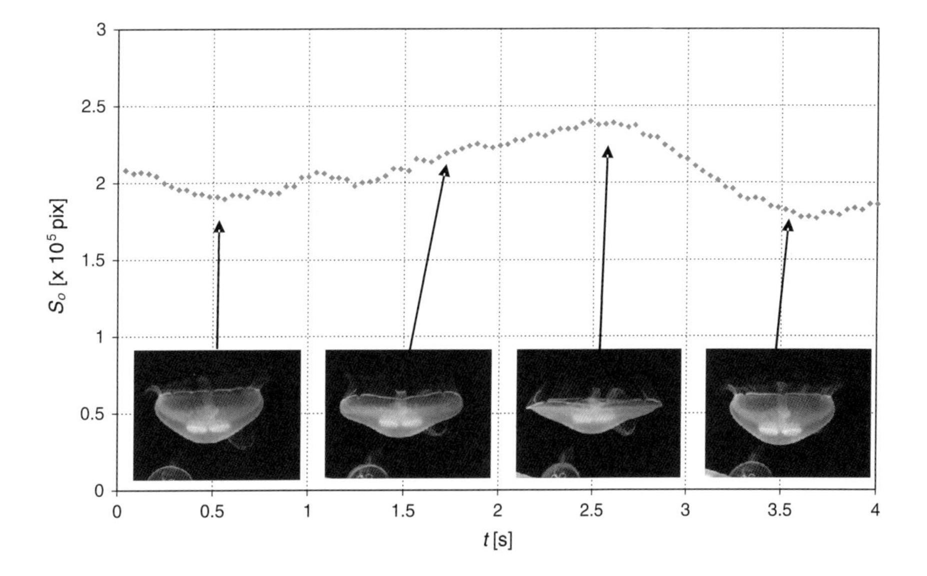


**Fig. 7** Alterations of the outer surface area of the bell during a cycle of the bell pulsation

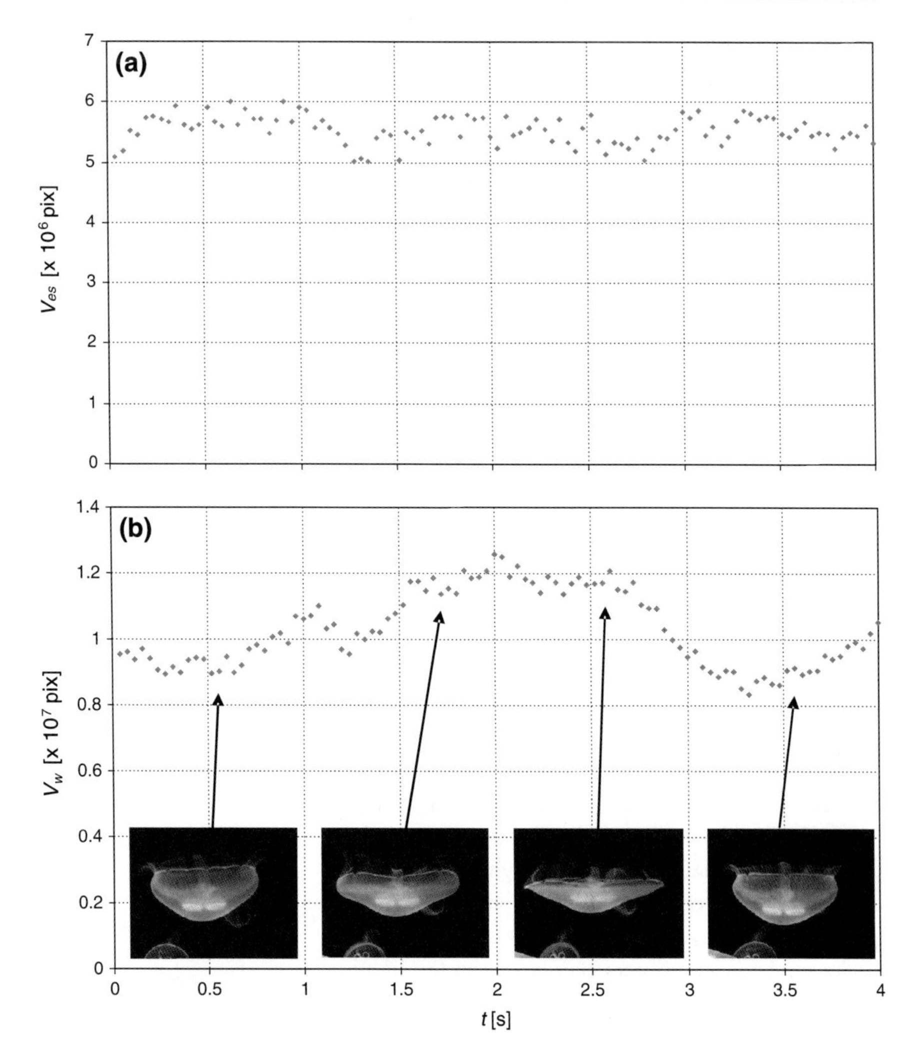


**Fig. 9** Volume during the pulsation cycle: (**a**) volume between the exumbrellar and subumbrellar (outer and inner) surfaces (bell volume); (**b**) volume of the water in the subumbrellar cavity

and hence the uncertainty of the inner bell contour at some places.

There was, however, a significant alteration in the subumbrellar cavity volume (i.e. the volume of fluid below the bell) (Fig. 9b). This was expected, since changes in this volume are needed to ensure propulsion of the medusa. The volume increased during the bell relaxation phase (time period between ca. 0.5–2.5 s, Fig. 9b). The ratio of the highest to the lowest values of the subumbrellar cavity volume was ca. 1.5.

The outer bell surface area ($S_o$) and the volume of water in the subumbrellar cavity ($V_w$) appeared to change in a similar way during the bell contraction cycle, since both diagrams (Figs. 7 and 9b) had a similar shape. Figure 10 shows the ratio between these two quantities ($V_w/S_o$).

Both $S_o$ and $V_w$ were calculated from the same images. Although the laws governing the changing of $S_o$ and $V_w$ during the bell contraction cycle appeared to be similar, the ratio $V_w/S_o$ was not constant (Fig. 10). The ratio increased when $V_w$ increased. Detailed consideration of the evaluation of $S_o$ and $V_w$ showed, however, that there was a certain time delay between the calculated signals (i.e. between time series of $S_o$ and $V_w$). To estimate the time delay between $S_o$ and $V_w$, the cross-correlation between the two signals was calculated using the following formula (Sach, 1997):

$$r(\Delta t) = \frac{\sum_i [(S_o(i + \Delta i) - \bar{S}_o) \cdot (V_w(i) - \bar{V}_w)]}{\sqrt{\sum_i (S_o(i + \Delta i) - \bar{S}_o)^2} \cdot \sqrt{\sum_i (V_w(i) - \bar{V}_w)^2}}, \quad (5)$$

where $r$ denotes the cross-correlation coefficient (between 0 and 1), $i$ is the current number of the

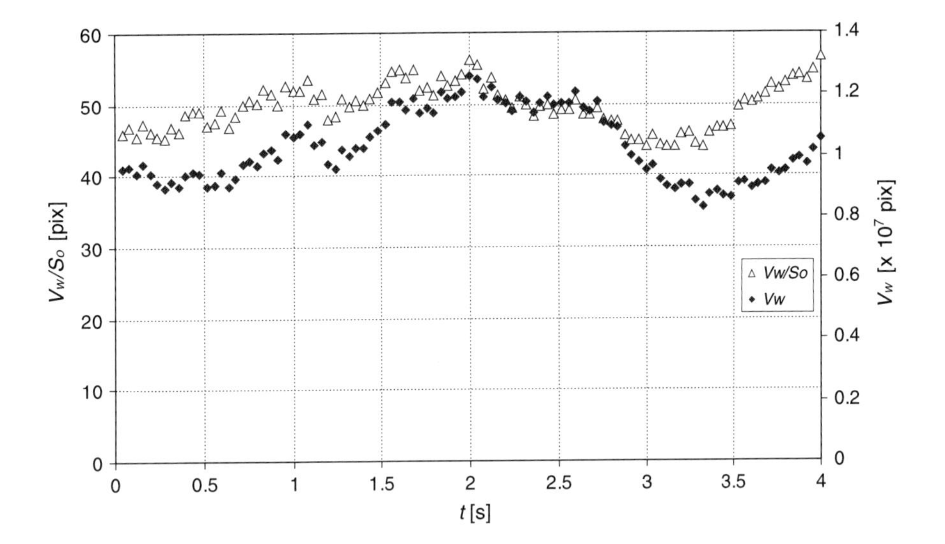


**Fig. 10** Ratio between the water volume in the subumbrellar cavity $V_w$ and the exumbrellar surface area $S_o$

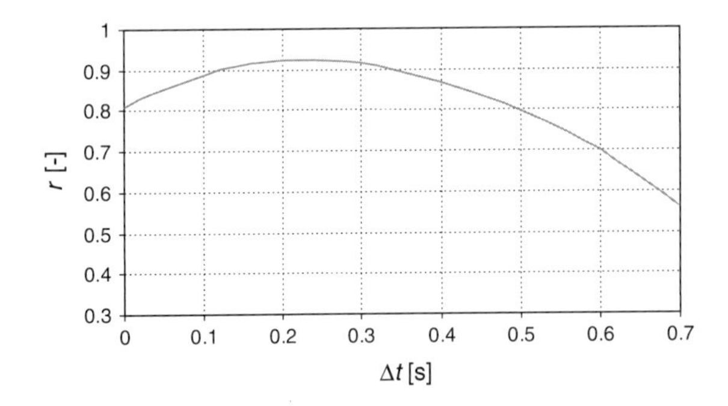


**Fig. 11** Correlation coefficient $r$ for different time delays $\Delta t$ between calculated time series of $S_o$ and $V_w$

image in the sequence and $\bar{S}_o$ and $\bar{V}_w$ denote the mean values of $S_o$ and $V_w$ over the whole sequence of $i$ images. $\Delta i$ is the sequence delay (i.e. the time delay $\Delta t$) between series of $S_o$ and $V_w$, where $\Delta i = 0, 1, 2,$ etc. coresponds to the time delay $\Delta t = 0$ s, 0.04 s, 0.08 s, respectively. The positive value of $\Delta t$ means that $S_o$ follows $V_w$.

The two time series were most closely correlated at a time delay of ca. 0.2 s, where the coefficient $r$ had its maximum value of ca. 0.923 (Fig. 11). The change in exumbrellar surface area thus lagged behind the change in subumbrellar cavity volume by ca. 0.2 s.

## Discussion

Kinematic characteristics of the scyphomedusa *Aurelia sp.* were obtained from reconstruction of video images using a new, relatively simple computer-aided visualization. This method enabled us to examine changes in the medusa outer bell surface and the subumbrelar cavity volume in the course of a bell pulsation (contraction/relaxation) cycle. During the contraction phase, in addition to reduction of bell diameter (aperture), average distance between two neighbouring points on the outer bell surface decreased by up to 25%, most notably at the bell margin (i.e. Fig. 6). Similarly, Ford & Costello (2000) noted that in oblate hydrozoan genera bell contraction occurred primarily at the bell margin. While the bell volume between exumbrellar and subumbrellar surfaces (bell mesoglea) did not change significantly during the pulsation cycle, the ratio of the highest (relaxation) to the lowest (contraction) subumbrellar cavity volume amounted to about 1.5. The recoil phase during which the bell refills with water is accomplished by elastic energy stored in the mesoglea during bell contraction (DeMont & Gosline, 1988; Megill, 2002). Our analysis indicates that on average this phase lasts nearly twice as long as the contraction phase; this result is comparable to the values reported for *Aurelia* by Costello & Colin (1994) and found also for some other medusae species (Glatfelter, 1973; Ford et al., 1997; Shorten et al., 2005).

The bell radius (point T5 with respect to bell top, Fig. 6b)) oscillated with a range (peak-to-peak), which was about 36 ± 5% of the maximum radius value ((90 ± 10 pixels)/(250 ± 10 pixels)). The relative change of the volume of the subumbrellar cavity with respect to the minimum cavity volume

was ca. 50%. This agrees well with the results of other investigators (Dabiri & Gharib, 2003), confirming again the reliability of the method.

Previous estimates of bell volume and the aperture area (Dabiri & Gharib, 2003) of jellyfish were grounded on insertion of a half-spline curve over the side image of jellyfish and the 3D bell was reconstructed by the revolution of a cubic spline around the axis of revolution. However, boundary points (pixels) on digital frames that separate the ambient fluid from the medusae bell have been selected by using the binary threshold filter with the conveniently set-up treshold level. The method here presented does not require this step. The boundary between the bell and the ambient is based on the exploration of the absolute gradients of pixels' intensity (Sobel, 1978), by sliding the $3 \times 3$ convolution mask over the digital image and the threshold value of the gradient magnitude is much easier to set, since the gradients in intensities of pixels at the boundary are much larger than elsewhere. The same method was applied in this work exploring movements of bell parts along the length of the *Aurelia* bell. This differs from the method of Colin & Costello (2002) who made the morphological measurements of six co-occuring hydromedusae directly from video recordings calibrated with scale bars. Subsequently, they applied the same method to assess morphological traits of three upstream foraging medusae (Colin et al., 2006). Scale bars are not needed in the method presented here and our study demonstrates the ability of this approach to make better determinations of the bell shape parameters leading in turn to a better estimate of jellyfish dynamics. The major disadvantage of our fast method of image analysis lies in the relativity of scale: all space dimensions are scaled in pixels. As it was shown in this work, there are (nondimensional) quantities which could be successfully extracted with this method. Moreover, the pixel dimensions could be scaled with ingested objects of known dimensions (fluorescent beads).

## Conclusions

We have demonstrated a relatively simple method of analysing the shape of the bell of *Aurelia* sp. that leads to an estimate of the velocities of parts of the outer bell profile. The obtained results, such as the duration of power and recovery stroke, were in agreement with the results of observations by other authors. This analysis also showed that alteration of the bell shape due to flexion, which conserves volume was a minor part of the bell's kinematics, while the changes of volume of the subumbrellar cavity play the dominant role during the relaxation/contraction of the bell. It can be concluded from the results of the analysis that the changes of subumbrellar cavity volume are manifested first by changes in the shape of the bell and second by changes in its outer surface area. These are, however, preliminary results that have to be further investigated.

It remains for the method presented here to be expanded on a larger population of *Aurelia* to confirm these interesting conclusions and to explore further how the swimming style is optimized according to the body size and shape. Knowledge of the kinematics of organisms with a complex shape is a prerequisite for the proper study of their dynamics, which is closely related to the hydrodynamics of the fluid that surrounds them. The latter will also be the focus of our future research.

## References

Attrill, M. J., J. Wright & M. Edwards, 2007. Climate-related increases in jellyfish frequency suggest more gelatinous future for the North Sea. Limnology & Oceanography 52: 480–485.

Brodeur, R. D., C. E. Mills, J. E. Overland, G. E. Walters & J. D. Schumacher, 1999. Evidence for a substantial increase in gelatinous zooplankton in the Bering sea, with a possible link to climate change. Fisheries Oceanography 8: 296–306.

Colin, S. P. & J. H. Costello, 2002. Morphology, swimming performance and propulsive mode of six co-occurring hydromedusae. The Journal of Experimental Biology 205: 427–437.

Colin, S. P., J. H. Costello & H. Kordula, 2006. Upstream foraging by medusae. Marine Ecology Progress Series 327: 143–155.

Costello, J. H., S. P. Colin & J. O. Dabiri, 2008. Medusan morphospace: phylogenetic constrains, biomechanical solutions, and ecological consequences. Invertebrate Biology, doi:10.1111/j.1744-7410.2008.00126x.

Costello, J. H. & S. P. Colin, 1994. Morphology, fluid motion and predation by the scyphomedusa *Aurelia aurita*. Marine Biology 121: 327–334.

Costello, J. H. & S. P. Colin, 1995. Flow and feeding by swimming scyphomedusae. Marine Biology 124: 399–406.

D'Ambra, I., J. H. Costello & F. Bentivegna, 2001. Flow and prey capture by the scyphomedusa *Phyllorhiza punctata* von Lendenfeld 1884. Hydrobiologia 451: 223–227.

Dabiri, J. O., S. P. Colin & J. H. Costello, 2006. Fast-swimming hydromedusae exploit velar kinematics to form an optimal vortex wake. Journal of Experimental Biology 209: 2025–2033.

Dabiri, J. O., S. P. Colin, J. H. Costello & M. Gharib, 2005. Flow patterns generated by oblate medusan jellyfish: field measurements and laboratory analyses. Journal of Experimental Biology 208: 1257–1265.

Dabiri, J. O. & M. Gharib, 2003. Sensitivity analysis of kinematic approximations in dynamic medusan swimming models. Journal of Experimental Biology 206: 3675–3680.

Daniel, T. L., 1983. Mechanics and energetics of medusan jet propulsion. Canadian Journal of Zoology 61: 1406–1420.

DeMont, M. E. & J. M. Gosline, 1988. Mechanics of jet propulsion in the hydromedusan jellyfish, *Polyorchis penicillatus*. I. Mechanical properties of thelocomotor structure. Journal of Experimental Biology 134: 313–332.

Ford, M. D. & J. H. Costello, 2000. Kinematic comparison of bell contraction by four species of hydromedusae. Scientia Marina 64(Suppl 1): 47–53.

Ford, M. D., J. H. Costello & K. B. Heilderberg, 1997. Swimming and feeding by the scyphomedusa *Chrysaora quinquecirrha*. Marine Biology 129: 355–362.

Gladfelter, W. B., 1972. Structure and function of the locomotory system of the Scyphomedusa *Cyanea capilata*. Marine Biology 14: 150–160.

Gladfelter, W. B., 1973. A comparative analysis of the locomotory systems of medusoid Cnidaria. Helgolnder wiss. Meeresunters 25: 228–272.

Graham, M., 2001. Numerical increases and distribution shifts of *Chrysaora quiquecirrha* (Desor) and *Aurelia aurita* (Linne) (Cnidaria: Scyphozoa) in the northern Gulf of Mexico. Hydobiologia 451: 97–111.

Graham, M., D. L. Martin, D. Felder, V. L. Asper & H. M. Perry, 2003. Ecological and economic implications of a tropical jellyfish invader in the Gulf of Mexico. Biological Invasions 5: 53–69.

Hays, G. C., A. J. Richardson & C. Robinson, 2005. Climate change and marine plankton. Trends in Ecology and Evolution 20: 337–344.

Lauder, G. V. & E. D. Tytell, 2006. Hydrodynamics of undulatory propulsion. Fish Physiology 23: 425–468.

Malej, A. & A. Malej, 2004. Invasion of the jellyfish Pelagia noctiluca in the Northern Adriatic: a non-success story. In Dumont, H., T. Shiganova & U. Niermann (eds.), Aquatic Invasions in the Black, Caspian, and Mediterranean Seas. Kluwer Academic Press, Dordrecht: 273–285.

Malej, A., V. Turk, D. Lučić & A. Benović, 2007. Direct and indirect trophic interactions of *Aurelia* sp. (Scyphozoa) in a stratified marine environment (Mljet Lakes, Adriatic Sea). Marine Biology 151: 827–841.

Matanoski, J. C. & R. R. Hood, 2006. An individual-based numerical model of medusa swimming behaviour. Marine Biology 149: 595–608.

Megill, W. M., 2002. The biomechanics of jellyfish swimming. Ph.D. Dissertation, Department of Zoology, University of British Columbia, 116 pp.

Mills, C. B., 1981. Diversity of swimming behaviours in hydromedusae as related to feeding and utilization of space. Marine Biology 64: 185–189.

Purcell, J. E. & M. N. Arai, 2001. Interactions of pelagic cnidarians and ctenophores with fish: a review. Hydobiologia 451: 27–44.

Sach, L., 1997. Angewandte Statistik: Anwendung statistischer Methoden. Springer-Verlag, Berlin.

Shorten, M., J. Davenport, J. E. Seymour, M. C. Cross, T. J. Carrette, G. Woodward & T. F. Cross, 2005. Kinematic analysis of swimming in Australian box jellyfish, *Chiropsalmus* sp. and *Chironex flecheri* (Cubozoa, Cnidaria: Chirodropidae). Journal of Zoology 267: 371–380.

Širok, B., T. Bajcar & M. Dular, 2002. Reverse flow phenomenon in a rotating diffuser. Journal of Flow Visualization and Image Processing 9: 193–210.

Sobel, I., 1978. Neighborhood coding of binary images for fast contour following and general array binary processing. Computer Graphics and Image Processing 8: 127–135.